MOLECULAR SYSTEMATICS

EDITED BY

David M. Hillis
THE UNIVERSITY OF TEXAS

AND

Craig Moritz
UNIVERSITY OF QUEENSLAND

Sinauer Associates, Inc.
Publishers
SUNDERLAND, MASSACHUSETTS, U.S.A.

To Ann and Fiona

MOLECULAR SYSTEMATICS

Library of Congress Cataloging-in-Publication Data

Molecular systematics / edited by David M. Hillis and Craig Moritz.
      p.   cm.
   Includes bibliographical references.
   ISBN 0-87893-279-8. — ISBN 0-87893-280-1 (pbk.)
    1. Chemotaxonomy—Research.   2. Molecular biology—Methodology.
I. Hillis, David M., 1958- . II. Moritz, Craig.
QH83.M665   1990
574.8'8—dc20
                                        89-48277
                                        CIP

Printed in U.S.A.

5   4   3   2

Contents in Brief

Contents

PART II. MOLECULAR TECHNIQUES   43

Chapter 4
Proteins I: Isozyme Electrophoresis   45
Robert W. Murphy, Jack W. Sites, Jr., Donald
G. Buth, and Christopher H. Haufler

Chapter 5
Proteins II: Immunological Techniques   127
Linda R. Maxson and R. D. Maxson

Chapter 6
Chromosomes: Molecular Cytogenetics   156
Stanley K. Sessions

Chapter 7
Nucleic Acids I: DNA–DNA Hybridization 204
Steven D. Werman, Mark S. Springer, and Roy J. Britten

Chapter 8
Nucleic Acids II: Restriction Site Analysis 250
Thomas E. Dowling, Craig Moritz, and Jeffrey D. Palmer

Chapter 9
Nucleic Acids III: Sequencing   318
David M. Hillis, Allan Larson, Scott K. Davis, and Elizabeth A. Zimmer

PART III. ANALYSIS   371

Chapter 10
Intraspecific Differentiation   373
Bruce S. Weir

Chapter 11
Phylogeny Reconstruction   411
David L. Swofford and Gary J. Olsen

Chapter 12
An Overview of Applications of Molecular Systematics   502
David M. Hillis and Craig Moritz

Preface

The need for a book on molecular systematics has been evident for many years. However, no one person can possibly become a practitioner and at the same time remain current in all of the molecular techniques used in systematic biology; the technology changes too quickly. Thus, we decided in 1987 to organize a multiauthored book on the subject. Because we were concerned about the possibility of uneven treatment by the various authors, we structured the chapters carefully and enforced the structure rigidly. We organized the book into three main sections that correspond to the three parts of every molecular systematic study: sampling design and execution, collection of molecular data, and data analysis. Our hope is that this book can guide beginners all the way through a molecular systematic study, and at the same time provide established investigators with new ideas, techniques, and approaches.

We use the term *systematics* in its broad sense to include the comparative study of biotic diversity at any level. The goals of molecular systematics are also the goals of systematics in general; this book deals specifically with molecular approaches because of the unique problems of collecting and analyzing molecular data. We hope the book will also be useful to nonmolecular systematists by describing the principles, applications, and limitations of molecular techniques.

A book of this type must rely heavily on cooperation from expert reviewers, and we have been fortunate to have extraordinary cooperation from the research community. John Avise, John Gillespie, Morris Goodman, Mark Kirkpatrick, Irv Kornfield, Mike Miyamoto, Colin Patterson, Vincent Sarich, and Allan Wilson sent us detailed comments on several chapters each, and we thank them for their considerable commitment of time. We also received very useful reviews of chapters from Loren Ammerman, James Archie, Robert Baker, Peter Baverstock, John Benzie, James Bull, Paul Chippindale, Joel Cracraft, Brian Crother, Ross Crozier, Llewellyn Densmore, Michael Dixon, Rafael de Sá, Herbert Dessauer, John Gold, Sheldon Guttman, James Hamrick, Richard Highton, John Kirsch, Mike Johnson, Linda Maxson, Steve Palumbi, James Patton, Craig Pease, Eric Pianka, Michael Ryan, Barbara Schaal, Charles Sibley, Montgomery Slatkin, Jerry Slightom, Carol Stepien, David Swofford, D. Tagle, Bruce Weir, and Gregory Whitt. We appreciate the time and effort that these reviewers have invested in this book.

Argye Hillis and Hamish McCallum provided invaluable statistical advice, and Michael Dixon and Loren Ammerman assisted with figure preparation. Thomas White provided advice and prepublication information on the polymerase chain reaction. We thank Linda Davis, Brad Garton, Diana

Hews, Beth Reid, and Vicki Young-Lehmeier for assisting with the correction, handling, and translating of computer files of the chapters. Andy Sinauer has contributed to every stage of the book, from planning and organizing to production; we thank him for his personal interest and concern for this book. The National Science Foundation and the Australian Research Council have provided generous support for our research in molecular systematics; this support provided us with the experience in a diversity of molecular techniques that we needed to edit this volume. Some of the travel involved in editing was generously supported by the University of Queensland.

Finally, our wives Ann Hillis and Fiona Hamer have assisted us and supported us throughout this project. We may never be able to repay them for all their help, encouragement, and extraordinary patience.

David M. Hillis
Austin, Texas, USA

Craig Moritz
Brisbane, Australia

Contributors

Peter R. Baverstock   University of New England, Northern Rivers, P.O. Box 157, Lismore, New South Wales 2480 AUSTRALIA

Roy J. Britten   Division of Biology, California Institute of Technology, Pasadena, California 91125 USA

Donald G. Buth   Department of Biology, University of California, Los Angeles, California 90024 USA

Charles J. Cole   Department of Herpetology and Ichthyology, American Museum of Natural History, New York, New York 10024 USA

Scott K. Davis   Faculty of Genetics, Texas A & M University, College Station, Texas 77843 USA

Herbert C. Dessauer   Department of Biochemistry and Molecular Biology, Louisiana State University Medical Center, New Orleans, Louisiana 70112 USA

Thomas E. Dowling   Department of Zoology, Arizona State University, Tempe, Arizona 85287 USA

Mark S. Hafner   Museum of Natural Science, Louisiana State University, Baton Rouge, Louisiana 70803 USA

Christopher H. Haufler   Department of Botany, University of Kansas, Lawrence, Kansas 66045 USA

David M. Hillis   Department of Zoology, University of Texas, Austin, Texas 78712 USA

Allan Larson   Department of Biology, Washington University, St. Louis, Missouri 63130 USA

Linda R. Maxson   Department of Biology, Pennsylvania State University, University Park, Pennsylvania 16802 USA

R. D. Maxson   Department of Biology, Pennsylvania State University, University Park, Pennsylvania 16802 USA

Craig Moritz   Department of Zoology, University of Queensland, St. Lucia, Queensland 4067 AUSTRALIA

Robert W. Murphy   Department of Ichthyology and Herpetology, Royal Ontario Museum, 100 Queen's Park, Toronto, Ontario M5S 2C6 CANADA

Gary J. Olsen   Department of Microbiology, University of Illinois, Urbana, Illinois 61801 USA

Jeffrey D. Palmer   Department of Biology, Indiana University, Bloomington, Indiana 47405 USA

Stanley K. Sessions   Department of Biology, Hartwick College, Oneonta, New York 13820 USA

Jack W. Sites, Jr.   Department of Zoology, Brigham Young University, Provo, Utah 84602 USA

Mark S. Springer   Division of Biology, California Institute of Technology, Pasadena, California 91125 USA

David L. Swofford   Center for Biodiversity, Illinois Natural History Survey, 607 E. Peabody Drive, Champaign, Illinois 68120 USA

Bruce S. Weir   Department of Statistics, North Carolina State University, Raleigh, North Carolina 27695 USA

Steven D. Werman   Department of Biology, Mesa State College, Grand Junction, Colorado 80501 USA

Elizabeth A. Zimmer   Department of Biochemistry, Louisiana State University, Baton Rouge, Louisiana 70803 USA

CHAPTER 1

MOLECULAR SYSTEMATICS: CONTEXT AND CONTROVERSIES

Craig Moritz and David M. Hillis

HISTORICAL CONTEXT OF MOLECULAR SYSTEMATICS

For centuries, naturalists have tried to detect, describe, and explain diversity in the biological world; this endeavor is known as systematics. The formalization of a hierarchical system of nomenclature by Linnaeus (1758) established a framework for describing and categorizing biological diversity. This hierarchical system was initially independent of evolutionary theory, but later workers (e.g., Darwin, 1859; Haeckel, 1866; reviewed by Mayr, 1983) developed the notion that classification should be based on phylogenetic relationships (Figure 1). The twentieth century has seen major conceptual and operational advances in the estimation of phylogeny (e.g., Hennig, 1966; Wiley, 1981; Felsenstein, 1982, 1988), as well as in the analysis of microevolutionary change (see Futuyma, 1986). These two aspects of evolution traditionally have been studied by different kinds of biologists: phylogeny by systematists, and microevolutionary change by population geneticists. However, the two fields have expanded toward a common middle ground, and an understanding of evolution requires a synthesis of the two disciplines. We view systematics in the broad sense of the term, as encompassing the study of both intraspecific and interspecific diversity.

Investigations of microevolutionary change and phylogeny require sources of heritable variation. Until the 1960s, systematics was based largely on analysis of morphological and behavioral variation, approaches that continue to be used with increasing sophistication. However, with the elucidation of the molecular basis of inheritance, biological macromolecules have assumed an increasingly important role in evolutionary studies. As evidenced by the chapters in this book, nucleic acids (DNA and RNA), pro-

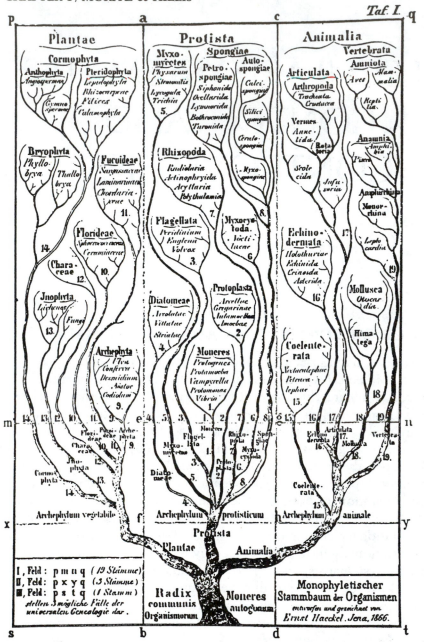

FIGURE 1. The phylogeny and classification of life as proposed by Haeckel (1866).

teins, and chromosomes can provide a broadly applicable set of heritable markers to examine genetic structure of populations or to estimate relationships among taxa. Conversely, such molecular studies have generated massive comparative data bases, providing important insights into the evolution of the molecules themselves (reviewed by MacIntyre, 1985; Nei, 1987).

Early applications of molecules to systematic problems were largely concerned with proteins. Immunological approaches (Chapter 5) were pioneered by Nuttall (1904) and achieved prominence with studies of relationships and times of divergence among hominoids (Goodman, 1961, 1963; Sarich and Wilson, 1966, 1967). Early applications of electrophoresis and specific histochemical staining (Chapter 4) revealed a wealth of protein variation within and among species (e.g., Hubby and Throckmorton, 1965; Hubby and Lewontin, 1966; Lewontin and Hubby, 1966). This approach, isozyme electrophoresis, has generated a massive comparative data base and is currently the most widely used approach in molecular systematics (see reviews by Avise, 1974; Lewontin, 1974; Avise and Aquadro, 1982; Nevo et al., 1984; Buth, 1984). Comparisons of amino acid sequences provided the first indications of a molecular clock (Zuckerkandl and Pauling, 1962) and have been used extensively to estimate phylogeny (e.g., Goodman et al., 1987). Variation in chromosome structure and number (Chapter 6) also provides a valuable source of genetic markers within and between species (see White, 1973).

Major advances in the manipulation and analysis of nucleic acids in the past decade have led to the widespread study of DNA and RNA variation. The sequences assayed have come from the nucleus, the mitochondrion, and the chloroplast. Approaches used include analysis of hybridization and dissociation of DNA (Chapter 7), use of restriction endonucleases to detect base substitutions and rearrangements (Chapter 8), and comparisons of the primary sequences (Chapter 9). Methods for collecting data on nucleic acid differences continue to improve; the challenge now is to further develop methods of sampling (Chapters 2 and 3) and data analysis (Chapters 10 and 11) to deal with the influx of molecular information.

CONTROVERSIES IN MOLECULAR SYSTEMATICS

The collection of molecular data and their use in systematics have led to several controversies, some of which have generated more heat than light. These controversies include arguments about the relative value of molecular versus morphological data, the constancy of rates of molecular evolution, the neutrality of molecular variants, the types of data that should be collected, the various philosophical approaches to analyzing data, and the meaning of "homology" in relation to molecular characters. Some of these debates are specific to molecular data, whereas others are general to all types of evidence used to estimate phylogeny. Each of these debates is reviewed at length elsewhere; here we merely outline the principal arguments and their implications for molecular systematics.

Molecules versus Morphology

There has been considerable debate over whether molecular or morphological features are inherently better sources of information for estimating phylogeny (Patterson, 1987). Some have claimed that molecular characters are relatively weak (e.g., Kluge, 1983), whereas others have claimed that morphological characters are likely to be misleading (e.g., Frelin and Vuilleumier, 1979; Sibley and Ahlquist, 1987a). Closer examination shows this to be an empty argument (Hillis, 1987). Comparative studies have shown that morphological change and molecular divergence are quite independent, responding to different evolutionary pressures and following different rules (Wilson et al., 1974, 1977). However, the real concerns for the practicing systematist are whether the characters examined exhibit variation appropriate to the question(s) posed, whether the characters have a clear and independent genetic basis, and whether the data are collected and analyzed in such a way that it is possible to compare and combine phylogenetic hypotheses derived from them.

The conflicts between molecular and morphological evidence have been overemphasized. The development of molecular systematics has not resulted in widespread refutation of phylogenetic hypotheses generated by morphologists, although the molecular approach is potentially powerful for generating and testing competing phylogenetic hypotheses. However, many systematists have emphasized the conflicts and deemphasized the concordance between inferences based on molecules and those based on morphology. The conflicts among molecular studies and among morphological studies are at least as great as those occurring between the two fields. Each approach has distinct advantages and disadvantages. For example, most (but not all) molecular data have a clear genetic basis and the total data set is limited only by the genome size; on the other hand, morphological data can be obtained from ancient fossils (e.g., Gauthier et al., 1988) and extensive preserved collections and can be interpreted in the context of ontogeny (Kluge and Strauss, 1986). In general, studies that incorporate both molecular and morphological data will provide much better descriptions and interpretations of biological diversity than those that focus on just one approach. Furthermore, it is possible to address some systematic problems only with morphological data and other problems only with molecular data (see Hillis, 1987; Fernholm et al., 1989). This book is concerned only with molecular variation because many issues are unique to molecular data and are inadequately covered elsewhere, not because we view molecules as inherently superior to morphological characters as markers of evolution.

Constancy of Evolutionary Rates

Early indications of a strong correlation between estimates of sequence divergence and of divergence time (Zuckerkandl and Pauling, 1962) raised the exciting possibility that molecular comparisons could provide indications of the time of divergence for taxa where no fossils exist. Although most biologists now accept a broad correlation between amount of molecular divergence (at least for proteins and DNA) and time, it is far from established that rates are constant (reviewed by Wilson et al., 1977, 1985; Thorpe, 1982; Gillespie, 1986c; Kreitman, 1987; Nei, 1987). Indeed, recent evidence (Goodman, 1981, 1985; Gillespie, 1984, 1986a,b; Britten, 1986; Vawter and Brown, 1986) indicates sufficient rate heterogeneity that one should not assume that rates are equal on an a priori basis. For instance, in a recent review of DNA sequences, Gillespie (1987) has calculated the ratio of the variance of the number of substitutions to the mean number of substitutions that occur along a lineage as ranging from 1 to 35 for amino acid substitutions and 1 to 19 for silent substitutions, indicating considerable fluctuations in evolutionary rate. This has significant implications for molecular systematics. Constancy of rates is a cornerstone of the neutral theory of molecular evolution (see below), is an assumption of some methods for estimating phylogeny (Chapter 11), and is widely assumed in estimating time since divergence (Chapter 12).

To some extent, the arguments over the molecular clock stem from different expectations. The utility of such a clock depends on the quality of information needed to test a specific hypothesis; if a clock indicates 3:20 PM, but the real time could be anywhere from 12:20 PM to 6:20 PM, the clock is useful only if one needs to know if it is morning or afternoon. In some cases, e.g., hominoid divergences (Sarich and Wilson, 1967), molecular estimates of divergence time have led to a substantial reevaluation of fossil evidence. However, most purported tests of hypotheses about divergence time have ignored problems associated with calibration and few have calculated appropriate confidence intervals. These confidence intervals can be so large in some cases that the term "clock," or even "sloppy clock," becomes meaningless (see Chapter 12).

Neutrality of Molecular Variants

A frequently voiced concern is that molecular characters are not neutral and that selection will bias analyses of intraspecific variation and estimates of phylogeny. This relates to a much broader argument over the evolutionary significance of molecular variation, the "neutralist–selectionist" controversy, that has been a major concern of molecular population genetics since Kimura's (1968) seminal paper on the neutral theory (reviewed by Lewontin, 1974, 1986; Kimura, 1983b, 1986; Gillespie, 1987). There can be no doubt that many protein, chromosome, and DNA variants are acted on by selection; it also appears that much molecular variation is consistent with predictions of various modifications of the neutral theory (Ohta, 1977; Sarich, 1977; Gillespie, 1987). Thus, the debate is reduced to whether or not most molecular variants are selectively neutral (or nearly neutral), and whether neutrality or selection should be considered the null hypothesis. The current lack of a general testable theory of molecular evolution based on selection dictates that neutrality must usually serve as the null hypothesis (but see Gillespie, 1986a). However, given the poor fit of many molecular data to the neutral theory (Gillespie, 1987), one should make a conscious distinction between testing for neutrality and simply assuming that it exists.

The impact of selection on systematic studies depends on the proportion of markers (loci) affected, the extent and sign of correlations among loci, and the robustness of the method of analysis to departures from neutrality (Chapters 2, 10, and 11). Where deviations from neutrality are likely to significantly bias analyses, the assumption of neutrality should be made explicit, preferably in a way that can be tested. However, because most departures from neutrality are thought to be locus specific, it is widely assumed that selection will have relatively minor effects on the overall analysis if numerous loci are examined.

Data Quality and Presentation

Population genetic or phylogenetic estimates can be only as precise as the primary data themselves. It is widely assumed that techniques that manipulate DNA sequences (DNA hybridization, RFLP analysis, and sequencing) are inherently more accurate than protein-based methods (allozyme electrophoresis and immunological techniques). Similarly, among the DNA methods, direct comparison of DNA sequence information is supposed to be the most informative. An assumption of these arguments is that the techniques are applied with equal rigor. Obviously, it is just as important to confirm DNA sequences using the complementary strand or overlapping primers and repeated runs (Chapter 9) as it is to use internal controls in allozyme electrophoresis (Chapter 4), immunological experiments (Chapter 5), and DNA hybridization (Chapter 7).

Inevitably, it is up to the investigator(s) to decide whether the data are of sufficient quality. However, the data should also be presented in such a way that peer reviewers and readers can judge the technical quality and extent of the data themselves. Unfortunately, once techniques have become established in the literature, there has been a tendancy on the part of editors and authors alike to dispense with the primary data, i.e., photographs of gels, chromosomes, or raw experimental data. In practice, this can lead to unnecessarily acrimonious debates over data quality and interpretation (e.g., Cracraft, 1987; Lewin, 1988; Sibley et al., 1988; Sarich et al., 1989). This is a poor reflection on the field as a whole and it is up to practitioners of molecular systematics to insist on rigorous standards of data quality and presentation.

Types of Characters and Methods of Analysis

The techniques of molecular systematics produce two fundamentally different types of information: distance data, where differences among molecules are measured as a single variable (e.g., immunological methods, Chapter 5; DNA hybridization, Chapter 7), and character data, where differences are measured as a series of discrete variables (characters), each with multiple states. Character data can be converted to distances, but distances cannot be converted into character data. Character data have some advantages for data collection and analysis. It is relatively easy to add information on new taxa to the data set (see Chapter 2) and data obtained from different sources (other molecules or other types of attributes) can be readily combined for analysis (e.g., Miyamoto, 1983b, 1985; Kluge, 1989).

Philosophical arguments about different approaches to phylogeny estimation, e.g., whether and how to apply parsimony, the justification for maximum likelihood approaches, and the successive weighting of characters based on their fit to an initial hypothesis (Felsenstein, 1978a, 1988; Wiley, 1981; Sober, 1983; Farris, 1983; Kluge, 1984; Lake, 1987; Carpenter, 1988; see Chapter 11) currently dominate discussions of phylogenetic analysis. Some molecular techniques inevitably restrict the range of applicable methods of analysis. This is not a problem so long as the remaining options can reliably estimate phylogeny, which in turn depends on the frequency with which assumptions specific to the method are violated and how sensitive the method is to those deviations. Considerable effort is therefore being given to examining the robustness of alternative methods for phylogenetic analysis (Chapter 11) and estimation of population genetic parameters (Chapter 10). Development of new methods of analysis and their implementation in computer algorithms also constitute a very active field (Chapters 10 and 11). Nonetheless, the greatest obstacle to the incorporation of the new flood of molecular data in systematics is a lack of adequate algorithm development and implementation (cf. the problem of alignment of multiple sequences discussed in Chapters 9 and 11).

Homology and Similarity in Molecular Systematics

The uses and misuses of the word **homology** frame a complex subject. Difficulties in its use arise as a result of differences in meaning between molecular and morphological (classical) biology (Patterson, 1988) and within molecular biology (Reeck et al., 1987; Aboitiz, 1987; Dover, 1987; Wegnez, 1987), and as a result of different meanings depending on context. In general, homology means inferred common ancestry, although it is commonly misused to mean similarity (Fitch, 1966, 1970; Reeck et al., 1987). Similarity is an empirical observation and can be quantified, whereas homology usually must be hypothesized and is an all-or-none condition. Therefore, two proteins cannot be "95% homologous," but instead are 95% similar (or 95% **isologous**; Wegnez, 1987), which may be used to infer that they are homologous. However, there are several reasons that the proteins may be similar, including common ancestry (homology), convergence, and gene conversion (Patterson, 1988).

There are also several types of homology that must be distinguished. Homologous sequences can diverge through speciation, in which case they are properly termed **orthologous** (Fitch, 1970). If, on the other hand, the homologous sequences diverged after gene duplication, they are properly referred to as **paralogous** (Fitch, 1970). Homologous sequences can also arise via lateral gene transfer (through retroviruses, for instance), in which case the sequences are said to be **xenologous** (Gray and Fitch, 1983). The distinction is necessary because only orthologous sequences can be used to infer phylogeny of species. Confusion of paralogous and orthologous sequences can result in a correctly estimated phylogeny for the molecules that differs markedly from that of the organisms from which they were sampled. Consider the example in Figure 2. A gene duplication event in the ancestor of species 1, 2, and 3 gave rise to the two paralogous sequences A and B.

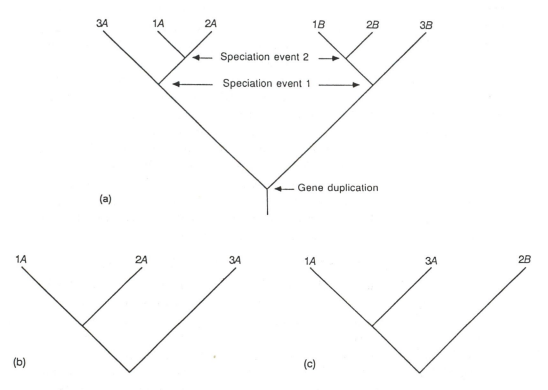

FIGURE 2.   The consequences of using orthologous versus paralogous genes to infer phylogeny. (a) The phylogeny of a set of homologous genes in three species (1–3); a gene duplication event in the ancestor of the three species gave rise to two sets of paralogous genes (A and B), and two subsequent speciation events gave rise to orthologous genes in each of three species. (b) The phylogeny inferred from comparison of either set of orthologous genes (notice that this is the correct species phylogeny). (c) The phylogeny inferred from comparison of two orthologous and one paralogous sequences (this is the correct gene phylogeny, but not the correct species phylogeny).

Subsequently, two speciation events gave rise to the three species, such that species 1 and 2 shared a more recent common ancestor (Figure 2a). One could potentially recover the phylogeny of the three species by examining the orthologous *A* sequences in each species, or by examining the orthologous *B* sequences in each species (Figure 2b). However, examination of paralogous sequences (e.g., *A* in species 1 and 3 and *B* in species 2) would result in incorrect inferences about species phylogeny, but correct inferences about gene phylogeny (Figure 2c). Thus, for problems of species phylogeny, the sequences examined must be orthologous.

Unfortunately, the problems do not end here, because "homologous" is used in at least two additional senses in molecular biology. In cytogenetics, it is standard to refer to the respective chromosomes in a chromosome pair of a diploid organism as "homologs" and to refer to the homologous pair of chromosomes in another species as "homoeologs" (Chapter 6), even though this is quite different from the use of "homology" in classical morphology (where "homonomy" is used to refer to a repeated structure in a single organism). In addition, a molecular probe is said to be "homologous" if it is used to study the same species from which it was derived, and "heterologous" if it is used to study a homologous sequence in another species (e.g., see Chapter 5). Although these multiple uses of the term homology can be confusing, their widespread use in the primary literature dictates that they be explained and incorporated in this book.

SCOPE AND USE OF THIS VOLUME

This volume aims to provide an overview of molecular methods currently used to analyze diversity within and among species. The primary goal is to provide new workers in this rapidly expanding field with sufficient technical and theoretical information to enable them to select one or more appropriate methods, to design and implement a study, and to analyze the resulting data, all with maximum efficiency. In selecting an appropriate technique for obtaining data, the basic questions to be considered are (1) will it produce a type of information compatible with the desired method of analysis; (2) is the signal-to-noise ratio likely to be sufficiently high to address the question(s) posed; and (3) is it cost effective and feasible given the available facilities and expertise? For practicing molecular systematists, the chapters in this volume may suggest alternative strategies for collecting and analyzing data and new perspectives on limitations and assumptions of familiar techniques.

The book has three major sections, each representing an important phase of a study: sampling design and methods (Chapters 2 and 3); methods for detecting and analyzing variation in proteins, chromosomes, and nucleic acids (Chapters 4 to 9); and methods for analyzing the data (Chapters 10 and 11). We have attempted to include a balance of viewpoints concerning different methods of data collection and analysis. One obvious omission is amino acid sequencing, a technique that is of great historical importance in molecular systematics (Goodman et al., 1987). However, amino acid sequencing has been largely replaced by nucleic acid sequencing, at least for most systematic applications. Despite automation, amino acid sequencing is technically much more difficult and time-consuming than nucleic acid sequencing, and amino acid sequences can be easily deduced from nucleic acid sequences. Nonetheless, comparison of deduced amino acid sequences with existing systematic data bases continues to be a highly productive endeavor and suggestions for phylogenetic analysis of amino acid sequences are included in Chapter 11. Otherwise, the coverage of techniques is fairly comprehensive. To facilitate comparisons, each of the molecular technique chapters is arranged into sections on (1) principles and comparisons of methods (including a discussion of assumptions), (2) applications and limitations, (3) laboratory setup, (4) protocols, and (5) interpretation and troubleshooting. A glossary of terms peculiar to molecular systematics and a list of common abbreviations are given after Chapter 12. Words and phrases included in the glossary are in bold type at their first appearance in the text.

For the most part, protocols are basic and well proven. Emphasis also has been placed on highlighting recent developments that appear particularly promising. However, for each approach, there is a wide range of alternative protocols not described here. This volume therefore is designed to complement existing manuals that focus on a single approach (see General References). These should be referred to for additional background and alternative methods once a particular approach has been adopted.

GENERAL REFERENCES

Proteins

Harris, H., and D. A. Hopkinson. 1976 et seq. *Handbook of Enzyme Electrophoresis in Human Genetics.* North-Holland, Amsterdam.

Richardson, B. J., P. R. Baverstock, and M. Adams. 1986. *Allozyme Electrophoresis.* Academic Press, Sydney.

Klein, J. 1982. *Immunology: The Science of Self–Nonself Discrimination.* Wiley, New York.

Chromosomes

Darlington, C. D., and L. F. La Cour. 1969. *The Handling of Chromosomes.* 5th ed. Allen & Unwin, London.

MacGregor, H., and J. Varley. 1983. *Working with Animal Chromosomes.* Wiley, New York.

Sharma, A. K., and A. Sharma. 1972. *Chromosome Techniques: Theory and Practice.* Butterworth, London.

Nucleic Acids

Ausubel, F. M. (ed.). 1989. *Current Protocols in Molecular Biology.* Wiley, New York.

Berger, S. L., and A. R. Kimmel (eds.). 1987. Guide to molecular cloning techniques. Methods Enzymol. 152:1–812.

Davis, L. C., M. D. Dibner, and J. F. Battey. 1986. *Basic Methods in Molecular Biology.* Elsevier, Amsterdam.

Erlich, H. A. (ed.). 1989. *PCR Technology: Principles and Applications for DNA Amplification.* Stockton Press, New York.

Hames, B. D., and S. J. Higgins (eds.). 1985. *Nucleic Acid Hybridisation: A Practical Approach.* IRL Press, Oxford.

Innis, M. A., D. H. Gelfand, J. J. Sninsky, and T. J. White (eds). 1990. *PCR Protocols: A Guide to Methods and Applications.* Academic Press, San Diego.

Sambrook, J., E. F. Fritsch, and T. Maniatis. 1989. *Molecular Cloning: A Laboratory Manual,* 2nd ed. 3 Volumes. Cold Spring Harbor Laboratory, Cold Spring Harbor, NY.

I
SAMPLING

CHAPTER 2

SAMPLING DESIGN

Peter R. Baverstock and Craig Moritz

INTRODUCTION

Molecular systematic studies require particularly careful planning because they are usually relatively expensive, and they frequently involve destructive sampling of the organisms (Chapter 3). The aim should therefore be to maximize the information obtained per specimen; too few specimens may lead to an inconclusive result; too many is sheer waste. Despite this requirement, molecular systematic studies seem especially prone to poor planning. Too often projects are well advanced before it is realized that the sampling strategy is inappropriate, the wrong tissues have been collected, the tissues have been stored inappropriately, the wrong technique has been chosen, or far too many or far too few specimens have been included.

Molecular systematic studies typically involve the following stages:

1. define the problem
2. conduct a pilot study
3. determine the appropriate sampling strategy
4. collect samples
5. analyze the samples
6. analyze the data

This chapter deals mainly with step 3, establishing the most efficient sampling design. However, steps 1 and 2 will have a profound influence on step 3 and are therefore given consideration. The remaining chapters of this book concern steps 4–6.

The cost of a project should take account not only of the cost of chemicals and other consumables, but also the cost of time, which includes both the collecting and the screening phases of the project. Thus, the sampling design should aim to minimize the number of specimens and their handling in a way that remains compatible with the biological questions being asked. Indeed, the most important first step is to clearly define the biological questions being asked. The questions should be stated in as specific and detailed a way as possible. It will be particularly useful to construct formal hypotheses as a guide to further steps in the analysis.

STATISTICAL CONSIDERATIONS

At the very outset, one needs to decide on the level of error that is acceptable. There are two types of errors: A type I error occurs if the null hypothesis is rejected when it should be accepted. A type II error occurs if the null hypothesis is accepted when it should be rejected. Type II errors are difficult to define for most biological systems, because the expected difference between two populations is usually unknown. They are usually expressed as the power of the test, i.e., 1 minus the probability of a type II error. The level of type I error one is prepared to accept depends on the consequences of being wrong. In biological studies it is usually set at 5%, but this limit should not be accepted blindly. For example, a researcher may be testing the hypothesis that a particular species of commercial fish has a population structure characterized by isolated demes. The corresponding null hypothesis is that the entire species is a single panmictic unit. Before launching into a full-scale study, the researcher should do a pilot study to see if there is any suggestion at all that the fish population shows evidence of genetic substructuring. Here the type I error might be set at 20%, since the consequence of being wrong (i.e., rejecting the null hypothesis when it is true) is that a fuller study is carried out. However, in the full-scale study, the type I error might be set at 1%, since here the consequence of being wrong is that inappropriate management procedures are adopted for the fish species.

The sample sizes required for a given level of type I error and a given power depend on the sampling variance. Many biological data follow a normal distribution, for which the mean and the variance are independent. However, genetic data such as allele frequencies determined by allozyme electrophoresis (Chapter 4) or RFLP studies (Chapter 8) may follow a binomial distribution where the variance can be estimated from the mean:

$$V_p = p(1 - p)/n$$

where p is the allele frequency and n is the sample size. For nuclear loci in a diploid population this distribution is appropriate if the genotype frequencies conform to **Hardy–Weinberg equilibrium** (HWE), otherwise resampling procedures should be used to estimate variances of allele frequencies (Chapter 10).

MOLECULAR SYSTEMATICS

There are three main applications of molecular systematics: studies of population structure (e.g., geographic variation, mating systems, heterozygosity, and individual relatedness), identification of species boundaries (including hybridization), and estimation of phylogenies. Each of these requires different approaches to virtually every phase of the study from project planning through pilot studies to sample sizes, sampling strategies, methods of data collection, and finally data analysis. It is therefore necessary to have a clear idea of the aims of the study very early in the planning stage. Determining the relationships of specific individuals (e.g., testing parentage) requires direct comparison of allozymes (Chapter 4) or, preferably, hypervariable sequences (Chapter 8) between putative relatives. Sampling methods and potential pitfalls are reviewed by Sensabaugh (1982) for allozymes and by Lynch (1988) for hypervariable sequences. The remaining applications are discussed below.

Studies of Population Structure

Background   The genetic structure of a population is perhaps the most fundamental piece of information for a species that requires management. For some species, the entire population may consist of a single panmictic unit (the panmictic model); others may consist of a series of small subpopulations, each largely isolated from other subpopulations (the stepping-stone or island models); still others may consist of a continuous population, but individuals within it exchange genes only with geographically proximate individuals (the isolation-by-distance model). Deciding which model best approximates the population structure of a particular species is usually the first step in understanding its population biology.

Population structure of animals has traditionally been studied using mark–release–recapture techniques. However, this approach has several limitations: dispersal may not be a good indicator of gene flow (Endler, 1979; Levin, 1981), it is unsuitable where the probability of recapture is low, and it is relatively expensive and time consuming. Gene flow in plants may involve both pollen and seeds and is particularly difficult to study by ecological methods. A molecular genetics approach can be used to help circumvent these problems. Moreover, the genetic approach reveals the net result of past historical demographic processes (e.g., Larson et al., 1984; Easteal, 1985), and is, therefore, a better predictor of the long-term population structure of the species.

The three different models of population structure result in different patterns of genetic differentiation within and between geographic localities. Therefore an analysis of the genetic structure of a species can give the investigator important clues to the population structure (e.g., Daly, 1981; Richardson, 1983; Crawford, 1984; see also Chapters 4 and 8).

When using the genetic approach it is desirable to sample as many loci as possible to minimize random variables and locus-specific effects such as selection (see Chapter 10). The need to examine a large number of loci is evident from the observations that the reliability of estimates of summary statistics such as heterozygosity, genetic distance, and F_{ST} depends more on the number of loci than on the number of individuals (Nei, 1978; Nei and Chesser, 1983; Chakraborty and Leimar, 1987).

To date, allozyme electrophoresis (Chapter 4) has been the genetic technique most widely used to study the genetic structure of populations, whereas restriction enzyme analysis of mitochondrial and nuclear DNA is being used with increasing frequency (Chapter 8). The distribution of variation within and among populations revealed by these methods may differ. Uniparentally inherited loci (e.g., mitochondrial DNA, Y-linked loci, most chloroplast DNA) are generally expected to show less variation within populations and more between populations than are biparentally inherited loci such as autosomal nuclear loci (see Chapter 8). Similarly, repeated sequences subject to strong concerted evolution (Chapter 8) may also have reduced levels of variation within populations. Any such alteration in the distribution of variation has important implications for sampling design.

Pilot Studies The pilot study usually has three major aims: (1) to find genetic markers (i.e., polymorphic loci), (2) to determine whether the polymorphic markers are suitable in a practical sense, and (3) to establish the feasibility of a large-scale sampling program.

Establishing Markers. Samples should be obtained from multiple populations to identify locally polymorphic markers as well as those with widespread variation. It is at this point that the distribution of variation within versus among populations should be assessed. For small numbers of populations, the sample size needed to detect a given level of differentiation at a diploid locus among populations (using G_{ST}; Nei, 1973) at least 50% of the time (i.e., a power of 0.5) with a type I error of 5% is approximately $2N > 1/G_{ST}$(Chakraborty and Leimar, 1987).

A suitable approach for the pilot study may be to collect relatively large samples (e.g., $N > 20$) from two localities and smaller samples ($N \approx 5$) from several other localities. This represents a trade-off between the need to assess within-population variation (particularly for diploid nuclear loci) and among-population variation (particularly for **mtDNA** and **cpDNA**). These samples should be assayed for as many loci as possible, preferably including different genetic systems (e.g., allozymes as well as mtDNA or cpDNA; see Chapters 4 and 8). At this point, it may be appropriate to approach other laboratories with experience in specific loci, rather than spending resources establishing loci that turn out to be uninformative (e.g., monomorphic).

The actual number of genetic markers required will depend to some extent on the subtlety of the population substructuring encountered and the variance among loci (see Chapter 10). One marker is clearly insufficient because the effects of selection and substructuring cannot be distinguished. Even two loci are insufficient because selection or linkage may give the same pattern for each. Thus, at the very least, three loci with common polymorphisms should be used. If fewer are found, some other approach to the problem should be explored. Ideally, six or more markers should be used.

Suitability of Markers. Because thousands of samples may need to be screened in the main study, markers should be inexpensive and easily scored. Moreover, for diploid (or polyploid) loci, it is essential that heterozygotes can be clearly and consistently distinguished from both homozygotes. This can cause difficulties with some allozyme markers (Richardson et al., 1986), but should not be a problem with **RFLPs** if a small number of loci is examined per gel (see Chapter 8). At this stage it is prudent to experiment with different tissues, different tissue treatment, different storage regimes, etc. (see Chapters 3 and 4) to improve the resolution of loci that are polymorphic but are proving difficult to score. These things should be sorted out before the main sampling program begins.

Feasibility of Population Sampling Program. The pilot study also gives the opportunity to test the feasibility of the main sampling program: have all the logistic problems been sorted out?

Sample Sizes and Strategies Sampled localities may have different allele frequencies for a polymorphic locus, but the difference may go undetected because of the small sample sizes used (a type II error). The smaller the actual difference in allele frequencies, the larger the sample sizes needed to reliably detect them. Thus, the first consideration in setting sample sizes is the magnitude of the allele frequency differences one expects to encounter. The only other considerations are the level of type I and type II errors one is prepared to accept. Again both can be reduced by increasing the sample size. Table 1 gives the minimum sample sizes required to detect given levels of allele frequency differences (assuming HWE) for various levels of type I error and various powers of test. For example, let us assume that we have two diploid populations, both polymorphic at a locus with allele frequencies of 0.5/0.5 and 0.55/0.45, respectively. Our null hypothesis is that the two populations do not differ in allele frequency. Let us assume also that we have decided on a type I error of 5% (i.e., we will reject the null hypothesis only when the data have less than a 5% probability of occurring if the null

Table 1. The number of diploid individuals in each of two samples required to detect given differences in allele frequency (dp)[a]

Power	dp	*p*				
		0.55	0.70	0.80	0.90	0.95
50%	0.05	760	645	492	276	146
	0.10	190	162	123	69	50[b]
	0.20	48	40	31	25	50[b]
	0.50	6[b]	9	13	25	50[b]
80%	0.05	1554	1319	1006	564	299
	0.10	389	332	252	141	76
	0.20	99	82	64	27	50[b]
	0.50	16	14	13	25[b]	50[b]
90%	0.05	2081	1766	1345	756	400
	0.10	520	444	337	189	102
	0.20	132	110	85	50	50[b]
	0.50	22	20	14	25[b]	50[b]

[a] A χ^2 test for homogeneity with the probability of a type I error set at 5% with powers of 50%, 80%, and 90% is used. Because the sample sizes required depend on the actual allele frequency (p), sample sizes required are given for various values of p (from Richardson et al., 1986).
[b] The χ^2 homogeneity test requires a minimum expected frequency of 5 in each cell—values marked have been set to meet this requirement.

hypothesis were true), and a power of 90% (i.e., we want to be sure that if we accept the null hypothesis it has a 90% chance of being correct). Then the two samples would each need to consist of at least 2081 individuals!

An alternative way of viewing the problem of sample size is to look at what level of discrimination will be achieved for a given sample size. For example, with sample sizes of 100 in each of two diploid populations, only differences in allele frequency of at least 0.1 to 0.2 (depending on the actual allele frequencies) will be detected with a probability of a type I error set at 5% and the power set at 80%. Smaller differences in allele frequencies will usually go undetected, even with such large sample sizes.

The number and geographic pattern of localities that ultimately need to be sampled will depend to a large extent on the actual scale of substructuring, which may not be apparent until after the first round of sampling. For example, if following the first round of sampling the entire population conforms to a panmictic unit, it may be decided to do no further sampling. But if substructuring is indicated, it may be at a level too fine for analysis from the first round of sampling and additional sampling at various hierarchical levels of geographic organization will be required (see Richardson, 1981; Johnson and Black, 1984). Alternatively, different genetic systems are likely to show heterogeneity at different levels (see Chapter 8). The budget for the program should foreshadow the possibility of additional rounds of sampling. At some localities, repeat samples should be taken to test for intralocality effects, an important consideration for species such as fish that may occur as different schools (Richardson et al., 1986).

Studies of Species Boundaries and Hybridization

Background There has been considerable debate in the literature concerning the most appropriate definition of a species. A common modern view is the evolutionary species concept, according to which a species is "a single lineage of ancestral-descendant populations which maintains its identity from other such lineages and which has its own evolutionary tendencies and historical fate" (Wiley, 1978). For sympatric sexually reproducing species, this reduces to the biological species concept (Mayr, 1969), according to which a species consists of a group of individuals capable of exchanging genetic material with each other but which are reproductively isolated from all other such groups.

There are at least five situations in which morphological data alone will be inadequate for defining species boundaries. First, two species may be **sympatric** (overlapping) or **parapatric** (abutting), but be so similar in morphology that their specific status goes undetected (e.g., Donnellan and Aplin, 1989). Second, two **allopatric** (geographically separate) populations may be morphologically different, but their status as species is questionable. Third, two parapatric populations may be morphologically distinct, but show clinal variation or broad hybridization. Fourth, two morphologically distinct forms may represent polymorphisms within a single interbreeding population (e.g., Titus et al., 1989). Fifth, asexual species may have morphologically similar forms that arose independently from sexual species.

Of the various molecular genetic approaches that may be brought to bear on the problem, allozyme electrophoresis (Chapter 4) appears to remain the most generally applicable and efficient, although cytogenetic (Chapter 6) and restriction site (Chapter 8) analysis can often be useful as well (see Chapter 12).

Different species usually have a fixed allelic difference at some of the loci screened in electrophoretic studies. Thus, for predominantly outcrossing species, the presence of sympatric cryptic species can be tested for by looking for variable loci that lack heterozygotes, while the status of sympatric morphotypes can be evaluated by testing for significant differences in genotype or allele frequencies (Chapter 10). For allopatric populations and asexual populations the aim is to assess the extent of genetic divergence between the populations being tested in relation to geographic variation within species. In all cases it is more important to maximize the number of loci screened than the number of individuals examined.

Pilot Studies, Sample Sizes, and Sampling Strategies   For sympatric out-crossing species, very small samples are adequate so long as they include both species. For example, let us assume that a sample of 10 individuals is collected. The null hypothesis under test is that all specimens belong to a single random-mating population. At one locus, six specimens are homozygous for one allele and four specimens are homozygous for a different allele; there are no heterozygotes. The best estimates of the allele frequencies at this variable locus are $p = 0.6$ and $q = 0.4$. The expected proportion of heterozygotes, based on the Hardy–Weinberg equilibrium for a random-breeding population, is $2pq$ which here is 0.48. The probability of an individual not being a heterozygote is therefore $1 - 2pq = 0.52$. The probability of all 10 individuals not being heterozygous is $(0.52)^{10} = 0.0014$. Clearly, the null hypothesis is under serious challenge. If an additional locus is found that shows the same pattern of fixed differences involving the same individuals, then the null hypothesis is under still more serious challenge; a reasonable alternative hypothesis is that two species are involved. Other hypotheses that deserve consideration are that the species is actually asexual, or that it is haploid, or that it has a very high level of self-fertilization.

In practice, a minimum of two loci showing patterns of fixed differences that are consistent between individuals should be aimed for. This is necessary because an apparent lack of heterozygotes at a locus can result from other effects. For example, variation may not be under simple genetic control (e.g., *Ldh-R* in *Mus domesticus*; Shows and Ruddle, 1968), or it may exhibit ontogenetic variation (e.g., hemoglobin in vertebrates), or, at least in theory, there may be very strong selection against heterozygotes.

The above argument rests on the assumption that fixed differences will be found. Clearly, the more loci that are screened, the greater the chance of finding such loci if two genetically distinct species really are represented. Consequently part of the aim of the pilot study should be to try different tissues, different treatments, etc. on a limited number of specimens with a view toward increasing the number of loci screened.

If the aim is to test whether two previously identified sympatric groups (e.g., distinctive morphotypes or chromosome races) are reproductively isolated, loci with shared polymorphism may also be useful. These can be examined one locus at a time, testing for significant differences in allele frequencies (Table 1) or significant deficiencies of heterozygotes (Chapter 10). Alternatively, several loci can be examined simultaneously using disequilibrium statistics (e.g., Ryman et al., 1979; see Chapter 10). Such analyses will usually require much larger sample sizes than is the case if loci with fixed differences are used (Brown, 1975). Moreover, disequilibrium statistics should be interpreted with caution because significant disequilibrium can result from many factors other than the presence of two reproductively isolated groups (Hartl and Clark, 1989).

Once the presence of two species has been indicated, a follow-up study is usually required in order to find morphological features diagnostic for the species, usually involving multivariate analyses. Here it is unlikely that the original sample will be sufficiently large for a full multivariate morphometric analysis, especially when the possible effects of age and sex are taken into account. However, additional specimens need to be "typed" only for the diagnostic loci. The sample sizes required for this part of the study will depend on the subtlety of any morphological differences between the species.

Methods for assessing whether allopatric populations represent distinct species are controversial. For example, some have suggested that a certain amount of divergence is indicative of being on separate evolutionary trajectories and can therefore be used to define species (e.g., Baverstock et al., 1977; Highton et al., 1989), although this approach has been strongly criticized (Frost and Hillis, 1990). A more appropriate approach may be to compare genetic divergence between two allopatric populations suspected of representing distinct species with that between similarly separated populations within each form (Jackson and Pounds, 1979). We have argued elsewhere that for studies of species boundaries and relationships, the proportion of fixed differences between two samples is the most appropriate measure of genetic divergence (Richardson et al., 1986; see also Frost and Hillis, 1990). Using this approach, shared polymorphisms with very different allele frequencies may be incorrectly scored as fixed differences, but, from the point of view of assessing genetic divergence between allopatric populations, very different allele frequencies indicate high genetic divergence and are operationally equivalent to fixed differences. Once again, however, because the variance of the estimate of between-population variation depends mainly upon the number of loci (Nei, 1978), every attempt should be made to maximize the number of loci screened.

The pilot study should consist of screening about five individuals of each of the two geographic forms. If no fixed differences are found, there is no point in screening additional individuals or additional populations since increasing the population sample size can only reduce the estimate of fixed differences. Any additional effort should focus on examining additional loci. Only where potentially diagnostic differences are found should additional sampling be contemplated. Here, small samples (about five individuals for diploid organisms) from each of several geographically widespread populations of each of the morphological forms should be screened.

Hybrid Zones   Although it is true that some information can be obtained from polymorphic markers, the population genetics of hybrid zones is most readily investigated if fixed genetic differences are found between the parental taxa involved in hybridization. Consequently every effort should be made in the pilot study to discover such fixed differences rather than rely on allele frequency differences.

Three additional features of hybrid zones are salient to the project planning stage (e.g., D. D. Shaw et al., 1987). First, genetic markers frequently show introgression over much broader geographic areas than might be predicted from morphology alone. Second, different genetic markers frequently show different levels of introgression. Third, uniparentally inherited nonrecombining markers (such as mtDNA and cpDNA) provide information of a different kind from diploid nuclear markers (Chapter 8).

As a consequence of these considerations, it is useful to screen both nuclear diploid and uniparental haploid loci for fixed differences. Moreover, the pilot samples should be taken from localities well away from the hybrid zone itself, and should involve several populations of each of the parental taxa.

Phylogenetic Relationships

Background   The single most important component of the project planning stage of a phylogenetic analysis is the decision as to which method(s) or sequence(s) are appropriate to the phylogenetic question at hand. The method chosen must yield sufficient variation to be phylogenetically informative, but not so much variation that convergences and parallelisms overwhelm informative changes (see Chapter 1).

There is a considerable body of evidence suggesting that the rate of evolution at the molecular level is at least roughly the same across all groups for a particular gene or set of genes (e.g., Wilson et al., 1977; cf. Gillespie, 1986c; see Chapter 1). As a consequence, the method chosen will depend to a large extent on the time frame over which divergence has occurred for the study group (see Chapter 12).

When the phylogenetic study begins, the time scale for the group in question will probably be largely unknown. Guesses based exclusively on morphology of extant forms are likely to be quite misleading because rates of morphological evolution vary enormously between groups (Cherry et al., 1982; Baverstock and Adams, 1987). Fossil data must also be interpreted with caution (Wilson et al., 1977). Therefore the prime purpose of the pilot study should be to determine which molecular technique or techniques are appropriate to the study group. Moreover, this should be done before extensive collection of tissues because the technique or techniques to be used may require that specific tissues be collected, and that they be stored in particular ways (Chapter 3).

Pilot Study   The aim of the pilot study is to obtain some idea of the extent of genetic divergence in the study group and thus to identify the most appropriate method(s) and sequence(s) for analysis. To reiterate, these should have low diversity within groups relative to the differences among groups. For the pilot study, it is desirable to sample individuals from taxa representing the two extremes of differentiation, i.e., two pairs of closely related taxa and two pairs of distantly related taxa. Again, it is desirable to evaluate the distribution of variation within and between groups for different types of loci (e.g., allozymes, mtDNA, rDNA sequences). It may be that some approaches work at higher taxonomic levels and others work at lower levels (Chapter 12). The number of specimens examined per group can be quite small (even one), unless shared polymorphisms among species are a likely possibility (e.g., closely related species). If shared polymorphism is a reasonable possibility, then at least two larger samples ($N \approx 10$) should be included. Multiple populations should be examined for closely related pairs.

In principle, these specimens could be subjected in the first instance to any one of the treatments discussed in Chapters 4 to 9 to obtain some idea of the appropriate method to be used in the main study. However, it would be most efficient to start with a method already available in the laboratory or to begin with a relatively cheap and fast method. If no technique is available locally, it would be wise to see if another laboratory with one of the techniques already established will run the pilot specimens rather than establish a method de novo that ultimately turns out to be inappropriate for the major study.

The pilot study will determine which technique(s) are most appropriate for the group, and hence how many specimens are needed for the main study, which tissues should be collected, and how they should be stored. It should be possible at this stage to estimate the cost of the study in terms of both consumables and time.

Sample Sizes and Sampling Strategy   The number of specimens needed per group to resolve relationships among groups depends critically on the amount of polymorphism relative to the extent of divergence. If, for a given method or sequence, the pilot study indicates that virtually all of the variation occurs among groups, then it is appropriate to use small samples per group (e.g., the *Anolis roquet* group, Gorman and Renzi, 1979). However, even here it is necessary to include multiple populations of closely related species, particularly if nonrecombining sequences such as mtDNA or cpDNA are being used (see Neigel and Avise, 1986). Thus, it may be most efficient to conduct the sampling and analysis in two steps: the first to identify clades of closely related taxa and the second to add additional geographically remote populations for each of the members of such clades. If the chosen method and sequence reveal appreciable polymorphism (relative to divergence) in some or all taxa, then larger sample sizes will be needed to estimate phylogeny (Archie et al., 1989). In this case correct choice of method of analysis (see Chapter 11) becomes even more important. For example, different methods of coding polymorphisms as character states are subject to very different levels of sampling error (Swofford and Berlocher, 1987).

The methods discussed in Chapters 4 to 9 fall into two broad categories—those that, by their very nature, yield distance data (e.g., immunological methods, DNA hybridization), and those that can yield character-state data (e.g., allozymes, chromosomes, restriction sites, and nucleic acid sequences). Methods that yield distance data alone require a different project strategy from those that yield character-state data. For character-state data, the cost (of both time and chemicals) goes up linearly with the number of taxa, whereas for distance data (where a matrix is required) the cost goes up with the square of the number of taxa. Therefore for methods that yield distance data, a sensible strategy might be to divide the project into several matrices, one providing the major branches for the group and others dealing with lower-order relationships.

At least one and preferably several outgroup taxa should be included in the analysis (Maddison et al., 1984). In the absence of a suitable outgroup, the data for the ingroup can be used to produce an unrooted tree, which provides valuable information, but it is usual to aim for a rooted tree (see Chapter 11). The use of more than one outgroup will be useful for jackknifing the final data set (see Chapter 10). The outgroups should be as closely related as possible to the ingroup (preferably including at least one member of its sister group), but one must be certain that the outgroup taxa are indeed outside of the group under study.

CONCLUDING REMARKS

We have attempted to highlight the necessity of proper project planning in the use of molecular methods in systematics. Some of the common pitfalls can be avoided by careful project planning and the judicious use of pilot studies. A particularly common pitfall is to attempt to include all three applications of molecular methods in systematics—population structuring, species boundaries, and phylogenetic reconstruction—into a single study. Yet these three applications, although using similar techniques, have such different strategies that attempts to combine them are almost certain to be inefficient, or at worst fail on all three.

CHAPTER 3

COLLECTION AND STORAGE OF TISSUES

Herbert C. Dessauer, Charles J. Cole, and Mark S. Hafner

INTRODUCTION

Research in molecular systematics requires tissue samples in which proteins and nucleic acids are maintained in the structurally intact physiologically active state. In field work to obtain such material, the collector is confronted with unique challenges. Field kits often must include unusual items, such as liquid nitrogen tanks or dry ice, because freezing has proved to be the most effective method for preserving the widest variety of tissue constituents. Even after obtaining the samples, the collector may encounter difficulties transporting the materials. For example, airlines are often reluctant to accept liquid nitrogen as baggage. Also, collectors working in foreign countries may find that customs officials require special permits for the importation of biologically active materials. In this chapter we offer advice on meeting the challenges of collecting, packaging, and preserving tissues, and we emphasize the need to develop synoptic collections of these materials.

REGULATIONS GOVERNING ACQUISITION OF SPECIMENS

Collectors should become familiar with local, state, national, and international laws and regulations, and they should allow adequate lead time to obtain the necessary permits. Regulations concerning the collection and transport of biological materials are designed to safeguard the general public health, protect domestic crops and livestock, and control illicit trafficking in endangered and threatened species. Scientists intending to import frozen tissues and other specimens must adhere to all applicable wildlife regulations for the countries involved.

Scientific collecting permits are usually necessary for sampling from natural populations; these are obtained from state and national fish and wildlife, forestry, or conservation offices (for examples of forms used by United States institutions see Dessauer and Hafner, 1984). Six months or more may be required to obtain permits for collecting; this may be extended to a year or more if international travel and endangered species are involved. Different documents may be required to collect, to travel in certain areas, to export from the country of origin, and to import into the researcher's country. To verify the animal's good health, some countries require a quarantine period following importation of certain living organisms.

REMOVING AND PRESERVING TISSUES IN THE FIELD

General Procedures

The collection and handling of tissues for use in future molecular studies can be carried out by individuals with minimal training (Dessauer and Hafner, 1984; Johnson et al., 1984). Although different groups of organisms present different problems, many procedures are common to all groups. Tissues should be sampled while the organism is alive or as soon after its death as possible. Even a tiny amount of tissue is valuable. For example, electrophoretic evidence on 14 protein loci was obtained from individual lice weighing less than 1 mg (Hafner and Nadler, 1988). As little as 1 μl of blood or a few nanograms of muscle was sufficient to obtain immunological evidence on the evolution of taxa for which antisera existed (Maxson and Szymura, 1984). Nanogram quantities of DNA amplified with the polymerase chain reaction (Mullis and Faloona, 1987) yield sufficient material for sequence studies of systematic interest (Pääbo et al., 1989). The kinds and quantities of tissues preserved will depend on the needs of the individual investigator (see Chapter 2); however, we recommend that collectors maximize quantities of as many tissue types as is feasible to help in the development of synoptical collections.

Packaging   Collectors should be aware of the importance of keeping instruments, containers, and reagents clean, and of placing tissue samples in a cold environment and away from light as quickly as possible. Tissues should be packaged in plastic cryotubes (Figure 1a), plastic bags (Figure 1b), or wrapped tightly in aluminum foil. Packages should have space for tissue expansion but otherwise have as little unused space (air pockets) as possible to minimize drying of tissues and denaturation of macromolecules. Plastic bags are suitable for storage in electric ultracold freezers, but generally are not suitable for immersion in liquid nitrogen. If tissues are placed in liquid nitrogen, care must be taken to exclude nitrogen from the package. Otherwise, when removed from the dewar flask, aluminum foil packages expand and may burst; similarly, plastic tubes may shatter.

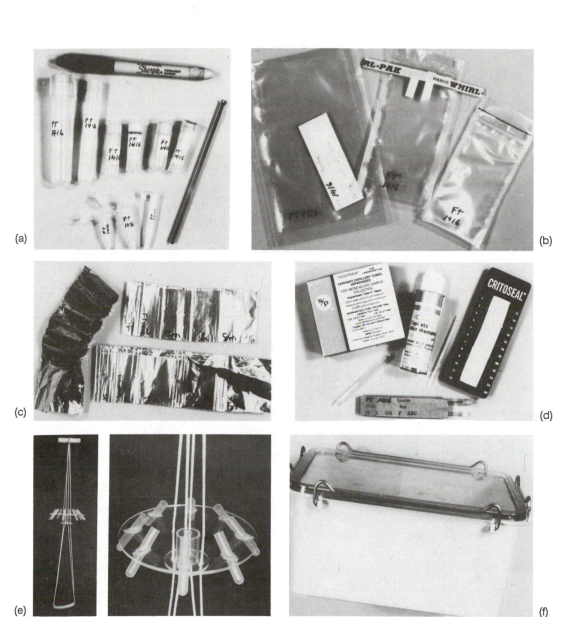

FIGURE 1. Materials and supplies that are useful for preparing and storing frozen tissues. (a) A plastic "French Straw" (right), assorted plastic containers, and an ink marker (top). The marker works on paper, plastic, and aluminum foil, and withstands freezing even in liquid nitrogen, so it can be used for both external and internal (backup) labels. The plastic straw and tubes can be used in liquid nitrogen if the lids are sealed well, but the small tubes with pop-off lids (bottom row) should be packaged in tightly wrapped foil or a larger tube for maximum security. (b) Plastic bags; these are acceptable for storage in electric ultracold freezers but are not recommended for use in liquid nitrogen. (c) Packets made by folding aluminum foil (see text). After tissues are sealed and folded within these packets (H, heart; L, liver; In, intestine and stomach; K, kidney; Sm, skeletal muscle), the package is wrapped in extra-heavy-duty aluminum foil, labeled by pressing the foil lightly with a ball point pen, and dropped into liquid nitrogen. (d) Glass tubes for collecting blood, sealant, and example (bottom) of capped tubes being slid into labeled piece of corrugated cardboard (see text for freezing instructions). (e) Hand centrifuge for field work in areas without electricity (Dessauer et al., 1983). (f) Plastic box (with gasket) for long-term storage of samples in an electric ultracold freezer.

To store tissue samples that do not fit conveniently into manufactured plastic tubes, we place them in subdivided packets of aluminum foil (Figure 1c). A rectangular piece of heavy duty foil is folded in half lengthwise, and open subdivisions are then created with two or three small folds made back on each other at regular intervals: we make as many attached packets as necessary for the different tissues being sampled from a specimen. After adding samples, we press out the air, seal by folding tightly, and wrap the package within a sheet of extra heavy duty aluminum foil. Such packets are readily customized to individual needs, they can be folded by assistants in advance of field work, and they transport efficiently in the flat (unopened) state.

Documentation of Samples   Careful documentation of samples is critical in all phases of work. A sample in a cryotube or other package is essentially useless if it has lost its label. It is important to (1) label samples and specimens so that no information is lost in wrapping, transport, storage, and entering of data into permanent records; (2) cross-reference the tissue sample with field collection data for the original specimen; (3) label containers, laboratory notebooks, and experimental samples during study; and (4) list specimens examined in research publications. Ideally, this citation will include the museum catalogue number for the voucher specimen (e.g., study skin, skull, preserved or dried body, pressed leaves) housed in a permanent repository. Although the museum or herbarium number is usually assigned long after the tissue sample was collected, all records pertaining to the sample (e.g., field catalogue data, notes, and photographs) should be cross-referenced with the permanent voucher number.

We recommend that individuals collecting tissue specimens in the field continue to use traditional collector's catalogues (Remsen, 1977; Herman, 1980). These should be organized so that each specimen receives a unique number preceded by the collector's initials. The entry should indicate the type(s) of tissues sampled. The package or tube containing the tissues should be marked clearly with the collector's initials and field number. The name of the specialist who provided identification of a specimen is an important part of the documentation.

Great care should be taken in labeling tubes and packages containing tissue samples. The following items have proved reliable: (1) high quality bond paper and a drafting pen with permanent, nonsmearing ink; (2) felt-tip pen with permanent ink that adheres to plastic tubes or packages; and (3) a ball-point pen that leaves a clear impression in aluminum foil. Experiment by exposing your materials to liquid nitrogen or other ultracold conditions followed by thawing. In addition, use a backup system (e.g., number written in ink and etched on the tube; labels both inside and outside the package).

On occasion, it is not possible to cross-reference a tissue sample with a permanently preserved voucher specimen, such as when a blood sample is taken from an individual in a zoo or from an endangered species that will be released after temporary restraint in the field. Under such circumstances, photograph the individual to document its identification and record its tag, band, or other identifying number, if known.

Preservation   As soon as possible after collection and packaging, most tissues should be dropped directly into liquid nitrogen or covered with dry ice. Field workers should be aware that liquid nitrogen is potentially hazardous (see pamphlet "Precautions and Safe Procedures-Liquid Atmospheric Gases" available from Linde Division, Union Carbide Co., 270 Park Ave., New York, NY, 10017 U.S.A.). Quick-freezing in liquid nitrogen generally shatters fragile glass hematocrit and microtubes filled with tissue fluids; such tubes must be frozen slowly before being subjected to ultracold temperature. In emergencies, a salt–ice mixture will substitute as a temporary refrigerant. Fragile capillary tubes and microtubes can be inserted into the slots of corregated cardboard (Figure 1d) or into a plastic straw such as those used for sperm storage (Figure 1a) for protection during long-term storage.

Fresh, unfrozen tissue gives the highest yields of animal mtDNA (Chapter 8). Tissues have been maintained successfully for 7–10 days unfrozen, immersed in a mannitol–sucrose buffer containing 100 mM EDTA. Soft tissues and especially oocytes, which contain 100 times the number of mitochondria per cell as somatic cells, are the best sources of mtDNA (Lansman et al., 1981; J. C. Avise, personal communication).

Cryopreservation is not required for tissues collected for certain purposes. Although not recommended for long-term preservation, immersion of tissues in an aqueous solution containing 2% 2-phenoxyethanol preserves the physicochemical, catalytic properties of many enzymes for at least 3 weeks (Nakanishi et al., 1969). Plasma albumin, which is stable in alcohol (Schwert, 1957), can be isolated from samples stored in phenoxyethanol at room temperature for at least a year. Lenses of vertebrate eyes, collected for sequence studies of α-crystallin, can be preserved in a saturated solution of guanidine hydrochloride. Hair and feathers of dry museum study skins represent valuable sources of keratins.

Cryopreservation of tissues is not absolutely required if the collection is solely for studies of DNA. Recent studies even show that some fragments of highly degraded DNA are recoverable from dried or traditionally preserved museum specimens (see section on "Stability of Macromolecules during Long-term Storage"; Houde and Braun, 1988; Pääbo et al., 1989). Preservation of nucleic acids depends primarily on the denaturation or inhibition of tissue nucleases with ethanol or EDTA, respectively.

The following procedure is recommended for the noncryogenic preservation of tissues for DNA analysis (Sibley and Alquist, 1981b; C. G. Sibley, personal communication). Place the tissue in a shallow dish and cut it into pieces of about 2–4 mm diameter, the smaller the better. Cover the minced tissue with twice its volume of ethanol (preferably 95%, but 75% can be used). After allowing the alcohol to diffuse through the tissue for 1 or 2 hr, replace the original alcohol, which has been diluted with tissue water. Do not use a blender or tissue homogenizer, as such treatment causes excessive degradation of DNA. After soaking in strong alcohol for at least 2 days, place the moist tissue, covered with about twice its volume of ethanol, in a plastic bag or other container for shipment and storage. Isopropanol, which is usually available as rubbing alcohol, and propanol can substitute for ethanol. The addition of EDTA to the alcohol will further stabilize the nucleic acids. In the laboratory, alcohol-preserved tissues may be stored at room temperature, but they are stable for longer periods if kept cool. Techniques for preserving DNA have been summarized by Arctander (1988).

Procedures Unique to Animal Tissue Collection

Animals collected should be handled with care consistent with appropriate guidelines for their welfare. Quick-freezing the entire organism is the fastest method. Unfortunately, this often diminishes the general usefulness of the tissues and their stability during long-term cryopreservation. For example, subsequent dissection of individual organs from the frozen specimen is difficult, blood plasma cannot be retrieved free of red cell proteins, and proteolytic enzymes and bacteria in digestive organs may cause destructive changes in biopolymers. We recommend that tissues be collected and packaged individually when larger animals are sacrificed. With vertebrates, blood, heart, liver, kidney, stomach, intestine, a sample of skeletal muscle, and perhaps other organs should be collected. Red blood cells of all vertebrates except mammals and some amphibians are excellent sources of DNA; the **buffy coat** of white blood cells is a good source of mammalian DNA. Avoid contaminating the tissues with bile salts, which are surface tension-reducing agents that could adversely influence tissue stability. However, bile may be saved since it has been useful for systematic work (Haslewood, 1967; Tammar, 1974).

Valuable material for molecular studies such as blood, hemolymph (Boyden, 1967), eggs (Sibley et al., 1974), snake venom (Russell, 1980), feather pulp (Marsden and May, 1984), and muscle biopsies can be obtained without sacrificing the animal. Methods for collecting tissues from most types of invertebrates are described in Wright (1978) and in papers listed in a bibliography of immunotaxonomic literature (Leone, 1968).

Anesthesia   The collector often requires anesthetic drugs to immobilize or euthanize animals. Background information on anesthetics, dose levels for a wide variety of both homeotherms and heterotherms, and ways to assess depth of anesthesia are given in McDonald (1976) and Lumb and Jones (1984). Inhalation anesthetics such as halothane and ether are valuable for use under laboratory conditions, but injectable drugs are more convenient in the field and, for most purposes, in the laboratory. Of these, ketamine, pentobarbital (Nembutal™), and tricane are most widely used. Ketamine is the drug of choice for many procedures involving a wide variety of species: it is easy to administer, induction of anesthesia is smooth, and dosage has a wide margin of safety.

Dosage levels of these drugs vary widely depending on the species, the size, sex, metabolic rate, and body temperature of the individual organism, and with the mode of injection of the drug. In general, small animals require higher doses relative to body size than large ones. Induction and recovery from anesthesia are slower in heterotherms than in homeotherms. Recovery of heterotherms from anesthesia may be sped up by raising their body temperature to increase metabolic rate. In selecting an anesthetic, remember that all are poisons that may influence subsequent molecular experiments. For example, rotenone, which would appear to be an excellent drug for collecting fish for molecular investigations, kills by inhibiting cellular oxidation involving NAD; this inactivates such enzymes as lactate dehydrogenase and malate dehydrogenase (Stecher et al., 1968). Fortunately, ketamine, pentobarbital, and tricane seem to have no detrimental effects on the proteins and nucleic acids commonly studied by molecular biologists.

We recommend the following procedures for the use of anesthetic drugs in the field. Anesthetic doses of either ketamine or pentobarbital for a wide selection of homeotherms and heterotherms fall between 20 and 30 mg/kg body weight when given intraperitoneally; doses of tricane for tetrapods fall between 100 and 200 mg/kg body weight. To anesthetize fish and other aquatic species, tricane may be added to water in a ratio of 1:2000 to 1:5000 parts drug to water. To lessen the risk of death from any of these drugs, we recommend treating the specimen with half the anticipated dose, followed by additional drug if needed. To euthanize animals with these drugs usually double or triple the dose levels required for anesthesia.

Blood and Hemolymph Collection   To prevent clotting we recommend the use of heparin or EDTA. One milligram of heparin or about 1 mmol of EDTA is sufficient to prevent clotting of 100 ml of blood. Blood diluted with STE prior to freezing preserves DNA and many proteins. Small samples can be obtained with only minor discomfort to the animal by "milking" blood from the tail or toe into a tube coated with anticoagulant (Figure 1d).

Microliter to milliliter samples have also been collected from the retro-orbital sinus of mammals (Riley, 1960). Larger samples can be collected directly from the heart, or from large, easily located vessels such as the femoral or jugular veins. Immunologists commonly collect rabbit blood from vessels in the ear. Before the needle is inserted, the vessels are caused to swell by wiping the ear with a mild irritant such as xylene. This approach probably will work with other mammals with large, highly vascularized ears. Blood of small birds is easily sampled from a wing vein (Arctander, 1988), and blood of some turtles may be sampled from a cervical sinus. Blood of large reptiles can be taken from caudal vessels (Gorzula et al., 1976). With the animal on its back, the needle is inserted through the skin at a point slightly distal to the vent at the midline. When the needle contacts a vertebra, caudal vessels are entered and blood is drawn into the syringe. We have obtained as much as 20 ml of blood from crocodilians with this method.

Heart puncture, particularly for heterotherms such as snakes and crocodilians, is easily accomplished without causing injury to the animal. The position of the heart may be located by pulsations visible on the antero-ventral body wall. A doppler ultrasonic device is also useful for locating the position of the heart in difficult cases (Brazaitis and Watanabe, 1982). Cardiac puncture is more complex in turtles, but one can approach the heart laterally, directing the needle through the soft tissues between the plastron and carapace. A more common practice is to tap the heart through a hole in the anteromedial corner of the right abdominal scute of the plastron.

Blood cells should be separated from plasma prior to freezing. Commercial hand centrifuges or a lightweight plastic centrifuge (Figure 1e; Dessauer et al., 1983) are useful for separating blood cells from plasma under field conditions. Methods for collecting hemolymph are described in bulletins 2, 5, and 15 (crustaceans), 2 (molluscs), and 37 (*Limulus*) of the Serological Museum (Boyden, 1948–1978).

Venom Collection   Collecting venom from snakes and other organisms is a dangerous task that should not be taken lightly. The fangs of the snake are inserted through a rubber or plastic membrane stretched across a collecting container. Many snakes discharge venom upon piercing the film. Additional venom may be obtained by gently massaging the region over the venom glands, avoiding undue compression, which may injure the glands and cause bleeding. If carefully handled, snakes can be "milked" repeatedly at 3-week intervals (Russell, 1980).

Procedures Unique to Plant Tissue Collection

Leaves, pollen, seeds, fern spores, and tubers of vascular plants have been preserved successfully for subsequent use in studies at the molecular level (Jensen and Fairbrothers, 1983). Many papers in the Serological Museum Bulletin include valuable information on collecting and handling plant material (Boyden, 1948–1978). As with animal tissues, plant tissue samples must be properly packaged and labeled with the collector's field number.

Seeds, pollen, and fern spores should be harvested only when mature. Viability of mistletoe seeds was greatest when collected during their period of dehiscence (Nickrent et al., 1984). Fairbrothers (in Dessauer et al., 1984) recommended the following protocol for preserving seeds from most plants: (1) remove fleshy portion, (2) dry, (3) place in vacuum-sealed container, and (4) store in the dark at or below freezing. Pollen and fern spores may be treated in the same manner after screening to remove debris.

Young, actively growing vegetative tissues are more valuable for molecular study than are mature leaves (e.g., ferns; see Werth et al., 1985a). Leaves of certain taxa show a "senescence" phenomenon wherein their proteins disappear during seasonal aging. Immediately on collection, leaf cuttings should be rinsed in distilled water, packaged, and frozen rapidly. Rapid freezing is particularly important when collecting leathery leaves that tend to rot rather than dry. Proteins in leaves ranging from ferns to oaks have survived freezing for at least 3–4 years.

For maximum yield of DNA, leaves should be pressed, dried overnight at 42°C, and then stored at room temperature (see also Chapter 8). The most important role of drying plant tissues may be the prevention of rotting, rather than the preservation of DNA. Although this method appears to preserve the integrity of DNA for several months, dried samples should be frozen at −70°C for long-term storage (Doyle and Dickson, 1987).

Chemical treatment to remove lipids and phenolic substances from freshly collected vegetative tissue prior to long-term storage probably should be avoided. Doyle and Dickson (1987) found that treatment with preservatives used in anatomical studies, or with solvents such as ethanol and chloroform–ethanol, tended to cause degradation of DNA. Similarly, Coradin and Giannasi (1980) found that chemical treatment interfered with subsequent analyses of flavanoids.

Collecting Cell Lines

Cryopreservation of living cells requires special collecting, freezing, and storage procedures if cells are to survive intact (Stowell, 1965; Watson, 1978; Hay, 1979; Hay and Gee, 1984). Cell damage is most likely to occur during the freezing and thawing process. Some cells will survive freezing and thawing if pretreated with a cryoprotectant such as glycerol or DMSO. For every species, tissue, and freezing system there is an optimum cryoprotectant concentration and freezing rate (Mazur, 1970). The cryoprotectant must be concentrated enough to protect the cells from freeze damage, yet dilute enough to avoid chemical injury to cells. Ideally, the rate of freezing should be controlled precisely, with rate depending on variables such as species, tissue type, size of the sample, cryoprotectant used, and the system used to freeze the sample. Generally, a cooling rate that is too fast results in death due to formation of ice crystals within the cells; too slow a rate causes death from the chemical consequences of solute concentration. Nevertheless, it is possible to store and recover low yields of viable cells without a highly specialized freezing procedure.

Semen samples and tissue biopsies are easily obtained under field conditions without permanent injury to the donor animal. The equipment and supplies needed to establish proper freezing conditions in the field are not elaborate: alcohol, freezing medium, and liquid nitrogen in an appropriate tank (Maure, 1978). Plastic "French Straws" have been useful as containers for storage of semen. If the freeze rates used are less than optimal, manipulation during the thawing process may increase the yield of viable cells (Mazur, 1970).

The following tissue biopsy protocol is recommended by Hay and Gee (1984) for use in the field. The tissue is collected aseptically, minced into fragments of about 1 mm diameter, and placed in a culture medium containing 10% serum, antibiotics, and 10–12% DMSO. The tissue is allowed to equilibrate with this "freeze medium" for 2–3 hr in an ice bath or refrigerator if available. The temperature is then lowered slowly to $-50°C$ (approximately 1°C/min) after which time the tissue container is dropped into liquid nitrogen. The gradual cooling procedure may be carried out in the nitrogen dewar by suspending the sample in the cold space above the liquid nitrogen. Successful cultures of some tissues have been established using samples equilibrated in the freeze medium and frozen immediately in liquid nitrogen (Taylor et al., 1978; R. J. Baker, personal communication).

TRANSPORT OF TISSUES FROM FIELD TO LABORATORY

Shipping Regulations

Tissues are usually transported either in styrofoam boxes packed with dry ice or in liquid nitrogen containers. Small samples are conveniently carried in personal baggage in thermos bottles containing refrigeration packs. Dry ice (referred to as "carbon dioxide, solid") and liquid nitrogen are classified as Restricted Articles by the International Air Transport Association, and the shipper must be aware of all pertinent regulations (see Dangerous Goods Regulations, 30th Edition, effective 1 January 1989).

Dry ice containers are usually accepted as baggage by the airline agent if the "Shipper's Certification for Restricted Articles" is attached to the package. Dry ice is designated as Hazard Class ORM-A, and packages containing dry ice must be so marked. No more than 200 kg of dry ice may be shipped in a single package.

Liquid nitrogen in nonpressurized, metal dewar flasks is authorized for shipment by air also. No more than 50 kg per flask can be shipped on an aircraft carrying passengers. The flask must be marked "Nitrogen (Liquid, Nonpressurized)," and must be further labeled to discourage loading or handling in any position other than upright. The upright position should be indicated prominently by arrows and the wording KEEP UPRIGHT placed at 120 degree intervals around the container. It must also be prominently marked DO NOT DROP—HANDLE WITH CARE. For shorter trips, the liquid nitrogen can be poured out and the tank checked as baggage. Most standard dewars are so well insulated that they will maintain a large mass of tissues frozen for many hours even in the absence of liquid nitrogen. If a tank contains few specimens, plastic tubes filled with water should be added to the tank to provide supercooled ice before pouring off excess nitrogen. A "Dry Shipper," which contains an absorbent that keeps nitrogen from spilling, alleviates problems of air transport of nitrogen in dewars; storage in some models is effective for up to 3 weeks.

Sources of Liquid Nitrogen and Dry Ice

Scientists conducting field work often have difficulty locating sources of liquid nitrogen and dry ice; useful contacts are universities, hospitals, welding supply companies, and mining operations. A partial listing of such sources in different areas of the world is given in Dessauer and Hafner (1984).

STORAGE OF TISSUES ON RETURN FROM THE FIELD

The majority of tissue samples are stored in an electrical ultracold freezer (−70 to −150°C), on dry ice, or in liquid nitrogen. Ease of access of samples within the freezer or nitrogen tank is of special importance. Samples can be stored in numbered moisture-proof boxes (Figure 1f). For easy retrieval, all specimens of a given taxon should be stored together; color codes on the outside of boxes facilitate the identification of contents as material in freezers is moved about. Each sample is identified by the collector's field number. A listing of the holdings in each freezer, complete with box number, contents of the box, and location in the freezer is maintained and routinely updated as samples are moved, used, granted, or discarded.

Ultracold storage space is very expensive to purchase and maintain, and it is important that materials be stored in a space-efficient manner. Thus, access and inventory procedures for frozen tissue collections should be extremely well organized. Freezers should be opened as rarely as possible; ultracold freezers are sensitive to even brief periods of temperature warmup, and every second that a freezer door is open while one searches for a particular sample is energy consuming and could eventually contribute to freezer failure. One must know exactly where each sample is located before opening the freezer. A map on the door of the freezer helps locate individual boxes efficiently. A "working freezer" should be used for storage of tissues that are currently being studied.

On a cost-per-sample basis, an ultracold freezer provides the most convenient and efficient method for long-term storage of large numbers of samples. Of the two designs (chest and upright models), chest models maintain more constant temperatures during use and are less prone to mechanical failure; upright models occupy less floor space and freezer boxes are more easily retrieved. Freezers may be equipped with a system that will sound an alarm in the event of electrical or mechanical failure, and they should be monitored at least once each day. During holiday periods, special arrangements must be made to ensure that freezers are monitored daily. Readily visible notices should be posted in freezer areas, indicating persons (and telephone numbers) to be contacted in case of freezer malfunction. Some form of backup storage system (other freezers, liquid nitrogen, or dry ice) should be available in the event of freezer failure.

Ideally, liquid nitrogen is better than ultracold freezers for long-term storage because of its much colder temperature ($-196°C$); however, if large numbers of samples are stored in the collection, it may be difficult to retrieve specific samples. Also, continual replenishment of evaporated liquid nitrogen may become costly. A small number of samples stored in liquid nitrogen tanks is relatively easy to organize for efficient retrieval. Liquid nitrogen freezers are available that are large enough to organize up to 15,000 2-ml cryotubes with easy access to any tube.

Household refrigerators and freezers may be used to store freeze-dried blood fractions, **acetone powders,** seeds and pollen, and enzymes in strong salt solutions or glycerol. Isolated samples of DNA and bacterial cultures containing DNA for cloning purposes may also be stored in a household freezer (Sambrook et al., 1989). However, frost-free appliances should be avoided, because of the danger that biomolecules may degrade during warming cycles.

STABILITY OF MACROMOLECULES DURING LONG-TERM STORAGE

Many proteins are far more stable than is generally assumed (Sensabaugh et al., 1971a,b; Dessauer and Menzies, 1984). For example, remnants of blood samples used in Nuttall's (1904) classical immunological study of mammalian relationships served as experimental material for an important test of protein stability (Keilen and Wang, 1947). Keilen and Wang found that hemoglobin, carbonic anhydrase, catalase, glyoxylase, and choline esterase had remained 75–85% active in blood samples that had been kept in the dark under aseptic conditions at room temperature for approximately 42 years.

Certain proteins retain activity for surprisingly long periods, even in tissues exposed to adverse conditions. Proteins in tissues with high proteolytic activity, such as those in viperid snake venoms, retained their activities for extended periods (Russell et al., 1960; Russell, 1980). Of 17 proteins commonly examined in electrophoretic studies, only alcohol dehydrogenase was undetectable in mammalian tissues stored at room temperature for 12 hr after death (Moore and Yates, 1983). Sensabaugh and colleagues (1971a,b) found that 8 of 11 proteins tested in an 8-year-old sample of dried blood retained at least some activity. Plasma albumin and esterase activity were detectable in mammal skins stored up to 16 years as standard museum preparations (J. L. Patton, personal communicaton). Albumin in muscle tissue from a mammoth frozen in northern Siberia for approximately 40,000 years (tissue that had probably undergone numerous freeze–thaw cycles) maintained sufficient immunological specificity to demonstrate that the species is properly classified with the elephants (Lowenstein et al., 1981). Certain proteins in acetone powders and freeze-dried tissues retain their structure and activity at room temperature for short periods of time and at freezer temperatures for much longer periods. Endocrinologists and botanists find that acetone powders retain many activities of the parent tissue. **Lyophilization** is commonly used to preserve immunoglobulins. Mellor (1978) suggests that vacuum drying of frozen tissues over a desiccant such as silica gel is less likely to damage sensitive proteins than is the usual method of freeze drying. Dried seeds of many taxa have remained viable for up to 10 years when stored in the dark, at or below 0°C. Macromolecules in dried pollen and fern spores are stable for at least 4 years when stored at or below −30°C (Anderson et al., 1978).

Decreases in activity traceable to changes in chemical composition and molecular conformation may arise during storage. These are often detected as modifications in electrophoretic banding patterns (McWright et al., 1975). Common modifications include oxidation of sulfhydryl residues (Chilson et al., 1965; Hopkinson, 1975; Harris and Hopkinson, 1976), oxidation of ferrous iron of heme proteins (Jiminez-Marin and Dessauer, 1973), deamination of asparagine residues (Wulf and Cutler, 1975), rearrangements of subunits of multimeric enzymes (Chilson et al., 1965), and formation of conformation isomers (Kitto et al., 1966; Dawson et al., 1967).

The addition of sulfhydryl reagents (e.g., dithiothreitol or mercaptoethanol), coenzymes, and activating ions to homogenates of previously frozen tissues stabilizes certain enzymes and may reverse some of the adverse effects of long-term storage (Chilson et al., 1965; Harris and Hopkinson, 1976). Sucrose, mannitol, or glycerol is often added to homogenizing fluids and diluents to stabilize proteins; **BSA** may be added to minimize surface denaturation of proteins in the highly dilute solutions of antigens and antibodies used in microcomplement fixation studies of nonmammalian albumins (Champion et al., 1974).

DNA is an extremely stable molecule, far less reactive chemically than proteins. Ordinary handling of fresh tissue, such as freeze–thaw processing or alcohol preservation, usually causes only minimal shearing. DNA isolated from such tissue samples is valuable for hybridization (Chapter 7), analysis of restriction fragment variants (Chapter 8), gene cloning, and sequencing (Chapter 9). For example, Sibley and colleagues (1988) used DNA isolated from alcohol-preserved tissues successfully in an extensive series of solution hybridization studies. From reptilian red cells stored frozen for over 20 years, we have isolated high-molecular-weight DNA that yielded restriction fragment fingerprints comparable to those obtained with DNA of fresh blood. McBee and colleagues (1987) isolated DNA from mammalian brain, skeletal muscle, and liver tissues maintained at 24°C from death through 72 hr after death. They found that DNA was most stable in brain and least stable in liver tissue.

Highly degraded segments of DNA have been isolated from mummies, museum specimens preserved dry and in ethanol, and even from formaldehyde-fixed and parafin-embedded tissues. For example, fragments yielding useful systematic data have been retrieved from the dried brain of a 7000-year-old human (Pääbo et al., 1988), from skin of the extinct quagga (Higuchi et al., 1984), and from 1000-year-old kernels of maize (Rollo et al., 1988). Unfortunately, DNA is not recoverable from all museum specimens. Tissues treated with picric acid or mercuric chloride or fixed for long periods in unbuffered or excess formaldehyde are unfavorable candidates (Goelz et al., 1985). Specimens preserved in alcohol are poor sources if the nucleases were not inactivated. The recovery of DNA from long-dead specimens is about 5–20% of that extractable from fresh tissue. The majority of the DNA is single stranded and degraded to between 200 and 300 base pairs, which is too small for restriction fragment analysis; however, some fragments of kilobase size may be present (Houde and Braun, 1988). A relatively high proportion of fragments appears to be derived from DNA that exists in multiple copies, such as mtDNA (Pääbo et al., 1988) and **Alu-repeats** (Pääbo, 1985).

In summary, to minimize detrimental influences of freezing on macromolecules (as contrasted to viable cells), tissues should be frozen quickly and thawed rapidly. Thawing tissues prior to the time to use them should be avoided when possible. Finally, proteins in homogenates are stabilized by the presence of coenzymes, sulfhydryl reagents, glycerol, DMSO, and the presence of other proteins.

DEVELOPMENT AND SUPPORT OF SYNOPTIC TISSUE COLLECTIONS

Collections of tissues in which macromolecules are structurally intact and/or physiologically active are important resources for investigations in many areas of biology; several natural history museums and herbaria world-wide are accumulating such collections (Dessauer and Hafner, 1984). The goal is to develop collections that are as synoptic and diverse as traditional skin, skeletal, dried, and spirit collections.

Disposition of Tissues for Long-Term Preservation

Tissues collected in the field may be handled in one of three ways, depending on whether the researcher maintains a formal frozen tissue collection and how soon the samples are to be used. If the collector does not maintain a tissue collection and does not plan to use the material immediately, the tissues (with complete documentation) should be sent to a recognized tissue repository. If the collector does not maintain a tissue collection but intends to use the samples immediately, all unused tissues, antisera, DNA clones, and other tissue isolates (with complete documentation) should be sent to a recognized tissue repository on completion of the research. If the collector or institution maintains a formal tissue collection, incoming tissue samples are incorporated into the permanent collection.

Curatorial Problems Unique to Tissue Collections

Curators of tissue collections face several unique curatorial problems; important among these are the acquisition and deacquisition of specimens and data base management.

Acquisition Policies   Curators of frozen tissue collections should seize every opportunity to acquire tissues from rare, unusual, and exotic species. Zoos, botanical gardens, and biomedical institutions often are sources of valuable specimens. Samples from rare species are valuable, even if they are not in optimal condition or lack complete data. Positive identification should be the primary concern for acceptance of materials into the collection; but details concerning the source of the specimen, including the location of the traditional voucher specimen, are desirable (see Leviton et al., 1985 for listing of museum symbolic codes). Examples of organisms from which tissues may be accepted without vouchers include large ungulates, proboscideans, and marine mammals such as whales; photographs can substitute as vouchers in such cases.

Deacquisition Policies   Unlike materials in conventional systematic collections, those stored in tissue collections are usually consumed as they are analyzed. When discussing transfer of these materials between collections and researchers, the word "loan" should be replaced with "grant." Because tissues are only quasipermanent in the collection, a premium is placed on their judicious dissemination and use. An efficient inventory system is required to keep track of their use and distribution. It is the joint responsibility of the donor and the recipient to be sure that transfer of tissue samples is legal (i.e., the recipient may require special permits to handle the material). The repository and person(s) who originally collected the tissues should be properly acknowledged in any publication resulting from their use.

Data Base Management   Many of the unique curatorial problems posed by frozen tissue collections can be minimized by a computerized inventory system. Programs designed for traditional museum collections may be adapted for use with frozen collections. The computer data base should include all field data and cross-references to the field catalogue number and voucher specimen number. When a sample is depleted, prompt deletion from the collection records avoids costly searches.

Guiding Principles   Museums and herbaria have maintained synoptical scientific collections for use in systematic research for decades, and, in some cases, for centuries. In contrast, collections of frozen tissues are relatively recent. Because individuals collecting tissues in the field generally save more material than is necessary for their immediate research, small collections of tissues are common worldwide (Dessauer and Hafner, 1984). Specimens in these collections, available to the general community of systematists, have helped answer many research questions.

The principle of obtaining large quantities of tissues when opportunity permits (rather than a small sample for a few experiments by a single scientist) greatly enhances the value of collecting efforts. The following principles (Dessauer et al., 1988) are intended to guide the development of synoptical collections of tissues representing the world's biota:

1. When collecting, do not limit efforts to only the specific material needed. Within the limits of permits, make general collections when opportunities arise, with emphasis on the unusual, the difficult to obtain, and on filling gaps in existing collections. Individuals who seek grants of tissue from large collections should also help to develop such collections.
2. When preparing specimens, discard as little material as possible. Obtain tissue samples for the research, select appropriate anatomical material to document the specimen by traditional means, then freeze as much of the specimen as possible for a synoptical frozen tissue repository.
3. In the laboratory, use only as much of any sample as is necessary to complete the experiments. Conserve as much of the sample as possible for future use.

4. After completing the experiments, place the remaining samples in a formal synoptical tissue repository.
5. Personal collections should be discouraged, as they are often lost to science. If a personal research collection is maintained, arrange for its conservation and timely transfer to an appropriate institutional collection in the event of disability or death.

Currently, individual scientists distribute samples of frozen tissues throughout the research community on an informal basis. It is hoped that, in time, a network of institutions will accept the responsibility for long-term maintenance, development, and distribution of such material (a list of collections is given in Dessauer and Hafner, 1984). Unique curatorial problems posed by frozen tissue collections have been addressed at workshops organized by the Association of Systematics Collections in 1983 (Dessauer and Hafner, 1984) and in 1988, and a set of guidelines has been proposed for their curation (Dessauer et al., 1988).

II

MOLECULAR TECHNIQUES

CHAPTER 4

PROTEINS I: ISOZYME ELECTROPHORESIS

Robert W. Murphy, Jack W. Sites, Jr., Donald G. Buth, and Christopher H. Haufler

INTRODUCTION

Protein **electrophoresis,** the migration of proteins under the influence of an electrical field, is among the most cost-efficient methods of investigating genetic phenomena at the molecular level. Since the origin of starch gel electrophoresis (Smithies, 1955) and the histochemical visualization of enzymes on gels (Hunter and Markert, 1957), and the classic studies of Harris (1966), Hubby and Lewontin (1966), and Lewontin and Hubby (1966), a major revolution in understanding micro- and macroevolutionary processes has occurred. Using enzymatic and nonenzymatic proteins, numerous investigations have focused on enzyme efficiency, estimating and understanding genetic variability in natural populations, gene flow, hybridization, recognition of species boundaries, and phylogenetic relationships, among other problems. The frequency of such investigations has not waned in recent years, but rather has increased as refinements and new methods have been developed.

Two general forms of protein data can be gathered simultaneously using electrophoretic methods. One is derived from **isozymes,** which are all functionally similar forms of enzymes, including all polymers of subunits produced by different gene loci or by different **alleles** at the same locus (Markert and Moller, 1959). The other data set consists of **allozymes,** a subset of isozymes, which are variants of polypeptides representing different allelic alternatives of the same gene locus. Both forms of data are important in molecular systematics, and both involve proteins that can be separated on the basis of net charge and size.

Here we provide step-by-step instructions on how to establish a horizontal starch gel electrophoresis laboratory, perform protein electrophoresis, stain for specific enzymatic and nonenzymatic proteins, and interpret the resultant gels. Although other methods of protein electrophoresis exist, using media such as cellulose acetate gels (Richardson et al., 1986), we have chosen to detail advances in horizontal starch gel methods because of their widespread use and efficiency. Ways of avoiding or recovering from common pitfalls are described. The electrophoretic principles and methods described are applicable to all organisms.

Where possible, we provide inexpensive alternatives to costly equipment and methods, but not at the expense of increased health risk. As with most molecular methods used in systematics, some aspects of the data gathering pose extreme health risks, both acute and chronic. Therefore, the appropriate level of caution, as known to us, is always given the highest priority.

PRINCIPLES AND COMPARISON OF METHODS

General Principles

Proteins are composed of amino acids joined together by covalent peptide bonds to form polypeptides. These sequences, or "primary structures," are genetically determined. Each of the 20 different amino acids has a unique side chain, characterized by its shape, size, and charge. The side chains on five of these amino acids are either basic and thus positively charged (NH_3^+; lysine, arginine, and histidine), or acidic and negatively charged (COO^-; aspartic acid and glutamic acid). Charged side chains are responsible for the movement of the proteins through a matrix during electrophoresis. The net charge of each protein varies with pH; at a low pH the amino groups become positively charged, and at high pH the carboxyl groups become negatively charged. Most proteins have a point at which the effect of positive and negative charges are equal, the **isoelectric point.** Isoelectric proteins do not move in an electrical field because they are attracted to neither the (positive) **anode** nor the (negative) **cathode.**

Uncharged amino acids are either nonpolar and hydrophobic or polar. These amino acids can become hydrogen-bonded to one another resulting in folding (β-structure) or helical (α-helix) configurations, termed secondary structure. Depending on the primary and secondary structure, the molecule usually undergoes additional folding to form its tertiary structure. The shape and size of a protein may also have an effect on protein migration, depending on the pore size of the electrophoresis matrix. To some extent the shape of a particular protein is determined by the relative charges of adjacent amino acids because of the effect of like charges repelling and different charges attracting. Finally, many proteins contain more than one polypeptide chain (subunit) bound together by hydrogen bonds, van der Waals forces, ionic bonds, disulfide bridges, and/or hydrophobic interactions. Proteins having more than one polypeptide (**multimeric**) have a quaternary structure (Darnell et al., 1986).

Some forms of electrophoresis separate proteins on the basis of net protein charge Q, shape as measured by radius r, strength of the electrical field d, and viscosity of the suspension medium n as given by the following equation:

$$u = \frac{Qd}{4\pi r^2 n}$$

Under appropriate conditions, the rate of movement, u, increases with net charge and strength of electrical field and decreases with the size of the molecule. The actual situation is usually more complex than this simple equation indicates (e.g., some proteins occur as relatively simple strands, and others in globular form) and indeed much remains to be learned about the physics of electrophoresis itself.

All electrophoretic techniques consist of an electrical power supply, a support matrix (cellulose acetate gel or strips, starch gel, etc.), and ionic buffers. Electrical current is applied to opposite ends of the suspension medium via the ionic buffers. Molecules (e.g., proteins) having a net positive charge (**cations**) migrate to the cathode, and negatively charged proteins (**anions**) migrate to the anode. Following electrophoresis, the proteins may be visualized by a number of different methods, the most frequently used being specific histochemical staining (Appendix 1 and references cited therein). After electrophoresing a protein sample in the gel matrix, the individual proteins are selectively stained. Most of the stains provide a specific substrate for the enzyme, allow it to catalyze the particular reaction involved, and then develop a dye that can be visualized in normal light or by fluorescence under UV light. Thus, from the hundreds or thousands of enzymes in the crude extract, proteins with the same substrate utilization can be identified.

As detailed later, numerous ionic buffers are available for electrophoresis. The buffers serve several functions. The primary function is to buffer against pH change, which occurs during electrophoresis: acid is produced at the anode and base at the cathode. The extent of pH change is directly related to the duration of electrophoresis, the voltage, and the current generated. Buffers also form an ionic connection between the electrical supply (electrodes) and suspension medium (gel) and reduce the interaction of charged groups on the protein with any charged groups in the matrix, and may modify the net charges of proteins, carry enzyme stabilizers (e.g., disodium EDTA), or provide enzyme catalysts (e.g., Mg^{2+}).

The amino acid sequences of proteins are changed by mutations in the encoding DNA locus. Such mutations may alter shape and net charge, as well as catalytic efficiency and stability (Shaw, 1965). Protein electrophoresis aims to reveal as many of these changes as possible. However, considering the principles of electrophoresis and the properties of protein side chains, it is unlikely that all allelic variants will be identified using a single buffer pH, buffer system, or concentration of gel, or even by using a single method of electrophoresis. Most laboratories concentrate on manipulating the first three of these four variables because of the difficulty in mastering the alternative technologies and the added expense of additional equipment and expendables.

Assumptions

The correct application of isozyme data requires that banding patterns observed on gels are correctly interpreted. The most basic assumption that evolutionary biologists make in using isozyme data is that changes in the mobility of enzymes in an electric field reflect changes in the encoding DNA sequence. Thus, if the banding patterns of two individuals differ, it is assumed that these differences are genetically based and heritable (see Matson, 1984). Also, it is assumed that enzyme expression is codominant, i.e., that both alleles at a locus are expressed. To interpret these banding patterns, one must know something about the number of subunits in the enzyme. Interpretation and nonheritable aspects of gel isozyme patterns are detailed below and elsewhere (e.g., Richardson et al., 1986). In addition to biochemical components of gel interpretation, one must be aware of compartmentalization of enzymes, or enzyme activity, in particular organs or organelles. For example, livers may have different enzymes than spleens or brains (see Murphy and Matson, 1986). In plants and animals, some enzymes are restricted to the cytosol whereas others are found only in mitochondria or chloroplasts.

For most enzymes, the genetic controls are well enough known to allow genetic inferences to be made from gel isozyme patterns. The distribution of isozymes per cell or tissue can be reliably predicted, homozygous and heterozygous individuals can be identified, and conclusions about genetic polymorphism, the breeding system of individuals, and population structuring can be drawn. It is necessary to conduct breeding studies only when banding patterns depart from expectations. Cell fractionation studies can demonstrate whether the enzymes are housed in the cytoplasm or one of several separate organelles (Weeden, 1983).

Population geneticists have developed statistical models for interpreting genetic population structure (Chapter 10). The most relevant of these to gel interpretation is the **Hardy–Weinberg equilibrium** principle. This states that in the absence of selection, drift, and migration, the frequencies of alleles in a randomly mating population will maintain a stable equilibrium with genotype frequencies of $AA = p^2$, $Aa = 2pq$, and $aa = q^2$, where p is the frequency of allele A, and q of the alternative allele a. Nonconformity to the prediction of Hardy–Weinberg equilibrium indicates that the phenotypic variation has a nongenetic basis or that one or more of the Hardy–

Weinberg assumptions is not met in the population. Thus, for example, the individuals may not be randomly mating, or some natural selective force may be acting on the species, or genes from neighboring populations may be migrating into the study site. If these principles of biochemistry, genetics, and gel interpretation are followed, electrophoresis can yield many valuable insights for the evolutionary biologist.

A major assumption in the use of allele frequency data to infer population structure is that alternate alleles at a given locus are selectively equivalent, or neutral (Kimura, 1983a,b). Exceptions to this assumption are known (see below), and accepting neutrality for most protein polymorphisms also requires accepting largely untested or poorly tested null hypotheses. However, in the absence of evidence for selection at a particular locus, it has been suggested that studies begin with neutrality as a working assumption (Allendorf and Phelps, 1981).

Comparison of the Primary Methods

The four primary methods of electrophoresis differ by the nature of the support medium: starch gel (including both horizontal and vertical systems), polyacrylamide gel, agarose gel, and cellulose acetate gel. Each method will be briefly described and discussed in terms of specific advantages and limitations. Less frequently used methods of resolving protein variants are not discussed herein; these include paper electrophoresis (Freifelder, 1982), isoelectric focusing, immunoelectrophoresis, and two-dimensional electrophoresis (Harris and Hopkinson, 1976; Hames and Rickwood, 1981). Advantages of the four basic methods are compared in Table 1, although choice of method will often be determined by availability of equipment and expertise.

Starch Gel Electrophoresis (SGE)   Hydrolyzed starch is heated in an ionic buffer solution and allowed to cool, thereby forming a gel. The ratio of starch to buffer can be varied to alter the size of the gel pores. Pore size allows for a sieving effect in the gel. Thus, these gels can separate on the basis of both size and charge.

Two forms of SGE exist: horizontal and vertical. In horizontal SGE, a poured gel is allowed to cool in a gel mold without further preparations. Vertical starch gels are poured into double-sided molds having a "gel comb" or "well former" that makes the "gel wells" for holding tissue extracts (Brewer, 1970; see also Chapter 8). In general, the vertical system requires a greater amount of starch and larger quantities of tissue extract, allows for fewer samples to be run per gel, and is thus more costly. The advantages to using vertical SGE include the avoidance of the phenomenon known as **electrodecantation**; as proteins migrate on horizontal gels, enzymes of high molecular weight tend to drop toward the bottom of the gel. This may make

Table 1. Comparison of the attributes of the four primary methods of protein electrophoresis on gel support media[a]

Attribute	SGE	PAGE	CAGE	AGE
Separates by charge[b]	yes	yes	yes	yes
Separates by size	yes[b]	yes[b]	no	no
Number of slices per gel	>6[b]	1	1	1
Toxic	no[b]	yes	no[b]	no[b]
Running time	4-24 hr	4-6 hr	0.3-3 hr[b]	3-4 hr
Minimum amount of sample required per gel	2 µl	2 µl	0.5 µl[b]	1 µl
Maximum amount of sample possible per gel	>50 µl[b]	>50 µl[b]	5 µl	>50 µl[b]
Amount of stain required	5-50 ml	10-50 ml	1-3 ml[b]	10-50 ml
Electroendosmosis	yes	no[b]	yes	yes
Voltage required (V/cm)	1-10	5-10	<3[b]	20
Cooling required	yes	at times	no[b]	yes
Gel easily handled	usually[b]	no	yes[b]	yes[b]
Simultaneously resolves cationic and anionic proteins	yes[b]	no	yes[b]	yes[b]
Allows counter-staining of adjacent slices	yes[b]	no	no	no

[a]SGE = starch gel electrophoresis; PAGE = polyacrylamide gel electrophoresis; CAGE = cellulose acetate gel electrophoresis; AGE = agarose gel electrophoresis.
[b]Perceived advantage.

slices from the upper regions of the horizontal gel inferior or inadequate for resolving these proteins. Nevertheless, the method of horizontal SGE is used almost exclusively in our laboratories and in the vast majority of other laboratories; the vertical system will not be discussed further (for more information, see Harris and Hopkinson, 1976; Siciliano and Shaw, 1976).

Polyacrylamide Gel Electrophoresis (PAGE)   Polyacrylamide gels are formed by the catalytic polymerization of monomeric forms of acrylamide and bisacrylamide. It allows the separation of proteins on the basis of both size and charge (Chrambach and Rodbard, 1971). The pore size of acrylamide gels can be controlled by altering concentrations of acrylamide and/or bisacrylamide. This sieving attribute has made PAGE one of the methods of choice in molecular biology laboratories examining nucleic acid sequences because, unlike most other forms of gel electrophoresis, it allows for the accurate, controlled separation of charged particles on the basis of molecular weight (Chapters 8 and 9). General references to this system are found in Hames and Rickwood (1981).

Cellulose Acetate Gel Electrophoresis (CAGE)   Electrophoresis can be carried out on preformed cellulose acetate gels or strips. The gel form of cellulose acetate is preferred because of repeatability of experiments (Harris and Hopkinson, 1976). A major advantage is that electrophoresis can be carried out with very small quantities of tissue homogenate. Although the gel itself is premade, it must be soaked in the appropriate buffer prior to electrophoresis. Due to the large pore size, CAGE has no sieving effect; proteins are separated on the basis of net charge only, although this does not appear to greatly reduce the number of variants identified (Richardson et al., 1986). In addition, the large pores also cause **electroendosmosis,** a "back wash" of buffer solution caused by gel charge groups and that accelerates the mobility of cationic isozymes but retards or reverses the anionic isozymes. Although this problem occurs with SGE and AGE, it is more pronounced with CAGE (Harris and Hopkinson, 1976). CAGE has been discussed in detail recently by Richardson et al. (1986).

Agarose Gel Electrophoresis (AGE)   Agar and agarose gels are prepared much in the same way as starch gels. Pure "agar" gels have a relatively high concentration of acidic groups (carboxyls and sulfates), resulting in considerable electroendosmosis and occasional adsorption of proteins, although adsorption problems may be overcome by use of highly purified agarose (Harris and Hopkinson, 1976).

APPLICATIONS AND LIMITATIONS

Intraspecific Applications

Population Structure   Protein polymorphisms in natural populations have been used to describe allele frequency changes in both time and space. For example, Mihok et al. (1983) monitored intermittent genetic changes at the transferrin locus in the rodent *Clethrionomys gapperi* over a 13-year period (1966–1978), and found relatively minor fluctuations in allele frequencies. A temporal study carried out by McClenaghan et al. (1985) on mosquitofish in the Savannah River drainage showed that allele frequencies were generally stable through time.

On a spatial scale, population genetic structure has been inferred from the geographic distribution of allele frequencies. For example, monarch butterflies (*Danaus plexippus*) are migratory, and Eanes and Koehn (1978) found that the considerable genetic subdivision in nonmigratory populations built up during summer residency in northeastern North America disappeared during migration. Among prairie dog (*Cynomys ludovicianus*) populations genetic variability was partitioned at several hierarchical levels that correlated with life history parameters (Chesser, 1983). In contrast, Shaklee (1984) surveyed damselfish populations (*Stegastes fasciolatus*) collected from Midway Island to Hawaii, a linear distance of about 2500 km.

Allele frequencies at all eight polymorphic loci were remarkably uniform (F_{ST} = 0.003; see Chapter 10) and, with the exception of a few very rare alleles, all populations possessed the same alleles at all loci.

Assessments of allozyme variability may be used to infer historical events that have significantly influenced the genetic structure of populations, such as bottlenecks (Bonnell and Selander, 1974). Studies of three introduced species of birds, *Passer domesticus* (Parkin and Cole, 1985), *P. montanus* (St. Louis and Barlow, 1988), and *Acridotheres tristis* (Baker and Moeed, 1987), revealed that founder populations are often more structured than source populations, and have fewer alleles and/or are less heterozygous. Easteal (1985) conducted a similar study and estimated N_e from both allozyme frequency and ecological data for founder populations of the giant toad (*Bufo marinus*) with known introduction histories in Hawaii and Australia, and concluded that the former gave more accurate estimates.

Extremely low levels of allozyme heterozygosity in broad geographic surveys may imply the occurrence of one or more recent severe bottlenecks, especially when genetically similar sister species possess much higher levels of variability (e.g., Menken, 1987). However, for a number of theoretical reasons, heterozygosity (H) estimates may not always be good indicators of past population bottlenecks (Nei et al., 1975; Chakraborty and Nei, 1977; Motro and Thomson, 1982; see Turner, 1984, for a possible empirical example). From a conservation perspective, estimates of allozyme variability may be used to make genetically based management recommendations for the recovery of endangered species (Vrijenhoek et al., 1985; A. F. Echelle et al., 1989).

Inbreeding, Outcrossing, and Dispersal   Allozyme studies can mesh genetic and ecological information to strengthen inferences about specific aspects of population structure, especially breeding structure and effective gene flow (e.g., Loveless and Hamrick, 1984; Chepko-Sade and Halpin, 1987; Ryman and Utter, 1987). Knight and Waller (1987) showed that low levels of gene flow in the annual *Impatiens capensis* correlated with pronounced population substructuring (F_{ST} = 0.46) and the largest range of fixation indices (F_{IS} = 0.26–0.94) reported for any plant species. Golenberg (1987) used several indirect estimators of gene flow and neighborhood size (Wright, 1943; Slatkin, 1985) in the selfing annual *Triticum dicoccoides* to show a very sharp decrease in gene flow beyond a distance of 5–7 m from any given parent plant. Among animal species, one or more indirect approaches have been used to infer gene flow patterns from the geographic distribution of allozyme frequencies (reviewed by Slatkin, 1985, 1987; see also Chapter 10). The importance of obtaining independent estimates of dispersal can be illustrated by K. L. Brown's (1985) study of the demographic and genetic characteristics of dispersal in mosquitofish (*Gambusia* *affinis*) in a thermally heated pond on the Savannah River Reservation, South Carolina. Genotype frequencies of the dispersers were nonrandomly distributed throughout the pond, and associations of genetic distance values and geographical distances between collection sites indicated that the dispersers did not constitute a random intermixing of refuge groups. Counter to intuitive expectations, dispersal in these populations resulted in an increase in allelic differentiation between sites, and an increase in mean levels of intrapopulation *H*.

Very few plant or animal species have adequate demographic data for estimating effective genetic dispersal (Endler, 1979) and N_e. In the house sparrow (*Passer domesticus*), Fleischer (1983) used demographic data to predict F_{ST} (Wright, 1943), and then tested the prediction with allozyme data and concluded that these birds approximated a stepping stone model (Kimura and Weiss, 1964) of genetic structure.

Some population genetic surveys have revealed striking examples of heterozygote deficiency, which could result from (1) strong selection against heterozygous genotypes, (2) inbreeding, or (3) a Wahlund (1928) effect (the inclusion of two or more genetically distinct units into a single population sample). In a number of studies, only the inbreeding explanation seems to be valid (see O'Brien et al., 1983, 1985b, 1987; Knight and Waller, 1987; Crouau-Roy, 1988).

Paternity Studies Allozyme studies combining ecological data on dispersal with genotypic data that establish paternity of offspring have allowed assessment of the relative importance of several factors affecting genetic structure in some mammals. An experimental removal and colonization study of pocket gophers (*Thomomys bottae*; Patton and Feder, 1981) revealed that migration (i.e., recolonization) was nearly random with respect to the available source populations. This movement depressed between-field genetic heterogeneity, but this was restored within a single generation due to a high variance in male reproductive success. Juvenile dispersal apparently was responsible for the maintenance of intrapopulation variability in highly socially structured breeding groups of a Neotropical cave-dwelling bat (McCracken and Bradbury, 1977) and colonies of yellow-bellied marmots (Schwartz and Armitage, 1980). Genetic markers revealed that inbreeding was avoided by the near total dispersal of male offspring from their natal colonies in both of these species.

Paternity of specific groups of offspring has been studied in several groups using detectable allozyme markers. For example, Tilley and Hansman (1976) collected female dusky salamanders (*Desmognathus ochrophaeus*) and their broods, and showed that at least 7% of all individual clutches were sired by more than one male. Insemination and fertilization are therefore effectively uncoupled allowing the opportunity for sperm competition, which has been shown in controlled laboratory matings in *D. ochrophaeus* (Houck et al., 1985). Similar studies include Evarts and Williams (1987), Harry and Briscoe (1988), and Quellar et al. (1988).

Species Boundaries   Frequently, evolution at the morphological and molecular levels is uncoupled (Wake, 1981), so protein studies often reveal discordant geographic patterns between levels of allozyme divergence and taxonomic boundaries inferred from morphological criteria. This is especially true for geologically old and morphologically conservative lineages (see Wake et al., 1983; Wake and Larson, 1987). For example, in a recent survey of protein variation within and between 17 populations of the salamander *Desmognathus fuscus,* Tilley and Schwerdtfeger (1981) found several populations that were consistently fixed for the same alternate **electromorph** (relative to *D. fuscus*) at 3 of 18 loci. This form was eventually described as *D. santeetlah* (Tilley, 1981), and no evidence was found for gene exchange between these two species. Similar cases in salamanders have been described by Larson and Highton (1978), Highton (1979), Hanken (1983), Wake and Yanev (1986), Wake et al. (1986), Good et al. (1987), and Good (1989). Examples from other taxa include Arulsekar et al. (1985), Crouau-Roy (1986), Ranker and Schnabel (1986), Adams et al. (1987), J. Shaw et al. (1987), Gastony (1988), and Donnellan and Aplin (1989). These examples highlight the power of this approach as a straightforward method for a quick determination of the genetic cohesiveness of morphologically defined species.

Ecological Genetics   The neutral theory of molecular polymorphism (Kimura, 1968; King and Jukes, 1969) has questioned the primacy of natural selection as an agent of molecular evolution (Lewontin, 1974; Nei and Koehn, 1983). Several statistical studies have concluded that most alternative alleles are selectively equivalent and may represent transient stages of replacement, with fixation probability being a function of mutation rates and effective population sizes (see Kimura, 1983a,b). However, these studies are based on several largely untested assumptions, and may not distinguish among various processes of neutrality and selection (Watt, 1985). Alternatively, several elegant multidisciplinary studies have investigated the selective basis for specific enzyme polymorphisms.

Watt (1985, 1986) outlined a bioenergetic approach for investigating possible functional and ecological differences between alternate allozymes. Documentation of adaptive differences in allozymes requires demonstration of (1) differences in a catalytic function, (2) allozyme-based catalytic differences having physiological effects, and (3) fitness differences in natural environments between physiological effects (see also Koehn, 1978). Watt (1977) identified four common GPI alleles in several natural butterfly

(*Colias eurytheme*) populations, and demonstrated genotypic differences in survivorship, flight activity, and mating success (Watt, 1983; Watt et al., 1985, 1986). Similar studies have been carried out on bivalves (Koehn et al., 1980, 1988; Koehn and Immermann, 1981; Koehn and Siebenaller, 1981; Hilbish et al., 1982; Hilbish and Koehn, 1985a,b), fishes (Powers et al., 1979; DiMichele and Powers, 1982a,b; Allendorf et al., 1983; Leary et al., 1984), and *Drosophila* (Heinstra et al., 1986; Barnes and Laurie-Ahlberg, 1986); these studies suggest that fitness differences do exist between allozymes segregating at single loci.

Interspecific Applications

Phylogenetic Systematics   Allozyme data (and to a lesser extent, isozyme data) have been used extensively to investigate phylogenetic relationships. Some recent reviews of phylogenetic applications include M. W. Smith et al. (1982) for vertebrates, Matson (1984) and Johnson et al. (1984) for birds, Crawford (1983) for plants, and Kilias (1987) for lichens. Buth (1984) reviewed the application of isozyme and allozyme data to systematic problems in general and Swofford and Berlocher (1987) provide a recent evaluation of methods for phylogenetic analysis (see also Chapter 11). The methods of data analysis used for phylogenetic analysis of these data vary widely and are highly controversial. Because each method of analysis has its limitations, and some are simply invalid (Buth, 1984; Chapter 11), the citations in these reviews (and below) should not necessarily be considered exemplary. Here, we restrict our comments to the use of allozymes; isozyme characters such as the presence of duplicate loci, patterns of gene expression, and the ability to form heteropolymers are considered below (Gene Expression and Gene Duplication).

Allozyme characters are subject to many of the same limitations as other forms of systematic data. For example, morphologically distinct species may show very low levels of divergence, and so differ by few phylogenetically informative characters even when many loci are screened (e.g., Murphy et al., 1983). At the other extreme, allozyme divergence may have proceeded to the point where too few electromorphs are shared, and many of those that are shared are convergent (e.g., Sites et al., 1984; Baverstock et al., 1985; Dimmick, 1987; Derr et al., 1987).

There are many groups, however, for which allozyme divergence provides information appropriate for analysis of intra- or intergeneric relationships. Allozyme electrophoresis is appropriate to analyzing intergeneric phylogeny in birds (Gutierrez et al., 1983; Patton and Avise, 1983; Johnson et al., 1988) and, occasionally, other groups as well (e.g., Hafner and Nadler, 1988). Most studies have focused on intrageneric relationships (see reviews cited above), and these are most informative when the individual loci are analyzed as discrete characters (Buth, 1984; Chapter 11). Among other things, this has the advantage that the evidence for each node is made explicit and can therefore be related back to the primary data (e.g., Avise et al., 1980; Miyamoto, 1983b; Patton and Avise, 1983; Hillis, 1985; Sites et al., 1990).

Paleobiogeography   Phylogenetic hypotheses formed from allozyme data can be applied to answering questions of paleobiogeography. A primary method is "Brooks parsimony analysis" (BPA: Wiley, 1988a,b), also known as cospeciation analysis. In the first step of this method, cladograms are constructed for the taxa in question (Brooks, 1981, 1990). Next, geographic areas in which the species occur are designated as if they were taxa. Using geological evidence, an "area cladogram" is constructed showing the historical connections among the study areas. Next, the taxa are treated as if they comprised a completely polarized multistate transformation series in which each taxon and each internal branch of the tree are numbered. The taxa are then coded using nonredundant linear coding (O'Grady and Deets, 1987) and the species names are replaced by their area names. A new area cladogram is constructed based on the phylogenetic relationships of the species, and this new area cladogram is presumed to represent a "picture" of the historical involvement of areas in the evolution of the species. Although this is the preferred method of paleobiogeographic analysis, it has not been applied to allozyme data (see also Kluge, 1988). A less preferable alternative is described below.

Rates of Evolution   Questions of relative rates of evolution are important considerations, especially if applying the **"molecular clock"** (see Chapter 12) or examining relative rates among different kinds of data from the same taxa, for example, allozymes versus morphology. Rosen and Buth (1980) provided a protocol for examining relative rates of evolution using allozyme data that included the calculation of "ancestral genetic distance" between all examined taxa and their common hypothetical ancestor. Murphy and Crabtree (1985a) applied this method to rattlesnakes and found that rates of divergence had been equal. However, they were unable to confidently calibrate the clock. Comparisons of relationships revealed by methods that assume equal rates (e.g., UPGMA) and those that do not (e.g., distance-Wagner trees; Chapter 11) frequently reveal marked variation among lineages in rates of change (e.g., Baverstock et al., 1979; Hillis, 1985). In general, the evidence for an allozyme clock is weak (Avise and Aquadro, 1982; Chapter 12).

Rate questions may also be applied to (1) estimated dates of dispersal or vicariance events, (2) the relative arrival times of taxa on oceanic islands, (3) relative roles of colonization, extinction, and historical factors in island biogeography, and (4) prediction of the time of origin of populations or geographic areas, such as islands, in the absence of supporting geological data or radioisotope dating, such as ^{14}C. For example, Murphy (1983a) used genetic distance data from presumed sister taxa of a number of reptile taxa presumably isolated by the same geological vicariant events in Baja California, Mexico and found a good correlation between geological dates and genetic similarity. Genetic distance data were then used to predict the age of one island in the Gulf of California, Isla Santa Catalina. Unfortunately, most of the sister taxa were "presumed," and not tested using cladistic methods. Genetic similarity data were extended to test the applicability of the MacArthur and Wilson (1963, 1967) theory of island biogeography as it relates to reptiles on islands in the Gulf of California (Murphy, 1983b), and to the colonization of islands by some rattlesnakes (Murphy and Crabtree, 1985a).

Hybridization   Ideally, studies of interspecific hybridization should incorporate three features, including (1) phylogenetic analysis of the taxa involved to allow inferences to be drawn about the origin of the hybrid zone (primary versus secondary; e.g., Hillis, 1985), (2) identification of autapomorphic electromorphs in each of the hybridizing species, which provides the most unambiguous genetic markers for gene flow inferences (Murphy et al., 1984), and (3) identification of at least three unlinked markers (fixed or nearly fixed electromorph differences) between hybridizing taxa (see Chapter 2). With three or more markers, F_1 individuals (heterozygous for parental electromorphs at all markers) can be clearly distinguished from F_2 or backcross classes, which will be heterozygous for some but not all markers (e.g., Hall and Selander, 1973). Few of the studies conducted to date satisfy all of these criteria, but most have contributed to a better understanding of the structure and dynamics of hybrid zones. Least informative are studies in which hybridizing populations are not characterized by any fixed allozyme differences (Greenbaum, 1981; Frykman and Bengtsson, 1984; Halkka et al., 1987). Hybridizing taxa distinguished by several fixed differences (e.g., Gorman and Yang, 1975; Patton et al., 1984; Szymura and Barton, 1986) offer greater potential for inferring the extent and symmetry of introgressed nuclear genes. Several recent studies have used allozyme markers to infer the extent of introgression of other classes of genetic markers (Lamb and Avise, 1986; Arnold et al., 1987b; Harrison et al., 1987; Nelson et al., 1987; Baker et al., 1989). Isozyme studies have also been used to study developmental stability as manifested by morphological asymmetry (Graham and Felley, 1985), and the origin and distribution of rare or unique alleles (called **hybrizymes** by Woodruff, 1989; see also Hunt and Selander, 1973; Sage and Selander, 1979; Greenbaum, 1981; Barton et al., 1983; Case and Williams, 1984; Murphy et al., 1984; Kocher and Sage, 1986; Gollmann et al., 1988). Future studies of hybridization that merge ecological and molecular genetic approaches in appropriate phylogenetic and biogeographic contexts offer great potential for understanding processes involved in genome divergence (Hewitt, 1988).

Parentage of Unisexual Biotypes   Allozyme electrophoresis is a powerful method for identifying the bisexual parent taxa involved in the hybrid origins of various unisexual taxa. Most carefully examined unisexual vertebrates appear to be of hybrid origin (reviewed in Dawley and Bogart, 1989). Typically, unisexual taxa are characterized by higher levels of multilocus heterozygosity than either parental form because of their hybridity and the absence of segregation. Laboratory studies have confirmed patterns of clonal inheritance of fixed heterozygosity in some unisexual lizards (Dessauer and Cole, 1986). In cases of multiple ploidy levels among different unisexuals of hybrid lineages, allozymes frequently show different staining intensities due to alleles that are represented unequally in the genome (Dessauer and Cole, 1984; Dawley et al., 1985). Ideally, an analysis of suspected hybridization events should be carried out in a phylogenetic context that will permit the identification of uniquely derived (autapomorphic) markers in the parental species; this will eliminate ambiguities arising from the use of shared ancestral (symplesiomorphic) alleles to define bisexual taxa involved in hybridization events (Wagner, 1983; Murphy et al., 1984; Funk, 1985; Moritz, 1987; Sites et al., 1990).

Allozyme data are useful in the estimation of clonal diversity within gynogenetic or parthenogenetic populations that arise through recurrent hybridization (Moritz et al., 1989c; Vrijenhoek, 1989), mutation (Parker and Selander, 1976; Spinella and Vrijenhoek, 1982), limited recombination (Asher, 1970; Bogart et al., 1987), or some combination of these factors. The matrilineal clones are frequently not a random representation of the possible genotypic diversity (Turner et al., 1983); interclonal selection may produce habitat or trophic specialists, or hybridogens with different life history characteristics, and these kinds of differences may be "frozen" during the origin of new clones (Vrijenhoek, 1989).

Origin of Polyploid Plants   As in studies of hybridization, isozymes have been valuable in identifying the parents of polyploid plants. Isozymes have supported hypotheses based on other lines of evidence (Roose and Gottlieb, 1976; Werth et al., 1985a) and differentiated among alternative hypotheses (Holsinger and Gottlieb, 1988; Gastony, 1986). Allozymes have shown that some polyploids have a single origin (Werth et al., 1985b) and that some are autopolyploids (Soltis and Rieseberg, 1986). Diversification and speciation in polyploid lineages have occurred through gene silencing (Werth and Windham, 1987; Haufler and Sweeney, 1989). However, if gene silencing regularly leads to diploidization of entire polyploid genomes (Haufler, 1987), the value of isozymes for assessing ploidy (especially in phylogenetically ancient groups) must be questioned.

Gene Expression and Gene Duplication

The expression of gene products is subject to both temporal (ontogenetic) and spatial (cells/tissues) variation in organisms. The predominance of products of different L-lactate dehydrogenase loci in different tissues of vertebrates (e.g., *Ldh-A* in skeletal muscle, *Ldh-B* in heart) is a classic example of this phenomenon (reviewed by Markert, 1983). An example of the evolutionary consequences of regulatory divergence in gene expression is that of the third L-lactate dehydrogenase locus (*Ldh-C*) in the bony fishes. Fishes of several morphologically primitive orders express *Ldh-C* in many tissues, whereas in most advanced teleosts *Ldh-C* expression is limited to eye or liver tissue (Shaklee et al., 1973; Whitt et al., 1975; Shaklee and Whitt, 1981). To ensure relevant comparisons of homologous gene products, extracts from homologous tissues/organs must be prepared and specimens at similar developmental stages compared.

The duplication of genes via aneuploidy, polyploidization, and regional gene duplications (Ohno, 1970; MacIntyre, 1976; Turner et al., 1980) can produce isozyme loci that often diverge markedly in their developmental expression (Whitt, 1981). Differences in gene number can serve as characters useful in systematic studies (Gottlieb, 1982b; Whitt, 1983, 1987; Buth, 1984). These differences can arise through gains of new genes (duplication) or losses (gene silencing); both conditions may be considered as derived relative to an ancestral state. For example, many groups of fishes (Buth, 1983) and plants (Gottlieb and Weeden, 1979; Gottlieb, 1982a) have extra loci encoding enzyme systems, suggesting that gene duplication events have played an important role in their evolution, perhaps in the acquisition of novel gene functions (Ohno, 1970; Markert et al., 1975; Fisher et al., 1980). In fishes, tetraploidization is followed by a shift from tetrasomic (pairing of homologous chromosomes in tetrads) back to disomic (pairing in diads) patterns of inheritance. During this "rediploidization," some 50% of the duplicated loci are silenced either by fixation of new mutations or the deletion of some codons (Allendorf and Utter, 1973; Ferris and Whitt, 1977a,b; Li, 1980). Patterns of malate dehydrogenase (MDH) meiotic segregation during rediploidization in the recently evolved tetraploid frog *Hyla versicolor* suggest polymorphic (tetrasomic, disomic, and tetrasomic–disomic) inheritance thought to be a transitory phase between complete tetrasomy and complete disomy (Danzmann and Bogart, 1982a). A phylogenetic evaluation of the catostomid fish *Moxostoma lachneri* suggests that "retetraploidization" of the second glucose-6-phosphate isomerase locus, *Gpi-B,* is due to reactivation, or postpolyploidization regional duplication (Buth, 1982a).

Isozyme staining intensities may be used to investigate ploidy levels. Danzmann and Bogart (1982b) and Dessauer and Cole (1984) found that gene dosages, and thus ploidy levels (*2n, 3n,* or *4n*), could be inferred

Table 2. Evolutionary patterns of creatine kinase gene expression in fishes

	Character state	
Character	Ancestral	Derived
1. Number of loci	2 (A, C)	4 (A, B, C, D)
2. Tissue specificity	Widespread	Restricted
3. Interlocus heterodimer formation	Present	Absent
4. Intralocus heterodimer formation	Present	Absent

Summarized from Ferris and Whitt (1978b) and Fisher and Whitt (1978, 1979).

accurately from staining intensities because subunit interactions were additive.

As discussed earlier, many enzymes are multimeric, composed of subunits that must be assembled in order for the enzyme to function. Multiple isozymes of multimers can be produced by combining different kinds of subunits in heterozygotes (heteromers) and by the interactions among multimers of duplicated genes in a multilocus isozyme system producing interlocus **heteropolymers.** Heteropolymer formation may be nonrandom because regulatory differences may suppress the formation of some or all of the possible heteromers, e.g., the heterotetramers of L-lactate dehydrogenase of some lizards (Gorman, 1971; Sites et al., 1986), fishes (Buth et al., 1980), and snakes (Murphy, 1988).

The "isozyme characters" (sensu Whitt, 1983, 1987; Buth, 1984; Murphy and Crabtree, 1985b) of gene number, tissue specificity of expression (gene regulation) and posttranslational modification, and heteropolymer assembly can be of systematic value only if they vary at a taxonomic level useful to the investigator. These characters may be useful for intraspecific, intrageneric, or intrafamilial comparisons depending on the group (Buth, 1984). However, the few studies of enzyme systems reveal certain limited group trends. Studies of creatine kinase (CK) expression in fishes by Ferris and Whitt (1978b), Fisher and Whitt (1978, 1979), and others permit the generalizations for CK isozyme characters listed in Table 2. Three of the four evolutionary patterns in Table 2 appear to hold for amphibians and reptiles (Buth et al., 1985). In contrast, LDH expression in sea snakes and cobras does not correlate with phylogenetic relationships (Murphy, 1988).

Limitations

Taxonomic Limits   Studies of population structure, breeding biology, and other intraspecific applications require sufficient levels of intraspecific variability. Allozymes are not sufficiently variable in some organisms, making other molecular methods such as mtDNA RFLP studies (Chapter 8) more appropriate. For example, DeSalle et al. (1987b) examined the distribution of mtDNA haplotypes in populations of *Drosophila mercatorum* distributed along a short altitudinal transect near Kamuela, Hawaii, and found statistically significant spatial and temporal heterogeneity in the absence of isozyme divergence. Intraspecific studies of birds are often hampered by very low levels of isozyme polymorphism (Barrowclough et al., 1985), yet Quinn and White (1987b) demonstrated extensive genomic DNA RFLP variability in the snow goose (*Anser c. caerulescens*). Similarly, Sites and Davis (1989) found far more restriction-site markers in both mtDNA and nuclear ribosomal DNA than in allozymes among central Mexican chromosome races of the lizard *Sceloporus grammicus*. These and other studies (Johnson et al., 1977; Avise et al., 1979a,b; Wetton et al., 1987) show a definite lower taxonomic limit to the resolving power of protein electrophoresis, but this limit may vary from group to group (e.g., Kessler and Avise, 1985b).

At the opposite extreme, some taxa have diverged to the extent that they share virtually no alleles. For example, Sites et al. (1984) surveyed 17 genera of batagurine turtles, and found the taxa to be so divergent and homoplasy so extensive that they could not recover well-corroborated branches for most basal stems of the cladogram. Divergence among these lineages approached or exceeded the limits of resolution of isozyme electrophoresis. Nei (1987: 251–252) proposed as a general rule that if genetic distance, D (Nei, 1972, 1978), is greater than 1.0, then the frequencies of back/parallel mutations will be high, and the variance of D large, even if numerous loci are assayed. The hierarchical taxonomic level at which phylogenetic utility is lost ($D \geq 1.0$) will vary with taxonomic assignments and taxon-specific rates of molecular evolution—birds appear to be decelerated (Avise and Aquadro, 1982)—but generally the greatest phylogenetic utility for isozymes will be at the level of species or closely related genera (Nei, 1987).

Sampling Limitations   Several kinds of limitations of isozyme techniques are recognized, including limits to the number of (1) loci resolved, (2) alleles per locus, and (3) individuals required for population or phylogenetic studies. The total number of loci that can now be visualized with histochemical staining techniques is in excess of 100 (Wright et al., 1983; Morizot and Siciliano, 1984), but this is still only a very small sample of the total genome. However, given the size of most eukaryotic genomes (summarized in Cavalier-Smith, 1985b; Loomis, 1988), this is a constraint common to most molecular techniques and will not be elaborated further here.

Limits to Detection of Segregating Alleles.   Hubby and Lewontin (1966) recognized that gel bands represented enzyme "phenotypes," and not necessarily all underlying allelic variation. King and Ohta (1975) introduced the term "electromorph" to label allozymes of the same mobility as different classes of alleles. Allendorf (1977) stressed that electromorph identity did not mean identity in DNA base sequence; homology is a "conditional" concept for isozyme phenotypes.

Because accurate estimation of allelic variation has important implications for many evolutionary questions (Coyne, 1982), the problem of "hidden heterogeneity" (Johnson, 1977) fostered several studies to determine how accurately conventional electrophoretic techniques estimate genetic variability. Singh et al. (1976) used a sequential assay of four different electrophoretic conditions, termed **sequential electrophoresis,** and heat stability tests, to examine *Xdh-A* variation in *D. pseudoobscura*. They resolved 37 alleles where only 6 had previously been identified by conventional protocols. Other approaches to detecting **cryptic alleles** include thermostability analysis (e.g., Chambers et al., 1981), peptide mapping (Ayala, 1982), and the use of polyacrylamide gels of varying pore sizes to produce a sieving effect for separation by size or molecular weight (Johnson, 1976, 1979).

Although these methods show that conventional isozyme electrophoresis may underestimate variability, they do not reveal what proportion of alleles may still remain undetected. Ramshaw et al. (1979) examined several human hemoglobin variants of known amino acid sequence using both standard and sequential acrylamide electrophoresis (varying conditions of pH and pore size). Three experiments determined what types and proportions of substitutions could be resolved by these methods. First, 8 and 17 hemoglobin variants out of 20 were detected by the two procedures, respectively. Second, groups of variants with the same amino acid substitutions in different parts of the molecule were screened by two approaches and revealed 77 and 90% of the known variants, respectively. Third, four of five pairs of hemoglobins differing by charge-equivalent substitutions in the same positions were separated by both procedures. There was no class of commonly indistinguishable substitutions, and Ramshaw et al. (1979) concluded that the standard protocol of electrophoresis was a powerful method for identifying most variants.

McLellan (1984) examined 14 whale myoglobins of known sequence by sequential polyacrylamide electrophoresis (five pH values) and was able to separate 13 of the 14 variants. No further resolution was obtained by altering concentration or composition of the gels, or by screening with other techniques such as urea denaturation or isoelectric focusing (McLellan and Inouye, 1986).

Aquadro and Avise (1982a) employed several starch and acrylamide conditions, gel sieving, isoelectric focusing, and thermal stability tests, to screen for cryptic alleles at three loci (*sAat-A, sMdh-A,* and *Est-1*) in five populations of *Peromyscus maniculatus. sAat-A* (*Got-1*) was previously known to segregate for two alleles across most of the range, *sMdh-A* (*Mdh-1*) was essentially monomorphic throughout the range, and *Est-1* was highly polymorphic, having eight alleles resolved in earlier studies. None of the employed techniques uncovered any additional variation in either *sAat-A* or *sMdh-A*. In contrast, sequential electrophoresis (five additional starch gel conditions) resolved 23 variants in *Est-1,* which were further resolved into 35 variants by heat denaturation, although the allelic nature of the latter group was not determined. Aquadro and Avise (1982b) also uncovered additional sMDH isozymes among 10 orders of birds using multiple buffers.

Thus, hidden heterogeneity is pervasive, and one cannot always rely on any single method to resolve all alleles. Equally important, however, are findings that (1) some loci are much more likely than others to harbor cryptic alleles, especially systems originally resolved as highly polymorphic by conventional methods, and (2) conventional methods will resolve most or all variation at the more conservative loci. Further, a number of classes of studies will be largely unaffected by this phenomenon (Coyne, 1982). Fixed differences between populations or species detected by conventional methods are real and can be increased only by resolution of additional alleles. Similarly, between-population allele frequency heterogeneity is also real, regardless of any underlying heterogeneity in electromorphs, because such differences in electromorph classes should also reflect the same deviations of cryptic alleles. Other kinds of studies (e.g., absolute estimates of heterozygosity) may be more affected by cryptic allelic variability, but to an unknown degree. Obviously, any problem addressed with isozyme techniques will be better understood by more accurate descriptions of allelic variation, and, where time and resources permit, we suggest that at least loci showing extensive variation under standard conditions be screened sequentially with additional buffers to maximize separation.

Null Alleles and Isoloci.   Other phenomena cause deviation from codominant expression of allozymes. **Null alleles** (those with reduced or absent expression of a protein product) are detected by reduced staining intensity of some single isozymes on the same gel; complete absence of activity may indicate null homozygotes (see Utter et al., 1987). These interpretations are often ambiguous and require confirmation by breeding studies (e.g., Stoneking et al., 1981). In the absence of breeding studies, quantification of a null allele cannot be made reliably. Apparent heterozygote deficiencies may be due to null heterozygotes being scored as active homozygotes (Foltz, 1986). Heterozygotes for null alleles are more readily detected if they either form partial heteropolymer isozymes in polymorphic single-locus systems

(Burkhart et al., 1984), or are expressed in multilocus, multimeric proteins (e.g., Utter et al., 1987; Engel et al., 1973; Allendorf et al., 1984). In both cases, reduced intensities of one or more multiple bands provide additional visual clues to the presence of null alleles.

Another difficulty may occur when isozymes having identical electrophoretic mobilities represent the products of two different loci of the same multilocus enzyme system (Utter et al., 1987). These **isoloci** may present rather complicated isozyme patterns, and, if allelic variation is present, determination of which locus is polymorphic, or if both are, may not be possible. Isoloci may be individually identifiable if their respective encoded loci are synthesized at different levels in different tissues, but this appears to be uncommon (Allendorf and Thorgaard, 1984). Under some circumstances, different staining intensities are expected (see Utter et al., 1987), but often such distinctions are difficult or impossible to make. Changing electrophoresis buffers often results in the separation of isoloci.

Other Sources of Phenotypic Variation of Isozymes   The phenomena described above may either limit the resolving ability of isozyme electrophoresis or complicate isozyme interpretations. Several "non-Mendelian" factors may also complicate isozyme phenotypes via in vivo or in vitro environmental conditions, or through the action of modifier loci.

Posttranslational Modification of Enzymes.   Polypeptide synthesis entails (1) translation, (2) polymerization, (3) termination, and (4) processing of the final protein product. Only the first step involves the direct coding of nucleotide sequences into primary protein structure, while the others are **posttranslational** processes that give a final structure to the product (Uy and Wold, 1977). These latter processes change the 20 primary amino acids specified by the genetic code as monomeric building blocks in polypeptide assembly into about 140 amino acids and derivatives in completed proteins (Uy and Wold, 1977). On gels, a number of these **epigenetic** events may produce **conformational isozymes,** or multiple forms of a single gene product that differ in secondary or tertiary structure (also called **secondary isozymes** or **subbands,** Richardson et al., 1986) and/or variants that differ in thermal stability (Lebherz, 1983). In some cases, modifying genes have been shown to be polymorphic for alleles that differ in their influence on electrophoretic mobilities of the protein products (Cochrane and Richmond, 1979; Womack, 1983; Dykhuizen et al., 1985). In other cases, altered mobilities appear to be restricted to specific tissues (Murphy and Crabtree, 1985b), or to be a function of environmental conditions and/or the physiological state of the organism (McGovern and Tracy, 1981; van Tets and Cowan, 1966), and in a cryptic species of the freshwater clam genus *Corbicula* the synthesis of an enzyme seems to be a seasonal event in an entire population (Hillis and Patton, 1982).

Mobilities of some proteins are also susceptible to protease degradation associated with repeated freezing and thawing (Harris and Hopkinson, 1976; Richardson et al., 1986), or long- and short-term aging of the sample (Walter et al., 1965; Kobayashi et al., 1984). Moore and Yates (1983) showed that many of the loci frequently screened in population and systematic studies were resistant to mobility modification when kept at room temperature up to 12 hr after death. Posttranslational effects can frequently be determined by evaluating relative intensity of isozyme staining; alternate segregating alleles usually give constant patterns of expression, while breakdown effects are likely to give a full range of expression of relative strengths (Richardson et al., 1986).

Intracistronic Recombination. Recombination between alternative nucleotide sequences within a single gene locus has been investigated theoretically by a number of workers with respect to its potential evolutionary importance (Watt, 1972; Strobeck and Morgan, 1978; Morgan and Strobeck, 1979; Golding and Strobeck, 1983). The mechanism can generate new alleles at rates several orders of magnitude above standard gametic mutation rates if some minimum level of variability is already present at a given locus. This hypothesis, that "polymorphism generates polymorphism," was empirically demonstrated by Ohno et al. (1969) in test crosses of 26 pairs of Japanese quail (*Coturnix c. japonica*) heterozygous for different combinations of four alleles for phosphogluconate dehydrogenase. They recovered 11 mutation-like events in enzyme phenotypes scored from 1011 test-cross progeny, including new electromorphs, novel-combination genotypes, and one case of the inheritance of three alleles. If these results are due to intracistronic recombination, and if this is a general phenomenon, then it may form the basis for the well-known "rare allele" observations in many hybrid zones (see above, and Woodruff and Thompson, 1980; Murphy et al., 1984). This type of recombination differs qualitatively from that between separate loci by producing new gene products, which may then be transmitted in a Mendelian fashion, rather than new genotypic combinations which will be disrupted in later generations.

LABORATORY SETUP

To carry out starch gel electrophoresis, a number of pieces of equipment are essential and others are highly desirable. In many cases, more expensive alternatives are more cost efficient in the long run because of time saved. Table 3 lists both the necessary and desirable equipment for SGE but does not include a detailed list of many supplies (see Werth, 1985). Table 4 provides a list of chemicals required for the stain and buffer recipes (Appendices 1 and 2, respectively).

Table 3. Basic equipment and non-chemical supplies necessary and desirable for starch gel electrophoresis[a]

Equipment	Description	Quantity needed	Quantity desirable	Protocol number
Major				
Freezer	Manual defrost	1	1	6
Refrigerator	>12 cu ft	1	1	6,8
Analytical balance	0.1 mg–100 g	1	1	1,2,6
pH meter	0.01 pH, with "Tris" probe	1	1	2,6
Fume hood		1	1	2,4,6
Water deionizer and filter		1	1	1,2,4,6
Power supplies	0-500 V, 0-100 mA	1	10+	4
Refrigerated, high-speed centrifuge	>10,000 g	0	1	1
Centrifuge rotor	Fixed angle, 24-36 place	0	1	1
Refrigerated chamber/walk-in refrigerator		0	1	4
Ultracold freezer	−70° C	0	1	1
CO_2 (or LN_2) backup for ultracold freezer		0	1	1
Incubator		0	1	6
Tissue homogenizer	High speed	0	1	1
Sonicator/cell disrupter		0	1	1
Water bath	≈25 x 35 cm	0	1	6
Microwave oven		0	1	2,6
Pipetters, set	1 μl–5 ml adjustable	0	>1	1,3,6
Single lens reflex camera	with macro lens and yellow filter	0	1	8
Ice machine	In lieu of blue ice	0	1	1,4
Minor				
Gel molds		1	>20	2,4
Buffer wells (trays)		1 pr.	>20 pr.	4
Desiccators		2	4	6
Spatula (stainless steel)	Large and small	12 ea.	>12	6
Magnetic stirrer	Preferably with hot plate	1	1	6
Magnetic stirring bars	Various sizes	1 pkg.	1 pkg.	6
Aspirator/vacuum line		1	1	2
Aspiration safety shield		1	1	2
Bunsen burner	5000 BTU	1	1	2
Heat gloves		1 pr.	2 pr.	2

(Continued)

Table 3. (*Continued*)

Equipment	Description	Quantity needed	Quantity desirable	Protocol number
Gel slicer		1	1	5
Polystyrene stain boxes		10	>200	6
Hazardous chemical disposal container(s)		1	2	6
Aluminum trays/blue ice		1	>10	4
Timer	5–60 min	1	3	6
Forceps	Small, straight tips	2	>4	3
Forceps	Small, curved tips	2	>4	3
Dissecting kit	Scalpel, scissors, etc.	1	2	1
Copy stand	Adjustable height	1	1	8
Light box		1	1	6,8
Ultraviolet lamp	Long wave (340 nm)	1	1	8
Ultraviolet light face shield		1	1	8
Pipetter	Ranging 1–10 ml fixed volume, bottle top	0	5	6
Liquid dispenser	Adjustable volume 10–60 ml	0	1+	6
Erlenmeyer flasks	125–1000 ml	Numerous		2,6
Glass bottles, narrow mouth, amber color	200–4000 ml	Numerous		6
Graduated cylinders	10–1000 ml	5	10	2,6
Beakers	3–3000 ml	1 ea.	12	6
Funnels	Large and small	2	>2	2,6
Wash bottles	250 ml	1	6+	1,5,6
Pasteur pipettes		As needed		2,6
Disposable rubber/Vinyl gloves		As needed		2,6
Disposable dust mask(s)		As needed		2,6

[a]See text for discussion and alternatives.

The room in which electrophoresis is to be carried out should be equipped with sufficient counter space and electrical outlets, a large sink, gas line (propane or natural gas), and a certified, working fume hood. If a fume hood is not available, then all procedures involving the use of 2-mercaptoethanol and some alcohols must be avoided; the former liquid is highly volatile and the fumes may be lethal. Water pressure must be sufficiently high to allow for a faucet aspirator, or a vacuum pump or vacuum line must be available. There should be an abundant supply of distilled water and a water filter/deionizer.

The starch gels must be cooled during electrophoresis. Ideally this is accomplished by performing electrophoresis in a walk-in refrigerator, chromatography chamber (dairy case), or standard refrigerator; horizontal gels are further cooled using ice-filled aluminum trays. If a refrigerated chamber is not available, the ice levels must be checked at least every 2 hr, making overnight electrophoresis runs difficult. In addition, tissue homogenates are kept cool on ice while gel loading takes place. Thus, an ice machine is highly desirable. If crushed ice is not readily available, then Blue Ice™ packs can be used during electrophoresis without having deleterious effects.

Most staining gels are placed in an incubator set at 37°C. Alternatively, gel staining can be carried out in dark cabinets or drawers, the only effect being a longer stain reaction time.

It may be necessary or desirable to construct some of the equipment, especially gel molds, buffer wells, gel origin guide, gel slicer, slicing tray, and aspiration shield. Plans and examples of equipment are provided in Figures 1–5 and detailed assembly instructions will be provided upon request to R. W. Murphy. Buffer well plans are designed to prevent accidental electrocution (see Spencer et al., 1966). Gel molds, buffer wells, and gel origin guides are constructed from high-quality acrylic plastic (transparent polymethyl methacrylate) sheets, such as Plexiglas™ G. The pieces of plastic are glued using either methylene chloride, chloroform, or a commercially available solvent (methylene chloride containing dissolved plastic).

Table 4 lists the chemicals necessary to establish an allozyme electrophoresis (specifically, SGE) laboratory having a capacity to use many different buffer combinations for running and staining of most enzyme systems that have been adapted to eukaryotes.

PROJECT PLANNING

The problems to be solved in preliminary studies are (1) what is the optimal buffer system?, and (2) how does expression vary among tissues and which tissues are best for analysis? This technical development phase can be combined with a pilot study (see Chapter 2) to determine the efficiency of the approach. The optimal gel buffer systems for particular proteins vary among taxa. Also, impurities in water can affect differences in electrophoretic conditions, making interlaboratory protocols vary. Unless multiple gel buffer systems are initially tried for each enzyme or general protein system, much of the variation may be "hidden," or unresolved (see above). Before the isozyme data are gathered, it is highly desirable, if not essential, to independently determine/confirm which of the various gel buffer systems

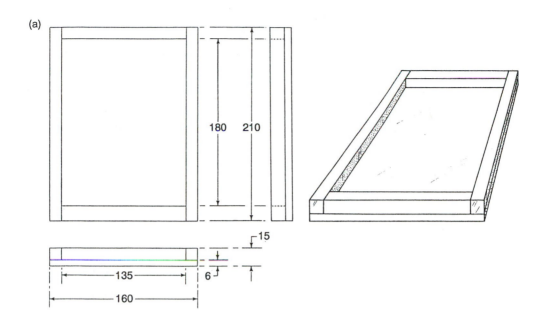

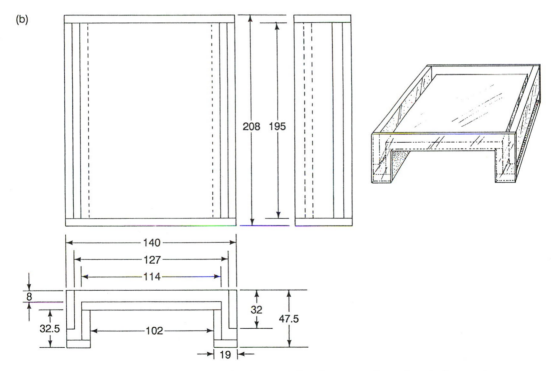

FIGURE 1. Plans for two types of gel molds used in horizontal starch gel electrophoresis. (a) Simple gel mold that requires the use of a sponge wick; (b) a wickless gel mold. The construction material is 1/4 in. acrylic plastic. All measurements are in millimeters.

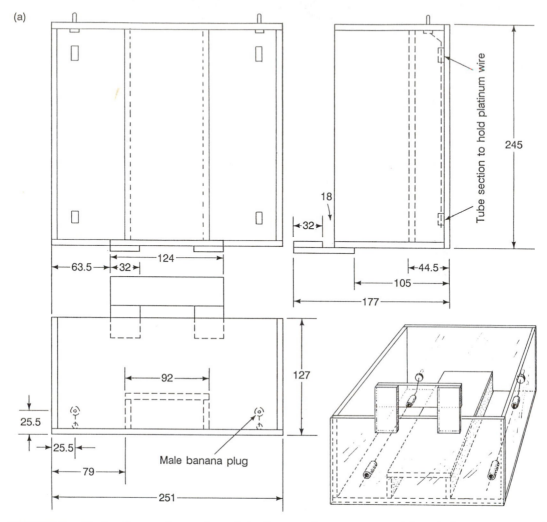

FIGURE 2.   Design for an electrophoresis buffer tray that prevents accidental electrocution. (a) Base; (b) cover. Construction material is 1/4 in. acrylic plastic. All measurements are in millimeters.

are useful. Therefore, we have avoided suggesting buffer and stain combinations.

With five gel setups, in a few days it is possible to determine optimal electrophoretic conditions by surveying a few specimens for a wide array of enzymes on virtually all commonly used gel buffer systems. Each of the five gels is made from a different buffer and can be cut into 5 × 5 minislices, allowing the rapid survey of 30 or more enzyme systems. Five minigels representing five different buffers are simultaneously stained in the same stain box making the protocol cost and time efficient. The specimens examined can represent the taxonomic diversity to be studied (see Chapter 2), a range of different tissue types, or both.

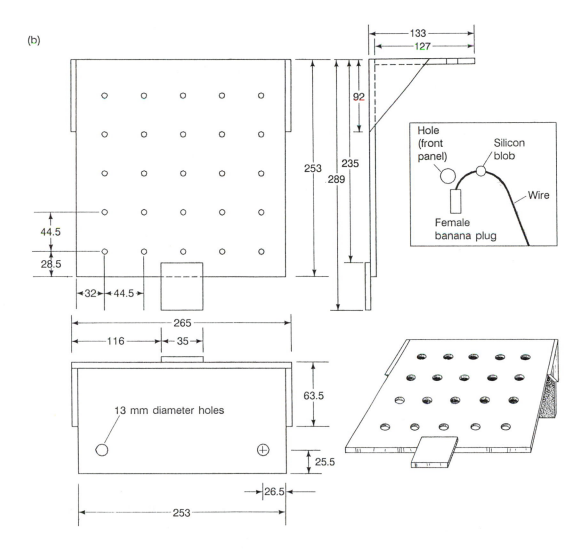

It may be important to have a mix of relatively rapidly and slowly evolving loci, especially if one study is to be compared to another, or if different hierarchical taxonomic levels are being examined. Some enzymes, such as those involved with glycolysis, tend to be relatively conservative in both number of loci and amount of allelic variability within loci (Gillespie and Kojima, 1968; Gottlieb, 1982a).

The final stage of planning involves the electrophoresis of allozymes from numerous individuals on established buffer systems and from known tissues to generate data on allozyme variation. Gel runs must be well planned in advance to avoid unnecessary reruns. Richardson et al. (1986) have detailed many variables that should be taken into consideration in the planning stages. Some of the more important considerations are as follows:

1. Enzyme systems sensitive to freezing and thawing (e.g., HBDH, G3PDH, IDDH, etc.) should be resolved first, preferably before freezing the tissues or extracts.
2. Tracking dye (Appendix 2) or a blank space should be used about every 10 specimens but without separating populations or taxa.
3. Initial specimen alignment on the gels should allow the first sample(s) to be repeated occasionally, or at least on the end of the gel, especially if more than two taxa or populations are being surveyed. To differentiate among different alleles, side-by-side comparisons are necessary.
4. Some enzyme systems require, or are better resolved with, the addition of known activators or cofactors such as EDTA and Mg^{2+} (e.g., PGM), or coenzymes such as NAD (e.g., ADH) or NADP (e.g., G6PDH, PGDH) (Harris and Hopkinson, 1976). These are added to specific gels before cooking (EDTA, Mg^{2+}) or following aspiration (coenzymes).

PROTOCOLS

1. Tissue homogenization
2. Preparation of starch gels
3. Gel loading
4. Electrophoresis
5. Gel slicing
6. Histochemical staining
7. Drying agar overlays
8. Documentation of results

Many of the chemicals used in the various protocols are extremely hazardous; Table 4 briefly summarizes the known acute and chronic health hazards. For more information, "Material Safety Data Sheets" may be available from suppliers free of charge on request when ordering chemicals. Contact with these organic and inorganic substances should be avoided as many are readily absorbed through the skin. Protocols should not be performed without using protective laboratory coats, rubber gloves, dust masks, and eye goggles whenever appropriate. Food, drink, and tobacco should not be allowed in the laboratory. Other safety precautions, such as eye wash stations, showers, chemical spill clean-up kits, and first aid kits, should be readily available. Operator safety must be accorded priority over all other considerations.

Protocol 1: Tissue Homogenization

(Time: 2 min/specimen)

Tissue extract preparation may precede or follow preparation of the starch gels depending on (1) the necessity for extremely fresh extracts or (2) desirability of preparing gels a day in advance. The homogenization of tissue samples far in advance of their use may result in significantly reduced levels of enzyme activity. Many extraction buffer

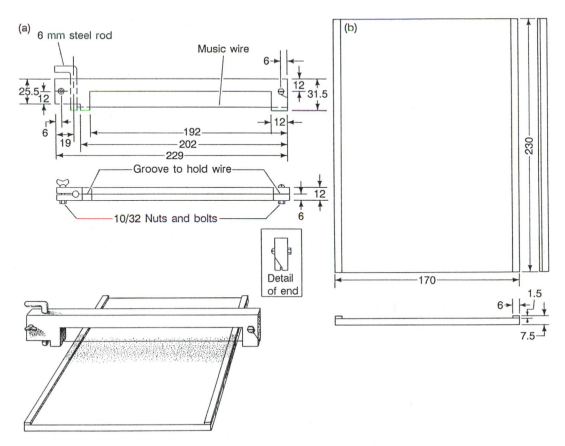

FIGURE 3.   Gel slicing apparatus. (a) Bow slicer (constructed from 1/2 in. aluminum bar); (b) gel slicing tray (constructed from 1/4 in. acrylic plastic). All measurements are in millimeters.

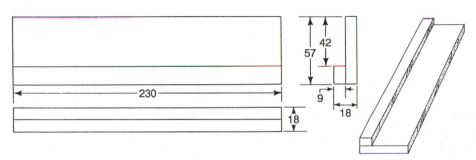

FIGURE 4.   Gel origin guide (constructed from 3/8 in. acrylic plastic). All measurements are in millimeters.

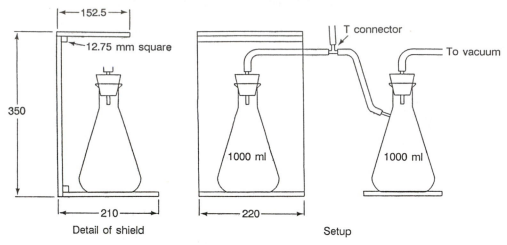

FIGURE 5.   Gel aspiration setup. Plastic implosion shield is made from 1/4 in. or thicker acrylic sheet.

recipes use 2-mercaptoethanol, a sulfhydryl reducing agent, to reduce subbands. However, at least in reptiles, this ingredient significantly reduces the activity levels of many enzyme systems.

Phenolic compounds in many plant tissues form complexes with proteins on homogenization. The addition of polyvinylpyrrolidone to the extraction solution usually reduces this problem; some plants also require other ingredients (see Werth, 1985).

There are several ways of extracting enzymatic proteins from cells including (1) simple maceration of tissue(s) with scissors followed by freezing, (2) use of a hand-held ground-glass homogenizer, (3) hand grinding with a glass test tube or rod sanded on its base and a porcelain spot plate (Werth, 1985), (4) motorized "plastic" (e.g., Teflon™) pestle and plastic (centrifuge tube) mortar, or (5) a high-speed tissue homogenizer with "generator" blade (Figure 6). Homogenization using devices designed to not disrupt cell membranes (Figure 6) may require that the samples be subjected to sonication or refreezing for 10 min at −20°C. All of the methods work very well, even without sonication; the latter, and initially most expensive method, is the fastest. If samples are not to be used immediately, refreeze, preferably in an ultracold freezer.

1. Dissect out desired tissues or retrieve previously dissected tissue samples from the freezer and place them in a clean grinding tube.
2. Dilute the samples 3- to 5-fold with grinding solution. The ice-cold grinding solution may be either distilled, deionized water or one of many solutions described in the literature (e.g., Selander et al., 1971; Harris and Hopkinson, 1976; Werth, 1985). If enzyme activity levels are to be surveyed, the tissue samples must precisely be weighed and diluted (Klebe, 1975; Kettler and Whitt, 1986; Kettler et al., 1986).

Table 4. Chemicals required for electrophoresis, use [enzyme system(s) and/or buffer(s)], location of storage, and health hazard information[a]

Chemicals (and stains involved)	Location[b]	Reference number[c]	Health & Safety[d]
Acetic acid, glacial	s		D
Acetone (RDH)	s		D
Cis-Aconitic acid (ACOH)	f	A-3412	E, S, R, I
Adenosine (ADA)	r	A-9251	E, S, R
Adenosine 5'-diphosphate (AK, ARK, CK, ENO, PK, TAT)	f	A-8146	N
Adenosine 5'-monophosphate (AK)	f	A-1877	E, S, R, I
Adenosine 5'-triphosphate (AK, GLAL, GUK, HK, PFK, PGAM, PGK, SUDH, UK)	f	A-6144	N
Agar (general)	s	A-7002	E, S, R, I
DL-Alanine (ALAT, ALPDH)	s	A-7502	N
DL-Alanyl-DL-methionine (PEP)	f	A-2128	N
Aldolase (GAPDH, PFK)	r	A-1893	A, I
Amaranth (tracking dye)	s	A-1016	T, E, S, R, I
3-Amino-9-ethyl carbazole (PER)	s	A-5754	C
N-(3-Aminopropyl)-diethanolamine (buffer)	s		F, D
N-(3-Aminopropyl)-morpholine (buffer)	s	A-9028	F, D, I
Ammonium hydroxide (general)	s		N
Ammonium molybdate (GLAL)	s	M-0878	N
Arsenic acid (Na salt) (ADA, GAPDH, TPI)	s	A-6756	F, C, E, S, R, I
Ascorbic acid (GLT)	s	A-1417	E, S, R
L-Ascorbic acid (GLAL, NTP)	s	A-0278	E, S
L-Aspartic acid (AAT)	s	A-9256	I
1,3-Bis(dimethylamino)-2-propanol (buffer)	s		N
Black K salt (CAP, ACP)	f	16318X[f]	N
Boric acid (buffer)	s	B-0252	E, S, R
Brilliant blue G (CBP, GP)	s	B-1131	E, S, R, I
Calcium chloride (GUK, PER)	s	C-3881	E, S, I
Citric acid (anhydrous, free acid) (buffer)	s	C-0759	E, R, A
Citric acid (dihydrate) (buffer)	s	C-7254	E, S
Citric acid (monohydrate) (buffer)	s	C-7129	E, S, R
4',6-Diamidino-2-phenylindole (DAPI)	f	D-1388	E, S, R, I
o-Dianisidine dihydrochloride (PEP)	r	D-3252	C, E, S, R, I
Dichlorophenol-indophenol (CBR, DDH, DIA, GR)	s	D-1878	E, S, R, I
Dihydroxyacetone phosphate (TPI)	r	D-7878	N
Dowex (ion exchange resin) (TPI)	s	50x4-200R	E, S, R, I
N,N-Dimethyl formamide (PER)	s	D-4254	D
Dimethylsulfoxide (βGA, βGALA)	s	D-5879	C, E, S

(Continued)

Table 4. (*Continued*)

Chemicals (and stains involved)	Location[b]	Reference number[c]	Health & Safety[d]
Ethanol (ADH, ODH)	s	xxxx[g]	D, E, R
Ethylenediaminetetraacetate (EDTA) free acid (buffer)	s	EDS	E, S, R, I
EDTA-dihydrate (PGAM, SUDH, buffer)	s	ED2SS	E, S, R, I
Fast blue BB (salt) (ALAT, AAT, EST)	f	F-0250	E, S, R, I
Fast blue RR (salt) (ALP)	f	F-0500	F, P
Fast garnet GBC (salt) (GLUR)	f	F-0875	M
Flavine adenine dinucleotide (GR)	f	F-6625	P, I
Fluorescein diacetate (Ca-2)	f	F-7378	E, S, R, I
Formaldehyde (FDH)	s	F-1635	F, D, S, R, I
D-Fructose-1,6-diphosphate (ENO, FBA, FBP, ALD, GAPDH, PK)	f	F-4757	N
D-Fructose-6-phosphate (GPI, PFK, buffer)	f	F-3627	N
Fumaric acid (FUMH)	s	F-4633	E, S, R, I
D-Gluconic acid lactone (HADH)	s	G-4750	E, S, R, I
D(+)-Glucose (AK, ARK, CK, GCDH, HK, PK)	s	G-8270	N
α-D-Glucose-1,6-diphosphate (PGM)	f	G-5875	N
α-D-Glucose-1-phosphate (PGM)	f	G-7000	N
D-Glucose-6-phosphate (G6PDH)	f	G-7250	N
Glucose-6-phosphate dehydrogenase (AK, CK, GPI, HK, MPI, PGM, PK)	r	G-5760	A, I
L-Glutamic acid (GLAL, GTDH)	s	G-1626	N
L-Glutamic dehydrogenase (ALAT, TAT)	r	G-2626	I, A, E, S, R
Glutathione, oxidized (disodium salt) (GR)	f	G-4626	N
Glutathione, reduced (FDH, HAGH, LGL)	r	G-4251	M, P, I
Glyceraldehyde-3-phosphate dehydrogenase (ALD, FBA, PF, PGAM, PGK, TPI)	r	G-0763	A, I
DL-Glyceric acid (GLYDH)	s	G-5626	N
Glycerol (fixitive)	s		E, S, H, B
DL-α-Glycerophosphate (G3PDH)	r	G-6014	N
α-Glycerophosphate dehydrogenase (PGK)	r	G-6751	A, I
Glycine (buffer)	s	G-6761	E, S, R, D
Glycolic acid (HAOX)	s	G-1884	F, E, S, R
Glycyl-L-leucine (PEP)	f	G-2002	N
Glyoxalase I (HAGH)	r	G-4252	A, I, E, S, R
Guanine (GDA)	s	G-0506	E, S, R, I
Guanosine-5'-monophosphate (GUK)	r	G-8252	E, S, R, I

(*Continued*)

Table 4. (Continued)

Chemicals (and stains involved)	Location[b]	Reference number[c]	Health & Safety[d]
1-Hexanol (ADH)	r	H-4504	D, E, R
Hexokinase (AK, ARK, CK, PK)	f	H-5000	A, I
L-Histidine HCl monohydrate (gel buffer)	s	H-8125	E, S
Hydrochloric acid	s		F, D, V
Hydrogen peroxide (CAT, PER)	r	H-1009	F, D
DL-β-Hydroxybutyric acid (Na salt) (HBDH)	r	H-6501	N
Hypoxanthine (XDH)	s	H-9377	E, S, R, I
Inosine (PNP)	s	I-4125	E, S, R, I
Inosine triphosphate (NTP)	f	I-0885	P, I
DL-Isocitric acid (IDH)	r	I-1252	N
Isocitric dehydrogenase (ACOH)	f	I-1877	A, I
α-Ketoglutaric acid (ALAT, AAT, TAT)	r	K-1750	E, S, R, I
DL-Lactic acid (buffer, LDH)	s	L-1250	D
L-Lactic dehydrogenase (AK, ALAT, CK, ENO, GUK, HAGH, PK, UK)	r	L-2500	A, I
L-Leucyl-L-alanine (general PEP)	r	L-9250	N
L-Leucylglycylglycine (PEP)	f	L-9750	N
L-Leucine-β-naphthylamide HCl (CAP)	f	L-0376	C, I
L-Leucyl-L-leucyl-L-leucine (PEP)	f	L-0879	N
Lithium hydroxide (buffer)	s	L-4256	F, D, I
Magnesium acetate (CK)	s	M-0631	P, I
Magnesium chloride (general)	s	M-0250	E, S, R, I
Magnesium sulfate (ALP, PK, buffer)	s	M-7506	E
Maleic acid (buffer)	s	M-0375	F, D
DL-Malic acid (buffer, MDH, MDHP, ME)	s	M-0875	D, E, S, R
Malic dehydrogenase (AAT, FUMH)	r	M-9004	A, I
D-Mannose-6-phosphate (MPI)	f	M-8754	N
2-Mercaptoethanol (FBP, NTP, PFK)	r	M-6250	F, D, E, S, R
Methylglyoxal (HAGH, LGL)	r	M-0252	E, S, R, I
Methyl alcohol	s		D, P, E, S, R
4-Methylumbelliferyl acetate (EST)	f	M-0883	E, S, R, I
4-Methylumbelliferyl-N-acetyl-β-D-galactosamide (βGALA)	f	M-9129	N
4-Methylumbelliferyl-N-acetyl-β-D-glucosamide (βGA, HA)	f	M-2133	N
4-Methylumbelliferyl-α-L-arabinoside (αARAB)	f	M-8880	N
4-Methylumbelliferyl-α-D-galactoside (αGAL)	f	M-7633	N
4-Methylumbelliferyl-β-D-galactoside (βGAL)	f	M-1633	N
4-Methylumbelliferyl-α-D-glucoside (αGLUS)	f	M-9766	N

(Continued)

Table 4. (*Continued*)

Chemicals (and stains involved)	Location[b]	Reference number[c]	Health & Safety[d]
4-Methylumbelliferyl-β-D-glucoside (βGLUS)	f	M-3633	N
4-Methylumbelliferyl-β-D-glucuronide (βGLUR)	f	M-9130	N
4-Methylumbelliferyl-α-D-mannopyranoside (αMAN)	f	M-4383	N
MTT (tetrazolium salt) (general)	r	M-2128	M, E, S, R, I
β-NAD (Nicotinamide adenine dinucleotide) (general)	f	N-7004	I
β-NADH (general)	r	N-6005	F, C, T, E, S, R
β-NADP (general)	f	N-3886	F, C, T, E, S, R
β-NADPH (GSR)	f	N-6505	F, C, T, E, S, R
Naphthol AS-BI β-D-glucuronic acid (βGLUR)	f	N-1875	E, S, R, I
Naphthol blue black (amido black) (GP)	s	N-9002	I
α-Naphthyl acetate (EST)	f	N-8505	E, S, I
β-Naphthyl acetate (EST)	f	N-6875	E, S, I
α-Naphthyl acid phosphate (ACP, ALP)	f	N-7000	E, S, R, I
β-Naphthyl acid phosphate (ALP)	f	N-7375	P, I
Nitro blue tetrazolium (NBT) (general)	r	N-6876	E, S, R, I
Nucleoside phosphorylase (ADA)	r	N-3003	A, I
1-Octanol (ADH, ODH)	s	O-4500	A, E, S, R
D-Octopine (OPDH)	f	O-4875	M, I
1-Pentanol (ADH)	s	P-8274	E, S, R, D
Peroxidase (PEP)	f	P-8125	A, E, S, R, I
Phenazine methosulfate (PMS) (general)	f	P-9625	M, E, S, R, I
Phenolphthalein diphosphate (ACP)	f	P-9875	E, S, R, I
L-Phenylalanyl-L-proline (PEP)	f	P-6258	N
Phosphocreatine (CK)	f	P-6502	N
Phospho(enol)pyruvate (AK, GUK, PK, UK)	f	P-7127	N
6-Phosphogluconic acid (Ba salt) (PGDH)	f	P-7627	P
6-Phosphogluconic acid (trisodium salt) (PGDH)	f	P-7877	N
Phosphoglucose isomerase (FBP, MPI) (=Glucose-6-phosphate isomerase)	r	P-5381	A, I
3-Phosphoglyceric phosphokinase (PGAM)	r	P-7634	A, I
Phosphoglycerate mutase (ENO)	r	P-8252	A, E, S, R, I

(*Continued*)

Table 4. (*Continued*)

Chemicals (and stains involved)	Location[b]	Reference number[c]	Health & Safety[d]
D(+)-2-Phosphoglyceric acid (ENO, PGAM)	f	P-0257	N
D(-)-3-Phosphoglyceric acid (ENO, PGK)	f	P-0769	N
Phospho-L-arginine (ARK)	f	P-5139	M, I
Polyvinylpyrrolidone (ALP, AAT)	s	PVP40	E, S
Potassium acetate (CK)	s	P-1147	E
Potassium bicarbonate (buffer)	s	P-9144	N
Potassium chloride (AK, ENO, GUK, PK, buffer)	s	P-4504	E, S, R, I
Potassium hydroxide (XDH)	s	P-1767	D, V
Potassium iodide (CAT, LGL)	s	P-8256	A, E, S, R
Potassium phosphate (dibasic—anhydrous) (buffer)	s	P-8281	N
Potassium phosphate (dibasic—trihydrate) (buffer)	s	P-5504	N
Potassium phosphate (monobasic) (buffer)	s	P-5379	N
Potassium sulfate (UK)	s	P-0772	E, S, R, I
Pyrazole (GLYDH, HADH, general)	s	P-2646	E, S, R
Pyridoxal-5'-phosphate (AAT, TAT)	f	P-9255	E, S, I
L-Pyroglutamic acid (PCDH)	s	P-3634	N
Pyruvate kinase (salt free) (AK, CK, ENO, GUK, UK)	f	P-9136	A, I
Pyruvic acid (ALPDH, HAGH, GLYDH)	r	P-2256	N
Retinol (RDH)	f	R-7632	E, I
Shikimic acid (SKDH)	s	S-5375	M, C
Sodium acetate (ACP)	s	S-8625	E, S, R, I
Sodium chloride (ALP, HBDH)	s	S-9625	E, R
Sodium hydroxide (buffer)	s	S-5881	F, D, I
Sodium phosphate (Na_2HPO_4) (buffer, AAT)	s	S-0876	E, S, I[e]
Sodium thiosulfate (buffer, CAT, TST)	s	S-8503	E, S, I
D-Sorbitol (buffer, IDDH)	s	S-1876	E, S, R, I
Starch (potato), hydrolyzed	r	S-4501	N
Succinic acid (free acid) (buffer)	s	S-7501	E, S, R
Succinic acid (disodium salt) (SUDH)	s	S-2378	N
Sucrose (buffer)	s	S-9378	E, S, R, I
5-Sulfosalicylic acid (CBP, GP)	s	S-2130	E, S, I
Sulfuric acid (GLT)	s		D, V
Trichloroacetic acid (CBP, GP)	s	T-4885	F, D
Triethanolamine (buffer)	s		E, S, R, C
Triosphosphate isomerase (PFK)	r	T-2507	A, I
Trizma base (buffer)	s	T-1503	E, S, I
L-Tyrosine HCl (TAT)	s	T-2006	S, R
Uridine 5'-monophosphate (UK)	f	U-1752	N

(*Continued*)

Table 4. (Continued)

Chemicals (and stains involved)	Location[b]	Reference number[c]	Health & Safety[d]
Venom—*Crotalus atrox* (PEP)	f	V-7000	F, A, E
Xanthine (XDH)	s	X-7375	E, S, R, I
Xanthine oxidase (ADA, GDA, PNP)	r	X-1875	E, I
Zinc chloride (HAGH)	s	Z-4875	D, V

[a]Sigma Chemical Company (St. Louis, Missouri, U.S.A.) catalog numbers are provided to distinguish among multiple forms of some chemicals; however, reagent-grade chemicals may be obtained from most major suppliers.
[b]s, = room temperature, shelf, f, = freezer ($-20°C$), r, = refrigerated (0-4°C).
[c]Sigma catalog numbers are provided as a reference and do not necessarily represent endorsement.
[d]Note: All of these chemicals may be harmful if inhaled, swallowed and/or absorbed through skin. A, allergen (esp. respiratory tract & skin); B, affects spermatogenesis, testes, epididymis, sperm ducts, male fertility index and post-implantation mortality; C, carcinogen; D, high concentrations are extremely destructive to tissues (mucous membranes, respiratory tract, eyes and skin); E, eye irritant; F, potentially fatal (if inhaled, swallowed, or absorbed through skin), at acute level; H, headache, nausea and/or vomiting; I, not thoroughly investigated; M, mutagen; N, no hazards known; P, poisonous; R, respiratory irritant; S = skin irritant; T, teratogen; V, corrosive.
[e]See chemical safety data sheet for specific hazards.
[f]K and K Lab, ICN Biomedicals Inc., Plainview, N.Y. USA.
[g]Denatured alcohol should not be used. We have achieved greatest ADH enzyme activity by using "gold" tequila as a source of ethanol; other liquors may work equally well.

3. Mechanically homogenize the mixture of tissue and grinding solution. The mixture should be kept ice-cold during homogenization.
4. Just prior to use, centrifuge the homogenate, preferably at >10,000 g for 15 to 30 min, to separate extracted proteins from cellular debris. Although highly desirable, centrifugation is not always necessary for some tissues and taxa.

Protocol 2: Preparation of Starch Gels

(Time: 2–3 hr/gel)

Gel cooking involves either the boiling of hydrolyzed starch in gel buffer (below) or the addition of starch to hot gel buffer (e.g., Micales et al., 1986). Hydrolyzed potato starch may be either made following the method of Smithies (1955) or purchased. Although relatively expensive, Connaught Medical Laboratories (Toronto, Ontario, Canada) starch has a long-standing reputation for consistently producing very high quality gels. Electrostarch Co. (Madison, Wisconsin) starch is relatively inexpensive, but highly variable in quality, and sometimes requires the addition of Connaught starch to make it usable. Recently started, Starch Art Corp. (P.O. Box 268, Smithville, Texas 78957) has produced highly satisfactory gels in test runs and is moderately priced. As with Electrostarch, a free sample is available on request. Other sources of hydrolyzed potato starch include various chemical (e.g., Sigma) and biological supply companies; these are invariably the most expensive and usually obtain their stock from the sources above.

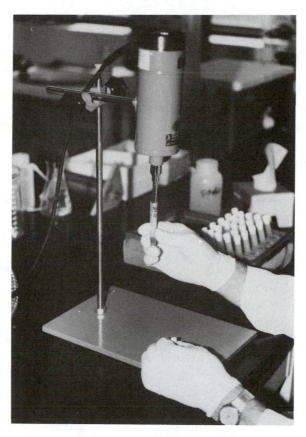

FIGURE 6.   Homogenization of tissue extracts using a high-speed homogenizer. See text for other methods.

Typically, starch gels are made in concentrations of 9–18% (w/v) starch in gel buffer, depending on the quality of starch, preferred texture of the gel, and desired sieving effect obtained during electrophoresis. The appropriate concentrations are determined by trial (and error).

Three problems may occur during gel preparation: undercooking, overcooking, and burning. Undercooking can be recognized by soft, wet gels that are difficult to handle following slicing; undercooking is rare. Overcooking is easily recognized during four stages: aspiration, cooling, loading, and slicing. During aspiration, overcooked gels may boil out of the flask. Vigorous shaking during aspiration may be required if the gel is to be saved, although this is sometimes ineffective. During cooling, deep crevasses or circular or octagonal patterns may be formed in the surface. Overcooked gel mixtures tend to stick to the gel molds, often splitting during loading or removal for slicing following electrophoresis; gel slices are tacky and sometimes impossible to handle. Burning can occur without overcooking. It results from not swirling the mixture vigorously enough during cooking and can be recognized by brown-black, burned starch on the bottom of the flask and/or dark flecks in the gel. Burning frequently results in tacky gels. Improperly cooked gels should be discarded.

Most types of gels can be cooked, poured, left overnight, and run the following day. However, Tris-citrate/borate, Tris-citrate III, lithium- borate/Tris-citrate, and Tris-HCl gels tend to crack during electrophoresis if used after this period of storage.

Finally, there are a number of peculiarities associated with some gel buffers. Tris-borate-EDTA II gels tend to stick to the flask after cooking. The problem can be overcome by lowering the percentage of starch by 0.5–1% and/or preparing an extra 20 ml of gel. Tris-citrate/borate and lithium-borate/Tris-citrate gels tend to split apart at the origin during running (see Protocol 4). Borate gels tend to be difficult to aspirate; slight undercooking and/or vigorous shaking during aspiration reduce these problems (see also Protocol 4).

1. Locate a stable, horizontal surface to hold gel molds until gels are cool enough to move ($\approx$ 1 hr). The surface should be near the aspirator.

2. Prepare gel molds for receiving hot starch: unglued wick molds (e.g., Micales et al., 1986) must have the edges clamped; use masking tape and seal the open portions of the legs of wickless molds. Place molds on the table or bench on top of a paper towel. Label the paper towel (or masking tape) noting the type of gel buffer to be poured and the date.

3. Weigh out 40 g (or appropriate weight) starch, place into a 1000 ml glass Erlenmeyer flask (narrow mouth, heavy duty rim), and add 400 ml gel buffer. Swirl contents until starch is well emulsified.

4. Cook gel by using one of the following methods:

   a. While wearing eye protection and insulated glove(s) continuously swirl flask above 5000 BTU Bunsen burner (Figure 7a). The mixture will become viscous and then quite rapidly much less viscous. As boiling begins (after about 3–4 min) stop heating.

   b. Use a magnetic stirring hot plate and large magnetic stirring bar to heat the starch–buffer mixture until the mixture becomes too viscous for the stirring bar to swirl. Remove the flask and occasionally swirl by hand until the mixture becomes less viscous once again, in about 1 min. Return the flask to the stirrer and continue heating until boiling as above. This procedure takes about 20 min.

   c. Cut the bottom out of a microwave oven having a stainless-steel interior in order to accommodate a magnetic stirring plate. Heat the starch–buffer mixture while stirring mixture until it becomes less viscous. Stop heating. (We have not used this method.)

5. Using an insulated glove quickly transfer molten gel to the aspiration shield, set flask on a heat pad, and cover the open hole of the T-connector to apply vacuum for about 15 sec (Figures 5 and 7b). The mixture will resume boiling. Swirling of the flask may be required during the first few seconds to avoid aspirating the gel out of the flask. Slowly release the vacuum.

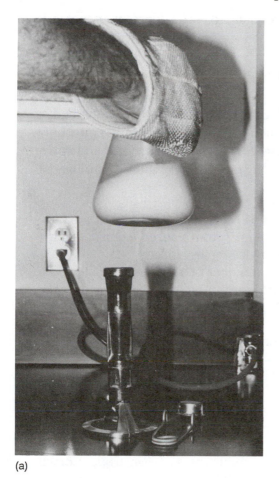

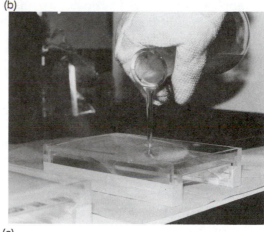

(a)   (c)

FIGURE 7.   (a) Cooking, (b) aspirating, and (c) pouring a starch gel.

6. Rapidly pour the hot mixture into a gel mold filling evenly and almost overflowing (Figure 7c); avoid dribbles.
7. Immediately (within 1 min) remove remaining air bubbles, if any, using a Pasteur pipette and pipette bulb.
8. Flush used flask in hot running water before remaining mixture solidifies.
9. After cooking all gels, and while they are cooling, fill buffer wells (trays).
10. Allow the gel to cool to ambient temperature, about 45–60 min, and gently cover with plastic food wrap. With both hands, hold the wrap at one end. Allow the opposite free end to contact one edge of the gel. Slowly lower the wrap allowing it to drop on the gel. If bubbles begin to form, lift the wrap and lower it again; air bubbles induce malformations in the gel surface. Pulling/tugging of the wrap should be avoided as this can form a split in the forming gel matrix. Gently write the name of the gel buffer on the wrap.

11. Place gel in refrigerator for 1 hr or allow to continue to cool at ambient temperature for 2 hr.

Protocol 3: Gel Loading

(Time: 10–20 min per gel)

The inoculation of protein extracts into horizontal gels is generally accomplished by the use of sample wicks—rectangular pieces of filter paper (Whatman No. 3) measuring 2–4 mm in width and 1 mm taller than the gel mold. Wicks can be hand-cut or purchased. The following protocol is used for loading multiple gels, and for right-handed operators.

1. Before loading, make sure that the buffer wells have been filled and labeled.
2. If applicable, remove frozen homogenized samples from freezer and initiate thawing, and recentrifuge if desirable; keep thawed samples chilled.
3. Number a piece of filter paper from 1 to the number of samples being applied to a gel, including tracking dye. Tape the paper to the table to the right of the operator.
4. Make stacks of wicks on the numbered filter paper. Each stack should contain as many wicks as there are gels to be loaded.
5. Remove gels from refrigerator and fold the plastic wrap back onto itself parallel to the sample origin exposing half, or more, of the gel.
6. Cut the edges of the gel free from the mold using a microspatula.
7. Slowly cut gel origin vertically using a thin, stainless steel microspatula and a gel origin guide (Figures 4 and 8a). The guide must be firmly held against the gel mold in order to avoid slipping. Trial buffer gels should be cut near the middle of the gel, others nearer to one edge.
8. Thoroughly wet the first five stacks of wicks with the first five tissue extracts, respectively, using Pasteur pipettes or 1–200 μl pipetters. Avoid cross-sample contamination by disposing used pipette tips between samples.
9. Place the narrow side of the gel nearest to operator. Gently open the origin of the well about 5 mm by pushing the wide side of the gel away. Using narrow-tip forceps, pick up a damp wick from the first stack and place it vertically into the gel origin against the narrow side, 1 cm from the left side of the mold, and in contact with the bottom of the gel mold (Figure 8). Load the remaining four samples, spacing the wicks about 1.5–2 mm apart. Sequentially load any other gel(s).
10. Repeat steps 8 and 9 using the next series of samples. Using tracking dye or a blank space about every 10 specimens facilitates later gel interpretation; allow 3 mm of space on either side of a tracking dye wick. The last wick should be soaked in tracking dye and located about 5 mm from the edge of the gel mold.
11. Once all samples have been loaded, examine the wick placement from the bottom-side of the gel mold to be sure that all wicks are completely inserted into the well. DO NOT shift wicks laterally. Using a rolling action of the index finger remove any bubbles from the bottom of the gel by pushing them to the origin.
12. Cover gel with plastic wrap and perform electrophoresis as described below.

(a)

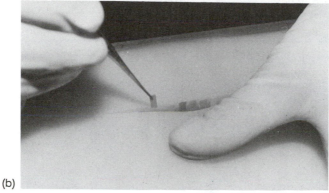

(b)

FIGURE 8.   (a) Cutting the gel origin using a gel origin guide and (b) loading a starch gel.

Protocol 4: Electrophoresis

(Time: 4–24 hrs)

Two primary types of buffer systems are used: continuous and discontinuous. In continuous systems, the gel buffer is usually a 10% or less dilution of the tray (electrode) buffer. In discontinuous systems (e.g., Tris-citrate/borate and Tris-HCl) the tray (borate tray buffer) and gel buffers are made of different electrolytes (Appendix 2); this system has the effect of "tightening" isozyme bands during electrophoresis (see Richardson et al., 1986). The tray buffer electrolytes can be observed to migrate through the gel.

Certain kinds of gels have peculiarities, especially the discontinuous buffer systems. In many buffer systems, e.g., Tris-HCl, the amperage (electrical current) drops as electrophoresis proceeds. Consequently, if running time is to be minimized the voltage should be progressively raised to the maximum level (Table 5) about every half hour but without exceeding 75 mA.

Tris-citrate/borate, and lithium-borate/Tris-citrate gels tend to split apart at the origin during electrophoresis, especially if the gels were cooked a day in advance or run overnight. Splitting typically occurs when the tray buffer electrolytes pass through the origin. There are three remedies to this problem. First, as previously mentioned, slightly overcook the gels during preparation. Second, push the gels halves together after the borate line has passed through the origin. Third, following 1–2 hr of electrophoresis, wedge plastic drinking straws or thin glass rods between the gel and inside edge of gel mold thereby forcing the gel halves together. These gels should be checked at the

Table 5. Recommended electrophoretic conditions for the wickless system described herein including electric potential in V/cm and average duration

Buffer combination	V/cm	Duration
Amine-citrate (morpholine)	4.2	overnight 14 hr
Amine-citrate (propanol)	4.2	overnight 14 hr
Borate (continuous)	3.9	overnight 18 hr
	8.3	6-7 hr
Borate (discontinuous)	3.3	overnight 14 hr
	5.5	7-8 hr
Histidine-citrate	12.0	6-7 hr
Lithium-borate/Tris-citrate	3.8	overnight 20 hr
	11.0	7-8 hr
Phosphate-citrate	2.2	overnight 20 hr
	4.4	10-12 hr
Tris-borate-EDTA I	2.7	overnight >18 hr
Tris-borate-EDTA II	3.3	overnight 18 hr
	11.0	6-7 hr
Tris-borate-EDTA-lithium	5.8	12 hr
Tris-citrate II	3.3	overnight >14 hr
	6.1	7 hr
Tris-citrate III	3.8	overnight 22-24 hr
Tris-citrate-borate	1.6	overnight 18 hr
	11.0	5-6 hr
Tris-citrate-EDTA	4.4	overnight 12 hr
	8.3	6 hr
Tris-EDTA	8.3	12 hr
Tris-HCl	9.7	5 hr
	2.2	20 hr
Tris-maleate-EDTA	4.8	18 hr

midpoint of electrophoresis to ensure that splitting has not occurred. Splits can be repaired by pushing the two halves of the gel back together.

It is frequently possible to run gels much more rapidly than recommended, a 4 hr minimum. Because of the sieving effect of starch gels, however, rapid running usually results in less-well defined protein bands following staining. Moreover, as gels begin to heat up, resistance increases and further heating will likely occur—to the extent of melting gels!

1. If wickless gel molds are used, remove the masking tape from the legs.
2. Place the gel mold in the buffer well box orienting the narrow end toward the cathode (negative, black terminal). If wick molds are used, a sponge-cloth must be used to complete the electrical circuit between the gel and buffer wells. While wearing rubber gloves, dip the sponge cloth into the well buffer and place it so that one end is in the buffer, and one on the gel surface 1 cm onto the gel and under the plastic food wrap.
3. Place either an aluminum tray filled with crushed ice or a frozen package of Blue Ice™ on the gel ensuring that the plastic wrap completely covers the gel and separates it from the ice pack. If Blue Ice™ is used, it is advisable to place a paper towel and glass plate between the gel and ice pack to prevent freezing of the gel surface.
4. Plug the well box top into the bottom, i.e., connect the buffer well electrodes to the power supplies (Figure 9).
5. Turn the power supply on, allow it to warm up for a few minutes, and adjust to desired voltage/amperage levels (Table 5). Amperage should not be allowed to exceed 100 mA, and preferably 75 mA, as overheating of the gel will likely occur.

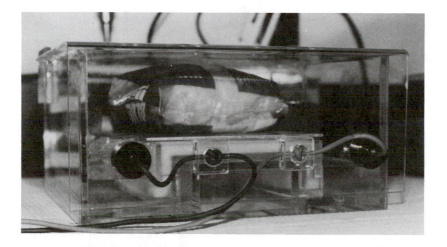

FIGURE 9.   Horizontal starch gel apparatus during electrophoresis. The electrophoresis buffer tray is a slightly more complex version of that shown in Figure 3. The gel is being cooled by using Blue Ice™.

6. After 15 min of electrophoresis, check tracking dye by examining edge of gel mold to ensure that gel was properly oriented in buffer well. If not, reverse polarity of the electrodes at the power supply.
7. Check ice levels every 2 hr if not running gels in refrigeration.
8. When tracking dye has reached the end of the gel, turn power supply off and remove gel (and gel mold) from buffer well box.

Protocol 5: Gel Slicing

(Time 5–10 min/gel)

Once electrophoresis has been completed, the gels need to be sliced and the slices placed in stain boxes. A number of methods have been developed including, among others, the use of "bow slicers" and slicing trays (Figures 3 and 10a), "multiple slicers" (Turner, 1980; Figure 10b and c), and nylon string (thread) (Micales et al., 1986). Gel

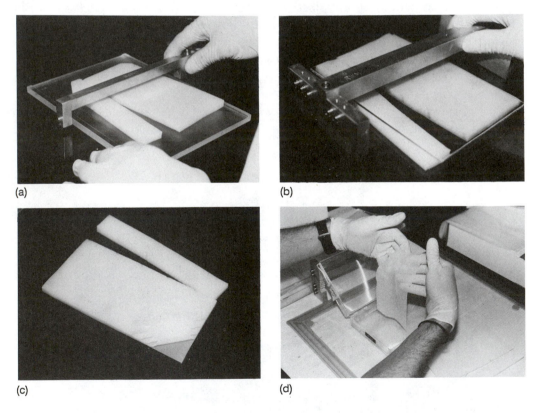

(a)

(b)

(c)

(d)

FIGURE 10. Gel slicing. (a) Use of a simple "bow slicer" (see Figure 4); (b) a "multiple slicer" (plans available on request); (c) a gel sliced with a multiple slicer; (d) gel slice handling. The multiple slicer does not require the use of a slicing tray. Note that when handling a gel slice, the fingers of both hands are touching to prevent stretching of the gel. Top glass (or plastic) plate in (a) has been removed for clarity.

slicing and handling should be carried out while wearing protective gloves, even though this increases the difficulty.

Several problems can occur during slicing, the most common of which is that of splitting or tearing the slices. Once a split has formed in a gel, it can be extremely difficult to transfer slices from the tray to the stain box; splits usually result from bending the gel too much while transferring it from one slicing tray to another. The easiest solution is to completely separate the split slice and transfer the two parts separately.

Improperly cooked gels are difficult to handle by hand. Transfer these slices by using plastic food wrap as a carrying medium.

1. Using masking tape, label stain boxes with the gel number, enzyme system or locus to be stained, gel buffer, and date. (This step is usually completed during electro-phoresis.)
2. Using a micro-spatula and gel origin guide, cut away the anodal and cathodal 1 cm of the gel (or legs of the wickless gel), 3–5 mm of the edges of the gel, and notch the left anodal and cathodal corners of the gel. Remove these edges and notch pieces from the mold leaving the greater portion of the gel in the mold.
3. Separate halves at the origin and remove wicks. The gel may be more difficult to move for some buffers (e.g., lithium-borate/Tris-citrate). Using a paper towel, gently dry the top of the gel, arrange the two pieces so that they form a "V" separated about 1 cm at one end, and cover with a piece of plate glass or a slicing tray.
4. Invert the sandwiched gel and gently dry the bottom of the gel. Choose the ap-propriate thickness of slicing tray (if applicable), center it up-side down on the bottom gel surface with the tray ridges aligned with the origin, and turn the gel right side up again.
5. Remove air bubbles between the gel and slicing surface. Failure to remove bubbles may result in holes in the gel slice and/or render the remaining gel incapable of being sliced. Re-cover top of the gel with second slicing tray (or glass plate).
6. Clean slicer wire with damp towel or steel wool.
7. Orient the gel so that the apex of the "V" is furthest away from the operator. Brace the (bottom) slicing tray to prevent it from moving toward the operator during slicing. Place the wire of the slicer on the raised ridges of the slicing tray, press downward on the slicer, and in one continuous operation slowly (about 3 cm per second) pull the wire through the gel. Gels usually move toward the operator slightly during slicing; do not stop pulling if this is observed, and do NOT press down on the top tray/plate.
8. Clean slicer wire with damp towel or steel wool. Do not immerse wire slicers in water.
9. Remove top tray/plate, carefully separate the gel from the bottom slice, and transfer the gel to the second slicing tray allowing for the "V" to have the opposite orientation (apex near operator). Similarly, move the anodal top slice but use both hands to support opposite sides of the gel. Lift anodal top gel slice to second tray forming a "V."
10. Open a plastic staining box and carefully transfer the anodal and cathodal bottom slices (Figure 10d) to the stain box. If an agar overlay, UV fluorescing, or limited

volume stain is to be applied, then it is important that no bubbles occur underneath the slice. Relatively expensive or critical stains should be made on slices cut from the bottom of the gel.

11. Repeat slicing. Always initiate subsequent slices from opposite ends of the gel to prevent uneven thinning. It may be necessary to repeat steps 4–5 if the remaining gel slides easily on the slicing tray. The top slice can be inverted and used, although it is preferable to stain with an agar overlay. Remaining portions of gels can be temporarily saved (24+ hr) by wrapping.

Protocol 6: Histochemical Staining

(Time: 2 min to 6 hr/stain)

The distance of migration of specific proteins through a starch gel is visualized by histochemical staining. These stains (Appendix 1) consist of a "substrate" on which a specific enzyme reacts, and a "detection mechanism" such as a dye or substance that fluoresces under long wave (340 nm) UV. The common mechanisms for detection include (1) the formation of a purple precipitate (formazan) by the reduction of NBT or MTT using PMS or DCIP as the intermediate electron carrier or reducer, respectively; (2) the nonfluorescence of NAD, which is formed from fluorescent NADH; (3) fluorescence of methylumbelliferone; (4) fast diazo dye (e.g., esterases); and (5) the oxidized form of O-dianisidine diHCl producing an insoluble brown precipitate. Many stains also contain cofactors, coupling enzymes, and other requisite molecules. Details of how these systems work are provided by Harris and Hopkinson (1976) and Richardson et al. (1986). A complete understanding of the concepts is desirable but not absolutely necessary, although such understanding greatly facilitates the resolution of staining problems when they occur.

Some stains (e.g., for PGM) are best applied to the gels in the form of an agar overlay, or an agar-based gel containing stain components; agarose may be preferred over agar because the latter inhibits the activity of some proteins through binding (Harris and Hopkinson, 1976). Most laboratories use agar because it is much less expensive. The overlays serve the function of containing the precipitating dye preventing it from either diffusing over a broad area of the gel or becoming too diffuse to be observed.

Several UV fluorescing stains (e.g., βGLU) may be applied to the gel slices as filter paper overlays, the overlays being cut from Whatman 1MM or other thin filter paper (Harris and Hopkinson, 1976). However, we have not noticed an advantage over simply applying these stains directly to the gel.

The quantity of specific chemicals in some recipes in Appendix 1 varies from amounts specified in other sources (e.g., Selander et al., 1971). These amounts are the minimum required to resolve these protein systems from the maximum diversity of taxa. Often these quantities can be reduced by applying less stain to a gel, especially once the region of activity has been identified. Most agar overlay stains can be easily accomplished using as little as 10 ml of stain solution.

Of the two dyes used in formazan-based stains, MTT is cheaper, more toxic, and precipitates more rapidly than NBT but tends to diffuse and is less stable. The two dyes can be used in concert. If NBT is yielding only faint bands initially, the addition of MTT during staining may help to intensify the isozymes.

For formazan stains, three components are particularly sensitive to light: PMS, MTT, and NBT. Therefore, the stock liquid solutions and staining gel slices must be kept out of light. Stock solutions should be stored in either amber glass bottles and/or bottles wrapped in aluminum foil.

All stains can be safely and conveniently prepared in Erlenmeyer flasks. Because some stains contain liquid components only (e.g., LDH), these may be mixed directly in the stain box so long as the stain buffer is applied first.

1. Dry chemicals should be weighed and placed in a 125-ml Erlenmeyer flask.
2. Add the liquid components. Liquid components are most safely handled using pipetting devices. In some cases adjustment of pH will be necessary (e.g., HADH). When mixing formazan-based stains, all powdered ingredients should be dissolved in the stain buffer and pH adjustments should be made before adding cofactors, PMS, and NBT (or MTT). Once completely mixed, pour the stain onto the gel and gently shake the box freeing the gel from the bottom. Agar overlays are prepared by bringing a 0.7% (w/v) mixture of agar/stain buffer to a boil, allowing it to set until all agar grains have disappeared, cooling to just below 50°C, adding remaining staining components, and pouring onto the gel slice. For the typical 50-ml stain, 35–40 ml of stain buffer is mixed with 0.35 g agar in a 125-ml Erlenmeyer flask; the remaining 10–15 ml of stain components are added to the warm agar just prior to covering the gel. Under ideal conditions, the agar is prepared in advance of slicing and staining by bringing the mixture to a boil in a microwave oven. The flask is corked or covered with aluminum foil and kept in a 50°C water bath until used. Cooling of hot, molten agar can be made rapid by the use of ice and an accurate thermometer. The molten agar forms a gel at around 42°C. Some fluorescent stains are prepared as small agar overlays. Do not view the UV light or fluorescing gel without the use of a UV light shield or protective glasses. Short wave lights are not necessary and should not be used because of the additional health hazard.
3. Most stains should be incubated at 37°C following staining.
4. Staining gel slices must be continuously monitored to prevent overstaining, which results in unresolvable, diffused, or smeared bands. Some stains must be scored and documented as soon as they are ready, sometimes within 5 min of staining. Stains using insoluble precipitates can be preserved (see below) and scored following the completion of all staining, even on the following day.
5. If the stain has been applied as a liquid, and not an agar overlay, siphon off the stain solution and save for appropriate hazardous waste disposal. Completely cover the gel slice with fixing solution (about 50 ml; Appendix 2) and refrigerate. If MTT is used as the dye, do not flood the gel slice with fixative or the formazan dye will wash out of the gel; apply only enough fixative to wet the gel slice (about 20 ml).

Troubleshooting   A number of problems may be encountered following application of the stain, the most common of which is the absence of enzyme activity on a gel. This may have several causes: (1) If the duration of electrophoresis is too long or short, the enzymes may have migrated off of the gel or remained in wicks in the origin, respectively. (2) It is possible that one (or more) of the stain components were omitted from the stain recipe. Successful staining may be possible by adding the missing component to the stain. (3) Very weak expression typically results from too little of a given component, or the use of a partially degraded solution of coenzymes. Under these conditions, it will be necessary to add additional stain components, or reorder the coenzyme. If more than one stain is resolving inadequately, check for common stain components, such as G6PDH. Coenzyme activity can be checked by electrophoresis and staining a small amount of the coenzyme along with tissue extracts where activity has been previously resolved. (4) A change in starch lot can result in the necessity to change the conditions of electrophoresis. (5) Shifts to a high pH can result in the conversion of NAD(P) to NAD(P)H. Check the pH of the final stain solution. (6) Finally, the addition of too much substrate or coenzyme can suppress enzyme activity.

Smeared isozymes may result from use of the wrong electrophoresis buffer conditions, too high a current (overheating), high concentrations of lipids in the tissue extracts, or (rarely) improper formation of the gel matrix.

Diffuse isozymes can indicate overstaining, less than ideal electrophoresis conditions, and/or that an agar overlay stain should have been applied. In the latter case, if light shaking of the gel results in disturbance of the formazan precipitate on top of the gel slice, then the stain should be applied as an agar overlay. Overstaining results in very dense isozyme banding patterns. Occasionally, background "ghost bands" may be observed. These bands result from the ability of an enzyme to act on an alternative substrate (e.g., LDH acting on DL-glyceric acid, the substrate of GLYDH), presence of sufficient substrate in the tissue extract, or contamination by bacteria, molds, and yeasts (e.g., ethanol and ADH). LDH, ADH, and other isozymes can be identified either by counterstaining, or by inclusion of the end product of the reaction, a procedure termed end-product suppression. For example, pyruvic acid suppresses (but does not stop) LDH, and pyrazole inhibits ADH. For some enzyme systems (e.g., GLYDH), use of one or more suppressers is required.

Protocol 7: Drying Agar Overlays

(Time: 6 hr)

Agar overlays can be dried on filter paper and saved as documentation as follows:

1. Cut filter paper (e.g., Whatman No. 1) to dimensions allowing it to fit into a stain box (12 × 17 cm) and label it with the enzyme system, gel number, and buffer conditions.
2. Decant or vacuum excess fixative from the stain box.

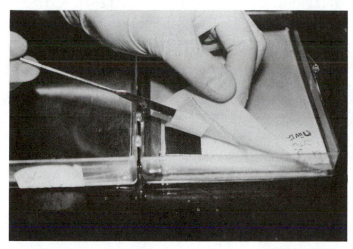

FIGURE 11.   The lifting of an agar overlay from a gel slice.

3. Cut the agar free from the edges of the gel slice using a microspatula.
4. Carefully overlay the filter paper on the agar overlay and then slowly lift the filter paper while separating the agar overlay from the gel slice using a microspatula (Figure 11).
5. Place the filter paper on a few paper towels agar-side up and allow to dry (several hours).
6. Once dry, curled overlays can be pressed flat and wrapped in plastic for safe handling. They should be stored in the dark and with light pressure to avoid recurling. Because the agar will retain dangerous chemicals for years, they should never be handled without wearing protective gloves and/or unless they are wrapped.

Protocol 8: Documentation of Results

(Time: 1–5 min/slice)

At the completion of staining, the isozyme patterns should be documented by photography or by drawing observed patterns on paper. Photography can be accomplished with a standard 35-mm camera, or more expensively with a Polaroid™ system. If 35-mm photography is used, a Y48 yellow filter should be fitted to the camera lens to increase the contrast between the stained isozyme patterns and the gel background; this filter is necessary for documenting UV stains. Photography is best carried out on a copy stand fitted with a light box. A polarizing filter helps to cut glare if the gel slice is illuminated from above but it should not be used when photographing UV stains. If using a UV lamp rather than a transilluminator, locate it close to the gel. When photographing, we place the stain box label on the gel to document each photograph.

INTERPRETATION AND TROUBLESHOOTING

The interpretation of the band patterns comprising the zymogram requires the knowledge of the subunit structure and the genetic control of the enzyme system. As discussed in the gene expression section, the tissue examined for enzyme activity may limit the number of gene products or subunits expressed. These variables may be manipulated by choosing a tissue that will express the desired gene products in the most scorable fashion.

The variables of subunit structure and genetic control are discussed below, followed by a brief introduction to some common problems faced in the interpretation of zymograms. For additional discussion see Harris and Hopkinson (1976), Rider and Taylor (1980), Moss (1982), and Richardson et al. (1986).

While many enzymes are **monomeric proteins** (i.e., made up of one polypeptide chain) the majority are multimeric (made of two or more polypeptide chains), most often dimers and tetramers. Harris and Hopkinson (1976), in a survey of the subunit structure of enzymes, found 28% monomers, 43% dimers, 4% trimers, 24% tetramers, and 1% octamers.

The simplest patterns of expression are single locus enzyme systems in diploid organisms. In a homozygous individual, only a single allelic product is formed. Even if the enzyme in question is a multimer, only one kind of homogeneous product will be assembled; this product is a **homomeric isozyme** seen as a single zone of activity on a gel. If the individual is heterozygous at this single locus, two kinds of allelic products are formed. In the case of monomeric enzymes, the two allelic products are produced in equal quantity, do not interact structurally, and are expressed equally in a given tissue of an individual (1:1 ratio). A zymogram illustrating triallelic variation involving a monomeric enzyme is shown in Figure 12. In the case of multimeric enzymes, the two allelic products are also produced in equal quantity in a given tissue but the products will usually randomly assemble to form all expected heteromers, in addition to homomers. It is usually the case that the subunits of multimeric enzymes form homomers and

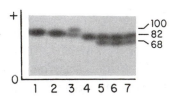

FIGURE 12. Zymogram exhibiting triallelic variation at the phosphoglucomutase locus (*Pgm-A*) in muscle extracts from the cyprinid fish *Luxilus cardinalis*. Specimens 1, 2, and 4 are homozygous expressing only the 82 homomer; specimen 3 is heterozygous expressing both 82 and 100 homomers; specimens 5, 6, and 7 are also heterozygous expressing both 68 and 82 homomers.

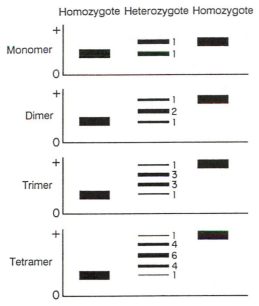

FIGURE 13. Diagram of isozyme patterns expected in homozygotes and heterozygotes for enzymes of common subunit composition. Modified from Harris and Hopkinson (1976). Ratios of intensity of isozyme activity in heterozygotes are indicated. See Table 2.

heteromers at random yielding banding patterns in predictable ratios. Because heteromers of similar composition can be assembled in several ways, the ratio of expected intensity of enzyme activity differs among isozymes according to the subunit structure of the enzyme (Figure 13). This variation is detailed in Table 6.

The situation is more complex for multilocus enzyme systems. Multimeric gene products of multiple loci in an enzyme system often retain their ability to form heteromers, and the number of isozymes formed can be considerable where heterozygosity occurs. Harris and Hopkinson (1976) provided the following equation for the computation of the expected number of isozymes (i) under these circumstances:

$$i = \frac{(L + h + n - 1)!}{n!(L + h - 1)!}$$

where L = the total number of loci, h = the number of heterozygous loci, and n = the number of subunits for this enzyme. The multilocus situation differs from its single locus counterpart in that whereas allelic products of a single locus can account for equal quantities of both products, the products of two different loci would rarely contribute equal quantities of both products in the same tissue. The predictable symmetrical ratios of isozymes in heterozygotes cannot be extended to the multiple loci unless, by chance, the two gene products are produced in equivalent proportions. Examination of enzyme expression in multiple tissues may aid in distinguishing single locus heterozygosity from

Table 6. Subunit structures of homomeric and heteromeric isozymes in heterozygotes[a]

	Monomer	Dimer	Trimer	Tetramer
Homomer	1	11	111	1111
Heteromers				1112
				1121
			112	1211
			121	2111
			211	
				1122
				1212
		12		1221
		21		2121
				2211
				2112
			221	1222
			212	
			122	2122
				2212
				2221
Homomer	2	22	222	2222

[a]Modified from Harris and Hopkinson (1976). Two alleles at this single locus determine polypeptide units 1 and 2, respectively. Random combination of subunits of multimeric enzymes is assumed.

similar isozyme patterns resulting from interactive multilocus products. However, the question of a two-allele, single-locus model versus that of interactive products of two homozygous loci can also be addressed by comparing the frequency of "heterozygotes" to the predictions of Hardy–Weinberg equilibrium (e.g., Ferris and Whitt, 1978a). The assumption of Hardy–Weinberg expectations for the distribution of allozyme products of a given locus is usually a safe one. Violation of this assumption suggests that additional study is necessary beginning with a reassessment of the scoring of that enzyme system. Scoring only "clear" bands and omitting "smeared" zones may overestimate the frequency of homozygotes. Frequently the report of 50% allele 1 and 50% allele 2 for a given locus in a table of allele frequencies is the result of incorrect scoring of an entire ($N > 5$) sample as heterozygotes.

Difficulty in scoring gels can occur when any of the other assumptions discussed previously are violated. Exceptions to expected subunit interactions and genetic control are often encountered. The random association of subunits of multimeric enzymes is sometimes restricted, yielding fewer zones of activity than expected. For example, creatine kinase is a dimer in all vertebrates but the heterodimer is not formed in heterozygotes at the Ck-A locus in teleost fishes (Ferris and Whitt, 1978b). The subunit structure of enzymes is often quite conservative across taxa; however, some enzymes have been reported to have a variety of structures in different groups of organisms (Manchenko, 1988). These reports may reflect real structural differences among taxa or the restriction of heteromer formation misinterpreted as structural differences. Rigorous testing (e.g., Ferris and Whitt, 1978b) should be applied in these cases. On rare occasions, allelic products have different catalytic properties and expected ratios of isozyme expression are not realized. Examination of a large series of individuals that resolve all heterozygous and homozygous categories should allow the correct interpretation of such variation. Epigenetic effects yield isozymes of different electrophoretic mobilities in different tissues and can suggest the action of more structural loci than are actually present. If only a single locus is active in this case, apparent heterozygosity or homozygosity should be correlated among tissues of an individual. The probability for unlinked multiple loci to covary in such a way can be addressed statistically (see Hartl and Clark, 1989).

In all studies that deal with questions of whether mobilities of electromorphs are equivalent or whether an individual is heterozygous at a locus, the resolution of discrete zones of enzyme activity on a gel is essential. If multiple buffer systems are not used or if tissue extracts no longer provide sufficient enzyme activity, the resolution may be inadequate. Interpretation of these suboptimal gels results in dubious data sets. For example, overstaining will obscure the subtle differences in relative activity of isozymes. In spite of the resolution of discrete zones of enzyme activity and efforts to limit enzyme expression to primary isozymes, some nongenetic subbanding may confound the interpretation of gels. The production of these secondary isozymes, or subbands, may vary by tissue location and age, enzyme system, and electrophoresis buffer employed. Resolution of this problem often comes by changing to another buffer or resolving variation at the locus. The subbanding problem is illustrated for dimeric glucose-6-phosphate isomerase (GPI) in Figure 14. This enzyme system is controlled by two loci, now known as *Gpi-A* and *Gpi-B*, in teleost fishes (Avise and Kitto, 1973); products of the latter locus predominate in muscle tissue and an interlocus heterodimer is usually formed. Isozymes of *Gpi-B* often yield two anodal subbands beyond each homomer or heteromer. This pattern might be confused with that of heterozygotes. However, as Figure 14 illustrates, heterozygotes serve to clarify the situation even if their homomers and heteromers are superimposed on some subbands obscuring the expected ratios of isozymes. The presence of two interlocus heterodimers in relevant individuals can provide an additional hint of heterozygosity in this example (e.g., specimens 6 and 9 in Figure 14 in which the variation at the *Gpi-A* locus is subtle).

Having noted the basics of gel interpretation we provide a few additional examples. The affect of using different buffers to resolve enzyme variability is shown in Figure 15. Alternative buffers can differentially affect both relative mobility and activity of isozymes. Although not shown, some buffers would result in smeared isozyme patterns. Figure 16

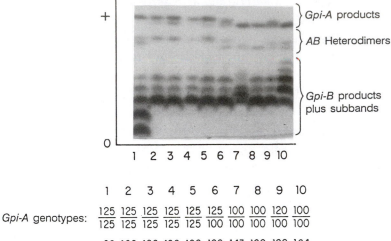

	1	2	3	4	5	6	7	8	9	10
Gpi-A genotypes:	$\frac{125}{125}$	$\frac{125}{125}$	$\frac{125}{125}$	$\frac{125}{125}$	$\frac{125}{125}$	$\frac{125}{100}$	$\frac{100}{100}$	$\frac{100}{100}$	$\frac{120}{100}$	$\frac{100}{100}$
Gpi-B genotypes:	$\frac{28}{100}$	$\frac{100}{100}$	$\frac{100}{100}$	$\frac{100}{100}$	$\frac{100}{100}$	$\frac{100}{100}$	$\frac{143}{100}$	$\frac{100}{100}$	$\frac{100}{100}$	$\frac{164}{100}$

FIGURE 14. Zymogram exhibiting variation at glucose-6-phosphate isomerase loci *Gpi-A* and *Gpi-B* in muscle extracts from the cyprinid fishes *Luxilus cardinalis* (lanes 1–6) and *Luxilus zonatus* (lanes 7–10). Genotypes are listed for each locus. Note the subbanding.

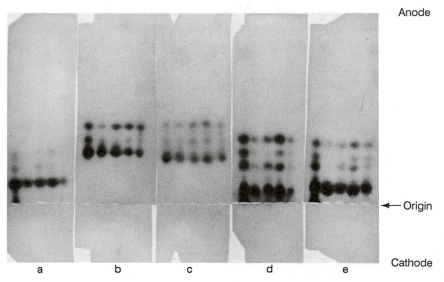

FIGURE 15. Zymogram demonstrating a gel-buffer screen from rattlesnakes for the enzyme system L-lactate dehydrogenase. (a) Tris-citrate III, pH 7.0; (b) Tris-citrate-borate, pH 8.2; (c) Tris-citrate II, pH 8.0; (d) Tris-citrate-EDTA, pH 7.0; (e) phosphate-citrate, pH 7.0. Increasing the pH also increases the net charge and relative mobility of the isozymes. Three or four isozymes are observed, depending on the buffer, but in no case are all of the anticipated five isozymes resolved (see Figures 13 and 16). Buffer (a) suppresses the activity of the more anionic system, products of the heart-predominating *Ldh-B* locus. Isozymes of the slower skeletal-muscle-predominating system *Ldh-A* cannot be adequately resolved on systems (d) and (e). Note that the minislices are uniquely notched.

Anode

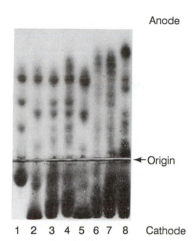

←Origin

1 2 3 4 5 6 7 8   Cathode

FIGURE 16.   Zymogram demonstrating intra- and interlocus variability of tetrameric L-lactate dehydrogenase (LDH) isozymes in spring peeper frogs, *Hyla crucifer.* The more anodal system is the heart-predominating locus, *Ldh-B,* and the cationic locus (on this buffer system) is muscle-predominating *Ldh-A.* Here resolution is inadequate for interpreting variation at the *Ldh-A* locus but is excellent for *Ldh-B;* the buffer revealing *Ldh-A* variation masks that of *Ldh-B.* Specimens 1, 2, and 5 are homozygous at both loci although 1 has a different allele expressed at *Ldh-A;* these specimens show the five expected isozymes (Figure 13). Lanes 3, 4, and 8 are heterozygous at *Ldh-A* but homozygous for two different alleles at *Ldh-B.* Lanes 6 and 7 are heterozygous at both loci.

demonstrates that multiple buffers may be required to resolve allelic variants at different loci of the same enzyme system. Diagnosis of a posttranslational modification is shown in Figure 17, although the exact nature of this modification is unknown.

Some enzyme systems may appear as background upon staining for others. These may be either desirable, as in the case of observing superoxide dismutase (SOD) following staining for glycerol-3-phosphate dehydrogenase (G3PDH; Figure 18), or undesirable (Figure 19).

Figure 20 documents the necessity of choosing the correct array of tissues to be surveyed. Finally, Figure 21 shows unacceptable resolution of an isozyme system. The optimal resolution of enzyme activity will facilitate a correct genetic interpretation of the zymograms. Documentation of results through the publication of zymograms is recommended strongly.

APPENDIX 1: ENZYME STAINING FORMULAS
(Compiled by Donald G. Buth and Robert W. Murphy)

Formulas for enzyme stains are frequently modified and republished, often as compilations for specific groups of organisms or even for single species. Textbook treatments

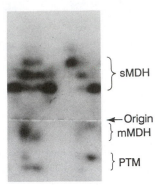

FIGURE 17. Zymogram of malate dehydrogenase (MDH) from salamanders (*Ambystoma maculatum*) showing variation in anionic isozymes (the supernatant locus *sMdh-A*) and cationic isozymes (the mitochondrial locus *mMdh-A*), and a posttranslational modification. *sMdh-A* is dimeric and heterozygotes appear similar to those of GPI (Figure 14). *mMdh-A* products are tetrameric and heterozygotes near the gel origin (arrow) appear smeared because the intralocus heteropolymers are too close together to be resolved. A posttranslation modification (PTM) of *mMdh-A* products results in more highly charged isozymes, which in this case lack the intralocus heteropolymers.

often provide a limited introduction to the vast array of stains available, whereas listings for specific groups of organisms are often limited to those systems well known or expressed only in those groups. Our listing is not meant to be all-inclusive; our selection is biased toward economical systems in use by botanists and zoologists. Our reference to stain sources usually extends only to the secondary literature wherein modifications are already noted.

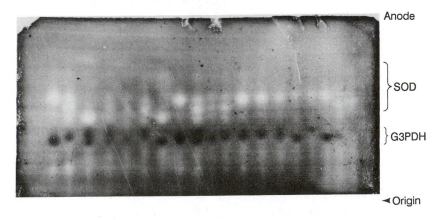

FIGURE 18. Zymogram showing the resolution of superoxide dismutase (SOD) isozymes (light bands) as background on a gel stained for glycerol-3-phosphate dehydrogenase (G3PDH) in spring peeper frogs, *Hyla crucifer*.

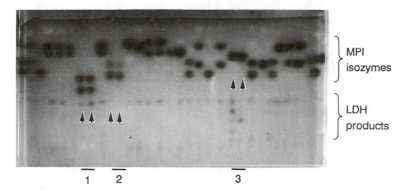

FIGURE 19. Zymogram showing extensive variability in mannose-6-phosphate isomerase (MPI) isozymes among some hylid frogs along with background resolution of L-lactate dehydrogenase products (LDH), which may be misinterpreted as a second MPI locus. All individuals except those indicated by arrows are *Hyla crucifer*. The more anodal isozymes in species 2, *H. cadaverina*, are *Ldh-B* products.

With few exceptions, the enzyme names and Enzyme Commission (EC) numbers used in this compilation are those recommended by the International Union of Biochemistry (IUBNC, 1984). Abbreviations of enzyme names are placed in capital letters; abbreviations are developed from the IUBNC (1984) recommended names and sometimes differ from common usage by the addition of letters for clarity. The listing of named

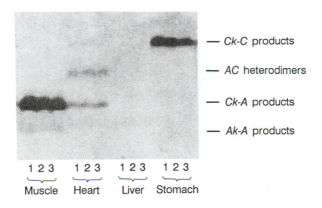

FIGURE 20. Zymogram of creatine kinase (CK) isozymes from the marine toad, *Bufo marinus*, showing differences in tissue expression and the presence of interlocus heteromers. In stomach, only products of the *Ck-C* locus are expressed whereas in skeletal muscle only *Ck-A* is seen. In heart tissue both locus products are expressed, albeit weakly, and the interlocus heteropolymer (heterodimer) is present; the pure locus products (homodimers) are not expressed in equal intensity eliminating the possibility of a heterozygotic state. CK is not expressed in liver. Adenylate kinase (*Ak-A* locus) activity is also resolved and limited to skeletal muscle tissue.

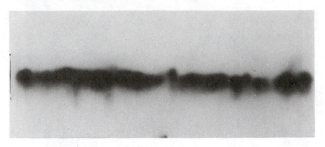

FIGURE 21. Unacceptable resolution of pyruvate kinase (PK) in spring peeper frogs, *Hyla crucifer*. Smeared bands could result from using an incorrect buffer system, overstaining, a bad gel, or bad samples (too high lipid content, denatured proteins, old samples, etc.). In addition, the sample wicks have not been spaced adequately.

loci controlling each enzyme system is beyond the scope of this appendix and other abbreviations are defined in the glossary.

The quaternary structures for enzymes listed herein were taken from those reported by Harris and Hopkinson (1976), Soltis et al. (1983), Richardson et al. (1986), Aebersold et al. (1987), Manchenko (1988), and personal communications from a number of researchers. For some enzymes, these structures are well documented and conservative across taxa. For others (e.g., catalase and glucose-6-phosphate dehydrogenase) several quaternary structures have been reported (Manchenko, 1988). Whether these and other enzymes actually exist in multiple structural forms or have a conserved single multimeric structure that is expressed as restricted subunit combinations remains to be investigated.

Many of the biochemicals used in enzyme stains are marketed in a number of forms. In some cases, ultrapurity is not required and considerable savings can be achieved through the purchase of a lesser grade. In some cases, the choice of a particular salt may be critical. We have listed (Table 4) the product number of many of these stain components keyed to the 1989 catalog of the Sigma Chemical Company (P.O. Box 14508, St. Louis, Missouri 63178 U.S.A.) to allow the reader to evaluate the kind and cost of these biochemicals. This choice does not necessarily represent our endorsement of these products.

Most of the stains below are based on a standard volume of 50 ml suitable for gel slices from most horizontal starch gel apparatus (scaled down from stain formulas for 100 ml volume used commonly for macroscale vertical apparatus of earlier studies). Some investigators have reduced the staining solution volume even further to conserve expensive stain components; many of these reductions are noted and others may be warranted.

The agar overlay method (see text) is recommended for most stains by some (e.g., Shaklee and Keenan, 1986). For 50 ml volume stains, we recommend 35–40 ml of buffer with the agar, 10–15 ml buffer with the stain components. Specific recommendations are made for stains using reduced volumes. See text for specific instructions and suggestions. Our staining methods reflect our own biases in choice of biochemical reagents and their means of storage and application. For example, we see no need to use the more expensive NADP in stains requiring glucose-6-phosphate dehydrogenase when an NAD-dependent form of G6PDH can be used (Buth and Murphy, 1980). NADP is used in solid form herein (see malate dehydrogenase NADP [MDHP]) but may be suitable in liquid stock for many other stains. We prefer NBT to MTT. Other individual preferences abound; for instance, Ayala et al. (1972) recommended adding phenazine methosulfate to stains after an hour or more of incubation with the other components. We urge researchers to experiment with the options listed among these "recipes" and to try other modifications of their own invention.

Unless otherwise indicated, the stained gel slices should be incubated in the dark at 37° C. Recipes for stain fixing solutions follow the stain recipes.

β-N-Acetylgalactosaminidase (βGALA) (EC 3.2.1.53)

Dimer Prepare this stain as an agar overlay (0.17 g agar in 10 ml water and combine with substrate mixture).

Dissolve the substrate:

4-methylumbelliferyl-N-acetyl-β-D-galactosamide	5 mg
dimethyl sulfoxide	0.25 ml

Add the dissolved substrate to

0.1 M phosphate-citrate buffer, pH 9.5	15 ml

Incubate and then view under UV light (long wavelength). Zones of activity will appear as bright areas. To enhance fluorescence, spray the gel with a concentrated solution of ammonium hydroxide. This stain was described by Aebersold et al. (1987).

N-Acetyl-β-glucosaminidase (βGA) (EC 3.2.1.30)

Dimer This enzyme was formerly refered to as hexosaminidase (HA). This stain may be prepared as an agar overlay (agar 0.17 g, H_2O 10.0 ml) that is added to the substrate below, or simply pour the following onto the gel surface:

4-methylumbelliferyl-N-acetyl-β-D-glucosaminide	5 mg
dimethyl sulfoxide	0.25 ml

Add the dissolved substrate to

0.1 M phosphate-citrate buffer, pH 4.5	15 ml

Incubate and then view under UV light (long wavelength). Zones of activity will appear as bright areas. To enhance fluorescence, spray the gel slice with a concentrated solution of ammonium hydroxide. This stain was described by Aebersold et al. (1987).

Acid Phosphatase (ACP) (EC 3.1.3.2)

Monomer or Dimer   Two stains are required in many vertebrates to resolve both "red cell" and "tissue" ACP isozymes. For dimeric "tissue" isozymes incubate the gel slice at ambient temperature for 30 min in a 0.05 M sodium acetate buffer, pH 6.0 (= 0.33 g sodium acetate in 50 ml H_2O with a minor pH adjustment). Drain, then add the following solution to the gel slice:

sodium acetate ($NaC_2H_3O_2 \cdot 3H_2O$)	0.33 g
α-naphthyl acid phosphate	0.15 g
black K salt	0.05 g
H_2O	50 ml

The pH of the staining solution is about 5.5 so further adjustment is usually unnecessary; the pH should be 5.0–6.0. The stain was modified from Shaw and Prasad (1970). Werth (1985) recommended the use of a stock solution of the substrate β-naphthyl acid phosphate in 70% acetone (use 1 ml of a 1% solution). Soltis et al. (1983) and Werth (1985) recommended the use of fast garnet GBC salt as a substitute for black K salt. Sigma fast black K salt has not provided satisfactory results in studies of reptiles. This stain was modified from Harris and Hopkinson (1976) and does not resolve "red cell" ACP in many vertebrates, including many reptiles. Monomeric "red cell" ACP isozymes, also known as erythrocytic acid phosphatase (EAP), may be stained as follows:

0.05 M citrate buffer, pH 6.0	50 ml
phenolphthalein diphosphate	0.2 g

Incubate for 1 hr, decant the staining solution, and spray the gel surface with a concentrated solution of ammonium hydroxide. Zones of activity will appear as pink bands. This same stain was described by Harris and Hopkinson (1976) who recommended 4 hr of incubation at 37°C. In some vertebrates "tissue" ACP isozymes can also be resolved.

Aconitase Hydratase (ACOH) (EC 4.2.1.3)

Monomer This enzyme was known formerly as aconitase (ACO or ACON). Mitochondrial and supernatant/cytosolic forms are known (Harris and Hopkinson, 1976). This stain may be prepared as an agar overlay.

0.2 M Tris-HCl, pH 8.0	50 ml
1.0 M MgCl$_2$·6H$_2$O	1.5 ml
(see note below)	
0.1 M cis-aconitic acid, pH 8.0	5 ml
isocitric dehydrogenase	3 units
NADP	0.01 g
5 mg/ml MTT	1 ml
5 mg/ml PMS	1 ml

This stain was modified from Harris and Hopkinson (1976) and Siciliano and Shaw (1976). Note that the magnesium chloride used is 1.0 M, not 0.1 M as is used in many other enzyme stains.

Adenosine Deaminase (ADA) (EC 3.5.4.4)

Monomer This stain may be prepared as an agar overlay.

0.2 M Tris-HCl, pH 8.0	15 ml
H$_2$O	35 ml
adenosine	0.04 g
arsenic acid	0.08 g
xanthine oxidase	0.4 units
nucleoside phosphorylase	1.8 units
5 mg/ml MTT	1 ml
5 mg/ml PMS	1 ml

This stain was modified from Spencer et al. (1968).

Adenylate Kinase (AK) (EC 2.7.4.3)

Monomer This stain may be prepared as an agar overlay.

0.2 M Tris-HCl, pH 8.0	50 ml
0.1 M MgCl$_2$·6H$_2$O	6 ml
adenosine 5'-diphosphate	0.03 g
D(+)-glucose	0.1 g
hexokinase	20 units
G6PDH	40 NAD units
10 mg/ml NAD	2 ml
5 mg/ml NBT	1 ml
5 mg/ml PMS	1 ml

This stain was described by Buth and Murphy (1980) as modified from Fildes and Harris (1966). A more sensitive, but more expensive, fluorescent stain modified from Harris and Hopkinson (1976) may also be used.

Alanine Aminotransferase (ALAT) (EC 2.6.1.2)

Dimer This enzyme was known formerly as glutamic-pyruvic transaminase (GPT). Prepare as an agar overlay (0.07 g agar in 5 ml buffer and then combine with the following):

0.2 M Tris-HCl, pH 7.0	5 ml
DL-alanine	0.04 g
α-ketoglutaric acid	0.02 g

NADH	0.02 g
L-lactic dehydrogenase	25 units

Monitor the development of expression under UV light (long wavelength). Enzyme activity is indicated by zones of defluorescence. This stain was modified from Harris and Hopkinson (1976) and Siciliano and Shaw (1976). See also Casillas et al. (1982). An alternative stain was described by Aebersold et al. (1987).

Alanopine dehydrogenase (ALPDH) (EC 1.5.1.17)

Monomer   Prepare as an agar overlay (0.07 g agar in 5 ml buffer).

0.2 M Tris-HCl, pH 8.0	10 ml
pyruvic acid	0.01 g
L-alanine	0.13 g
NADH	0.01 g

Incubate and view under UV light (long wavelength). Enzyme activity is indicated by zones of defluorescence. This stain was modified from Shaklee and Keenan (1986). See also Dando et al. (1981); this enzyme may be limited to certain invertebrate animals (Manchenko, 1988).

Alcohol Dehydrogenase (ADH) (EC 1.1.1.1)

Dimer Mix the substrate alcohol and the buffer thoroughly prior to adding the other stain components. An agar overlay may be appropriate in some circumstances.

0.2 M Tris-HCl, pH 8.0	40 ml
95 or 100% ethanol	5 ml
10 mg/ml NAD	1 ml
5 mg/ml NBT	1 ml
5 mg/ml PMS	1 ml

Products of some ADH loci may be resolved better using other alcohols [e.g., 0.2 ml 98% 1-hexanol or 0.3 ml amyl alcohol (1-pentanol]. Some related enzymes are named specifically (e.g., octanol dehydrogenase: EC 1.1.1.73) although the substrate, octanol, also may resolve other alcohols (e.g., ADH). This stain was modified from Brewer (1970). Werth

(1985) warned of fume contamination of nearby incubating gels and recommended sealing the ADH stain box with plastic wrap.

Alkaline Phosphatase (ALP) (EC 3.1.3.1)

Monomer or Dimer   First combine the following:

0.2 M Tris-HCl, pH 9.0	50 ml
NaCl	0.85 g
$MgCl_2 \cdot 6H_2O$	0.01 g

Then add:

α-naphthyl acid phosphate	0.15 g
polyvinylpyrrolidone (PVP)	0.15 g
fast blue RR salt	0.05 g

This stain was modified from Boyer (1961) and Ayala et al. (1972).

α-L-Arabinofuranosidase (ARAB) (EC 3.2.1.55)

Subunit Structure Unknown

0.1 M phosphate-citrate buffer, pH 4.0	5 ml
4-methylumbelliferyl-α-L-arabinoside	0.01 g

Incubate for approximately 30 min and then view under UV light (long wavelength). Zones of activity will appear as bright areas. To enhance fluorescence, spray the gel slice with a concentrated ammonium hydroxide solution. This stain was modified from Harris and Hopkinson (1976).

Arginine Kinase (ARK) (EC 2.7.3.3)

Dimer The following stain may also yield adenylate kinase (AK) gene products. A control slice from the same gel must be stained specifically for AK to ascertain, by a process of elimination, which zones of activity are ARK. Prepare as an agar overlay (0.09 g agar in 8 ml buffer).

0.2 M Tris-HCl, pH 7.0	12 ml
0.1 M $MgCl_2 \cdot 6H_2O$	1 ml
adenosine 5'-diphosphate	0.02 g
D(+)-glucose	0.04 g

hexokinase	20 units
phospho-L-arginine	0.02 g
G6PDH	40 NAD units
10 mg/ml NAD	1 ml
5 mg/ml NBT	1 ml
5 mg/ml PMS	1 ml

This stain was modified from Shaklee and Keenan (1986). This enzyme is present only in some invertebrate animals (Shaklee and Keenan, 1986).

Aspartate Aminotransferase (AAT) (EC 2.6.1.1)

Dimer This enzyme was known formerly as glutamate-oxaloactetate transaminase (GOT). Mitochondrial and supernatant/cytosolic forms are known (Harris and Hopkinson, 1976). For the resolution of this enzyme from extracts of relatively fresh tissues, dissolve the following:

0.2 M Tris-HCl, pH 8.0	50 ml
L-aspartic acid	0.23 g
α-ketoglutaric acid	0.10 g

Readjust the pH to 8.0 with 4.0 N NaOH. Then add:

| pyridoxal 5-phosphate | 0.01 g |
| fast blue BB salt | 0.10 g |

This stain was modified from Schwartz et al. (1963). A final pH of 8.0 is critical to the success of this stain (Soltis et al., 1983). Siciliano and Shaw (1976) recommended dissolving each component in specific order with a pH adjustment at the end. A more sensitive, but more expensive, fluorescent stain is found in Harris and Hopkinson (1976).

Calcium-Binding Proteins (CBP) [Nonspecific]

Monomer CBPs migrate rapidly toward the anode and are somewhat diffuse in appearance (Buth, 1979b, 1982b). Creatine kinase gene products (Ck-A) and other proteins that predominate in the tissue examined (e.g., those often scored as "general proteins") will also stain via this procedure.
Dissolve the dye in water:

| H_2O | 50 ml |
| brilliant blue G dye | 0.05 g |

Then add acids:

| trichloroacetic acid (CCl_3CO_2H) | 7.5 g |
| 5-sulfosalicylic acid ($2\text{-}HOC_6H_3\text{-}1\text{-}COOH\text{-}5\text{-}SO_3H$) | 2.5 g |

This stain was modified from Massaro and Markert (1968).

Catalase (CAT) (EC 1.11.1.6)

Tetramer?   Allow the gel slice to warm to ambient temperature, then add

| 0.06 M sodium thiosulfate ($Na_2S_2O_3.5H_2O$) | 15 ml |
| 3% hydrogen peroxide | 35 ml |

Incubate at ambient temperature for 1 min (or less if bubbling is observed), drain, and add

| 0.09 M potassium iodide (KI) | 50 ml |

Flush the gel slice quickly with water to remove KI as soon as the white zones of catalase activity are visible against the blue background. DO NOT place stained gel slice in gel fixitive! Photograph the developed gel slice immediately if possible. The gel slice may be stored in water in a dark refrigerator for a few days but the blue background stain will be lost eventually. This stain was modified from Brewer (1970). Siciliano and Shaw (1976) recommended a longer incubation period (15 min) for the initial solution. Soltis et al. (1983) and Werth (1985) noted that up to 1 ml of glacial acetic acid may have to be added to the KI solution to induce or improve staining. An alternative CAT stain was described by Harris and Hopkinson (1976) and Aebersold et al. (1987).

Creatine Kinase (CK) (EC 2.7.3.2)

Dimer This stain may be prepared as an agar overlay.

0.2 M Tris-HCl, pH 7.0	50 ml
0.1 M MgCl$_2$·6H$_2$O	1 ml
adenosine 5'-diphosphate	0.03 g
D(+)-glucose	0.05 g
hexokinase	40 units
phosphocreatine	0.05 g
G6PDH	40 NAD units
10 mg/ml NAD	1 ml
5 mg/ml NBT	1 ml
5 mg/ml PMS	1 ml

This stain was described by Buth and Murphy (1980) as modifed from Shaw and Prasad (1970). A more sensitive, but more expensive, fluorescent stain was described by Harris and Hopkinson (1976).

Cytosol Aminopeptidase ("CAP"/"AP"/"PEP") (EC 3.4.11.1)

Monomer This enzyme was known formerly as leucine aminopeptidase (LAP); the current IUBNC (1984) name and EC number may be changed as more is learned about peptidases. Richardson et al. (1986) refer to this enzyme as "Pep-E" (see Peptidase). A conservative approach would be to consider this enzyme under the category of generic aminopeptidases EC 3.4.-.-. Incubate the gel slice in the following solution for 30 to 60 min:

0.1 M KH$_2$PO$_4$ buffer, pH 7.0	50 ml
0.1 M MgCl$_2$·6H$_2$O	1 ml
10 mg/ml L-leucine-β-naphthylamide HCl	0.1 ml

Then dissolve the following salt in a small quantity of water and add it to the incubation solution:

black K salt	0.03 g

Continue incubation. This stain was modified from those of Brewer (1970), Shaw and Prasad (1970), and Ayala et al. (1972). Some of these staining methods involve a preincubation step in a boric acid solution which may not be necessary. Werth (1985) recommended fast garnet GBC salt as a possible substitute for black K salt.

Dihydrolipoamide Dehydrogenase (DDH) (EC 1.8.1.4)

Monomer or Dimer This enzyme was known formerly as diaphorase (DIA) and lipoamide dehydrogenase (EC 1.6.4.3; see Muramatsu et al., 1978).

0.2 M Tris-HCl, pH 8.0	50 ml
2 mg/ml 2,6-dichlorophenol-indophenol	1 ml
NADH	0.01 g
5 mg/ml MTT	1 ml

Zones of enzyme activity will appear pink/purple against the blue background of the gel. The blue DCIP color will clear overnight if the developed gel is kept refrigerated (dark) yielding a white gel with purple isozymes. This stain was modified from those of Kaplan and Beutler (1967) and Brewer (1970). Harris and Hopkinson (1976) used this stain to resolve "NADH diaphorase" recognizing this enzyme as a synonym of cytochrome-b_5 reductase (EC 1.6.2.2). Aebersold et al. (1987) noted that this stain may also resolve xanthine oxidase (XO) gene products as well as those of a variety of other enzymes.

Enolase (ENO) (EC 4.2.1.11)

Dimer Prepare this stain as an agar overlay.

0.2 M Tris-HCl, pH 8.0	50 ml
0.1 M MgCl$_2$·6H$_2$O	5 ml
adenosine 5'-diphosphate	0.05 g
D(-)-3-phosphoglyceric acid	0.05 g
NADH	0.04 g
pyruvate kinase	20 units
L-lactic dehydrogenase	125 units
phosphoglycerate mutase	125 units

Monitor the development of expression under UV light (long wavelength). Enzyme activity is indicated by zones of defluoresence. This stain was modified from Harris and Hopkinson

(1976). Two alternative agar-overlay, fluorescent stains have been described for ENO that call for slightly different sets of components that may be less expensive (Siciliano and Shaw, 1976; Shaklee and Keenan, 1986; Aebersold et al., 1987).

Esterase (EST) [Nonspecific]

Monomer or Dimer The following general esterase stain can resolve a number of gene products with broad substrate specificities; these enzymes might be identified specifically using other methods. The products resolved nonspecifically might be considered as generic carboxylic ester hydrolases EC 3.1.1.- (IUBNC, 1984).

0.2 M Tris-HCl, pH 7.0	50 ml
α-naphthyl acetate solution	3 ml
fast blue BB salt	0.05 g

Incubate at ambient temperature; dark is not required. To prepare the stock substrate solution (1% solution in 50% acetone), dissolve the α-naphthyl acetate in the acetone, then add the water. Clearer resolution of some esterase products can be achieved by using β-naphthyl acetate as the substrate. This stain was modified from Brewer (1970). A fluorescent stain for dimeric "esterase-D" (EC 3.1.1.-) has been described by several investigators (e.g., Harris and Hopkinson, 1976; Soltis et al., 1983; Shaklee and Keenan, 1986; Aebersold et al., 1987).

Formaldehyde Dehydrogenase (FDH) (EC 1.2.1.1)

Dimer

0.2 M Tris-HCl, pH 8.0	50 ml
37% formaldehyde (= reagent)	3 drops
glutathione, reduced	0.05 g
10 mg/ml NAD	3 ml
5 mg/ml NBT	1 ml
5 mg/ml PMS	1 ml

Prepare the stain and incubate the gel slice at 37°C in the dark in a fume hood. This stain was modified from an unpublished formula of G. P. Manchenko (personal communication). See also Manchenko (1988).

Fructose-bisphosphatase (FBP) (EC 3.1.3.11)

Dimer or Tetramer This enzyme was known formerly as hexosediphosphatase (HDP) and fructose-1,6-diphosphatase.

0.2 M Tris-HCl, pH 8.0	50 ml
$MgSO_4 \cdot 7H_2O$	0.25 g
D-fructose-1,6-diphosphate	0.02 g
glucose-6-phosphate isomerase (PGI = GPI)	50 units
G6PDH	40 NAD units
2-mercaptoethanol (1 drop in 10 ml H_2O)	1 drop
NADP	0.02 g
5 mg/ml NBT	1 ml
5 mg/ml PMS	1 ml

Prepare stain and incubate the gel slice at 37°C in the dark in a fume hood. This stain is modified from Shaw and Prasad (1970).

Fructose-bisphosphate Aldolase (FBA) (EC 4.1.2.13)

Tetramer This enzyme was known formerly as aldolase (ALD) and is also currently abbreviated as FBALD by some investigators.

0.2 M Tris-HCl, pH 8.0	50 ml
glyceraldehyde-3-phosphate dehydrogenase	200 units
D-fructose-1,6-diphosphate	0.08 g
10 mg/ml NAD	2 ml
5 mg/ml NBT	1 ml
5 mg/ml PMS	1 ml

This stain was modified from Ayala et al. (1972). Aebersold et al. (1987) recommended the addition of 0.01 g arsenic acid (Na_2HAsO_4) to an FBA stain of 50 ml volume.

Fumarate Hydratase (FUMH) (EC 4.2.1.2)

Tetramer This enzyme was known formerly as fumarase (FUM). The following stain may also yield malate dehydrogenase (MDH) gene products. A control slice from the same gel must be stained specifically for MDH to ascertain, by a process of elimination, which zones of activity are FUMH.

0.2 *M* Tris-HCl, pH 8.0	50 ml
fumaric acid	0.05 g
malic dehydrogenase	150 units
10 mg/ml NAD	1 ml
5 mg/ml NBT	1 ml
5 mg/ml PMS	1 ml

This stain was modified from Brewer (1970). The FUMH stain described by Siciliano and Shaw (1976) called for the use of the disodium salt of fumaric acid and MDH in greater concentration. Aebersold et al. (1987) recommended the addition of 0.01 g pyruvic acid to a FUMH stain of 50 ml volume to supress LDH.

α-Galactosidase (α GAL) (EC 3.2.1.22)

Dimer

0.1 *M* phosphate-citrate buffer, pH 4.0	5 ml
4-methylumbelliferyl-α-D-galactoside	0.01 g

Incubate for approximately 30 min and then view under UV light (long wavelength). Zones of activity will appear as bright areas. To enhance fluorescence, spray the gel slice with a concentrated ammonium hydroxide solution. This stain was modified from Harris and Hopkinson (1976).

β-Galactosidase (β GAL) (EC 3.2.1.23)

Monomer

0.1 *M* phosphate-citrate buffer, pH 4.0	5 ml
4-methylumbelliferyl-β-D-galactoside	0.01 g

Incubate at 37°C for approximately 30 min and then view under UV light (long wavelength). Zones of activity will appear as bright areas. To enhance fluorescence, spray the gel slice with a concentrated ammonium hydroxide solution. This stain was modified from Harris and Hopkinson (1976).

General Proteins (GP) [Nonspecific]

Various Quaternary Structures   Creatine kinase gene products (Ck-A) and other proteins that predominate will also stain via this procedure.

Prepare stock solution:

naphthol blue black (amido black)	1 g
stain fixing solution (1:5:5 mixture of glacial acetic acid, methanol, and water).	500 ml

Filter stain to remove undissolved dye. Stain gel slices in 50 ml of solution for 20 min at 20°C. Wash slices in fixing solution several times until background is pale. The stain may be reused. This stain was modified from Selander et al. (1971).

Glucose Dehydrogenase (GCDH) (EC 1.1.1.118)

Dimer?   The enzyme was known formerly as hexose-6-phosphate dehydrogenase (H6PDH).

0.05 *M* potassium phosphate buffer, pH 7.0	50 ml
D(+)-glucose	9 g
10 mg/ml NAD	2 ml
5 mg/ml NBT	1 ml
5 mg/ml PMS	1 ml

This stain was modified by Berg and Buth (1984) from that described by Harris and Hopkinson (1976).

Glucose-6-phosphate Dehydrogenase (G6PDH) (EC 1.1.1.49)

Dimer?   NADP (0.02 g in 400 ml) should be added to the gel before electrophoresis.

0.2 *M* Tris-HCl, pH 8.0	50 ml
0.1 *M* MgCl$_2$·6H$_2$0	3 ml
D-glucose-6-phosphate	0.3 g
NADP	0.03 g

5 mg/ml NBT	1 ml
5 mg/ml PMS	1 ml

This stain is modified from Brewer (1970).

Glucose-6-phosphate Isomerase (GPI) (EC 5.3.1.9)

Dimer   This enzyme was known formerly as phosphohexose isomerase (PHI) or phosphoglucoisomerase (PGI). This stain may be prepared as an agar overlay.

0.2 M Tris-HCl, pH 7.0	50 ml
0.1 M MgCl$_2$·6H$_2$0	5 ml
D-fructose-6-phosphate	0.04 g
G6PDH	40 NAD units
10 mg/ml NAD	2 ml
5 mg/ml NBT	1 ml
5 mg/ml PMS	1 ml

This stain was described by Buth and Murphy (1980) as modified from DeLorenzo and Ruddle (1969).

α-Glucosidase (α GLUS) (EC 3.2.1.20)

Tetramer

0.1 M phosphate-citrate buffer, pH 4.0	5 ml
4-methylumbelliferyl-α-D-glucoside	0.01 g

Monitor the development of expression under UV light (long wavelength). Zones of activity will appear as bright areas. To enhance fluorescence, spray the gel slice with a concentrated solution of ammonium hydroxide. This stain was modified from Harris and Hopkinson (1976). Aebersold et al. (1987) recommended a stain buffer pH of 8.0.

β-Glucosidase (βGLUS) (EC 3.2.1.21)

Subunit Structure Uncertain

0.1 M phosphate-citrate buffer, pH 4.0	5 ml
4-methylumbelliferyl-β-D-glucoside	0.01 g

Incubate for approximately 30 min and then view under UV light (long wavelength). Zones of activity will appear as bright areas. To enhance fluorescence, spray the gel slice with a concentrated ammonium hydroxide solution. This stain was modified from Harris and Hopkinson (1976).

β-Glucuronidase (β GLUR) (EC 3.2.1.31)

Tetramer   Some investigators abbreviate this enzyme as "GUS."

0.1 M phosphate-citrate buffer, pH 4.0	5 ml
4-methylumbelliferyl-β-D-glucuronide	0.01 g

Incubate for approximately 30 min and then view under UV light (long wavelength). Zones of activity will appear as bright areas. To enhance fluorescence, spray the gel slice with a concentrated ammonium hydroxide solution. This stain was modified from Harris and Hopkinson (1976). A more expensive, nonfluorescent ("positive") stain was described by Aebersold et al. (1987).

Glutamate-ammonia Ligase (GLAL) (EC 6.3.1.2)

Subunit Structure Uncertain   This enzyme was known formerly as glutamine synthetase (GS). Preparation of a substrate solution and a visualization solution is required.

Substrate solution: Dissolve 0.10 g MgCl$_2$·6H$_2$O in 2.0 ml H$_2$O, then add

0.2 M Tris-HCl buffer, pH 8.0	3 ml
L-glutamic acid	0.20 g
adenosine 5'-triphosphate	0.08 g
NH$_4$OH (concentrated)	0.2 ml

Raise the pH to 9.3 with 1.0 N NaOH. Incubate the gel slice in the substrate solution at 37°C for at least 1 hr then remove the substrate solution from the gel slice but do not rinse. Cover the gel slice with a 50% acetone solution and incubate it at ambient temperature for 15 min. Flush the gel slice with water and add the following visualization solution:

L-ascorbic acid   0.8 g
ammonium molybdate solution   6 ml
    (see below)

Incubate at 37°C. The ammonium molybdate solution can be prepared as a stock (2.5 g ammonium molybdate, 8 ml concentrated H_2SO_4, 92 ml H_2O). This stain was described by Morizot et al. (1983; D. C. Morizot, personal communication).

Glutamate Dehydrogenase (GTDH) (EC 1.4.1.2)

Tetramer?

0.1 M potassium phosphate buffer, pH 7.0	35 ml
1.0 M L-glutamic acid	15 ml
10 mg/ml NAD	3 ml
5 mg/ml NBT	1 ml
5 mg/ml PMS	1 ml

This stain was modified from Shaw and Prasad (1970). Brewer (1970) recommended a staining buffer of pH 9.0. Aebersold et al. (1987) recommended the addition of 0.014 g adenosine 5'-diphosphate and 0.001 g pyridoxal 5-phosphate to a GTDH stain of 50 ml volume.

Glutamate Dehydrogenase (NADPH⁺)(GTDHP) (EC 1.4.1.4)

Subunit Structure Uncertain

0.1 M potassium phosphate buffer, pH 7.0	35 ml
1.0 M L-glutamic acid	15 ml
NADP	0.02 g
5 mg/ml NBT	1 ml
5 mg/ml PMS	1 ml

Counterstain for GTDH. This stain was modified from Shaw and Prasad (1970). Brewer (1970) recommended a staining buffer of pH 9.0. Aebersold et al. (1987) recommended the addition of 14 mg adenosine 5'-diphosphate and 1 mg pyridoxal 5-phosphate to a GTDHP stain of 50 ml volume.

Glutathione Reductase (NAD(P)H) (GR) (EC 1.6.4.2)

Dimer   Prepare as an agar overlay (0.17 g agar in 10 ml H_2O, mix the stain components in the 13 ml buffer):

0.2 M Tris-HCl, pH 8.0	13 ml
2 mg/ml 2,6-dichlorophenol-indophenol	1 ml
glutathione, oxidized	0.02 g
flavine adenine dinucleotide	0.002 g
NADH	0.01 g
5 mg/ml MTT	1 ml

This stain was modified from Aebersold et al. (1987). Alternative stains are discussed by Brewer (1970). Dihydrolipoamide dehydrogenase (DDH; EC 1.8.1.4) may also be resolved as a second, relatively slower system (see Harris and Hopkinson, 1976) because it appears if glutathione is omitted from the stain. In addition, because Aebersold et al. (1987) noted that the DDH stain may also resolve xanthine oxidase (XO) gene products as well as those of a variety of other enzymes, this may also be true for GR. FAD may not be necessary if NADPH is used instead of NADH.

Glyceraldehyde-3-phosphate Dehydrogenase (GAPDH) (EC 1.2.1.12)

Tetramer   Prepare the substrate solution:

0.2 M Tris-HCl, pH 7.0	10 ml
aldolase	100 units
D-fructose-1,6-diphosphate	0.25 g

Incubate the substrate solution at 37°C for 30 min, then add the following:

0.2 M Tris-HCl, pH 7.0	40 ml
arsenic acid (Na_2HAsO_4)	0.08 g
10 mg/ml NAD	2 ml
5 mg/ml NBT	1 ml
5 mg/ml PMS	1 ml

This stain was modified from those of Ayala et al. (1972) and Siciliano and Shaw (1976). In order to minimize L-lactate dehydrogenase (LDH) staining if this is a problem, 0.1 g of pyruvic acid may be added to the staining solution (Harris and Hopkinson, 1976).

Glycerate Dehydrogenase (GLYDH) (EC 1.1.1.29)

Subunit Structure Uncertain   The following stain may also yield lactate dehydrogenase (LDH) gene products. A control slice from the same gel may be necessary.

0.2 M Tris-HCl, pH 8.0	50 ml
DL-glyceric acid	0.2 g
pyruvic acid	0.05 g
pyrazole	0.05 g
10 mg/ml NAD	2 ml
5 mg/ml NBT	1 ml
5 mg/ml PMS	1 ml

This stain was modified from Siciliano and Shaw (1976).

Glycerol-3-phosphate Dehydrogenase (G3PDH) (EC 1.1.1.8)

Dimer   This enzyme was known formerly as α-glycerophosphate dehydrogenase (αGPD or αGPDH).

0.2 M Tris-HCl, pH 8.0	50 ml
DL-α-glycerophosphate, pH 8.0	1 g
0.1 M MgCl$_2$	1 ml
10 mg/ml NAD	1 ml
5 mg/ml NBT	1 ml
5 mg/ml PMS	1 ml

This stain may require a higher pH of stain buffer (pH 9.5), 2 ml of NAD, and incubation for 1 hr before adding PMS. This stain was modified from Shaw and Prasad (1970).

Guanine Deaminase (GDA) (EC 3.5.4.3)

Dimer   This stain may be applied as an agar overlay (use 0.07 g agar in the 5 ml H$_2$O and mix the stain components in the 5 ml buffer). Otherwise simply mix the following:

0.2 M Tris-HCl, pH 8.0	5 ml
H$_2$O	5 ml
guanine	0.25 g
xanthine oxidase	2.6 units

5 mg/ml MTT	2 ml
5 mg/ml PMS	1 ml

If the enzyme activity is low, the amount of XO may have to be increased. However, Harris and Hopkinson (1976) recommended that less XO be used, 0.1 units. This stain was modified from Richardson (1983).

Guanylate Kinase (GUK) (EC 2.7.4.8)

Monomer   This stain may be prepared as an agar overlay (0.07 g agar in 5 ml of buffer and mix the stain components in the remaining 8 ml of buffer).

0.2 M Tris-HCl, pH 8.0	13 ml
0.1 M MgCl$_2$·6H$_2$O	2 ml
adenosine 5'-triphosphate	0.01 g
guanosine 5'-monophosphate	0.04 g
phospho(enol)pyruvate	0.01 g
NADH	0.01 g
potassium chloride (KCl)	0.07 g
calcium chloride (CaCl$_2$)	0.01 g
pyruvate kinase	10 units
lactic dehydrogenase	125 units

Monitor the development of expression under UV light (long wavelength). Enzyme activity is indicated by zones of defluorescence. This stain was modified from Harris and Hopkinson (1976). See also Morizot and Siciliano (1982).

Hexokinase (HK) (EC 2.7.1.1)

Monomer   The following stain may also yield glucose dehydrogenase (GCDH) gene products. Either a control slice from the same gel must be stained specifically for GCDH to ascertain, by a process of elimination, which zones of activity are HK, or 0.05 g of gluconic acid must be added to the stain to supress GCDH. HK and GCDH often have very different tissue-specific expression so the choice of tissue may make continued control testing unnecessary.

0.2 M Tris-HCl, pH 8.0	50 ml
0.1 M MgCl$_2$	1 ml
adenosine 5'-triphosphate	0.25 g

D(+)-glucose	5 g
G6PDH	80 NAD units
10 mg/ml NAD	2 ml
5 mg/ml NBT	1 ml
5 mg/ml PMS	1 ml

Incubate the gel slice in the dark at ambient temperature or at 37°C. This stain was modified from Shaw and Prasad (1970). Soltis et al. (1983) and Werth (1985) suggested that 0.02 g tetrasodium EDTA be added to a HK stain of 50 ml volume.

D-2-Hydroxy-acid Dehydrogenase (HADH) (EC 1.1.99.6)

Dimer   This enzyme has multiple substrate affinities, and may be confused with 3-hydroxybutyrate dehydrogenase (HBDH). ADH may appear with prolonged staining. Dissolve 2.0 g D-gluconic acid lactone in 25 ml H_2O. Adjust the pH to 12.5 with sodium hydroxide pellets ($\approx$0.5 g). Incubate this solution at ambient temperature for 30 min with occasional stirring. Then readjust the pH to just below 8.0 by adding, dropwise, 12 N HCl. Add 25 ml 0.2 M Tris-HCl, pH 8.0 (at this point, the pH of the substrate solution should be 8.0 but a minor adjustment with 4 N HCl may be necessary). Add the following to the substrate solution and apply to the gel slice:

pyrazole	0.05 g
10 mg/ml NAD	1 ml
5 mg/ml NBT	1 ml
5 mg/ml PMS	1 ml

This stain was described by Buth (1980) as modified from Tobler and Grell (1978).

(S)-2-Hydroxy-acid Oxidase (HAOX) (EC 1.1.3.15)

Tetramer   This enzyme was known formerly as glycolate oxidase (GOX) EC 1.1.3.1. Prepare as an agar overlay.

0.2 M Tris-HCl, pH 8.0	50 ml
glycolic acid	0.05 g

5 mg/ml MTT	1 ml
5 mg/ml PMS	1 ml

This stain was modified from R. L. Garthwaite (personal communication).

Hydroxyacylglutathione Hydrolase (HAGH) (EC 3.1.2.6)

Dimer   This enzyme was known formerly as glyoxalase II (GLO-II). This stain should be prepared as an agar overlay (0.17 g agar in 10 ml H_2O, mix the stain components in the 15 ml buffer):

0.2 M Tris-HCl, pH 8.0	15 ml
methylglyoxal	0.24 ml
glutathione, reduced	0.03 g
glyoxalase I	150 units

Incubate at 37°C for at least 30 min, then add

1.0 M zinc chloride ($ZnCl_2$)	80 µl
pyruvic acid	5 mg
L-lactic dehydrogenase	500 units
10 mg/ml NAD	3 ml
5 mg/ml MTT	1 ml
5 mg/ml PMS	1 ml

Incubate the gel slice at 37°C in the dark. This stain was modified from Aebersold et al. (1987).

3-Hydroxybutyrate Dehydrogenase (HBDH) (EC 1.1.1.30)

Dimer   This enzyme may be confused with 2-hydroxy-acid dehydrogenase (HADH) due to the multiple substrate affinities of the latter. A control slice from the same gel should be stained for HADH. Electromorphs in common between slices stained for HBDH and HADH are probably HADH. ADH products may appear with prolonged staining.

0.5 M potassium phosphate buffer, pH 7.0	25 ml
H_2O	20 ml
0.1 M $MgCl_2$	2 ml
NaCl	0.30 g
DL-β-hydroxybutyric acid	0.63 g
10 mg/ml NAD	3 ml

5 mg/ml NBT	1 ml
5 mg/ml PMS	1 ml

This stain is modified from Shaw and Prasad (1970).

L-Iditol Dehydrogenase (IDDH) (EC 1.1.1.14)

Tetramer   This enzyme was known formerly as sorbitol dehydrogenase (SDH or SORD).

0.2 M Tris-HCl, pH 8.0	50 ml
2.0 M D-sorbitol	5 ml
10 mg/ml NAD	1 ml
5 mg/ml NBT	1 ml
5 mg/ml PMS	1 ml

This stain was modified from Lin et al. (1969).

Isocitrate Dehydrogenase (IDH) (EC 1.1.1.42)

Dimer   The following stain is for NADP-dependent isocitrate dehydrogenase. Mitochondrial and supernatant/cytosolic forms are known (Harris and Hopkinson, 1976).

0.2 M Tris-HCl, pH 8.0	50 ml
0.1 M MgCl$_2$	3 ml
0.1 M DL-isocitric acid	3 ml
NADP	0.01 g
5 mg/ml NBT	1 ml
5 mg/ml PMS	1 ml

This stain was modified from those of Henderson (1965), Brewer (1970), Shaw and Prasad (1970), and Ayala et al. (1972). Note that Brewer (1970) erred in listing manganese chloride (MnCl$_2$) instead of magnesium chloride in this stain.

L-Lactate Dehydrogenase (LDH) (EC 1.1.1.27)

Tetramer

0.2 M Tris-HCl, pH 8.0	50 ml
1.0 M lithium lactate, pH 8.0 (see below)	8 ml
10 mg/ml NAD	1 ml

5 mg/ml NBT	1 ml
5 mg/ml PMS	1 ml

The stock substrate solution may be prepared using either DL-lactic acid or lactic acid solution; the pH should be adjusted to 8.0 with the addition of LiOH. This stain was modified from Shaw and Prasad (1970).

Lactoylglutathione Lyase (LGL) (EC 4.4.1.5)

Dimer   This enzyme was known formerly as glyoxalase I (GLO-I).

0.1 M potassium phosphate buffer, pH 7.0	12 ml
methylglyoxal (40% solution)	0.5 ml
glutathione, reduced	0.04 g

Incubate the gel slice in this solution at 37°C for 40 min. Remove the solution and prepare an agar overlay of the following (0.21 g agar in the 30 ml H$_2$O, add KI when cooled):

H$_2$O	30 ml
0.9 M potassium iodide (KI)	0.1 ml

Areas of activity will be seen as blue zones on the gel slice. This stain was modified from Harris and Hopkinson (1976).

Malate Dehydrogenase (MDH) (EC 1.1.1.37)

Dimer   The following stain is for NAD-dependent malate dehydrogenase. Mitochondrial and supernatant/cytosolic forms are known (Harris and Hopkinson, 1976).

0.2 M Tris-HCl, pH 8.0	50 ml
2.0 M DL-malic acid	5 ml
10 mg/ml NAD	1 ml
5 mg/ml NBT	1 ml
5 mg/ml PMS	1 ml

This stain is modified from Brewer (1970).

Malate Dehydrogenase (NADP$^+$) (MDHP) (EC 1.1.1.40)

Tetramer   The following stain is for NADP-dependent malate dehydrogenase.

The convention for name abbreviation of this kind of enzyme (+P to MDH) follows Aebersold et al. (1987). This enzyme was known formerly as malic enzyme (ME). Mitochondrial and supernatant/cytosolic forms are known (Harris and Hopkinson, 1976). NADP (0.02 g in 400 ml) should be added to the gel before electrophoresis.

0.2 M Tris-HCl, pH 8.0	50 ml
0.1 M MgCl$_2$	1 ml
2.0 M DL-malic acid, pH 8.0	5 ml
NADP (see note below)	0.02 g
5 mg/ml NBT	1 ml
5 mg/ml PMS	1 ml

This stain was modified from those of Ayala et al. (1972) and Cross et al. (1979). It is important that NADP be used in solid form in this stain. There is often sufficient breakdown of NADP to NAD in liquid stocks in prolonged storage that NAD-dependent MDH activity will be resolved in addition to MDHP. If there is any doubt as to the identity of MDHP, a control slice from the same gel should be stained specifically for MDH to ascertain, by a process of elimination, which zones of activity are MDHP. Aebersold et al. (1987) recommended adding 0.02 g oxaloacetic acid to an MDHP stain of 50 ml volume.

Mannose-6-phosphate Isomerase (MPI) (EC 5.3.1.8)

Monomer   This stain may be prepared as an agar overlay.

0.2 M Tris-HCl, pH 8.0	50 ml
0.1 M MgCl$_2$	1 ml
D-mannose-6-phosphate	0.05 g
glucose-6-phosphate isomerase (GPI = PGI)	50 units
G6PDH	40 NAD units
10 mg/ml NAD	2 ml
5 mg/ml MTT	1 ml
5 mg/ml PMS	1 ml

Products of L-lactate dehydrogenase (LDH; EC 1.1.1.27) may appear as faint bands

following staining. LDH activity can be supressed by adding 0.05 g of pyruvic acid. This stain was described by Buth and Murphy (1980) as modified from Nichols et al. (1973).

α-Mannosidase (α MAN) (EC 3.2.1.24)

Monomer or Dimer

0.1 M phosphate-citrate buffer, pH 4.0	5 ml
4-methylumbelliferyl-α-D-mannopyranoside	0.01 g

Incubate for 30 min and then view under UV light (long wavelength). Zones of activity will appear as bright areas. To enhance fluorescence, spray the gel slice with a concentrated ammonium hydroxide solution. This stain was modified from Harris and Hopkinson (1976).

Nucleoside-triphosphate Pyrophosphatase (NTP) (EC 3.6.1.19)

Dimer   This enzyme was known formerly as inosine triphosphatase (ITP). This stain requires the preparation of a substrate solution and a visualization solution.

Substrate solution:

0.2 M Tris-HCl, pH 8.0	15 ml
MgCl$_2$·6H$_2$O	0.1 g
inosine triphosphate	0.03 g
2-mercaptoethanol	120 μl

Mix these components and pour on the gel slice in a fume hood. Incubate at 37°C for at least 1 hr. A filter paper overlay may be desired to keep the gel slice moist (Harris and Hopkinson, 1976; Aebersold et al., 1987). After incubation, remove the substrate solution from the slice but do not rinse. Add the following visualization solution:

L-ascorbic acid	0.5 g
ammonium molybdate solution (see below)	12 ml

Incubate at 37°C in a fume hood in the dark. The ammonium molybdate solution can be prepared as a stock (2.5 g ammonium molybdate, 8 ml concentrated H_2SO_4, 92 ml H_2O). This stain was modified from Aebersold et al. (1987).

Octanol Dehydrogenase (ODH) (EC 1.1.1.73)

Dimer

0.2 M Tris-HCl, pH 8.0	50 ml
1-octanol	3 ml
95% ethanol	1 ml
10 mg/ml NAD	1 ml
5 mg/ml NBT	1 ml
5 mg/ml PMS	1 ml

This stain, modified from those of Ayala et al. (1972) and Shaklee and Keenan (1986), includes ethanol, which may also serve as a substrate for alcohol dehydrogenase (ADH). A control slice from the same gel should be stained for ADH using only ethanol as substrate to ascertain which zones of activity are ODH.

D-Octopine Dehydrogenase (OPDH) (EC 1.5.1.11)

Monomer   Prepare this stain as an agar overlay (0.17 g agar in 15 ml buffer, the stain components in the remaining 10 ml buffer):

0.2 M Tris-HCl, pH 8.0	25 ml
0.1 M $MgCl_2$	1 ml
D-octopine	0.01 g
10 mg/ml NAD	1 ml
5 mg/ml NBT	1 ml
5 mg/ml PMS	1 ml

This stain was modified from Shaklee and Keenan (1986). This enzyme is present only in certain molluscs (Shaklee and Keenan, 1986).

Peptidase (PEP) (EC 3.4.-.-)

Subunit Structure Variable   The terms dipeptidase (EC 3.4.13.11) and tripeptide aminopeptidase (EC 3.4.11.4) are recommended over the more generic term "peptidase" (IUBNC, 1984). However, the

multiple substrate affinities of these enzymes and problematic assignment of their homology makes the exact assignment of EC numbers difficult. Exceptions are those of proline dipeptidase (Pep-D; EC 3.4.13.9) and perhaps "cytosol aminopeptidase" (Pep-E; EC 3.4.11.1). Recommended substrates for the resolution of products of seven peptidase loci described from vertebrates follow Frick (1981, 1983, personal communication), Richardson et al. (1986), and/or Matson (1989). The tissue distribution of these gene products is often restricted (e.g., Frick, 1983; Matson, 1989).

Pep-A: glycyl-L-leucine [dimer]
Pep-B: L-leucylglycylglycine [monomer]
Pep-C: glycyl-L-leucine [monomer]
        [specific: DL-alanyl-DL-methionine]
Pep-D: L-phenylalanyl-L-proline [dimer]
Pep-E: see cytosol aminopeptidase herein
        [monomer]
Pep-F: L-leucyl-L-leucyl-L-leucine [?]
Pep-S: glycyl-L-leucine [tetramer?]

The following general stain for peptidases is modified from Merritt et al. (1978) and is best used as an agar overlay:

0.2 M Tris-HCl, pH 8.0	50 ml
di/tripeptide (see above)	0.04 g
snake venom (from Crotalus atrox)	0.01 g
peroxidase	0.02 g
o-dianisidine dihydrochloride	0.01 g

Snake venom is used in this stain as a source of L-amino acid oxidase. The substitution of a less purified but adequate source of this enzyme (via snake venom) was advantageous financially at one time but may no longer be so. Several stain formulas list specifically the venom of the eastern diamondback rattlesnake (Crotalus adamanteus) for use in peptidase stains (e.g., Siciliano and Shaw, 1976). We have tried the less expensive venom of a closely related rattlesnake, the western diamondback (C. atrox, recommended herein), and found that it yielded equivalent results.

Peroxidase (PER) (EC 1.11.1.7)

Subunit Structure Uncertain

3-amino-9-ethyl carbazole	0.04 g
N,N-dimethyl formamide	2.5 ml

Then add:

0.05 M sodium acetate buffer, pH 5.0	45 ml
0.1 M calcium chloride ($CaCl_2$)	1 ml
3% hydrogen peroxide (H_2O_2)	1 ml

Incubate the gel slice in a refrigerator, usually for 30–60 min. This stain was modified from Shaw and Prasad (1970) by Soltis et al. (1983). See Brewer (1970) and Siliciano and Shaw (1976) for additional PER stains.

6-Phosphofructokinase (PFK) (EC 2.7.1.11)

Dimer   This stain may be prepared as an agar overlay (0.14 g agar in 10 ml of water and stain components in the remaining 10 ml of buffer).

0.2 M Tris-HCl, pH 8.0	10 ml
H_2O	10 ml
D-fructose-6-phosphate	0.012 g
adenosine 5'-triphosphate	0.012 g
$MgCl_2 \cdot 6H_2O$	0.04 g
arsenic acid (Na salt)	0.20 g
2-mercaptoethanol	20 μl
aldolase	36 units
triosphosphate isomerase	500 units
α-glycerophosphate dehydrogenase	40 units
10 mg/ml NAD	1 ml

Prepare stain and incubate the gel slice at 37°C in a fume hood. Incubate for 30 min and then view under UV light (long wavelength). Zones of activity will appear as bright areas. To enhance fluorescence, spray the gel slice with a concentrated ammonium hydroxide solution. This stain was modified from Harris and Hopkinson (1976), who recommend the addition of 2-mercaptoethanol (final conc. 10 mM) and ATP (final conc. 0.2 mM) to gel before degassing, but this may not be necessary.

Phosphoglucomutase (PGM) (EC 5.4.2.2)

Monomer   This stain may be prepared as an agar overlay.

0.2 M Tris-HCl, pH 8.0	50 ml
0.1 M $MgCl_2$	5 ml
α-D-glucose-1-phosphate	0.1 g
G6PDH	40 NAD units
10 mg/ml NAD	2 ml
5 mg/ml NBT	1 ml
5 mg/ml PMS	1 ml

This stain was described by Buth and Murphy (1980) as modified from Spencer et al. (1964). The reaction requires glucose-1,6-diphosphate in addition to glucose-1-phosphate listed above. However, the former biochemical is expensive if purchased in purified form and only a trace is necessary. The Sigma G-7000 product is said to carry a trace of G-1,6-P (0.01–0.2%) which is usually enough to stain for PGM but may not be so in some cases. Soltis et al. (1983) and Werth (1985) recommended the Sigma G-1259 product, which, although more expensive than G-7000, has a sufficient amount of G-1,6-P (1%) as a contaminant to guarantee PGM activity.

Phosphogluconate Dehydrogenase (PGDH) (EC 1.1.1.44)

Dimer   This enzyme was known formerly as 6-phosphogluconate dehydrogenase (6PGD or 6PGDH). Turner (1974) identified a second gene product that sometimes appears with this stain as glucose-6-phosphate dehydrogenase. Limit the following staining solution to the immediate area of anticipated activity. NADP (0.02 g in 400 ml) should be added to the gel before electrophoresis.

0.2 M Tris-HCl, pH 8.0	5 ml
0.1 M $MgCl_2$	5 ml
6-phosphogluconic acid	0.01 g
NADP	0.01 g
5 mg/ml NBT	0.5 ml
5 mg/ml PMS	0.5 ml

This stain is modified from Shaw and Prasad (1970). Harris and Hopkinson (1976) recommended an agar overlay for PGDH.

Phosphoglycerate Kinase (PGK) (EC 2.7.2.3)

Monomer   This stain should be prepared as an agar overlay (0.14 g agar in 10 ml buffer and stain components in the remaining 10 ml buffer).

0.2 M Tris-HCl, pH 8.0	20 ml
0.1 M MgCl$_2$	1 ml
D(-)3-phosphoglyceric acid	0.03 g
adenosine 5'-triphosphate	0.05 g
NADH	0.02 g
glyceraldehyde-3-phosphate dehydrogenase	20 units
α-glycerophosphate dehydrogenase	5 units

Monitor the development of expression under UV light (long wavelength). Staining may occur rapidly; check after 5–10 min. Enzyme activity is indicated by zones of defluorescence. This stain was modified from those of Beutler (1969) and Harris and Hopkinson (1976). Aebersold et al. (1987) recommended using 10 times this amount of magnesium chloride and adding 300 units of triose-phosphate isomerase. We have found the use of α-glycerophosphate dehydrogenase to be unnecessary in amphibians, reptiles, and mammals. A stock partial substrate buffer solution may be prepared as follows:

PGK partial substrate solution:

M$_g$Cl$_2$·6H$_2$O	1 g
EDTA	0.20 g
0.2 M Tris-HCl, pH 8.0	20 ml
H$_2$0	80 ml

Store partial substrate solution refrigerated. Add 3-phosphoglyceric acid, NADH, ATP, and GAPDH to the partial substrate solution in the quantities described above, incubate, and view under UV light (long wavelength).

Phosphoglycerate Mutase (PGAM) (EC 5.4.2.1)

Dimer

0.2 M Tris-HCl, pH 8.0	10 ml
MgCl$_2$·6H$_2$O	0.02 g
D(+)2-phosphoglyceric acid	0.03 g
adenosine 5'-triphosphate	0.02 g
disodium EDTA	0.01 g
NADH	0.01 g
3-phosphoglyceric phosphokinase	400 units
glyceraldehyde-3-phosphate dehydrogenase	100 units

Monitor the development of expression under UV light (long wavelength). Enzyme activity is indicated by zones of defluorescence. This stain is modified from those of Harris and Hopkinson (1976) and Siciliano and Shaw (1976). An alternative stain is suggested by Richardson et al. (1986).

Purine-nucleoside Phosphorylase (PNP) (EC 2.4.2.1)

Trimer   This enzyme was known formerly as nucleoside phosphorylase (NP). This stain should be prepared as an agar overlay.

0.05 M potassium phosphate buffer, pH 7.0	50 ml
inosine	0.01 g
xanthine oxidase	0.4 units
5 mg/ml MTT	1 ml
5 mg/ml PMS	1 ml

Incubate the gel slice at 37°C in the dark. This stain was modified from those of Harris and Hopkinson (1976) and Ward et al. (1979).

Pyrroline-5-carboxylate Dehydrogenase (PCDH) (EC 1.5.1.12)

Subunit Structure Uncertain

0.2 M Tris-HCl, pH 8.0	50 ml
L-pyroglutamic acid	0.05 g
10 mg/ml NAD	1 ml
5 mg/ml NBT	1 ml
5 mg/ml PMS	1 ml

This stain was modified from Mulley and Latter (1980).

Pyruvate Kinase (PK) (EC 2.7.1.40)

Tetramer   This stain may be prepared as an agar overlay.

0.2 M Tris-HCl, pH 8.0	50 ml
0.1 M MgCl$_2$	6 ml
adenosine 5'-diphosphate	0.03 g
D(+)-glucose	0.09 g
hexokinase	20 units
phospho(*enol*)pyruvate	0.02 g
G6PDH	60 NAD units
10 mg/ml NAD	1 ml
5 mg/ml NBT	1 ml
5 mg/ml PMS	1 ml

This stain was described by Buth and Murphy (1980) as modified from Brewer (1970). This stain may also resolve adenylate kinase gene products. A fluorescent stain that develops very rapidly was described by Harris and Hopkinsion (1976)

Retinol Dehydrogenase (RDH) (EC 1.1.1.105)

Subunit Structure Uncertain

retinol	0.05 g
acetone	3.5 ml

Then add to:

0.1 M phosphate buffer, pH 7.0	50 ml
10 mg/ml NAD	1 ml
5 mg/ml NBT	1 ml
5 mg/ml PMS	1 ml

This stain recipe was supplied by R. L. Garthwaite (personal communication).

Shikimate Dehydrogenase (SKDH) (EC 1.1.1.25)

Subunit Structure Uncertain

0.2 M Tris-HCl, pH 8.0	50 ml
shikimic acid	0.05 g
NADP	0.01 g

5 mg/ml NBT	1 ml
5 mg/ml PMS	1 ml

This stain was modified from Soltis et al. (1983).

Succinate Dehydrogenase (SUDH) (EC 1.3.99.1)

Monomer

0.1 M phosphate buffer, pH 7.0	50 ml
adenosine 5'-triphosphate	0.04 g
succinic acid	0.2 g
disodium EDTA	0.2 g
10 mg/ml NAD	3 ml
5 mg/ml NBT	1 ml
5 mg/ml PMS	1 ml

This stain was modified from Brewer (1970).

Superoxide Dismutase (SOD) (EC 1.15.1.1)

Dimer and Tetramer   This enzyme was known formerly as indophenol oxidase (IPO) or tetrazolium oxidase (TO). Mitochondrial and supernatant/cytosolic forms are known (Harris and Hopkinson, 1976). The mitochondrial form is a tetrameric manganoprotein, whereas the cytosolic form is a dimeric cuprozinc protein (Healy and Mulcahy, 1979).

0.2 M Tris-HCl, pH 9.0	50 ml
5 mg/ml NBT	1 ml
5 mg/ml PMS	1 ml

Incubate the gel slice exposed to light at ambient temperature or at 37°C. The enzyme appears as light bands on a dark background and is frequently resolved on other enzyme systems (e.g., G3PDH, PNP, ACOH). This stain is modified from Johnson et al. (1970). In some cases, resolution may be improved if this stain is applied as an agar overlay; replace the NBT with 1.0 ml of 5 mg/ml MTT in this instance. Siciliano and Shaw (1976) recommended the addition of 0.05 g MgCl$_2$·6H$_2$O in an SOD stain of 50 ml volume.

Thiosulfate Sulfurtransferase (TST) (EC 2.8.1.1)

Subunit Structure Uncertain   This system was previously refered to as rhodanase (RDS).

0.2 M Tris-HCl, pH 8.0	50 ml
potassium cyanide	0.012 g
sodium thiosulfate ($Na_2S_2O_3$)	0.5 g
5 mg/ml MTT	1 ml
5 mg/ml PMS	1 ml

This recipe was given to us by R. D. Sage (personal communication).

Triose-phosphate Isomerase (TPI) (EC 5.3.1.1)

Dimer   This stain may be prepared as an agar overlay (0.17 g agar in 8 ml buffer, stain components in remaining 2 ml buffer).

0.2 M Tris-HCl, pH 8.0	10 ml
dihydroxyacetone phosphate solution (see below)	1 ml
glyceraldehyde-3-phosphate dehydrogenase	800 units
arsenic acid (Na_2HAsO_4)	0.1 g
10 mg/ml NAD	3 ml
5 mg/ml NBT	1 ml
5 mg/ml PMS	1 ml

This stain was modified from Brewer (1970). Preparation of the dihydroxyacetone phosphate (DHAP) solution follows D. C. Morizot (personal communication). Rinse 5 g Dowex-50 resin (supplied with DHAP) at least four times in distilled water. Add 0.25 g DHAP and 5 g washed Dowex-50 resin to 10 ml water; swirl for 30 sec. Filter into graduated cylinder and add water to 25 ml. Incubate at 38–40°C for 4 hr. Adjust pH to 4.5 with $KHCO_3$. Freeze in 1-ml aliquots. An alternative procedure that does not involve the very expensive DHAP is described by Ayala et al. (1972). The latter procedure requires 2 hr of substrate preincubation and two pH adjustments for each stain preparation.

Tyrosine Aminotransferase (TAT) (EC 2.6.1.5)

Subunit Structure Uncertain   The stain should be prepared as an agar overlay (0.14 g agar in 10 ml H_2O, mix the stain components in the 15 ml buffer).

0.2 M Tris-HCl, pH 8.0	15 ml
L-tyrosine HCl (1% solution in 0.2 N HCl)	1 ml
α-ketoglutaric acid	0.01 g
adenosine 5'-diphosphate	0.01 g
pyridoxal 5-phosphate	0.01 g
L-glutamic dehydrogenase	100 units
10 mg/ml NAD	1 ml
5 mg/ml NBT	1 ml
5 mg/ml PMS	1 ml

This stain was modified from Aebersold et al. (1987) and D. E. Campton (personal communication).

Uridine Kinase (UK) (EC 2.7.1.48)

Monomer   This enzyme was known formerly as uridine monophosphate kinase (UMPK). This stain should be prepared as an agar overlay (0.14 g agar in 10 ml buffer and mix the stain components in remaining 10 ml buffer).

0.2 M Tris-HCl, pH 8.0	20 ml
$MgCl_2 \cdot 6H_2O$	0.1 g
uridine 5'-monophosphate	0.04 g
adenosine 5'-triphosphate	0.04 g
phospho(enol)pyruvate	0.01 g
NADH	0.01 g
potassium sulfate (K_2SO_4)	0.35 g
pyruvate kinase	40 units
L-lactic dehydrogenase	250 units

Monitor the development of expression under UV light (long wavelength). Enzyme activity is indicated by zones of defluorescence. This stain was modified from Harris and Hopkinson (1976).

Xanthine Dehydrogenase (XDH) (EC 1.1.1.204)

Monomer or Dimer   There is some question as to whether stains such as the following resolve XDH activity or that of xanthine oxidase (XO; EC 1.1.3.22) as an altered form of XDH (Richardson et al., 1986). Hypoxanthine is quite insoluble; Brewer (1970) recommended heating the substrate in the buffer and then adding the other components when the solution has cooled. Richardson et al. (1986) recommended suspending the hypoxanthine in acetone (0.04 g hypoxanthine per ml of acetone). Another method to increase the amount of substrate in solution requires the preparation of the following:

0.2 M Tris-HCl, pH 8.0	5 ml
0.1 M KOH	5 ml
hypoxanthine (see note below)	0.2 g

Stir this solution for at least 10 min, then add:

0.2 M Tris-HCl, pH 8.0	40 ml
10 mg/ml NAD	1 ml
5 mg/ml NBT	1 ml
5 mg/ml PMS	1 ml

Adjust the pH to 8.0, if necessary. Cover the gel slice with the stain solution; make sure that any undissolved hypoxanthine covers the gel slice. Incubate the gel slice at 37°C in the dark. Note that for some species, as much as 1 g of hypoxanthine may be required for optimal resolution. For some organisms, xanthine may be a more suitable substrate. This stain was modified from those of Brewer (1970) and Shaw and Prasad (1970).

Stain Fixing Solutions

Many stain fixing solutions have been used. These include the following:

1. 1:5:5 glacial acetic acid:methyl alcohol:water

   Caution must be exercised when handling fixed gel slices; do not breathe vapors. Do not soak (completely immerse) gels stained with MTT in this fixative or significant fading will result. This fixative is from Selander et al. (1971).

2. 50% ethanol (in water)

   This should not be used for the fixation of AAT and PER stains; see Soltis et al. (1973).

3. 50% glycerol (in water)

   This fixative may be preferred for gels stained using MTT as a dye to reduce fading. Soak gels for several hours before wrapping if the gels are to be saved. This method of fixation may result in the resolution of faint LDH isozymes (see also Soltis et al., 1983; Werth, 1985).

APPENDIX 2: BUFFERS AND TRACKING DYE FOR ISOZYME ELECTROPHORESIS

(Compiled by Donald G. Buth and Robert W. Murphy)

The importance of buffers has reemerged as a critical issue for the resolution of enzymes in starch gels. However, the search for optimal buffers often leads to a trade-off of cost versus quality. The more buffers needed for optimal resolution of a large array of enzymes, the higher the cost in both effort and funds. To illustrate both extremes, Harris and Hopkinson (1976) recommended a different buffer, or slight modification thereof, for almost every enzyme system listed, whereas Siciliano and Shaw (1976) "worked to develop just a few buffer systems which would give clear resolution of many proteins" and reported "clear resolution" of enzymes using only two buffers: Tris-citrate, pH 7.0, and Tris-borate-EDTA II, pH 8.0. Many laboratories have geared their operations toward the latter extreme, which has resulted in less than "clear resolution" for many enzymes of many taxa. We encourage the initial screening of enzyme systems using several of the buffer systems listed herein followed by comparisons using similar buffers, if necessary, for optimal resolution of enzymes. Additional buffers are listed by Brewer (1970), Selander et al. (1971), Clayton and Tretiak (1972), Harris and Hopkinson (1976), Steiner and Joslyn (1979), Shaklee and Tamaru (1981), Conkle et al. (1982), Soltis et al. (1983), Cheliak and Pitel (1984), Werth (1985), Micales et al. (1986), Selander et al. (1986), Shaklee and Keenan (1986), and Aebersold et al. (1987). Several of those listed for use in cellulose acetate electrophoresis by Richardson et al. (1986:153–154) may be applicable to starch gel work. The reader must remain aware that buffer formulas are usually derived empirically and additional modification should be encouraged.

Our buffer accounts include a descriptive name of the system, molarities of components in solution, exact gram measures of components in 1 liter equivalents, formulas for stock solutions as well as dilutions for electrode chambers and the gels, and references. We have resisted listing the electrical potential for each of these buffer systems, although many other compilations of buffer formulas provide such information. Among these, only Brewer (1970) identified correctly the fact that such potentials are related to the length of the gel mold and should be expressed as "volts per linear centimeter of gel." We find most published voltages to be at or beyond the high end of applicability and improved resolution (as well as lab planning) can be gained with electrophoretic runs for longer duration at lower voltages.

Electrophoresis Tracking Dye

Amaranth	0.01 g
Brilliant Blue G	0.01 g
Ethanol	2.5 ml
H_2O	7.5 ml

This stain was modified from a recipe formerly prepared by Gelman Sciences, Inc., Ann Arbor, Michigan. It gives both blue and red markers. It is not necessary to use nondenatured ethanol, although this may be preferable to keep methyl and isopropyl alcohol out of the gel.

Amine-citrate (Morpholine)

Stock solution:

(0.04 M) citric acid monohydrate	8.4 g/liter

Adjust to desired pH by adding ≈10–15

ml/liter of N-(3-aminopropyl)-morpholine
Electrode: Undiluted stock solution
Gel: 1:19 dilution of stock solution

These gels are hazardous and should be handled only with protective gloves. This buffer was described by Clayton and Tretiak (1972). Werth (1985), Shaklee and Keenan (1986), and Aebersold et al. (1987) recommended its use at pH 6.1, 6.0, and 7.0, respectively. Its range of use may be pH 6.0–8.0 (D. E. Campton, personal communication). Aebersold et al. (1987) suggested the inclusion of 0.01 M EDTA in the stock solution.

Amine-citrate (Propanol)

Stock solution:

(0.04 M) citric acid monohydrate	8.4 g/liter

Adjust to the desired pH by adding ≈10–15 ml/liter 1,3-bis(dimethylamino)-2-propanol
Electrode: Undiluted stock solution
Gel: 1:19 dilution of stock solution

This buffer was described by Clayton and Tretiak (1972). It may be optimal at pH ≈7.5.

Borate (Continuous), pH 8.6

Stock solution:

(0.25 M) boric acid (H_3BO_3)   15.5 g/liter

Adjust to pH 8.6 with NaOH (pellets)
Electrode: Undiluted stock solution
Gel: 1:9 dilution of stock solution

This buffer was modified from Sackler (1966) who used it at pH 8.8.

Borate (Discontinuous)

Electrode:

(0.30 M) boric acid (H_3BO_3)   18.6 g/liter
(0.03 M) sodium chloride   1.75 g/liter
  (NaCl)

Adjust to pH 8.0 with 2 N NaOH

Gel:

(0.02 M) boric acid (H_3BO_3)   1.24 g/liter

Adjust to pH 8.6 with 2 N NaOH
This buffer was described by Brewer (1970).

Histidine-citrate (Discontinuous)

Electrode:

(0.41 M) citric acid dihydrate   120.6
  (Trisodium salt)   g/liter

Adjust to pH 7.0 or 8.0 with HCl.
Gel:

(0.005 M) L-histidine HCl   1.05 g/liter
  monohydrate

Adjust to pH 7.0 or 8.0 with NaOH
This buffer was described by Fildes and Harris (1966), Brewer (1970), and Harris and Hopkinson (1976).

Lithium-borate/Tris-citrate

Stock solution A:

(0.19 M) boric acid (H_3BO_3)   11.8 g/liter
(0.03 M) lithium hydroxide   1.26 g/liter
  (LiOH·H_2O)

Adjust to pH 8.1

Stock solution B:

(0.05 M) Tris   6.06 g/liter
(0.008 M) citric acid   1.68 g/liter
  monohydrate

Adjust to pH 8.4
Electrode: Undiluted stock solution A
Gel: 1:9 mixture of stock solutions A:B, final pH 8.3

This discontinuous buffer is the "lithium hydroxide" buffer described by Selander et al. (1971), which is a slight modification of that described by Ridgway et al. (1970). This buffer will often cause gels to separate as the citric acid buffer front passes through the origin.

Phosphate-citrate

Stock solution:

(0.214 M) potassium   37.3 g/liter
  phosphate dibasic
  (K_2HPO_4)
(0.027 M) citric acid   5.67 g/liter
  monohydrate

Adjust to pH 7.0
Electrode: Undiluted stock solution
Gel: 1:25 dilution of stock solution
This buffer was modified from Selander et al. (1971) who called for 0.214 M potassium phosphate dibasic (anhydrous) but provided the measurement (grams/liter) for this molarity of potassium phosphate monobasic (KH_2PO_4). If the latter is used, use 29.1 g/liter. Harris and Hopkinson (1976) described a similar buffer using 0.245 M sodium phosphate monobasic ($NaH_2PO_4·H_2O$) and 0.15 M citric acid monohydrate adjusted within the range of pH 5.9–7.5; dilute the stock 1:39 for the gel.

Tris-borate-EDTA I

Stock solution:

(0.90 M) Tris	109 g/liter
(0.50 M) boric acid (H_3BO_3)	30.9 g/liter
(0.02 M) disodium EDTA	6.7 g/liter

Adjust to pH 8.6 with NaOH (pellets)

Electrode:

Anode: 35 ml stock solution + 215 ml H_2O
   (1:6 dilution)
Cathode: 50 ml stock solution + 200 ml
   H_2O (1:4 dilution)

Gel: 1:19 dilution of stock solution This is a modification of the buffer described by Boyer et al. (1963) referred to as EBT by Wilson et al. (1973) and Shaklee and Keenan (1986), and as TBE by Aebersold et al. (1987). Shaklee and Keenan (1986) recommended the use of 7.4 g/liter tetrasodium EDTA in this buffer.

Tris-borate-EDTA II

Stock solution:

(0.50 M) Tris	60.6 g/liter
(0.65 M) boric acid (H_3BO_3)	40.2 g/liter
(0.02 M) disodium EDTA	6.7 g/liter
(= "versene")	

Adjust to pH 8.0
Electrode: Undiluted stock solution
Gel: 1:9 dilution of stock solution

This is the TVB (Tris-versene-borate) buffer of Selander et al. (1971) and Siciliano and Shaw (1976). Another version of this buffer is described by Brewer (1970): 0.21 M Tris, 0.15 M boric acid, and 0.006 M disodium EDTA adjusted to pH 8.0 for the electrode and 21 mM Tris, 20 mM boric acid, and 0.68 mM disodium EDTA adjusted to pH 8.6 for the gel. Werth (1985) described another system, termed "salamander B" and attributed to S. I. Guttman, which uses the same stock solution and concentrations for the electrode and gel buffers: 84 mM Tris, 7.9 mM boric acid, and 0.86 mM disodium EDTA adjusted to pH 9.1 with HCl.

Tris-borate-EDTA-lithium

Stock solution:

(0.9 M) Tris	109 g/liter
(0.4 M) lithium hydroxide (LiOH·H_2O)	16.8 g/liter
(0.5 M) boric acid (H_3BO_3)	30.9 g/liter
(0.1 M) EDTA free acid	29.2 g
H_2O	bring to 1000 ml

Adjust to pH 9.1 with NaOH

Electrode:

stock solution	225 ml
sucrose	100 g
H_2O	bring to 2000 ml

Gel:

stock solution	40 ml
sucrose	50 g
H_2O	bring to 1000 ml

This buffer system was described by Turner (1973).

Tris-citrate II

Stock solution:

(0.687 M) Tris	83.2 g/liter
(0.157 M) citric acid monohydrate	33.0 g/liter

Adjust to pH 8.0
Electrode: undiluted stock solution
Gel: 1:29 dilution of stock solution

This is the "continuous Tris-Citrate II" buffer of Selander et al. (1971). Soltis et al. (1983) listed other modifications of the Tris-citrate buffers of Shaw and Prasad (1970) including (1) 0.135 M Tris plus 0.032 M citric acid, pH 8.0, diluted 1:14 for the gel, (2) 0.135 M Tris plus 0.017 M citric acid, pH 8.5, diluted 1:14 for the gel, (3) 0.223 M Tris plus 0.086 M citric acid, pH 7.5, diluted 1:27.5 for the gel, and (4) 0.223 M Tris plus 0.069 M citric acid, pH 7.2, diluted 1:27.5 for the gel. Other variations include 0.13 M Tris plus

0.043 M citric acid, pH 7.0, diluted 1:14 for the gel (Siciliano and Shaw, 1976), 0.094 M Tris plus 0.0235 M citric acid, pH 8.6, diluted 1:5 for the gel (Harris and Hopkinson, 1976), and 0.22 M Tris (27 g/liter) plus 0.086 M citric acid monohydrate (18.1 g/liter), pH 5.8, diluted 1:27.5 for the gel (Shaklee and Keenan, 1986).

Tris-citrate III

Stock solution:

(0.75 M) Tris	90.8 g/liter
(0.25 M) citric acid monohydrate	52.5 g/liter

Adjust to pH 7.0 with NaOH (pellets)

Electrode:

Anode: 35 ml stock solution + 215 ml H_2O (1:6 dilution)
Cathode: 50 ml stock solution + 200 ml H_2O (1:4 dilution)

Gel: 1:19 dilution of stock solution

This buffer was described by Whitt (1970) and Rainboth and Whitt (1974).

Tris-citrate/borate (pH 8.7)

Electrode:

(0.30 M) boric acid (H_3BO_3)	18.6 g/liter
(0.06 M) sodium hydroxide (NaOH)	2.4 g/liter

Adjust to pH 8.2

Gel:

(0.076 M) Tris	9.21 g/liter
(0.005 M) citric acid monohydrate	1.05 g/liter

Adjust to pH 8.7

This is the "discontinuous Tris-citrate" or "Poulik" buffer described by Selander et al. (1971). This buffer will often cause gels to separate as the borate buffer front passes through the origin. Chippindale (1989) modified this system by using 26.2 g/liter Tris to bring the pH of the gel buffer to 9.5.

Tris-citrate-EDTA

Stock solution:

(0.135 M) Tris	16.4 g/liter
(0.045 M) citric acid monohydrate	9.46 g/liter
(1.3 mM) disodium EDTA	0.44 g/liter

Adjust to pH 7.0
Electrode: Undiluted stock solution
Gel: 1:14 dilution of stock solution

This buffer is modified from Avise et al. (1975). A modification of this system using different molarities of the components (pH 5.7–8.6) and limiting the disodium EDTA to the gel buffer was described by Harris and Hopkinson (1976).

Tris-EDTA

Stock solution:

(0.1 M) Tris	12.1 g/liter
(4.5 mM) disodium EDTA	1.5 g/liter

Adjust to pH 9.6
Electrode: Undiluted stock solution
Gel: 1:9 dilution of stock buffer

This buffer was described by Harris and Hopkinson (1976).

Tris-HCl

Electrode:

(0.30 M) boric acid (H_3BO_3)	18.6 g/liter
(0.06 M) sodium hydroxide (NaOH)	2.4 g/liter

Adjust to pH 8.2
Gel:

(0.01 M) Tris	1.21 g/liter

Adjust to pH 8.5 using concentrated HCl

This discontinuous buffer was described by Selander et al. (1971). Harris and Hopkinson (1976) recommended a continuous buffer modification using 0.1 M Tris-HCl with a 1:4 dilution for the gel or 0.3 M Tris-HCl with a 1:14 dilution for the gel. They recommended the use of this buffer in the range of pH 8.6–9.6.

Tris-maleate-EDTA

Stock solution:

(0.10 M) Tris	12.1 g/liter
(0.10 M) maleic acid	11.6 g/liter
(0.01 M) disodium EDTA	3.36 g/liter
(0.01 M) MgCl$_2$·6H$_2$O	2.03 g/liter

Adjust to pH 7.4 with NaOH

Electrode: Undiluted stock solution
Gel: 1:9 dilution of stock solution

This buffer was described by Brewer (1970) and Selander et al. (1971) at pH 7.6 and 7.4, respectively. Brewer (1970) noted that some of the reagents (e.g., EDTA) will not go into solution until the NaOH is added. Harris and Hopkinson (1976) recommended the use of this buffer in the range pH 6.5–7.4.

CHAPTER 5

PROTEINS II: IMMUNOLOGICAL TECHNIQUES

Linda R. Maxson and R. D. Maxson

PRINCIPLES AND COMPARISON OF METHODS

General Principles

Comparative immunological techniques have been used for phylogenetic inference since the pioneering work of Nuttall (1904). The reasons why the technique worked were not understood at that time. Today immunological methods are better understood, and specialized techniques have been developed to allow either qualitative or quantitative estimates of amino acid sequence differences between homologous proteins.

In both qualitative and quantitative immunological methods, **antibodies** are produced to an **antigen** using protocols that vary as to the purity of the antigens and the immunization schedule. The degree of reactivity between the antibodies and antigens from different species is then measured using various direct or indirect properties of the antibody–antigen binding process. The results depend on the **affinity** and **specificity** of the antibodies as well as how accurately the methods can measure the antibody affinities and specificities relative to different antigens.

When a foreign antigen is injected repeatedly into an animal (typically a rabbit), the animal responds by producing a spectrum of antibodies having varying binding affinities and specificities to the foreign material. After an appropriate immunization period, the animal is bled and the resultant serum contains the antibodies used in immunological studies. The degree of antibody binding affinity and specificity can vary with immunization protocol from low-affinity and narrow-specificity (low-diversity) antibodies that will recognize only major **antigenic sites**, to very high-affinity and broad-specificity antibodies capable of detecting single amino acid replacements within numerous antigenic sites. The antibodies raised against antigen X_A from species A can be used to test relative cross-reactivities of antigen X_A (the **homologous** antigen) to antigens X_{B-Z} from a series of related species B through Z (the **heterologous** antigens). The degree of immunological cross-reaction obtained with the heterologous antigens relative to that obtained with the homologous antigen can be used as a general measure of the genetic relationship between the species. Whether this measure is an estimate of sequence difference or is a measure of the binding avidity of the dominant antigenic site must be evaluated for each technique. The various immunological methods differ primarily in the manner in which antisera are produced and the means by which immunological cross-reactions are measured. As a rule, prolonged immunization with several antigen challenges increases the precipitating ability of the antiserum, increases its sensitivity to differences in the antigens, and broadens the antibody diversity (Prager and Wilson, 1971b; Cocks and Wilson, 1972).

In this chapter we provide a nominal introduction to several popular immunological techniques, but have emphasized the powerful and cost-effective quantitative **microcomplement fixation** (MC'F) assay.

Assumptions

In all of the assays it is assumed that the stronger the immunological reaction observed, the more similar the proteins. The relationship between immunological reactivity and sequence differences has been determined only for MC'F (Maxson and Maxson, 1986). Other immunological techniques have been shown to exhibit a general correlation with MC'F distances. The degree to which such correlations introduce additional errors to estimates of divergence is unknown; even a moderate variance in the correlation can compromise estimation of phylogeny. For example, the decay in binding energy of a single, major antigenic site as it incurs amino acid substitutions may be correlated with overall divergence, but it will not be linear with overall protein divergence, nor will the 5–10 amino acids comprising the antigenic site be representative of the entire protein.

Comparison of the Primary Methods

Precipitin and Immunodiffusion Tests   Nuttall (1904) introduced the precipitin test as a means of comparing members of different families of the animal kingdom and demonstrated the potential of the field of systematic serology (Boyden, 1942, 1964). Precipitin tests rely on mixing antibody and antigen in solution and measuring the resultant antibody–antigen precipitate (Klein, 1982). In general, the greater the amount of precipitate, the more closely related the two species.

Antisera used in these tests are usually produced over a relatively short period of time (4–6 weeks). Low-affinity, narrow-specificity antibodies

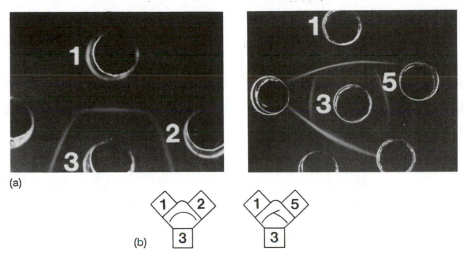

FIGURE 1   (a) Ouchterlony gels with anti-albumin antibody to species L in well 3 and plasma from species L in wells 1 and 2. Plasma from species M is in well 5. The line of identity between wells 1 and 2 indicates no detectable difference in the antigen. The spur with species L over M is due to numerous antigenic differences between albumins of species L and M (see Klein, 1982, for details). (b) Trefoil immunodiffusion, with antibody and plasma as in (a).

predominate in such antisera (Klein, 1982). The turbidity caused by the precipitating antibody–antigen complex initially was titered "by eye" in a series of serial dilution tests. More quantitative measurements of the amount of precipitate formed were obtained with the photronreflectometer (Libby, 1938). Subsequent variations and improvements were made on the precipitin tests, including diffusion of the antibody and antigen in agar (Ouchterlony, 1958). This eliminated the need for a spectrophotometer since the precipitate was visible in the clear agar (Figure 1a). Prager and Wilson (1971b) showed an excellent correlation between results obtained in precipitin assays and in quantitative MC'F tests. Using the same antisera, they concluded that MC'F was better able to discriminate between closely related lysozymes than precipitin tests, and that MC'F was more economical of material.

Immunodiffusion on trefoil plates was introduced by Goodman and Moore (1971) as an improvement over standard Ouchterlony immunodiffusion methods. These authors attempted to quantify degree of species relatedness from the size, position, and intensity of the precipitin lines and/or spurs formed by precipitating antigen and antibody complexes in agar (Figure 1b). The trefoil design had the advantage of concentrating antisera in a small volume of agar as well as permitting the simultaneous comparison of three species (antigens).

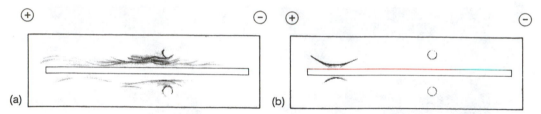

FIGURE 2   (a) Immunoelectrophoresis using antibody to whole plasma from species L in the center trough and whole plasma from species L in the top well and whole plasma from species M in the bottom well. (b) Immunoelectrophoresis testing for purity of the albumin preparation of *Hyla regilla*. Antisera to purified albumin is in the center trough and whole plasma is in the top well and the albumin preparation is in the bottom well. Had the antibody been made to more than just albumin, numerous arcs would have appeared.

Immunoelectrophoresis Immunoelectrophoresis compares a number of proteins of two different species by a combination of electrophoretic separation of the proteins followed by antibody precipitation. Plasma from each species is electrophoresed on an agar gel, followed by precipitation of antigen–antibody complexes (precipitin arcs) with antisera. Electrophoresis separates plasma proteins based on charge and size, and the antibodies recognize and precipitate proteins that are sufficiently similar to those of the species to which the antibodies were produced. The gel is then rinsed and the precipitin arcs are stained with a general protein stain to visualize all antibody–protein complexes (Figure 2a). In some instances a protein-specific staining method may be used (histochemical stains for esterases or radioactive iron for transferrin). The number of arcs is counted for both species and is used as a measure of species similarity (Frair, 1964; Minton and Salanitro, 1972). Generally, the identity of the specific antigens that are represented by precipitin arcs is not known, and in many cases arcs overlap and make individual protein discrimination difficult. Immunoelectrophoresis is especially valuable for testing the purity of an antigen and in determining whether or not the antisera is directed solely against a single antigen (Figure 2b).

Radioimmunoassay Radioimmunoassay (RIA) can measure nanogram to picogram amounts of protein and was the first molecular technique to extract useful genetic information from fossil proteins (Lowenstein, 1985a). Serial dilutions of antigen are bound to polyvinyl microtiter plate cups that are then washed and saturated with a solution of chicken egg protein. Rabbit antisera (to one protein, or to whole serum) are added and allowed to react with bound antigen for 24 hr. Unbound antisera is washed out and ^{125}I-labeled goat anti-rabbit γ-globulin (GARGG) is added to bind any rabbit antisera adhering to the microtiter cups. Unbound GARGG is washed out and radioactivity is measured in a scintillation counter. The amount of radioactivity in each microtiter cup is used as a measure of the antibody–antigen reaction and most likely measures the binding affinities of the strongest antigenic site. Refinements of the RIA involving competitive binding techniques can improve specificity (Lowenstein et al., 1981). RIA requires radioactive material, and is more complex and costly than other immunological assays. However, it can be used with much smaller amounts of protein.

Microcomplement Fixation MC′F measures reactions between a soluble antigen and antibodies in dilute solution under conditions in which only high-affinity antibodies react. Broad-specificity (high-diversity) antisera are prepared to a single purified antigen and are used to estimate the number of unmodified antigenic sites and thereby estimate the sequence differences between homologous proteins from different species. MC′F data can be obtained for two proteins until they differ in their amino acid sequences by roughly 35%. However, sequence differences can be estimated only until no identical antigenic sites remain in common (Maxson and Maxson, 1986), and this depends on the antibody diversity. For albumin, a large protein with at least 25 antigenic sites (Benjamin et al., 1984), this horizon is between 20 and 30% sequence divergence.

Homologous reactions are standardized and those with heterologous antigens are measured relative to the homologous reaction. In this way, antigens from all species are compared and the results converted to the same units of **immunological distance** (ID).

Complement plays a dual role in MC′F; it is a sensitive measure of the amount of antibody that is bound to antigen, and it assays this reaction by lysing sensitized sheep red blood cells. The fixation of complement requires tightly bound antibody and antigen. By working with very dilute concentrations of antisera and antigen, only high-affinity antibodies will be bound. Altered antigenic sites (those with amino acid replacements, AAR) do not bind antibody and are excluded from the antibody–antigen reaction. No albumin is excluded, however, because it has sufficient antigenic sites so that it does not matter if some of the sites are excluded. The fraction of antibody that is excluded causes a proportional drop in complement fixation because the complement only binds to the bound antibody. The number of excluded antigenic sites (Figure 3) is measured by monitoring the increase in antisera concentration needed to obtain equivalent "fixation" of complement (Figure 4a). In other words, all MC′F reactions, including heterologous reactions, are equivalent in having the same amount of bound complement, bound antigen, and bound antisera. The amount of extra antisera needed to compensate for the excluded antibody allows estimation of the number of excluded antigenic sites (Table 1; modified from Maxson and Maxson, 1986).

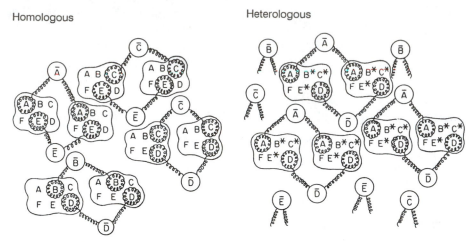

FIGURE 3   Antibody assortment-exclusion process. Albumin has at least 25–33 antigenic sites. Altered sites (*) are excluded in the heterologous reaction, as indicated by the antibodies ($\bar{A}$–$\bar{E}$) to altered sites *not* participating in the antibody–antigen reaction. In MC'F, complement is trapped in the lattices formed by reacting antibody and antigen.

Experimentally, antisera (antibodies) and serum (antigen) are combined in dilute solution in the presence of a precise amount of complement (see below). The reaction is incubated on ice overnight, during which time complement becomes bound as antibody–antigen lattices are formed. The next day the degree of antibody–antigen reaction is monitored by assaying any unbound complement. The amount of bound complement serves to

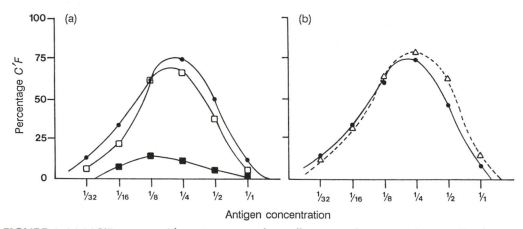

FIGURE 4   (a) MC'F curves with antisera to *Hyla regilla* (titer: 1/6,000) and *H. regilla* plasma (•) and with *H. cadaverina* plasma at titers of 1/6,000 (■) and 1/4,500 (□). The ID between these species is 100 log (6,000/4,500) = 12.5. (b) MC'F curves with *H. regilla* antisera at a titer of 1/6,000, using purified albumin as the antigen (Δ) and using whole plasma as the antigen (•).

Table 1. Derivation of the relationship between immunological distance (ID) and amino acid replacements (AAR) in antigenic sites of proteins

Definitions:

1. R = Heterologous antibody titer/homologous antibody titer
2. $ID = 100 \log R$ (Log is base 10)
3. $1 = e^{-x} + xe^{-x} + (x/2!)e^{-x} + \cdots$ (Poisson distribution)
4. S = number of antigenic sites (25 − 50 for albumin)
5. $X = AAR/S$ (the fraction of AAR per antigenic site)

Therefore:

$1 - 1/R$ is the measured fraction of excluded antigenic sites
$1 - e^{-AAR/S}$ is the theoretical fraction of excluded antigenic sites

Setting these two equations equal and substituting:

$ID = (43/S)AAR$

normalize the homologous and heterologous reactions, and the amount of antisera needed to obtain equivalent complement fixation allows an estimate of the degree of sequence similarity between the homologous proteins of the two species being compared.

Advantages of Each Technique In all the above techniques, the degree of similarity of the antigenic sites of homologous proteins is measured. As most studies use short-term antisera, a single highly conserved antigenic determinant can dominate the results. This is particularly true for many precipitin and RIA analyses. This greatly limits the amount of genomic information that can be measured since single antigenic sites may consist of as few as 5–10 amino acids. There is evidence that RIA results can be correlated with MC'F ID data over portions of their range of applicability (Prager et al., 1980; Lowenstein et al., 1981). The RIA reaction, however, differs significantly from MC'F in two respects. At least for comparisons of distantly related taxa, RIA is performed at higher concentrations of antisera where low-affinity antibodies can be encouraged to bind, whereas MC'F is done at very low antisera concentrations where all but the highest affinity antibodies can be prevented from binding. Since RIA uses the antigenic sites having the highest binding affinity, it can use low-specificity antisera, whereas MC'F requires high-specificity antisera to obtain usable results. The MC'F assay has been shown to utilize a large number of antigenic sites and to effectively "count" the number of sites that are eliminated by AAR. MC'F has the advantage of being rapid, relatively easy to perform, and extremely cost effective. Thus, for many applications, MC'F is better able to discriminate among closely related antigens and has the potential to estimate amino acid sequence differences between homologous proteins (Benjamin et al., 1984; Maxson and Maxson, 1986).

APPLICATIONS AND LIMITATIONS

Precipitin Tests

These immunological assays have been applied to numerous studies of phylogenetic relationship among plants and animals. Antisera are most commonly raised in rabbits over a 3- to 6-week period. Serum, tissue extracts, or specific proteins are used as antigens. Relatively large amounts (roughly 20 times that used in MC'F) of both antiserum and antigen are required for these assays, making it difficult to study small species or species from which only small samples can be obtained. Pairwise comparisons are completed and numerical estimates of difference are obtained from the amount of precipitate formed by the antibody reactions. All measurements are made relative to a homologous reaction.

Nuttall (1904) summarized results from 16,000 precipitin tests and Boyden (1942) evaluated all of this work in a treatise on systematic serology. In an attempt to provide objective data on the question of which invertebrate group is the sister group of vertebrates, Wilhelmi (1942) used precipitin tests making antibody to tissue extracts of echinoderms, annelids, arthropods, and chordates. He concluded that the echinoderms had the closest phylogenetic affinity to the chordates. However, it is unlikely that short-term antisera and precipitin tests can discriminate branching patterns at such great distances. Additional work with the photronreflectometric precipitin technique has been reported by Cei and his students, primarily studying amphibians (Cei, 1972) and reptiles (Cei and Castro, 1973). Prager and Wilson (1971b) have used quantitative precipitin assays to compare seven different bird egg white lysozymes, Sarich (1985) has studied rodent relationships, and Joger (1985) has examined lizard relationships using quantitative precipitin analysis.

Immunodiffusion and Immunoelectrophoresis

Extensive work on systematic relationships among animals and plants has been performed using the **Ouchterlony** immunodiffusion method (Ouchterlony, 1958). Whereas early studies of mammals (primates, in particular) used antisera made to plasma or serum, this was later refined to making antisera to a single purified protein, such as albumin or transferrin. Goodman and his students examined mammalian relationships using modifications of this approach (Baba et al., 1975; Dene et al., 1978). Studies of phylogenetic relationships among reptiles and crocodilians subsequently were carried out by Dessauer and his students (Schwaner and Dessauer,

1982; Densmore, 1983). Some early studies were done on plant material (Fairbrothers and Johnson, 1964).

In all instances, measures of similarity were derived based on interpretation of degree of antibody–antigen reactivity. Major limitations are the degree of subjectivity in interpreting patterns and degrees of cross-reaction. Attempts to objectively evaluate reactions have been made (Goodman and Moore, 1971), but in general one still is left with a subjective, qualitative measure of species resemblance. Moreover, if appropriate long-term antisera have been made, MC'F usually is better able to discriminate among very similar antigens and we have found that MC'F provides more data by enabling comparisons of species over a greater range of genetic distances (L. Maxson, unpublished data).

Immunoelectrophoresis has been used for systematic work in a variety of taxa. Application of this technique can be found in groups as diverse as insects (Stephen, 1974), snakes (Minton and Salanitro, 1972), and turtles (Mao and Chen, 1982). Antisera are produced to whole serum and the patterns of relationship determined from this multiple protein approach are generally concordant with those based on other biochemical analyses. However, the complex patterns seen in immunoelectrophoresis (Figure 2a) typically require subjective judgments and allow only qualitative assessment of species relationships.

Radioimmunoassay

In 1980, Lowenstein used an adaptation of a solid-phase double-antibody RIA method to detect albumin in a small piece of tissue from a 40,000-year-old Siberian mammoth (Prager et al., 1980). Since that time, RIA has been used to probe phylogenetic affinities of recent fossil material and to help resolve systematic questions concerning extinct and fossil species (Lowenstein et al., 1981; Lowenstein, 1985b). Lowenstein and his collaborators have used both standard RIA and competitive binding modifications of RIA to improve specificity. They report a high correlation between the RIA distance and MC'F ID when using the same antisera in both assays (Lowenstein et al., 1981). Although other immunological techniques can be used to compare some recent fossil material (Prager et al., 1980), RIA is the only assay that can work with degraded proteins at relatively large distances (Shoshani et al., 1985). RIA can continue to give measurable cross-reactions even when only a single modified antigenic determinant is present. It is this feature that allows RIA to extract information from fossil material, such as museum skins of the recently extinct Tasmanian wolf and the quagga (Lowenstein, 1985a).

Microcomplement Fixation

Early immunological methods provided qualitative estimates of overall protein similarity, which were interpreted as direct reflections of the underlying genome. Within certain limits, MC'F can also provide quantitative data that directly estimate amino acid replacements (AAR) between homologous proteins (Table 1).

Basically, MC'F is a rapid method for comparing sequence differences between homologous proteins (Prager and Wilson, 1971a; Wilson et al., 1977; Maxson and Maxson, 1986). Serum albumin has been the preferred molecule for MC'F study among the vertebrates because it is a monomeric protein, the product of a single gene, large (590+ amino acids; Benjamin et al., 1984), highly **immunogenic** , and evolves at a rate that is well suited to studies of vertebrate phylogeny. Albumin also is easily purified, relatively stable, and sufficiently abundant in vertebrates so that 0.5–1.0 ml of plasma usually contains enough albumin to produce sufficient antisera for thousands of comparisons. Enough albumin for several comparisons can be obtained from as little as 10 μl of plasma or a few milligrams of muscle. This enables albumin sampling (without killing) of rare or endangered species (Maxson and Szymura, 1984; Hutchinson and Maxson, 1987a). For example, MC'F can reliably compare albumins until they differ at about 20–30% of their sequence, corresponding to an estimated difference of 120–180 AAR between each pair of species studied (Maxson and Maxson, 1986). This is somewhat misleading, however, since MC'F counts only those AAR that prevent an antigenic site from participating in the MC'F reaction. As albumin has from 25–50 major antigenic sites (Benjamin et al., 1984; Maxson and Maxson 1986), MC'F can directly monitor only that many substitution events. For albumins evolving at roughly a 1% difference every 3 million years (Wilson et al., 1977), this still provides a sufficient data base for phylogenetic studies of species divergences throughout the Cenozoic and late Cretaceous. The range of divergence times over which MC'F can be used may vary among groups according to the rate of albumin evolution.

Among the first applications of MC'F to evolutionary questions were studies of lactate dehydrogenase in representative amphibians (Salthe and Kaplan, 1966) and egg-white lysozymes in birds (Arnheim et al., 1969). The first studies using antibody to highly purified serum albumin were the then controversial analyses of the close relationship of humans and chimpanzees by Sarich and Wilson (1966, 1967). Since that time, extensive phylogenetic work has been accomplished with MC'F analyses of a wide variety of organisms (bacteria, plants, and animals), using a variety of proteins.

Prokaryotes Understanding of the phylogenetic relationships among the prokaryotes has been difficult due to a paucity of reliable morphological characters across diverse taxa. A few MC'F analyses of bacteria have used alkaline phosphatase (Cocks and Wilson, 1972; Steffen et al., 1972) or azurin (Champion et al., 1980) as the antigen.

Plants Relatively little MC'F work has been accomplished in plants with the exception of studies with plastocyanins (Wallace and Boulter, 1976). A primary deterrent in this group may be the lack of a widespread molecule that is abundant, easily purified, stable, antigenic, and evolves at a suitable rate for MC'F comparisons.

Insects MC'F studies among insects were initiated by Fink and Brosemer (1973), who produced rabbit antibodies to glycerol 3-phosphate dehydrogenase to study the relationships among honeybees and bumblebees. Subsequent MC'F analyses involved work on *Drosophila* with glycerophosphate dehydrogenase (Collier and MacIntyre, 1977), acid phosphatase-1 (MacIntyre et al., 1978), and larval protein (Beverley and Wilson, 1982, 1985).

Vertebrates A wealth of systematic studies among all major vertebrate lineages has been carried out using MC'F. Early studies are described in Leone (1964) and Wright (1974). Subsequent studies of bird evolution have involved work with lysozyme, ovalbumin, and ovotransferrin (Prager and Wilson, 1976). Mammalian evolution has been probed extensively by Sarich and colleagues using both albumin and transferrin (Cronin and Sarich, 1975; Sarich and Cronin, 1976; Pierson et al., 1986). Studies of sciurids (Maxson et al., 1981), Australian rodents (Baverstock et al., 1986), and cetaceans (Lint et al.,1988) have also been reported. Reptilian phylogeny has been studied by Gorman et al. (1971); there have also been analyses of snake phylogeny (Dowling et al., 1983; Mao et al., 1983; Schwaner et al., 1985; Dessauer et al., 1987; Cadle, 1988), turtle relationships (Chen et al., 1980; Mao et al., 1987), and lizard phylogeny (Gorman et al., 1980; Shochat and Dessauer, 1981). To date, little work has been done on albumin evolution in fish (Davies et al., 1987). Extensive MC'F analyses of relationships among the Amphibia include studies on salamanders (Maxson et al., 1979; Busack et al., 1988), frogs (for examples see Maxson, 1984; Hutchinson and Maxson, 1987b and references therein), and caecilians (Case and Wake, 1977).

LABORATORY SETUP

Standard laboratory equipment for immunological analyses are indicated in Table 2, and specialized chemical reagents in Table 3. Access to approved animal care facilities and a reliable supplier of laboratory rabbits are essential. We prefer New Zealand White rabbits as they are easy to handle and produce sufficient antisera for foreseeable needs. In our experience, females are easier to work with and produce higher titer antisera than males.

Table 2. Laboratory equipment for immunological analysis[a]

Equipment	Use (protocols in parentheses)
Autoclave	Making Alsever's solution (4)
Balance, analytical	Weighing small samples*
Balance, top-loading	General weighing*
Cadmium-coated test tube racks for 24 and 40 ml (29 × 116 mm) centrifuge tubes	MC'F experiments (4)
Centrifuge, clinical tabletop with swinging bucket rotors for 48 (13 × 100 mm) tubes	MC'F experiments (4)
Centrifuge, microtube	Centrifuging small samples*
Centrifuge, high-speed refrigerated rotors for 40 ml round-bottom tubes	Antibody production (2)
Cold room (0–4°C)	Reagent storage *; MC'F (4)
Cornwall automatic pipettes: 2 ml and 5 ml	MC'F setup (4)
Cylindrical polyacrylamide gel electrophoresis unit	Antigen purification (1)
Dishpans to hold 1 set of MC'F tubes	MC'F experiment (4)
Drying ovens	Drying glassware *
Electrophoresis power supply	Antigen purification (1); Immunoelectrophoresis (3)
Freezer, upright; household	Storing antisera and antigens*
Freezer, ultracold	Long-term antisera storage*
Ice machine	MC'F setup (NE)
Immunoelectrophoresis gel trays, knives, and setup	Immunoelectrophoresis (3)
Luerlok glass syringes (5 ml)	Antibody production (2)
pH meter, with Tris electrode	Buffer preparation*
Pipettes, volumetric and serological	MC'F setup (4)
Rabbit care facilities	Antibody production (2)
Rabbit restraining cage	Immunization (2)
Refrigerator (0–4°C)	Reagent storage*
Scintillation counter	RIA (3)
Spectrophotometer (with sipper)	Reading MC'F experiments (4)
UV light	Antigen purification (1)
Water bath 45 × 90 cm	MC'F experiment (4)
40 ml Pyrex round-bottomed centrifuge tubes (29 × 116 mm)	MC'F experimental tubes (4)

[a]Not all items listed are needed for some of the applications described in this chapter. Therefore, the protocols for which each item is needed are listed; items that are essential for most protocols are marked with an asterisk (*). Items that are nonessential but facilitate a procedure are marked NE.

Table 3. Specialized reagents

Reagents	Use (protocols in parentheses)
Bovine serum albumin (fraction V)	MC'F (4)
Chicken ovalbumin	RIA (3)
Freund's complete adjuvant	Antibody production (2)
Freund's incomplete adjuvant	Antibody production (2)
Guinea pig complement	MC'F: setup (4)
Hemolysin (rabbit anti-sheep red blood cell antisera)	MC'F: preparing sensitized sheep red blood cells (4)
Heparin	Antibody production (2)
^{125}I-labelled goat anti-rabbit γ-globulin	RIA (3)
Ionagar	Immunodiffusion; immuno-electrophoresis (3)
Isobutyl alcohol	Antigen purification (1)
Isotris and Isosatris buffers	MC'F: setup and making cells (4)
Pentobarbital	Antibody production (2)
Phenoxyethanol	MC'F (4)
Polyoxyethylene 23-lauryl ether	Antigen purification (1)
Sheep red blood cells	MC'F: reading experiments (4)

For MC'F (Protocol 4), special tubes must be prepared for use in all experiments. Forty-eight pyrex, round-bottomed, 40-ml centrifuge tubes (29 × 116 mm with pour spout) and two cadmium-coated metal racks, each of which holds four rows of six tubes, constitute a "set" of MC'F tubes. Condition new tubes by soaking overnight in acid dichromate cleaning solution. Rinse repeatedly and then fill tubes with 1 N HCl and soak for 2 hr. Rinse thoroughly with deionized water and dry (upside down in 180°C oven). Run complement controls (Protocol 4, parts E–G) in all tubes twice. Properly conditioned tubes will all read the same at the end of an MC'F run. If the set does not give a uniform reading, fill all tubes with deionized water and place in 180°C drying oven for 2 hr, rinse, dry, and repeat complement controls. When tubes are conditioned, clearly label each set to avoid any confusion of MC'F sets in the lab. Once an MC'F set has been conditioned, the tubes should only be cleaned by thorough rinsing (three to four times) in deionized water and dried in an oven (180°C) after each experiment. Detergent should *never* be used.

PROTOCOLS

1. Antigen purification
2. Antibody production

3. Immunodiffusion, immunoelectrophoresis, and RIA
4. Microcomplement fixation

Protocol 1: Antigen Purification

(Time: Part A, 2 hr; Part B, 1.5 hr; Part C, 1 hr)

The choice of antigen will vary depending on the problem to be studied. The more closely related the species to be studied, the more rapidly evolving the protein should be to ensure sufficient events are monitored to allow resolution of species relationships. Purification methods will vary depending on the antigen selected. Our example is for serum albumin, which is extremely useful for studying vertebrate species that have diverged throughout the Cenozoic. To probe older divergences, more slowly evolving molecules need to be studied. To probe divergences among invertebrates, plants, or bacteria, suitable molecules (other than albumin) must be identified. Most antigens are purified for use as immunogens by standard methods and stored at high concentration in the cold (0 to $-15°C$; Champion et al., 1974).

PART A. PREPARATION OF TUBES AND SOLUTIONS

1. Wash 10-cm × 7-mm glass tubes in 5% NaOH for 30 min. Rinse with tap water and place tubes in polyoxyethylene 23-lauryl ether (3 drops/liter) for 30 min. Rinse with deionized water and dry thoroughly.
2. Dip each tube in 2% solution of Photo-Flo (Kodak) and dry upright in test tube rack.
3. Dilute electrophoresis buffer 1/10 for use.
4. Prepare ANS stain.

PART B. PREPARATION OF GELS FOR ELECTROPHORESIS

1. For 24 tubes, mix 25 ml acrylamide mix A (Appendix), 25 ml acrylamide mix B (Appendix), and 40 mg ammonium persulfate in a 250-ml sidearm vacuum flask. Degas with vacuum suction, swirling gently.
2. Pipette the acrylamide solution into clean glass tubes that have been sealed on the bottom with parafilm and placed upright in a test tube support.
3. Use a fine-tipped Pasteur pipette to layer a water-saturated solution of isobutyl alcohol atop each acrylamide gel to exclude air and allow the acrylamide to polymerize.
4. After 30 min, rinse the alcohol from the top of each gel in running tap water and remove the parafilm from the bottom of each tube. Place the tubes in a cylindrical gel electrophoresis unit.
5. Fill the bottom chamber with a chilled 1/10 dilution of the electrophoresis buffer. Carefully pipette a little buffer atop each gel and then fill the top chamber with buffer.
6. Turn power on for 15 min at 5 mA/gel at room temperature to electrophorese ammonium persulfate away from origin.

NOTE: For extensive purifications, the entire electrophoretic apparatus can be set in an ice bath so the buffer does not overheat.

PART C. PURIFICATION OF ANTIGEN

The following steps are for purifying albumin from serum or muscle extracts. Methods of preparing other purified proteins will vary. The key to success is preparing a highly purified antigen.

1. Add crystalline sucrose to plasma (10–15% of sample volume).
2. Layer the sample atop each of the buffer-covered gels with a long-tipped 0.2 ml pipette (0.03–0.08 ml/gel).
3. To purify antigens from two species at one time and prevent sample contamination, pour the top chamber buffer into a beaker while loading the samples and then carefully pour the buffer back into the chamber.
4. Electrophorese the gels for 30–45 min at 5 mA/tube and 100–120 V at room temperature or in an ice bath. The time will vary depending on the temperature and the sample being purified.
5. Remove each gel from the glass tube using a water-filled syringe equipped with a fine needle for slipping around the outside of the gel.
6. Place the gels on a plastic tray and stain them. For albumin, stain gels with a piece of filter paper soaked in ANS stain.
7. Remove the paper and examine the gels under ultraviolet light. Albumin will be visible as a leading fluorescent band (Hartman and Udenfried, 1969).
8. Cut the albumin out of the gel with a clean razor blade.
9. Place the pieces of albumin-containing gel in a small flask or test tube in isotris buffer (final concentration roughly 0.5–1 mg/ml).
10. Elute antigen with occasional shaking for 2–3 days at 4°C.

NOTE: This albumin preparation can be stored in the freezer indefinitely.

PART D. MODIFICATIONS

1. To purify albumin from PPS + blood, first centrifuge the sample and remove any hemoglobin precipitate. Treat the resultant supernatant as plasma, but omit sucrose as the PPS makes the sample sufficiently dense to layer on the gels.
2. To purify albumin from PPS + muscle, centrifuge the sample and concentrate the supernatant in Centricon filters (Amicon Corp.) at 2000 g at 4°C for 45 min. The concentrated supernatant is treated as described in Part C.

Protocol 2: Antibody Production

(Time: Part A, 13 weeks; Part B, 30 min for one rabbit; 1 hr for six rabbits)

Three young female rabbits are used for each antiserum. Rabbits are quarantined for at least 1 week prior to immunization. For MC'F, antisera should be made only to purified proteins. For a given antigen, rabbits are given four injections over a 13-week immunization period. Each rabbit receives approximately 1–3 mg of antigen during this time.

PART A. IMMUNIZATION SCHEDULE

1. Prepare a 1:1.2 emulsified suspension of antigen eluant (Protocol 1C, step 9) and Freund's complete adjuvant. For three rabbits, pipette 2.0 ml of antigen and 2.4 ml of Freund's into a 5-ml glass Luerlok syringe that has been attached with a "bridge" to a second similar syringe. (The "bridge" can be made from two 18-gauge stainless-steel needles, or comparable units can be purchased commercially). Expel excess air and repeatedly compress the two syringes, effecting emulsification. Leave emulsion in one syringe. Remove the bridge and attach a 23-gauge disposable needle to the syringe just before injecting the rabbits.

NOTE: The emulsion can be prepared prior to immunization and left in the refrigerator overnight.

2. With electric hair clippers, shave two sites over the rear haunches of each rabbit. Slip the needle into the skin and inject about 0.5 ml of emulsion, forming a small bump. Each rabbit should receive 1–1.5 ml of emulsion.
3. Seven weeks after the first intradermal immunization, inject a 1:1.2 emulsion of antigen eluant + Freund's incomplete adjuvant at two sites near the initial injection sites.
4 Three weeks later, inject 1 ml of antigen eluant in the marginal ear vein by placing the rabbit securely in a restraining cage, leaving the ears accessible. Shave the area over the marginal ear vein with a razor and slip a 25 gauge disposable needle into the vein, parallel to the margin of the ear. Slowly administer the eluant. Dab petroleum jelly on ear.
5. Administer a second intravenous injection of 1 ml of antigen eluant in the marginal ear vein 1 week later.

PART B. COLLECTION OF ANTISERA

1. One week after the last injection, anesthetize the rabbit with an injection of pentobarbital (1 mg/ml) in the marginal ear vein ($\approx$ 1–2 ml/rabbit). Bleed by heart puncture, using an 18-gauge heparinized needle and 40-ml disposable syringe containing 1 ml of heparin (1000 units/ml).
2. Remove the rabbit from the restraining cage and place it on its back. Feel for the beating heart, slip the needle in the heart, and pull back gently on the syringe. When the syringe is full, leave the needle in the heart and empty the syringe into a labeled centrifuge tube containing 2 ml of heparin (1000 units/ml isotris). Reattach the syringe and continue collecting blood. Cover the filled tube with parafilm and place it in an ice bucket. If blood flow ceases, remove the needle and take a fresh syringe and needle and begin again. Sixty to 100 ml of rabbit blood (the antiserum) is typically obtained.
3. Centrifuge the blood at 4000 g at 4°C for 20 min. Decant the supernatant into 20

ml labeled scintillation vials. Individual rabbit antisera are stored at −15°C and are stable indefinitely. Once the rabbit antisera are safely stored, euthanize the rabbits by approved procedures.

Protocol 3: Immunodiffusion, Immunoelectrophoresis, and RIA

(Time: Part A, 1 day; Part B, 3 days; Part C, 3 days)

PART A. IMMUNODIFFUSION (OUCHTERLONY AND TREFOIL PLATES)

For in-depth description of this method see Goodman and Moore (1971), Maxson and Daugherty (1980), Klein (1982), and Densmore (1983). Steps 1–4 describe the construction of Ouchterlony plates; Trefoil plates are available commercially.

1. Dissolve Ionagar in isotris buffer (1% agar solution) by heating the solution with constant stirring until it just begins to boil. The solution becomes clarified when the agar is dissolved.
2. Pipette 5 ml aliquots of hot agar solution into 60 × 15 mm plastic tissue culture dishes. Cover and allow to cool.
3. Wrap the filled plates in plastic bags and store in refrigerator until needed.
4. To use, make wells in solidified agar using a punch made from a 12-gauge stainless-steel needle or a commercially made instrument.
5. Label wells clearly. Use disposable microhematocrit tubes to fill antigen and antiserum wells. Leave plates at room temperature for 3–12 hr to allow precipitation to occur. Monitor plates at intervals until precipitation is complete.
6. Photograph or sketch results (see Figure 1).

PART B. IMMUNOELECTROPHORESIS

The procedure below can be performed with commercially available kits. For an in-depth description of the technique see Klein (1982) and Mao and Chen (1982). Immunoelectrophoresis also is used to test antisera for purity (Figure 2b) as described by Wallace et al. (1973).

1. Prepare adhesive agar (0.1% immunoelectrophoresis agar; Appendix).
2. Brush film of adhesive agar on a clean microscope slide on a slide holder on a level surface; allow to dry for 5–10 min.
3. Heat agar to 85–90°C.
4. Pipette warm immunoelectrophoresis agar onto slides in slide holder (10 ml/3 slides).
5. Cover and cool for 15 min.
6. Place slides in humid chamber for 15–30 min.
7. Use commercial punch to make antigen wells and center antiserum trough. Carefully remove agar from wells, using mild suction.
8. Add antigens to wells with 5-μl pipette and place slides in electrophoretic apparatus.

9. Electrophorese (using veronal buffer; Appendix) at 7.4 V/cm for 60 min.
10. Remove slides from electrophoretic apparatus. Carefully cut gel out from trough and add antiserum to center trough with 100 μl pipette. Place slide in a humid chamber for ≅20 hr at room temperature.
11. Remove slides and rinse in 1% NaCl for 6 hr.
12. Remove slides and reimmerse in fresh 1% NaCl for 16 hr.
13. Rinse slides in fresh deionized water for 1 hr.
14. Remove slides, fill wells and trough with deionized water.
15. Cover slides with wet blotter strips and dry.
16. Remove strips from gel; if they stick, wet with tap water.
17. Immerse slides in staining tank for 5 min. Use general protein stain or specific stain.
18. Rinse and destain gels. Dry gels and record results.

PART C. RADIOIMMUNOASSAY

For in-depth description of this method see Klein (1982) and Lowenstein (1985a,b).

1. Make serial dilutions of antigens in deionized water; concentration range must be experimentally ascertained (10^{-3}–10^{-4}).
2. Bind antigen dilutions to polyvinyl microtiter plates for 1 hr.
3. Wash microtiter plates with a 2% solution of chicken ovalbumin or BSA.
4. Add 0.2-ml aliquots of antiserum (10^{-3}–10^{-4} in deionized water) to microtiter plates and incubate at room temperature for 24 hr.
5. Aspirate out unbound antibody.
6. Add 1 mg ^{125}I-labeled goat anti-rabbit γ-globulin (GARGG) to the microtiter plates; let sit for 24 hr.
7. Wash out unbound GARGG with running tap water.
8. Cut up microtiter plates and count radioactivity remaining in individual wells in a scintillation counter.

Protocol 4: Microcomplement Fixation

(Time: Part A, 1 day; Part B, 1 hr; Part C, 1–3 days; Part D, 1 day; Part E, 1.5 hr; Part F, 45 min; Part G, 2 hr)

The general protocol for MC′F analyses has been described in Champion et al. (1974). It is important to standardize all experimental conditions and monitor experimental results to be sure repeatable MC′F curves are obtained for all the reported and analyzed data. Broad MC′F curves are one indication that the conditions for quantitative results are not being met.

PART A. TITERING AN ANTISERUM

The antiserum titer is defined as the dilution of antiserum that gives 70–75% fixation at the peak of the MC′F curve in standard experiments that are incubated for 18–22 hr at 0°C. In Figure 4a, the titer of the homologous antiserum to *H. regilla* albumin is 1/6000.

1. Prepare rabbit antisera as described in Protocol 2.
2. Set up a series of MC'F reactions (part D) with the homologous antigen and varying dilutions of antiserum. As titer is time dependent, it is convenient to standardize the incubation time for all experiments using the same antiserum. Titer appears to vary by roughly 5–10% per hour. Optimal MC'F curves for amphibian albumins usually are obtained at antiserum dilutions of 1/2000 to 1/6000. In our laboratory over the past decade, 20 hr antisera titers range from 1/1000 to 1/9000 (averaging 1/4400).
3. Pool individual antisera made to the same antigen (part B).
4. For convenience, make a stock solution of a 1/25 dilution of pooled antisera (part B). For very high titer antiserum, make a 1/100 stock solution.

PART B. POOLING ANTISERA

When there is insufficient antigen to immunize three rabbits, two rabbits or even a single rabbit have been used successfully (Hutchinson and Maxson, 1987b). Substantial work has been done in my laboratory (L.R. Maxson) with antisera from individual rabbits, as well as with an antisera pool of the same individuals. In these instances, the IDs obtained with individual antisera and with pooled antisera are within experimental error (Maxson, 1981). In rare instances in which one rabbit has an atypically low titer, both high- and low-titer antisera pools have been made and found to give the same IDs (Busack et al., 1988). This is true only where all antisera have been made as described in Protocol 2. In some cases, low-titer antisera do not permit distant comparisons. This is probably due to low antibody diversity. In such cases using individual antisera, rather than a lower titer antisera pool, permits more distant reactions to be measured.

1. Titer individual rabbit antisera by a standard MC'F experiment (parts E–G).
2. Pool individual antisera in inverse proportion to their titers to maximize antibody diversity.
3. Heat pooled antisera at 60°C for 20 min to inactivate rabbit complement.
4. Centrifuge the heated antisera at 27,000 g for 20 min at 4°C; decant the supernatant and store in 20-ml aliquots in scintillation vials at −15°C. The antisera pools are stable indefinitely.

PART C. TESTING ANTISERA FOR PURITY

1. Test pooled (or individual) antisera for purity by immunoelectrophoresis (Protocol 3; Figure 2b). When purified antigen is scarce, simple Ouchterlony tests are performed (Maxson and Daugherty, 1980).
2. If only a single precipitin line is formed with a sample of unpurified antigen, it is concluded that the antisera is directed primarily to albumin.
3. Perform MC'F tests (parts E–G) using both the immunogen and whole plasma or muscle extracts (Figure 4b) as antigens. If the antisera is directed primarily against albumin, the results from both MC'F experiments will be indistinguishable (Maxson and Roberts, 1985).

PART D. THE ANTIGEN

Stock solutions of antigen are typically 1/100 dilutions of plasma or 1/10 solutions of PPS + tissue. A series of doubling dilutions of antigen are made for each experiment to ensure that the "peak" of the curve (the point where maximum antibody–antigen binding occurs) is included in the six tubes comprising the experimental curve. For albumin, the peak of the MC'F curve is usually obtained at concentrations of albumin of 100 ng/ml.

PART E. SETTING UP AN EXPERIMENT

We describe an experiment that will titer an antiserum, derive a slope, and make one heterologous comparison. This is outlined in Figure 5. The homologous antisera are directed against albumin from species BV and the heterologous species is BX.

1. MC'F tubes: A typical experiment consists of 6 experimental curves of 6 MC'F tubes each (tubes 1–36; Figure 5), 6–10 tubes of antibody, antigen, and/or complement controls (tubes 37–46), a double complement control tube (tube 47), and a cell control tube (tube 48). Each MC'F experiment *must* include at least one homologous curve, and all antibody controls should be run in duplicate.
2. Add isotris buffer (Appendix) to all 48 MC'F tubes.
   a. Use an adjustable 5-ml automatic pipette to add 3 ml of buffer to tubes 1–36 (Figure 5).
   b. Add 4 ml buffer to antibody and antigen control tubes 37–44.
   c. Add 5 ml buffer to complement control tubes 45 and 46.
   d. Add 4 ml of buffer to tube 47 (double complement tube) and 6 ml of buffer to the cell control (tube 48).
   e. Cover the tubes with plastic wrap and set aside.
3. Antigen dilution tubes: For each antigen, label a series of six (10–20 ml) test tubes and fill with buffer (using the automatic pipette), such that the first antigen dilution tube has at least $2(n + 1)$ ml of buffer and the remaining five dilution tubes have $(n + 1)$ ml of buffer, where n is the number of curves using that antigen. In this example (with four homologous curves), the dilution series for BV will use 11 ml in the first tube and 5 ml in subsequent tubes. When antigen controls are run, as in this example, an additional 1 ml of buffer is needed in each first tube.
4. Antigen dilutions:
   a. Add a previously determined amount of the antigen stock solution (typically 0.01 ml of a 1/100 plasma solution/ml of buffer) to the first dilution tube. For BV, 0.10 ml of 1/100 plasma solution is added to 11 ml of buffer.
   b. Use a clean 5-ml pipette to make the serial dilutions. Remove 5 ml from the first (1/1) tube and mix it in the next (1/2) tube, then remove 5 ml from this tube and transfer it to the next (1/4) tube. Repeat until the final (1/32) tube contains 10 ml.
   c. Prepare similar antigen dilutions for all antigens.

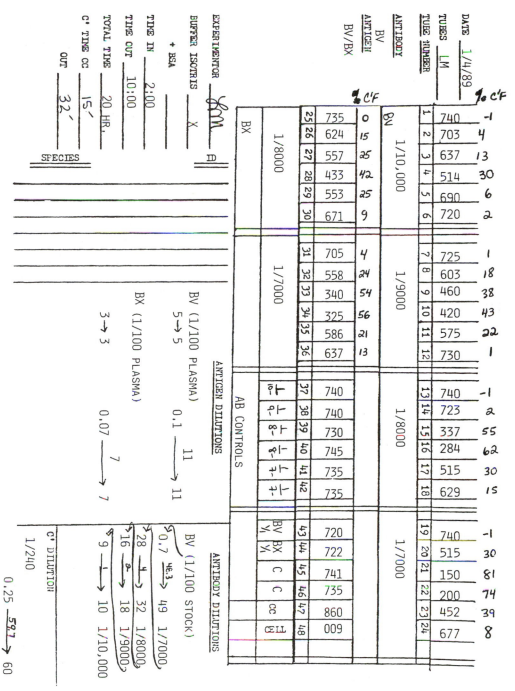

FIGURE 5 An MC'F flow sheet outlining an experiment titering species BV antibody and comparing species BX to species BV and depicting the results in absorbance and percentage complement fixed (%C'F).

5. Antibody dilutions:
   a. Use a well-labeled clean test tube or small flask for each antibody dilution.
   b. Carefully pipette the calculated amount of buffer into each tube, taking care to wipe off the outside of each pipette to remove any excess antibody.
   c. Make serial antibody dilutions.
6. Adding the antigen:
   a. Using a 1 ml-pipette, add 1 ml of the most dilute (1/32) antigen dilution for BV to tubes 6, 12, 18, and 24 of the MC'F set.
   b. Next add 1-ml aliquots of the 1/16 antigen dilution to tubes 5, 11, 17, and 23.
   c. Repeat for the remaining four tubes of the BV antigen dilution.
   d. For the antigen control, pipette an additional 1 ml of the most concentrated (1/1) antigen dilution into tube 43.
   e. Add the antigen dilution for the heterologous BX to tubes 25–36 in the same manner, beginning with the most dilute and ending with the most concentrated.
   f. Add 1 ml of the 1/1 antigen dilution to antigen control tube 44.
7. Adding complement: Optimize the complement concentration to achieve lysis of SSRBC in 25–35 min under standard conditions. Complement titer varies among different batches and different suppliers. Typical dilutions vary from 1/140 to 1/280. Stock solutions of complement are prepared weekly and stored in the freezer.
   a. Pipette buffer for complement into an Erlenmeyer flask in an ice bucket.
   b. Place the MC'F set (tubes 1–48) in an ice bath.
   c. Thaw complement and pipette precise amount into the chilled buffer.
   d. Use a 2-ml adjustable pipette to pipette 1 ml of the complement dilution into tubes 1–46 and 2 ml into tube 47 (the double complement control). No complement is added to tube 48 (cell control).
   e. Gently shake the tubes to rinse antigen and complement from the walls.

NOTE: It is critical to place the experimental tubes on ice *before* adding the complement and to add the complement *before* adding the antibody.

8. Adding the antibody: the addition of antibody marks the start of the experiment. Eighteen to 22 hr after this point, the experiment can be completed by adding SSRBC and monitoring cell lysis.
   a. Pipette 1-ml aliquots of diluted antibody into the experimental tubes, beginning with the antibody control tubes and then the curves, from most dilute (1/32) antigen to most concentrated (1/1) antigen. Thus, 1 ml of 1/7000 antibody dilution is added to tubes 41, 42, and then to tubes 36–31.
   b. Use a clean 1-ml pipette to add the remaining 1/7000 dilution to tubes 24–19.
   c. Use a clean pipette and add 1 ml of the 1/8000 antibody dilution to control tubes 39 and 40, and to experimental tubes 30–25 and (with a clean pipette) tubes 18–13.
   d. Add the 1/9000 dilution to tubes 38, and 12–7 with a clean pipette.
   e. Add the 1/10,000 dilution to tubes 37, and 6–1 with a clean pipette.
   f. Swirl the tubes gently to mix reagents.

g. Cover the experiment with plastic wrap and aluminum foil and place in cold room overnight.

h. Record "time in" on the flow sheet (Figure 5).

PART F. PREPARING SENSITIZED SRBC (SSRBC)

Prior to reading the experiment the next day, fresh SSRBC must be prepared. All work with SRBC is done in isosatris buffer (Appendix) to retard cell lysis. Preparation of SSRBC generally takes approximately 45 min.

1. Centrifuge an aliquot (5 ml) of sheep blood in Alsever's solution (Appendix) at ≈850 g for 3–4 min at room temperature.
2. Discard the supernatant and resuspend the packed cells in an equal volume of isosatris.
3. Repeat the washing cycle two to three times.
4. After the final centrifugation, aspirate off the supernatant and dilute 1 ml of the cell pellet in 18 ml of isosatris (this dilution may vary with the hematocrit of the sheep). To determine the appropriate concentration of the SRBC, lyse 0.1 ml of the cell dilution in 15 ml of deionized water. The absorbance at 413 nm should be between 0.800 and 0.850. If it is not, adjust the concentration of the cell dilution by adding more washed cells or more isosatris buffer and retest the resultant dilution.
5. In an Erlenmeyer flask, mix equal amounts of appropriately diluted cells and a 1/250 dilution of hemolysin (hemolysin is diluted 1/25 in isosatris and stored in the freezer; a fresh 1/10 dilution in isosatris of that dilution is made each day). Add the diluted hemolysin to the cells while gently swirling the flask.
6. Cover the flask with parafilm and incubate the mixture in a 35°C water bath for 15 min, shaking gently at 5 min intervals.
7. Dilute this mixture 1/10 with isosatris; these are the SSRBC used in MC'F. The SSRBC should be stored at 4°C and used within 16 hr.

PART G. READING THE MC'F EXPERIMENT

1. At the end of the 18–22 hr incubation period, remove the experiment from the cold room and add 1 ml of SSRBC to each tube (1–48) with a 2-ml automatic pipette.
2. Remove the entire experiment from the ice bath and place it in a 35°C water bath.
3. Swirl all the tubes and monitor the experiment, shaking all tubes at 5 min intervals.
4. The double complement tube (tube 47) will be the first tube to show complete lysis; this should occur within 14–16 min. Record this time on the flow sheet (Figure 5). Lysis can be seen by holding the tubes up to the light. Recognizing degrees of lysis can be facilitated by using a pattern on a light board and looking through the tube at the pattern. The pattern comes sharply into focus as lysis nears completion. Lysis in the antibody control and complement control tubes should begin within 25–35 min.
5. Remove the experiment at 80–85% of total lysis and place it in an ice bath and shake all the tubes to help quench the reaction. Record the time the experiment is placed on ice on the flow sheet (Figure 5). This facilitates determining the complement

concentration to be used in subsequent experiments. If the experiment is removed from the 35°C water bath before obtaining 80% lysis, the lower absorbance values will contribute to greater experimental error. If the experiment is removed too late, lysis exceeds 85% in the controls, and the SSRBC become limiting, making the titer appear lower.

6. Quickly pour the contents of the 48 experimental tubes into 48 numbered prechilled test tubes (13 × 100 mm).
7. Centrifuge (tabletop, clinical) these smaller tubes at ≅850 g for 10 min to sediment unlysed SSRBC.
8. Read the absorbance of the supernatant in each of the 48 tubes at 413 nm in a spectrophotometer. Record absorbances of each tube on the flow sheet (Figure 5).

NOTE: Antibody controls generally are run in duplicate. Double complement control tubes are used to estimate how long the entire experiment will take. Complement control tubes are used to judge whether antigens and antibodies may be pro- or anticomplementary (see Interpretation and Troubleshooting).

INTERPRETATION AND TROUBLESHOOTING
Immunodiffusion

Results Ouchterlony and trefoil immunodiffusion plates are incubated for a fixed period of time, which is generally dependent on the titer of the antisera used. Lines of precipitation are observed and recorded, taking care to obtain precise information on the size, position, and intensity of all spurs (Figure 1).

Data Reduction Immunodiffusion comparisons are usually scored in terms of the size, positions, and intensity of the spurs that are formed. Antigenic distances are estimated from pairwise comparison of all combinations of antigen and antisera that are made. When using antisera to a mixture of proteins, the identity of the proteins forming specific precipitin bands is generally unknown. With antisera to a single protein and/or staining with protein-specific stains, one can infer the similarity of the homologous antigens of two species as the more similar antigens will show the smallest spurs. In some studies, a scale of antigenic distance is derived based on subjective interpretation of the data and correlations with other measures of immunological relatedness (Goodman and Moore, 1971; Schwaner and Dessauer, 1982).

Radioimmunoassay

Results and Data Reduction The raw data from RIA consists of counts of radioactivity measured by a scintillation counter. The specificity of the antisera plays a major role in determining the final amount of radioactivity measured. Specificity can be improved by an additional competitive binding technique (Lowenstein et al., 1981). Correlations have been reported between radioimmunoassay results and distances measured by MC'F (Lowenstein and Ryder, 1985; Pierson et al., 1986). However, IDs measured by RIA are not equivalent to IDs measured by MC'F.

Microcomplement Fixation

Results   Peak complement fixation is obtained at concentrations of approximately 100 ng albumin/ml. In areas of antigen excess (e.g., tube 1 in Figure 5) or antibody excess (tube 6 in Figure 5), little or no complement appears to be fixed. These tubes read close to the antibody, antigen, and complement controls (tubes 37–46 in Figure 5) that do not fix complement. When absorbance readings are converted to percentage complement fixed (Figure 5) and plotted versus antigen concentration (Figures 4 and 6), bell-shaped curves result. The peaks of these curves are the percentage complement fixed values used to make all calculations.

Data reduction   The absorbance values are corrected for the cell and antibody controls by calculating percentage complement fixed (%C'F) as follows:

$$\%C'F = 100 \times (AC - ET)/(AC - CC)$$

where   AC = absorbance of the antibody control tube
ET = absorbance of an experimental tube
CC = absorbance of the cell control tube

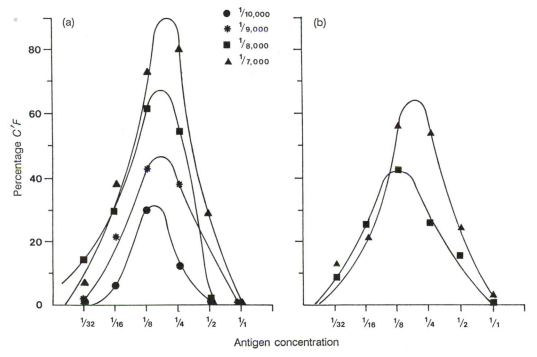

FIGURE 6   MC'F curves for (a) species BV at four concentrations of anti-BV albumin antibody and (b) species BX at two concentrations of anti-BV albumin antibody.

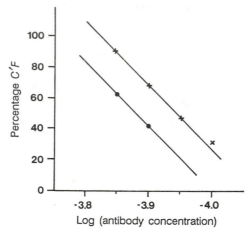

FIGURE 7 The slope determinations for anti-BV albumin antibody with the homologous and heterologous antigens. The slope is close to 400 for both antigens with this antibody (384 with BV and 370 with BX).

The %$C'F$ values can be plotted versus the log of the antigen concentration (from 1/32 to 1/1) and a curve fitted to the data to interpolate peak fixation (Figure 6). Immunological distance is calculated as the ratio of the amount of antisera it takes for the heterologous antigen to achieve the same complement fixation as the homologous antigen (Figure 4). When the peaks of both curves are close, but not identical, rather than adjusting the antisera concentrations to match peaks and repeating the entire experiment, a slope correction can be applied.

The slope of an antiserum (Champion et al., 1974) relates the change in the log of the antiserum concentration to the change in %$C'F$ (Figure 7). Slope is characteristic of an antiserum and various protocols, and does not change when heterologous antigens are used in place of the homologous antigen. Slope is determined when titering a new antiserum. In this example, the slope is 384. Knowing the slope permits calculation of immunological distance without precisely matching peaks of homologous and heterologous curves. In our laboratory, we use a slope correction only when peaks differ in heights by less than 15%$C'F$. For greater differences in peak heights, antibody concentrations are adjusted and the experiment repeated.

Immunological distance (*ID*) between two antigens is calculated as

$$ID = 100 \log \frac{[\text{HET AB}]}{[\text{HOM AB}]} (\times \text{ or } \div) \text{ slope correction factor}$$

where [HET AB] = the antibody concentration used in the heterologous curve
[HOM AB] = the antibody concentration used in the homologous curve

The slope correction factor is defined as

$$\text{antilog}([\%C'F \text{ homologous} - \%C'F \text{ heterologous}]/\text{slope})$$

When the peaks of the homologous and heterologous curves match precisely, the slope correction factor is 1. When the peak of the heterologous curve is less (more) than that of the homologous curve, the slope correction is multiplied (divided) to adjust the heterologous curve to that of the homologous curve.

The slope of the BV antibody in the example in Figure 7 can be calculated as 384 and the *ID* between the two species is

$$100 \ (\log\{[1/7000]/[1/8000]\} \times 1.04) = 7$$

Sources of Error   When interpreting an experiment, care must be taken to avoid using data that are either overfixed or underfixed. If too little antibody is present in an experimental curve (tubes 1–6 in Figure 5), very little complement is fixed, and large slope corrections are not very reliable. Likewise, if an experimental curve is overfixed (tubes 19–24 in Figure 5), the reaction becomes nonlinear and it is difficult to correctly interpolate the actual peak fixation.

In most instances, antigen, antiserum, and complement controls all read about the same in a given experiment. On occasion, some antisera or some antigens independently enhance (are procomplementary) or retard (are anticomplementary) the binding of complement. For this reason antibody controls are run in duplicate in all experiments and all antigens are tested the first time they are used. Anticomplementary antigens can be compensated for by additional purification, or by running an entire curve (six tubes) of antigen controls (1/1 to 1/32) for that species.

Some antisera may be anticomplementary at high antibody concentrations. To compensate for this, the concentration of complement can be increased and the experimental tubes at that antibody concentration plus the antibody controls are kept in the water bath until the control tubes exhibit 80% lysis, or for 90 min (beyond that time complement becomes unstable). If sufficient lysis is not obtained in the controls, the antibody cannot be used at these distances. It is critical to always run homologous curves in each experiment because the precise antibody titer will vary with experimental conditions. Even when working with anticomplementary antisera, the homologous reaction can be removed first and the heterologous reactions can remain in the water bath until their antibody controls are ready.

Finally, when working with heterologous antigens at IDs of over 100, a shift in the peak of the curve toward higher antigen concentration or a broadening of the curve may be noticed. Sometimes this shift towards higher concentration may be so pronounced as to place the peak in the first tube. Although the concentration of albumin in the starting tube can be increased and broad peaks can still be measured, we do not recommend pushing the limits of the MC'F reaction for reasons outlined in Maxson and Maxson (1986). Shifts in antigen concentration at peak fixation and broad curves are both indications that the antibody exclusion process has reached its limits.

Tests of Reciprocity   For estimating phylogenies, one must have antisera to each species to be placed on the tree. This permits reciprocal tests (anti-A vs B and anti-B vs A) that indicate how well the albumin sequence differentiation between species pairs is being estimated. Ideally, the only ID difference in reciprocal tests should be attributable to experimental error and would be of the order of 2 ID units. In reality, however, there are other sources of experimental error that affect reciprocal estimates of sequence differences. The standard deviation from reciprocity of most work performed in our laboratory over the years averages around 10%. We have defined standard deviation of reciprocity (Maxson and Wilson, 1975) as:

$$\sqrt{\frac{1}{n} \sum_{i=1}^{n} \left[\frac{(a_i - b_i)100}{a_i + b_i} \right]^2}$$

where: a_i = ID of anti-A to B
b_i = ID of anti-B to A
n = number of reciprocal pairs

Because of this definition, small deviations from reciprocity in IDs under 10 units greatly inflate the calculation. When working with very close species, we often report deviation from reciprocity in terms of ID units, rather than percentage standard deviation. This was most notable in studies of marsupial frogs where deviations averaged from 2.5 to 4.4 units (Scanlan et al., 1980; Duellman et al., 1988).

On occasion, the reciprocal data demonstrate a significant nonrandomness, where all IDs estimated with one particular antiserum run higher (or lower) than reciprocal IDs. All reciprocal data are tested for significant nonrandomness as suggested by Sarich and Cronin (1976), and reciprocal data are "corrected" as necessary. Both raw data and corrected data are reported and separately evaluated. In our experience, phylogenetic conclusions from both raw and corrected data are concordant.

Phylogenetic Analysis of Species with One-Way Tests   Species for which no antisera are prepared cannot be placed precisely in a phylogenetic tree. However, by comparing such species in one-way tests with all available antisera, the approximate placement of such species relative to the previously defined lineages can be inferred. When several species are compared to a panel of antisera, and where the panel defines a tree, "species group" relationships can often be identified by the patterns of cross reaction to all the antisera (Hutchinson and Maxson, 1987b; Duellman et al., 1988). Beverley and Wilson (1982) have described a method for adding species for which no antisera are available to a phylogeny.

APPENDIX: STOCK SOLUTIONS

Acrylamide mixture "A"

40 g acrylamide
1 g bisacrylamide (N,N'-methylene-
    bisacrylamide)
250 ml deionized water

Mix in dark bottle and store in refrigerator

Acrylamide mixture "B"

4.04 g Tris
1.308 g glycine
0.24 ml TEMED (N,N,N',N'-tetramethyl-
    ethylenediamine)
200 ml deionized water

Mix in dark bottle and store in refrigerator

Adhesive agar for immunoelectrophoresis

0.1 g Ionagar #2
25 ml veronal buffer
75 ml deionized water

Alsever's Solution

10.250 g dextrose
4.000 g sodium citrate, dihydrate
0.275 g citric acid, monohydrate
2.100 g sodium chloride

Dissolve the above reagents sequentially in 500 ml deionized water. Autoclave for 15 min at 15 psi. Remove immediately. Store under sterile conditions in refrigerator.

ANS Stain

10 ml 0.1 N sodium phosphate, pH 6.8
0.3 ml anilino naphthalene sulfonate (1
    mg/ml water; store covered in aluminum
    foil in refrigerator)

10× Electrophoresis Buffer

1 M Tris
0.5 M glycine

Immunoelectrophoresis agar (1%)

1 g Ionagar #2
25 ml veronal buffer
75 ml deionized water

Heat until agar is dissolved

Isosatris buffer

4.0 g BSA (bovine serum albumin, fraction
    V)
4.0 liters 1× isotris buffer

10× isotris stock buffer

327.0 g NaCl
48.4 g Tris
5.92 g $MgSO_4 \cdot 7H_2O$
0.9 g $CaCl_2 \cdot 2H_2O$
0.1 g merthiolate

Adjust pH to 7.45 with concentrated HCl. Bring volume to 4 liters with deionized water.

PPS (phenoxyethanol-phosphate-sucrose) preserving solution (Nakanishi et al., 1969)

85.6 g sucrose (reagent grade)
16.7 ml 1 M KH_2PO_4
83.3 ml 1 M K_2HPO_4
15.0 ml phenoxyethanol (Eastman #4861)

Bring volume to 1 liter with deionized water. Stir well to dissolve phenoxyethanol.

Veronal buffer

0.05 M sodium acetate
0.05 M sodium barbital

Adjust to pH 8.6 with 0.01 N HCl

CHAPTER 6

CHROMOSOMES: MOLECULAR CYTOGENETICS

Stanley K. Sessions

PRINCIPLES AND COMPARISON OF METHODS

General Principles

This chapter concerns the analysis of microscopically visible aspects of the molecular structure of chromosomes. The term "chromosome" was introduced in 1888 by Wilhelm Waldeyer, and the chromosomal theory of inheritance was put forward and elaborated by Theodor Boveri, Walter S. Sutton, and Thomas H. Morgan in the first part of this century. Ever since that time the study of chromosomes (known as either karyology or cytogenetics) has occupied a prominent place in genetics in both clinical and academic applications, as well as in comparative biology and phylogenetic studies. Comparative cytogenetics is thus an old field with diverse schools of interpretation concerning chromosome structure, function, and evolution.

The history of cytogenetic research can be divided into several eras, each coupled with major technological innovations that triggered revolutions in analytical approaches (Hsu, 1979). The modern era in cytogenetics, including the incorporation of molecular methods, was initiated by the development of three main technological breakthroughs (reviewed in Hsu, 1979): (1) the discovery that hypotonic treatment spreads metaphase chromosomes, allowing accurate assessments of chromosome numbers and morphology; (2) the development of chromosome banding techniques, which allows the identification of homologs (within karyotypes of the same species) and homoeologs (between karyotypes of different species); and (3) the development of techniques for **in situ hybridization** of nucleic acid probes to cytological preparations of chromosomes, by which specific DNA sequences can be localized to particular chromosomes and parts of chromosomes (Gall and Pardue, 1969; John et al., 1969).

The field of molecular cytogenetics is centered on the technique of in situ hybridization of nucleic acids, and this methodology will be discussed

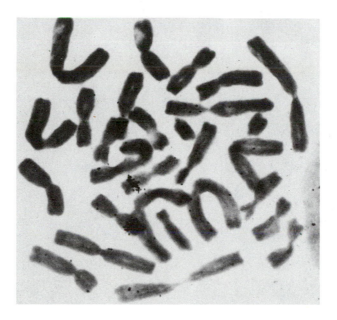

FIGURE 1 In situ hybridization of a ribosomal DNA probe to mitotic chromosomes of the salamander *Hydromantes italicus*. From Nardi et al. (1986); reproduced with permission.

in detail in this chapter. The basis of in situ nucleic acid hybridization is the annealing of mobile probe molecules and stationary target molecules to form base-paired duplexes. Comparative studies using in situ hybridization have primarily involved locating repetitive sequences, such as **satellite DNA**, ribosomal gene clusters, or the extensively reduplicated genes of **polytene chromosomes** using highly radioactive molecular probes (Figure 1). In situ hybridization can also be used to locate specific RNA transcripts on lampbrush chromosomes (Figure 2; Diaz et al., 1981; Varley et al., 1980), and techniques have been developed over the last few years for reliably locating single-copy DNA sequences on mitotic chromosomes (Figure 3; Harper and Saunders, 1984), and for using nonradioactive antibody probes visualized with enzymes or fluorescent dyes (Langer et al., 1981; Manuelidis et al., 1982; Pinkel et al., 1986; Frommer et al., 1988).

A novel and potentially powerful approach to the study of the comparative molecular structure of chromosomes uses **monoclonal antibodies** directed against nuclear proteins on polytene chromosomes of insects and lampbrush chromosomes of salamander oocytes (Ragghianti et al., 1988; Figure 4).

These advances, along with the availability of a wide range of probes and the increasing accessibility of techniques for constructing new probes, make molecular cytogenetics an increasingly powerful approach for the

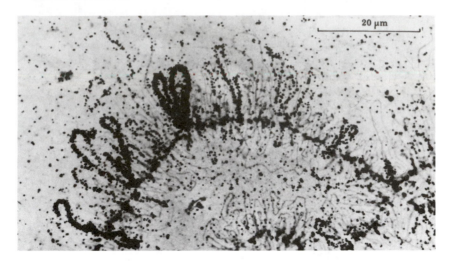

FIGURE 2   In situ hybridization of a repetitive DNA probe to RNA transcripts on lampbrush chromosomes of the European newt, *Triturus carnifex*. From Macgregor and Andrews (1977); reproduced with permission.

study of chromosomal evolution. In situ hybridization has become the simplest and most direct way to physically map genes or any DNA sequence to chromosomes (Henderson, 1982; McKusick, 1988).

The usefulness of in situ hybridization for physical mapping of specific genes and other kinds of sequences to particular chromosomes and regions of chromosomes depends not only on the quality of cytological preparations but also on the unambiguous identification of homologous and homoeologous chromosomes. For this reason, I have included some reliable methods for preparing various kinds of chromosomes, including polytene and lampbrush chromosomes as well as mitotic metaphase spreads, and have also included methods for obtaining different kinds of banding patterns for the identification of homologs. The banding patterns themselves reveal information about the molecular structure of chromosomes, especially when sequence-specific dyes are used.

In many cases, chromosomes can be identified on the basis of their morphology alone including relative size, centromere position, and secondary constrictions (Figure 5; Table 1). Some chromosomes, such as insect polytene and oocyte lampbrush chromosomes, have intrinsic, complex patterns of bands or other markers that facilitate the identification of homologs (White, 1973). Chromosome identification in most species, however, depends on an analysis of induced banding patterns. Usually it is necessary to band the same chromosome preparations that are used for in situ hybridization.

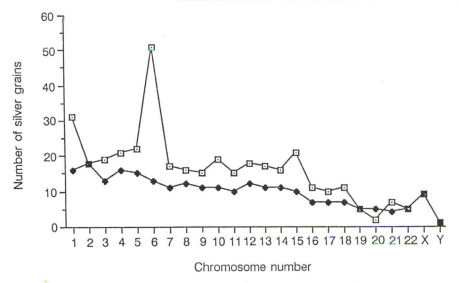

FIGURE 3 Distribution of silver grains over chromosome arms (long arms only) in 153 chromosome preparations in situ hybridized with a cDNA probe for *Tcp-1*, a gene coding for a testicular germ-cell protein. The gene is localized to the long arm of chromosome 6. Closed diamonds: expected (assuming random distribution); open squares: observed. After Fonatsch et al. (1987).

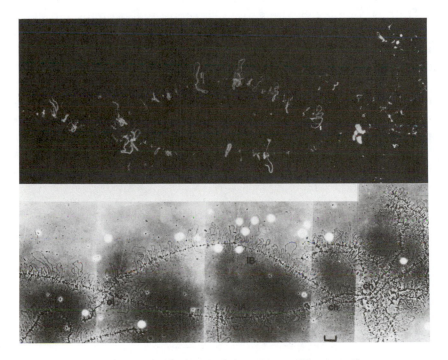

FIGURE 4 Monoclonal antibody directed against nuclear protein on lampbrush chromosomes of the salamander *Triturus vulgaris*. From Ragghianti et al. (1988); reproduced with permission.

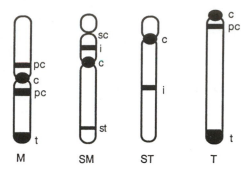

FIGURE 5   Diagram illustrating common terminology for chromosome morphology and positions of bands. Chromosome morphology: M, metacentric; SM, submetacentric; ST, subtelocentric; T, telocentric; band position: c, centrometric; pc, pericentric (or pericentromeric); i, interstitial; sc, secondary constriciton; t, telomeric (or terminal).

Assumptions

The primary assumption in the use of in situ hybridization is that chromosomal DNA can be denatured in such a way that it will anneal with reasonably high efficiency to complementary single-stranded nucleic acid probes to form hybrid duplexes. This is not a trivial matter, since chromosomal DNA is complexed with various chromosomal proteins and RNA. In fact, the efficiency of in situ hybridization is constrained by the difficulty in obtaining complete denaturation of chromosomal DNA, the loss of DNA during fixation and slide pretreatment, and the presence of chromosomal proteins (Henderson, 1982).

It is commonly assumed that banding patterns reflect differences in sequence organization (e.g., GC-rich or AT-rich regions) or repetitiveness. For example, G-bands are thought to represent AT-rich regions, and C-bands correspond to constitutive heterochromatin, generally assumed to be rich in highly repetitive sequences (Comings, 1978; Hsu, 1979; Schmid and Guttenbach, 1988).

Table 1. Terminology for chromosome morphology (centromere position)[a]

Centromere index	Terminology
0.00–0.12	telocentric = t
0.13–0.25	subtelocentric = st
0.26–0.38	submetacentric = sm
0.39–0.50	metacentric = m

[a]From Levan et al. (1964) and Green et al. (1980). Centromere index = length of short arm/length of whole chromosome.

Comparison of the Primary Methods

Preparation of Mitotic Metaphase Chromosomes   The preparation of useful spreads of mitotic metaphase chromosomes involves five steps: (1) selection of tissues with a high mitotic activity (or stimulation of such activity), (2) in vivo or in vitro treatment with a mitotic arresting agent (with or without cell cycle synchronization), (3) hypotonic treatment of tissues or cells, (4) fixing (and storing) tissues or cells, and (5) making permanent chromosome preparations on slides. Specific protocols for the production of mitotic chromosome spreads are given later in this chapter.

The simplest method for obtaining mitotic chromosomes is to select a tissue that has an intrinsically high rate of mitotic cell division in vivo. Root tips in plants, and developing embryos, larvae, or regenerating blastemas in animals are good in this regard. In adult vertebrates, high mitotic rates may be found in bone marrow, intestinal epithelium, corneal epithelium, kidneys, spleen, and gonads, depending on the species. Mitotic proliferation can sometimes be stimulated in vivo by subcutaneous or intraperitoneal injection of a mitogen such as phytohemagglutinin (PHA) or pokeweed mitogen (PWM), or even activated yeast suspension. Use of these tissues avoids the need for tissue culture, which can be expensive and unpredictably time consuming if working with a variety of species. The disadvantage of the in vivo technique is that specimens generally must be sacrificed to harvest the tissue.

In vitro methods involve culturing peripheral blood or cells from explants of various other kinds of tissues (see Freshney, 1987, for general tissue culture methods). There is a wide range of media and culture conditions that have been used for various species, and often the tissue culture requirements must be worked out for particular species of interest. One great advantage of in vitro culture is that it may be possible to synchronize cell cycles to increase the yield of chromosomes, decrease the variation in chromosome condensation between spreads, and decrease the required dose of the mitotic arrest agent (Watt and Stephen, 1986).

In general, both in vivo and in vitro chromosome cultures require a mitotic spindle inhibitor to block cells in mitotic metaphase. The most commonly used spindle inhibitors are colchicine, its synthetic analogue colcemid (deacetylmethylcolchicine), and vinblastin (Tjio and Levan, 1956; Macgregor and Varley, 1983; Watt and Stephen, 1986). For in vivo culture, relatively high concentrations of colchicine are injected directly into the body cavity or under the skin (colcemid is substantially more potent than colchicine and is used at lower concentrations). Some organisms, such as aquatic amphibian larvae, or tissues, such as plant root tips, can simply be immersed in a solution of colchicine. The optimal treatment time depends on the cell cycle of the species used and on the desired level of chromosome contraction. Cell cycle time is proportional to genome size (**C-value,** the haploid amount of nuclear DNA), at least in poikilotherms. The treatment time is short (1–2 hr) for organisms with small C-values (most vertebrates and invertebrates), but is much longer (24–72 hr) for species with very large C-values (e.g., lungfish, salamanders, and certain species of flowering plants). Much smaller amounts of colchicine (or colcemid) are needed for cells in tissue culture.

"Squash" and "Splash" Techniques   There are two main techniques for the production of permanent chromosome preparations suitable for molecular hybridization studies: the "squash" and the "splash" techniques. Both are designed to achieve optimal flattening and spreading of cytological material on microscope slides. In the squash technique, small pieces of tissue (thick cell suspensions can also be used) are macerated and/or finely minced on a slide and then firmly squashed beneath a siliconized coverslip. In the splash technique, a thick suspension of cells is splashed onto a slide from a pipette, usually from a distance, and spreading of cells and chromosomes occurs via surface tension. For both techniques it is critical that the tissues be exposed to a hypotonic solution (either distilled water or dilute KCl) before fixation, and then fixed in freshly prepared, ice-cold "3:1" fixative made with 3 parts absolute ethanol (for squashing) or methanol (for splashing) and 1 part glacial acetic acid.

A useful feature of the 3:1 fixative is that tissues (or cell pellets) can be stored in it for years so long as they are kept in tightly sealed vials at $-20°C$. If such long-term storage is necessary (or desirable) it is important to make sure that the tissues have been well fixed in at least two changes of 3:1 for at least 15 min (to remove all water) and then stored in fresh 3:1. For preparations that will eventually be used for in situ hybridization, it is advisable to keep the 3:1 fixative ice-cold to minimize hydrolysis. Storage at $-20°C$ is necessary because this fixative decomposes rapidly at room temperature. Cytological tissues fixed and stored in this manner have been used successfully not only for in situ hybridization but also for the extraction of high-molecular-weight DNA sequences (P. E. Barker et al., 1986).

To make slide preparations using the squash technique, small pieces of tissue are removed from the 3:1 fixative and briefly soaked in 45% acetic acid (this treatment hydrolyzes cytoplasmic components and can eventually reduce the tissue to a nuclear suspension). The softened tissue is then removed in a small drop of acetic acid to a very clean, subbed (gelatinized) microscope slide, minced as finely as possible, covered with a siliconized coverslip, and squashed firmly with the thumb on a cushion of absorbant paper. The preparation is now either examined directly, or made permanent by freezing on dry ice. The slides can then be stored indefinitely if kept desiccated at 4°C. The main advantages of the squash technique are that it is a very quick and efficient way to analyze a particular specimen without building up a large number of unnecessary slides. Also, excellent photo-micrographs of selected chromosome spreads can be obtained during the process of screening using phase contrast optics. A disadvantage of the squash technique is that the chromosomes can be damaged or lost during the process of making the slides permanent. Another common problem with the squash technique is that the preparation can be ruined by the slightest sideways movement of the coverslip during squashing or by the inclusion of large bits of tissue, lint, or air bubbles beneath the coverslip. The squash technique works best for organisms with very large chromosomes, especially salamanders, plants, and insect polytene chromosomes, and is not recommended for species with very small chromosomes such as mammals, birds, and reptiles because of the difficulty in obtaining sufficiently flattened chromosomes.

The splash technique involves preparing a suspension of cells fixed in 3:1 methanol:acetic acid. Methanol is used because it evaporates faster than ethanol. Cells are collected, dispersed, and incubated in a hypotonic solution (0.075 M KCl) either directly, or after centrifuging out of culture medium. The cells are then centrifuged out of the hypotonic solution and resuspended in fixative (preferably ice-cold), washed several times in fresh fixative, and finally resuspended in a small volume of fixative. The resulting concentrated cell suspension can either be stored at $-20°C$, or used immediately to make slides. The concentrated cell suspension is pipetteted (splashed) onto slides using various techniques designed to maximize spreading of cells and chromosomes. One commonly used method is to splash several drops of the cell suspension onto ice-cold slides wet with distilled water from a height of 0.5 m or more and drying them on a slide warmer (40°C). Another technique is to use a 1-ml pipetter and a plastic pipette tip to pipette a cell suspension up and down at several different spots on a slide that has been warmed to 60°C on a slide warmer. Each time the cell suspension is drawn back into the pipette, cells are left sticking to the slide in concentric rings. The suspension is drawn back into the pipette before moving to the next spot. An advantage of this latter technique is that several spots can be placed in controlled positions on a single slide, which facilitates screening the slides for good chromosome spreads. Splash slide preparations can be stored indefinitely if they are kept desiccated at 4°C.

A disadvantage of the splash technique is that it is sometimes difficult to see or photograph good examples of chromosome morphology until after the preparations have been stained and/or coverslipped (although chromosome spreads can be located using phase contrast optics or using a defocused condenser under bright field optics). Whichever technique is used to obtain chromosome preparations for in situ hybridization, it is important that the actual preparations are made near one end of the slide for ease of handling later, especially during autoradiography.

Cytological preparations for in situ hybridization should be made on slides that have been coated with a thin layer of gelatin ("subbed") to minimize loss of material during processing. Subbed slides are particularly important for autoradiography to prevent the nuclear track emulsion from slipping during development, fixing, washing, and staining.

Preparation of Polytene Chromosomes   Polytene chromosomes are somatic chromosomes that have undergone many rounds of endoreplication (DNA replication without division of the cell or nucleus) such that each chromosomal element consists of hundreds to thousands of unseparated **chromatids**. Polytene chromosomes are found in the cells of dipteran insect larvae, in collembolans (springtails), and in certain other invertebrates (White, 1973). The familiar bands of polytene chromosomes are formed by **chromomeres** (densely packed chromatid fibers) that are found along the length of each chromatid. Polytene chromosomes are particularly useful for gene mapping and comparative studies because of their large size and banding patterns.

Polytene chromosomes are very easy to prepare from dipteran larvae. Good examples are midges of the genus *Chironomus* and fruitflies of the genus *Drosophila*. In both of these organisms, larval growth occurs by increase in cell size rather than cell number, and involves endoreplication of their chromosomes. Polytene chromosomes are easiest to prepare from salivary glands, which are found near the anterior end of the larva. The glands are quickly exposed by removing the larva's head with watchmakers' forceps or needles and removing the adhering fat bodies; the glands can then be squashed on a subbed slide under a siliconized coverslip.

Preparation of Lampbrush Chromosomes   Lampbrush chromosomes represent bivalents at diplotene stage in female meiotic cells, and they are found in the oocytes of most animals (see Callan, 1986, for a recent review of lampbrush chromosome structure and function). Lampbrush chromosomes consist of two duplicated homologous chromosomes held together at regions of crossing over, or chiasmata (Figures 2 and 4). Each homologous chromosome of the bivalent consists of an axis formed by the two closely associated sister chromatids that connects a series of ellipsoid chromomeres. Lateral loops, each consisting of a single thread of DNA double helix, extend from many of the chromomeres and are sites of active RNA synthesis (Callan, 1986).

The largest and most easily studied lampbrush chromosomes occur in the oocytes of salamanders. A generalized method for obtaining lampbrush chromosomes from amphibian oocytes is included in the protocol section of this chapter (see also Callan et al., 1987). The procedure involves manually isolating and opening the nucleus ("germinal vessicle") of immature oocytes with needles and/or very sharp watchmakers' forceps in an unbuffered "isolation medium" consisting of a 5:1 mixture of 0.1 M KCl and 0.1 M NaCl. The nuclear contents (including the lampbrush chromosomes) are then transferred to and dispersed in an appropriate salt solution in an observation chamber on a specially designed slide. The optimal salt concentration varies among species and must be determined empirically.

The lampbrush chromosomes gradually sink to the bottom of the observation chamber and can then be observed and photographed with phase contrast optics. For in situ hybridization and/or immunocytochemistry the preparation must be centrifuged to firmly attach the lampbrush chromosomes to the coverslip at the bottom of the chamber. The plexiglass disc observation chamber is designed so that it will fit on an epoxy plug inside a large centrifuge tube (Figure 6). After centrifugation, the dispersal medium can be removed and the lampbrush chromosomes fixed and dehydrated.

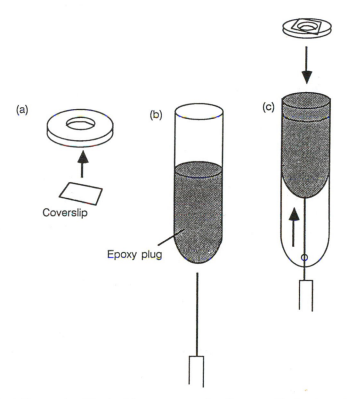

FIGURE 6 Centrifuge tube fitted with an epoxy plug for centrifuging lampbrush preparations (Protocol 11). (a) Dispersion chamber consisting of a plastic disk with a 7-mm hole bored in the center, and a coverslip attached to the bottom with paraffin; (b) polymerized epoxy plug in 30-ml centrifuge tube; (c) dispersion chamber is positioned on the epoxy plug by raising the plug with a probe inserted through a hole in the bottom of the centrifuge tube.

Chromosome Banding The three most common methods for banding chromosomes are **Q-banding, G-banding,** and **C-banding.** The simplest of these is Q-banding, which involves soaking the slides in a buffer and then staining with quinacrine mustard or quinacrine dihydrochloride. This produces fluorescent bands that are brightest in AT-rich regions of the chromosomes (Hsu, 1979) but that are also influenced by variation in protein composition of the chromosomes (Benn and Perle, 1986). The disadvantage of Q-banding is that the bands are visible only with UV optics and they fade quite rapidly. G-banding is also simple and involves brief treatment with trypsin or NaOH and staining with Giemsa (or similar dyes) in a phosphate buffer. The result is alternating light and dark bands, the latter representing primarily AT-rich regions and thus corresponding to most Q-bands.

Whereas Q- and G-banding require little or no pretreatment of the chromosomes, C-banding requires a stringent extraction step resulting in significant loss of chromosomal DNA (at least 60%; Pathak and Arrighi, 1973). For C-bands, chromosomes are treated with a strong base at an elevated temperature, incubated in a sodium citrate solution at high temperature, and stained in a concentrated Giemsa solution. This results in the extraction of almost all of the non-C-band chromatin, leaving only **constitutive heterochromatin,** which usually contains rapidly reassociating repeated sequences (Comings et al., 1973). Methods for Q-, G-, and C-banding are given in the protocol section.

Various specialized banding procedures have also been developed (see Rooney and Czepulkowski, 1986). Some of the most useful are fluorescence banding using various fluorochromes such as chromomycin A3 and DAPI, differential replication banding using bromodeoxyuridine (particularly useful for species that are difficult to G-band, such as salamanders; Figure 7), and **nucleolar organizer region (NOR)** banding using silver nitrate. Various restriction endonucleases have also been used to induce banding patterns (e.g. Ferrucci et al., 1987; Figure 8). Vast differences are often observed in the response of chromosomes of different organisms to these banding procedures, reflecting differences in chromosome organization. For example, banding with chromomycin A3 (counterstained with methyl green) or with distamycin A (counterstained with mithramycin) specifically stains small stretches of GC-rich chromosomal DNA and yields multiple chromosomal bands in mammals but stains almost nothing but nucleolar organizer regions in fishes, amphibians, and reptiles (Sessions and Kezer, 1987; Schmid and Guttenbach, 1988).

One potentially very useful (but little used) approach to the identification of homologs is the induction of **cold-induced constrictions,** or CICs (Callan, 1966; Sessions, 1982). These constrictions are chromosome specific, and can be induced in organisms with large chromosomes, such as

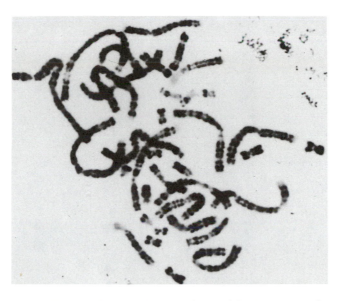

FIGURE 7   Differential replication banding in an embryo of the Japanese salamander, *Hynobius tokyoensis*. Reproduced with permission from Kuro-o et al. (1987).

certain plants and salamanders, by prolonged treatment of the specimens at 0.5–2.5°C in the presence of colchicine. The CICs apparently represent regions of incompletely condensed intercalary heterochromatin, and may be analogous to late-replicating fragile sites in human chromosomes (Laird, 1987).

Chromosome preparations can be banded either before or after in situ hybridization. For banding after autoradiography it is important to use a banding procedure that requires little or no pretreatment, such as G-banding or Q-banding. Prehybridization banding procedures requiring stringent pretreatments may have an adverse effect on the hybridization reaction due to loss of chromosomal DNA.

In Situ Hybridization   In situ hybridization of nucleic acid probes to chromosome preparations involves four general steps: (1) preparation and labeling of probe, (2) the hybridization reaction between probe and target, (3) removal of unbound or nonspecifically hybridized probe, and (4) visual detection of the sites of hybridization.

Any DNA (cloned or uncloned) can be used as a probe for in situ hybridization. DNA or RNA probes of any desired sequence can be obtained through recombinant DNA technology and/or by in vitro DNA amplification (see Chapter 9). Both RNA and DNA double-stranded or single-stranded probes can be used. RNA probes have the advantage that

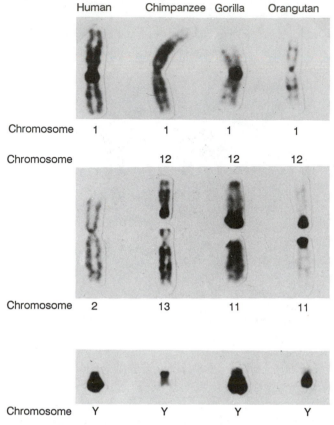

FIGURE 8   Restriction endonuclease banding in apes and human. Top row: human chromosome 2 and ape homologs; middle row: human chromosome 2 and ape homologs; bottom row: human Y chromosome and ape homologs. Reproduced with permission from Ferrucci et al. (1987).

they produce less background than DNA probes, and unbound or non-specifically hybridized RNA is efficiently removed by RNase digestion after the hybridization reaction. It is somewhat more difficult to reliably remove nonspecifically bound DNA probes. In general, a better signal-to-noise ratio is obtained when vector sequences are removed from the purified DNA, although in some cases the extra DNA provided by the vector may facilitate the formation of DNA networks that enhance the signal (see below).

There are three main sources of labeled nucleic acid probes used for in situ hybridization: (1) in vivo labeling of ribosomal RNA, (2) in vitro production of RNA from DNA templates using either *Escherichia coli* or SP6 polymerase, and (3) nick translation of double-stranded DNA using DNase I and DNA polymerase I (for other labeling techniques see Arrand, 1985; Schleif and Wensink, 1981; Berger and Kimmel, 1987; Ausubel,

1989; and Chapter 8). The radioisotopes most commonly used to label probes for nucleic acid hybridization are ^{32}P, ^{125}I, and ^{3}H; choosing one for in situ hybridization involves a trade-off between sensitivity and resolution. ^{32}P yields the highest specific activity and is extensively used in transfer-hybridization experiments (Chapter 8) but is not used in in situ hybridization because its high energy disintegrations result in poor resolution. Tritium is usually considered the best radioisotope for in situ hybridization because of the extremely low energy of the β particles emitted (0.018 MeV; Pardue, 1985). The low energy β particles emitted by ^{3}H travel only about 1 μm through autoradiographic emulsion, which results in close spatial correspondence between silver grains and hybridized target (Macgregor and Varley, 1983; Pardue, 1986). The disadvantage of using tritium is that it often necessitates long autoradiographic exposure times, depending on the specific radioactivity of the probe. Shorter exposure times are achieved with ^{125}I but the radiation emitted is significantly more energetic than that of tritium with the danger of less precise resolution and higher background (Pardue, 1986).

The highest specific radioactivities are achieved by in vitro labeling (Macgregor and Varley, 1983; see protocol section). In vitro transcription of RNA by *E. coli* polymerase for in situ hybridization involves the random transcription of double-stranded or single-stranded DNA to yield labeled RNA (a protocol is given below). In vitro transcription with SP6 RNA polymerase involves cloning a known DNA sequence adjacent to a promoter of the phage SP6 to produce a single-stranded tritium-labeled RNA probe. SP6 polymerase and cloning vectors are commercially available (see Pardue, 1985 and Chapter 9 for cloning protocols).

Nick translation of DNA is usually considered the most successful and generally useful method for obtaining labeled probes with high specific activity for in situ hybridization (Macgregor and Varley, 1983). The advantages of nick translation are that it is a relatively rapid, simple, and inexpensive reaction, it yields uniformly labeled DNA, and it can be used to produce probes with high specific activities (Ausubel, 1989). Furthermore, nick translation can be used to prepare either radioactive or nonradioactive (biotin, bromodeoxyuridine) probes. Nick translation involves the use of DNase I to create single-strand "nicks" in double-stranded DNA, followed by exposure of the nicked DNA to *E. coli* DNA polymerase I in the presence of labeled deoxyribonucleotides. The polymerase binds to a nick and removes nucleotides from the 5' side while adding labeled nucleotides to the 3' side as it moves down the DNA strand (Arrand, 1985). This action results in the translation (i.e., translocation) of the nicks along the DNA strand in a 5' to 3' direction and the replacement synthesis of radioactive DNA strands. Denaturation of nick-translated DNA thus yields uniformly labeled DNA fragments that can be used as probes. The length of the fragments produced as well as the extent of incorporation can be controlled by the amount of DNase I that is used. The specific activity of the probes is controlled by the number and specific activities of tritiated deoxyribonucleotides. The best results are obtained if all of the deoxyribonucleotides are tritiated, but at minimum the reaction should include [^{3}H]dCTP and [^{3}H]dTTP, since these are available at the highest specific activities. The detection of single copy sequences requires a specific activity of $2–4 \times 10^7$ cpm/μg, which can be achieved by the use of precursors with high specific activity (40–110 Ci/mmol; Harper and Saunders, 1984).

Labeled probes may also be obtained by in vitro DNA amplification: the polymerase chain reaction (PCR; Saiki et al., 1988 and Chapter 9). This technique involves the exponential amplification of a DNA segment by repeating cycles of polymerase-mediated oligonucleotide primer extension. The basic PCR cycle consists of (1) heating to denature the template DNA, (2) cooling to permit precise primer annealing, and (3) warming to facilitate polymerase-mediated extension. For labeled probes, radioactive or biotinylated deoxyribonucleotides are included in the reaction mixture. Since the copy number of the DNA segment doubles after each cycle, it can be amplified several million fold (2^n, where n = the number of cycles).

Nonradioactive labeled probes have been successfully used in in situ hybridization for the detection of reiterated sequences. The most successful nonradioactive technique involves the use of biotin-substituted nucleotides to label nucleic acid probes (Langer et al., 1981). The probe is detected by biotin-specific antibodies, avidin, or streptavidin conjugated to fluorescent or enzymatic reagents. The main advantage of nonradioactive probes is that they eliminate the need for autoradiography and long exposure times. Also, the probes are stable for long periods of time and may be stored at −20°C for many months or years. The disadvantages are that the detection methods are less sensitive than autoradiography, are not as easily quantified, and are often more difficult to photograph adequately with black and white film.

The Hybridization Reaction Although almost every laboratory has its own minor modifications, the hybridization reaction of nucleic acid probes to chromosomal preparations involves four general steps (see protocols): (1) removal of endogenous RNA from the cytological preparation (for DNA targets), (2) denaturation of the DNA of the target chromosomes (for DNA targets), (3) incubation of the cytological preparation with probe dissolved in hybridization buffer, and (4) removal of nonspecifically bound probe.

For hybridizing to DNA targets it is important to remove endogenous RNA because it can compete with target sequences for hybridization with the labeled probe. Endogenous RNA is removed before hybridization by treating the cytological preparations with a high concentration of pancreatic ribonuclease, or with a mixture of ribonuclease A and ribonuclease T1. This step is, of course, omitted if the target is RNA.

Chromosomal target DNA must be denatured for in situ hybridization. This is accomplished by briefly treating the slide preparations with 0.07 M NaOH at room temperature. Alternatively, the slide preparation can be exposed to 70% formamide at a higher temperature (formamide lowers the melting temperature of double-stranded DNA). Unfortunately, both procedures result in some degradation of chromosomal morphology. Alternatively, denaturation with 0.1 M HCl results in better chromosomal morphology but reduced hybridization due to depurination of the target DNA (Macgregor and Varley, 1983; Pardue, 1986).

The actual hybridization reaction involves first dissolving the probe in hybridization buffer with 50% formamide. Optionally, 10% dextran sulfate, and sheared, denatured, noncompetitive carrier DNA at 1000-fold greater concentration than the probe may also be added to the hybridization reaction mixture (see below). Formamide lowers the melting temperature of double-stranded DNA, and thus lowers the temperature required for the hybridization reaction. Formamide is especially important for RNA–DNA hybridization because it retards the rate of RNA degradation (Schleif and Wensink, 1981). The formamide should be of the highest quality, deionized, and stored at −20°C. The addition of dextran sulfate to the hybridization mixture accelerates the hybridization reaction approximately 10-fold by increasing the effective concentration of the probe, and favors maximum binding by enhancing the formation of networks of probe molecules on the target. The addition of dextran sulfate to the hybridization mixture is essential for the detection of single-copy sequences. Adding sheared, denatured, noncompetitive, unlabeled DNA to the hybridization mixture reduces nonspecific binding of the probe. For DNA probes, such "blocking" DNA can be extracted from *E. coli* (Pardue, 1986) or from salmon sperm (Malcolm et al., 1986). Unlabeled ribosomal RNA from *E. coli* can be used for RNA probes (Pardue, 1986).

Probes for in situ hybridization should be used at a concentration that will nearly saturate the target DNA without contributing to background silver grains. Large targets, such as localized clusters of repetitive sequences and the reiterated sequences of polytene chromosomes, require lower concentrations. The optimal probe concentration depends on the probe, its specific radioactivity, and the nature of the target. Therefore, it is generally best to use a range of concentrations (e.g., from 1 to 20 ng per slide; Malcolm et al., 1986). In general, the probe should be dissolved to provide a total of 3×10^6 cpm, as determined by a liquid scintillation counter (Macgregor and Varley, 1983).

The hybridization reaction is initiated by pipetting the probe mixture onto the center of the cytological slide preparation, covering with a coverslip, and incubating in a moist chamber at 37°C for approximately 12 hr.

After an appropriate incubation time, the hybridized cytological preparations must be washed to remove probe molecules and their degradation products that are not bound to complementary sequences on nuclei or chromosomes. This procedure is necessary to remove both weakly hybridized molecules and unbound or nonspecifically bound molecules and is essential to reduce background radioactivity. This step can involve different levels of stringency depending on the nature of the hybrid. Washing usually involves incubation in 2× SSC at a temperature that is slightly lower than that used for the hybridization reaction, in addition to treatment with 50% formamide or 5% cold TCA. If an RNA probe is used, washing includes a mild digestion with ribonuclease. Following the washing step, the slides are usually dehydrated in ethanol and air dried. The slides are now ready for autoradiography.

Autoradiography   Visualization of sites of hybridization between a radioactive probe and its cytological target is achieved by autoradiography. This procedure involves coating the slides with a thin layer of nuclear track photographic emulsion consisting of silver halide crystals suspended in a gelatin matrix. Radiation from light or from radioactivity sensitizes the crystals to form a "latent image," which is visualized when the crystals are reduced to metallic sliver by photographic developer. The resulting grain density is highest immediately over the source of radiation and decreases symmetrically on each side of the source with increasing distance. The rate of decrease of grain density from the source determines the resolution, and is dependent on the radioisotope that is used. The vast majority of silver grains produced by the β particles emitted by tritium will be located within 0.5 μm from the source, although 1–2% of the particles may travel up to 3 μm. ^{125}I produces a greater scatter of grains, up to 16 μm from the source, although approximately 90% of the grains will fall within a 3.5 μm radius and at least half of the grains will be at the same distance as those produced by tritium (Henderson, 1982).

Several different nuclear track emulsions are available with different sensitivities. The single most important property of the emulsion is the intrinsic background of silver grains formed in the absence of exposure. Therefore, it is necessary to test each batch of new emulsion as it arrives from the supplier by developing coated blank slides and examining them under a microscope. A background of less than 50 grains per field of view under a 100× oil objective is considered very good, but a grain count of over 100 is unacceptable (Macgregor and Varley, 1983). Unacceptable emulsion should be returned to the supplier for a replacement. Background can also be controlled by careful handling and storage of the emulsion.

After the slides are coated with emulsion, the preparations are exposed for a length of time that must be determined empirically. The objective is to obtain a sufficient number of silver grains to unambiguously detect hybridization but not so many that cytological detail is obscured. The best exposure time for a particular in situ hybridization experiment can be determined by including several replicates or cytologically suboptimal preparations that can be used as test slides. One test slide is developed at a given interval to determine whether exposure has been adequate. Exposure times can vary from days to months, depending on the concentration and specific activity of the hybridized probe molecules.

Detection of Hybridization by Other Methods   Nonradioactively labeled probes are detected using fluorescence (e.g., fluorescein isothiocyanate, FITC) or an enzyme (alkaline phosphatase or horseradish peroxidase) that can be reacted with a substrate to form a cytologically visible stain. Biotin-labeled probes can be detected by treating the hybridized preparations with an anti-biotin primary antibody (e.g., goat) followed by a secondary antibody (e.g., rabbit anti-goat) that is conjugated with horseradish peroxidase or FITC. The peroxidase is then visualized by reacting with diaminobenzidine tetrahydrochloride (DAB) and hydrogen peroxide, which results in a reddish-brown to black signal. Alternatively, the preparations are treated with alkaline phosphatase- or peroxidase-conjugated avidin (or streptavidin), proteins that bind to biotin very tightly (McInnes et al., 1987). Alkaline phosphatase is visualized by reacting with 5-bromo-4-chloro-3-indolyl phosphate (BCIP) and nitro blue tetrazolium (NBT), resulting in the deposition of a purple precipitate at sites of hybridization. Complexing the avidin with electron-dense colloidal gold allows the hybridized probes to be visualized with electron microscopy (Hamkalo and Hutchison, 1984).

Bromodeoxyuridine (BrdU)-labeled probes can be visualized with four-layered immunohistochemistry using mouse anti-BrdU monoclonal primary antibody, rabbit anti-mouse secondary antibody, peroxidase-conjugated swine anti-rabbit third antibody, and peroxidase antiperoxidase complex fourth antibody (Frommer et al., 1988).

Whereas the products of both alkaline phosphatase and peroxidase staining are stable for long periods of time, fluorescein has the disadvantage that it fades rather quickly. A solution of p-phenylenediamine is sometimes added to retard fading of fluorescence (Pinkel et al., 1986). Preparations labeled with fluorescein can be counterstained with two DNA-specific fluorescent dyes: 4,6-diamidino-2-phenylindole (DAPI) and propidium iodide (Pinkel et al., 1986). The propidium iodide fluoresces red and allows simultaneous observation of the yellowish-green fluorescein-labeled hybridized probe and total DNA. DAPI fluoresces blue and is used so that biotin-labeled and total DNA can be observed separately. Fluorescein and propidium iodide are excited at 450–490 nm and propidium iodide can be viewed separately at 546 nm. DAPI is excited in UV at 360 nm.

APPLICATIONS AND LIMITATIONS

Applications

Cytogenetic studies can contribute an array of information independent from morphological, biochemical, behavioral, and other characters that are used for phylogenetic analysis. As with biochemical data, cytogenetic information can reveal differences and similarities that may not be obvious at the morphological level. The inherent attractiveness of cytogenetics is that it encompasses several levels of biological organization ranging from the morphological to the molecular, depending on the applicable technology. At one extreme, the overall amount of DNA in a genome can have substantial phenotypic consequences at the "whole organismal" level in terms of cell size and cell cycle time and the effects these cellular parameters can have on organismal development rate (Sessions and Larson, 1987). These phenotypic correlates of genome size have been termed "nucleotypic" by Bennett (1972). Chromosomes can be studied as a morphological manifestation of the genome in terms of their microscopically visible size, shape, number, and behavior during meiosis and mitosis. The analysis of ploidy levels or meiosis can provide unique insights into changes in breeding systems or modes of inheritance (e.g., apomixis or translocation heterozygosity). At a lower level, banding studies reveal aspects of the general structural organization of chromatin along the lengths of individual chromosomes. At a still lower level, chromosomes may be probed with known DNA or RNA sequences using in situ hybridization to reveal finer details of chromosomal anatomy in terms of the spatial arrangements, or presence and absence, of particular kinds of sequences. It is now possible to go one step further and directly microdissect selected regions of chromosomes for sequence analysis (Pirrotta, 1986).

Most phylogenetic studies using cytogenetics have used mainly chromosome number and morphology and, less commonly, various kinds of chromosome bands. Most of the phylogenetically useful variation in chromosome number and morphology is attributable to **Robertsonian translocations** (fusions and fissions of chromosomes at their centromeres) and two different kinds of inversions (Figure 9): **pericentric inversions,** involving the centromere, and **paracentric inversions** occurring outside of the centromeric region. While pericentric inversions can often be documented by chromosomal morphology alone (i.e., shifts in centromere position), confirmation of Robertsonian translocations and paracentric inversions usually requires some kind of chromosome markers such as bands, NORs, or hybridized probes.

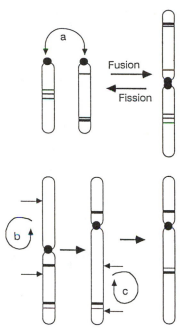

FIGURE 9. Major kinds of chromosome rearrangements. (a) Robertsonian translocations (fusion and fission); (b) pericentric inversion; (c) paracentric inversion (small arrows indicate location of breaks).

Although differences in chromosome structure are often correlated with taxonomic differentiation, the role of cytogenetic change in actual processes of speciation is controversial (Patton and Sherwood, 1983; Sites and Moritz, 1987). The fixation of structural rearrangements, which may involve shifts in heterochromatin, is probably far more important to speciation than changes in the amount of heterochromatin (Patton and Sherwood, 1982). Some organisms, such as many groups of salamanders, show little or no variation in cytologically visible aspects of chromosome structure despite extensive changes in morphology and biochemistry, suggesting that chromosomal morphology has been strongly constrained (Sessions and Kezer, 1987). The reasons for this cytogenetic stasis are unknown.

The easiest and most successful application of in situ hybridization concerns the localization of any moderately long sequence that is repeated more than 100 times at one place in the genome (Macgregor and Varley, 1983). Consequently, most comparative studies have focused on repetitive sequences such as those coding for ribosomal RNA, tRNA, and histones, as well as highly repeated satellite DNA sequences. Single-copy genes have always been easily detected in dipteran polytene chromosomes because all gene sequences are multiplied several hundred times and are localized and concentrated.

Although there have been many comparative studies of sequence local-ization using in situ hybridization, these data have rarely been used fore-stimating phylogeny. Phylogenetic analyses are possible using such char-acters as the location(s) of sequences (e.g., various repeat families) among and within chromosomes, sequence structure and copy number, spatial relationships among identified genes and other sequences and to specific bands or other markers, and the localization of functional versus nonfunc-tional NORs. These kinds of studies have been particularly important for the identification of homologies between chromosomes or parts of chromo-somes among distantly related species for phylogenetic analysis (e.g., *Dro-sophila,* Steinemann et al., 1984).

The use of such characters is predicated on our understanding of their evolution. Two different (but not mutually exclusive) views concerning the mode of evolutionary change in the molecular structure of chromosomes are the "repatterning" hypothesis (Mancino et al., 1977; Cremisi et al., 1988) and the "homosequentiality" hypothesis (Macgregor and Sherwood, 1979). According to the repatterning hypothesis, interspecific differences in the chromosomal location of certain repetitive DNA sequences (e.g., ribo-somal RNA genes) reflect the redistribution of chromosomal elements with-in karyotypes. A prediction based on this view is that evolutionary changes in sequence location should be relatively conservative (i.e., slow, unique, and irreversible). The homosequentiality hypothesis, on the other hand, postulates that differences in the apparent location of various sequences reflect localized amplification or diminution of sequences with fairly stable chromosomal locations. This view predicts relatively rapid and reversible changes in the cytologically visible chromosomal location of certain kinds of sequences. A likely cytogenetic mechanism for this kind of rapid change is unequal crossing over within repetitive sequences, which can quickly generate products of different lengths. A particular sequence could quickly become too small to be visualized by in situ hybridization or, alternatively, a diminutive sequence could rapidly expand in size and become detectable.

A testable model for evolutionary changes in chromosome size and molecular structure involving major groups of repetitive sequences is pre-sented by Macgregor and Sessions (1986). According to this model, highly repetitive DNA sequences originate and grow at particular chromosomal locations such as centromeres, or wherever they are tolerated. Nonhomol-ogous recombination at such sites would result in the rapid spread of the new sequence throughout the karyotype and subsequent concerted evolu-tion among nonhomologous chromosomes. Subsequent small rearrange-ments would result in the gradual stochastic breakup, dispersal, and de-gradation of these sequences as they are moved away from the centromere regions along the chromosome arms. A prediction based on this model is that recently evolved sequences should be homogeneous in structure, localized in large clusters (especially at or near the centromeres or telomeres), functionally inert, and taxonomically restricted in occurrence. Ancient sequences, on the other hand, should show more sequence complexity, should be found as small clusters in intercalary positions along the chromosome arms, may be transcribed, and are likely to have a wider taxonomic occurrence. Information that is consistent with this model is available from both salamanders (Macgregor and Sessions, 1986; Cremisi et al., 1988) and plants (Flavell, 1986).

There is growing evidence that much of the repetitive DNA in mammalian genomes may have originated from functional genes (e.g., tRNA genes) through a **retroposition** (reverse transcription) mechanism (Deininger and Daniels, 1986). This mechanism involves the copying of RNA molecules back into DNA with subsequent integration and amplification of these copies at new genomic sites. Retroposition may have been the dominant source of the major repetitive DNA families in mammals, but there are no known examples of high copy number retroposon families in non-mammalian genomes (Deininger and Daniels, 1986).

The picture that has emerged from comparative studies on diverse organisms is that genomes are almost incredibly dynamic in terms of position and/or structure of identified sequences, especially in terms of the number, kinds, and locations of various repetitive sequences. Cytogenetic mechanisms such as unequal crossing over, inversions, and translocations have clearly played a dominant role, and we are just beginning to understand the role of transposons, retroposons, and the phenomenon of gene conversion in chromosomal evolution (Doolittle, 1985; Deininger and Daniels, 1986). It is clear that we have very little understanding of the relationship between the molecular structure and function of chromosomes. For example, clusters of ribosomal sequences have been found on almost every single chromosome, in addition to a stable nucleolus organizer region (NOR) in the European newt, *Triturus vulgaris* (Andronico et al., 1985), and certain simple-sequence satellite DNA sequences are transcribed by lampbrush chromosomes in salamander oocytes (Varley et al., 1980). These results make it difficult to make testable predictions concerning rates, constraints, and directions of evolutionary change, and indicate that full and proper use of molecular cytogenetic information for phylogenetic analyses will require a better understanding of the molecular basis of chromosome structure and function.

Limitations

One of the most serious limitations of molecular cytogenetics concerns the reliability of chromosome identification. Ideally, this identification should be based on banding patterns or some other chromosome-specific markers, independent of the localization of particular sequences. The chromosomes of most mammals and various other organisms are readily G-banded and show complex, chromosome-specific banding patterns. Other organisms, such as amphibians, have seemingly G-band-resistant chromosomes, and unambiguous chromosome identification is more difficult and requires a variety of specialized banding techniques.

Limitations of in situ hybridization mainly concern the sensitivity and efficiency of the hybridization reaction. The sensitivity of in situ hybridization using radioactive probes depends on three main parameters: (1) the specific radioactivity of the probe, (2) the efficiency of the hybridization reaction, and (3) the autoradiographic procedure. For many years these parameters limited most in situ hybridization studies to repetitive sequences that can be localized with poorly defined probes of low specific activity and suboptimal hybridization conditions (Henderson, 1982). The specific radioactivity of a probe is limited only by the specific activity of the nucleotide precursors used in the synthesis of the probe. For clusters of repeated sequences, including polytene chromosomes, RNA probes labeled with [^{3}H]UTP at 50 Ci/mmol are sufficiently radioactive (Pardue, 1985). For smaller targets, the specific radioactivity of the probe can be increased by using additional ^{3}H-labeled nucleotides.

The efficiency of hybridization depends on numerous factors, including the concentration of the probe, the ionic strength of the hybridization mixture, the incubation temperature for the hybridization reaction, the type of chromosomes and the complexity of the site, as well as the method used to prepare the slides and the age of the slides. Ideally, all available complementary target sites will hybridize with the probe at saturation concentrations. This is precluded, however, by the nature of cytological preparations, including the difficulty in obtaining complete denaturation of chromosomal DNA, loss of chromosomal DNA during denaturation, and the possibility of steric hindrance by chromosomal proteins (Henderson, 1982). Overall, the efficiency of hybridization has been estimated to be 6–10% (Macgregor and Varley, 1983).

Some of these limitations of the hybridization reaction have been counteracted by using dextran sulfate in the hybridization reaction. Dextran sulfate is essential for the detection of single-copy chromosomal sequences (Harper and Saunders, 1984). It is possible that the signal can also be enhanced by vector sequences that are attached to cloned probes. These sequences are radiolabeled and free to participate in network formation, thus contributing to the overall signal at the hybridization sites.

The main limitation with nonradioactive probes is that they cannot detect target chromosomal sequences of less than 30–60 kb. Also, problems are often encountered in the accessibility of chromosomal target DNA to the reagents, and "halos" of signal are often seen around chromosomal targets that represent diffuse strands of DNA that are more accessible than the compact DNA in the interior of the chromosomes (Pinkel at al., 1986). These problems can be minimized by careful preparation and storage of the prehybridized slides, and (in the case of biotinylated probes) amplifying the signal by using multiple layers of avidin (Pinkel et al., 1986).

LABORATORY SETUP

The most essential piece of equipment for cytogenetic studies is a compound microscope equipped with high-quality phase-contrast objective lenses. Other equipment needed, for banding procedures, molecular techniques, and even tissue culture, are commonly found in most laboratories. Table 2 lists the major equipment necessary to set up a molecular cytogenetics laboratory. (For a complete listing of worldwide suppliers of equipment and supplies, see *The Biotechnology Directory 1989*, Stockton Press, New York.)

PROTOCOLS

1. "Subbed" slides
2. Mitotic chromosomes from gut epithelium
3. Mitotic chromosomes from plant root tips
4. Squash technique for mitotic and meiotic chromosomes
5. Yeast method for mitotic chromosomes from small vertebrates
6. Splash technique for slide preparations of mitotic chromosomes
7. Mitotic chromosomes from peripheral blood in vertebrates
8. Mitotic chromosomes from fibroblast cultures (reptiles)
9. Mitotic chromosomes from insect embryos
10. Polytene chromosomes from dipteran salivary glands
11. Lampbrush chromosomes
12. C-banding
13. Q-banding
14. G-banding
15. Fluorochrome banding with chromomycin A3
16. $AgNO_3$-banding for NORs
17. Differential replication banding with BrdU
18. Modification of BrdU banding for salamander embryos
19. Labeling probes via nick translation for in situ hybridization
20. Labeling in situ hybridization probes via transcription of cRNA using *E. coli* RNA polymerase
21. In situ hybridization procedures for reiterated sequences using a DNA probe
22. In situ hybridization using an RNA probe
23. Localization of single copy sequences on chromosomes
24. Autoradiography for detection of radioactive probes
25. Postautoradiography G-banding with Wright's stain
26. In situ hybridization with biotin-labeled DNA probes

Table 2. Equipment needed to set up a molecular cytogenetics laboratory[a]

Equipment	Use (protocols in parentheses)
Autoclave	sterilization*
Balance, analytical	weighing small samples*
Balance, top-loading	weighing large samples*
Camera attachments for microscopes	chromosomal photography*
Centrifuge, benchtop, with swinging rotors (preferably refrigerated)	preparing mitotic chromosomes (5, 6, 7, 8, 9, 18) and lampbrush chromosomes (11)*
Centrifuge, microtube	centrifuging small samples
Centrifuge, vacuum	probe labeling (19, 20)
Darkroom	autoradiography (24), general photography*
Fluorescence attachments for microscope	fluorochrome banding (15), fluorochrome labeled probes (27) and antibodies (29)
Freezers ($-20°C$ and $-70°C$)	storing chromosome tissues, probes, labels, and antibodies*
Incubator, tissue culture (20–37°C)	tissue culture (7, 8)
Microscope, compound, with phase contrast lenses (minimum: 10×, 20×, 40×, 100× oil immersion)	examining chromosomes*
Microscope, dissecting binocular, with substage lighting	polytene chromosomes (10), lampbrush chromosomes (11)
Microscope, inverted compound, with phase contrast lenses	checking tissue cultures, lampbrush chromosomes (11)
Ovens, 30–90°C	C-banding (12), AgNOR-banding (16), labeling probes (19, 20, 27), in situ hybridization (21–23, 26)*
pH meter	making buffers, etc.*
Refrigerators	storing samples and reagents*
Safelight filter with 25W bulb	autoradiography (24)
Scintillation counter and geiger counter	labeling probes (19, 20, 26)*
Tissue culture hood	tissue culture (8)
UV light source, 15W	R-banding (17, 18)
Waterbaths, 45–100°C	autoradiography (24), in situ hybridization (21, 28)
Water purification system	

[a] Items that are essential for most protocols are marked with an asterisk.

27. Detection of hybridized biotinylated probes with fluorescein-avidin
28. In situ hybridization with BrdU-labeled probes
29. Immunolocalization of antigens on lampbrush chromosomes

Protocol 1: "Subbed" Slides

(Time: ≈1 hr handling plus 24 hr incubation)

Microscope slides should be very clean; washing in hot water and detergent is recommended, but at minimum slides can be agitated in 95% ethanol to which several drops of glacial acetic acid have been added, and then wiped dry. Reference: Macgregor and Varley (1983).

1. Wash slides in hot water and detergent, and rinse copiously in hot water and then distilled water.
2. After a final rinse in distilled water, dip the slides into the subbing solution (Appendix).
3. Drain the slides and dry in a rack overnight at 60°C. Subbed slides can be stored indefinitely in a slide box.

Protocol 2: Mitotic Chromosomes from Gut Epithelium

(Time: incubation from 2 to 48 hr, depending on species, plus ≈1 hr handling)

This technique (from Kezer and Sessions, 1979) works best for amphibians.

1. Give healthy, well-fed animals an intraperitoneal injection of 1.0% aqueous colchicine, approximately 0.1 ml/g body weight.
2. Let animal incubate at a physiologically comfortable temperature for 1 hr (mammals), 4 hr (reptiles), 24 hr (frogs), or 48 hr (salamanders).
3. Kill the animal by overanesthesia (see Chapter 3).
4. Remove the stomach and intestines, squeeze out any contents, and open lengthwise using fine pointed scissors. Also remove the spleen, kidneys, and (if male) gonads and make small cuts with scissors.
5. Submerge tissues in a large volume (e.g., 50 ml) of distilled water in a flask or beaker and agitate vigorously for 10–15 min. The water should be changed if it becomes cloudy or full of debris.
6. Remove tissues from water, blot briefly on paper towels, and submerge in 50–100 ml of freshly prepared, ice-cold 3 parts ethanol, 1 part glacial acetic acid for at least 15 min on ice (this first volume of 3:1 fixative can be reused for all specimens during a particular fixing session if kept cold).
7. Transfer fixed tissues to a 20-ml vial with fresh, cold 3:1 fixative. Glass scintillation vials with plastic cone inserts in the caps are ideal for storing tissues fixed in 3:1 (do not use foil liners, as they will decompose into the fixative). The tissues can now be stored indefinitely (10 years or more) in an ordinary freezer, or used immediately to make slides using the squash technique (see Protocol 4).

Protocol 3: Mitotic Chromosomes from Plant Root Tips

(Time: 4–6 hr incubation plus ≈12 min)

Root tips may be obtained either from germinated seeds or bulbs, or from the surface of the potting soil when the plant is tipped out of its pot. For potted plants, it is best to water liberally 1 or 2 days before taking root tips. Seeds may be germinated on moist filter paper in a petri dish, and roots can be obtained from bulbs by suspending them with toothpicks over dishes of water so that they are partially submerged. Healthy growing root tips are brittle, translucent, and white, with more opaque, tapered tips. The most rapidly dividing cells are located proximal to the tip. If roots are not available, it is possible to use young leaves or the mitotically active ovary or ovule wall of developing flowers or fruits instead (Dyer, 1979).

1. Cut off the distal 1 cm of the root tips and immerse them in a solution (aerated with an airstone) of 0.05% colchicine at room temperature for 4–6 hr (germinated seeds can be left intact).
2. Fix severed root tips in freshly prepared 3:1 ethanol:acetic acid for at least 1 hr.
3. Macerate the tissues in 1 M HCl at 60°C for 5 min.
4. Soak the root tips in 45% acetic acid for 1–5 min.
5. Transfer a root tip to a drop of 45% acetic acid on a clean slide, cut off the terminal 1 mm of the root tip and discard the rest, and crush and mince thoroughly with a scalpel or razorblade.
6. Make squash preparations (Protocol 4).

Protocol 4: Squash Technique for Mitotic and Meiotic Chromosomes

(Time: <5 min per preparation)

There is a certain amount of "art" in making good squashes and practice is usually necessary. Once the technique is mastered it is very fast, and it is convenient to set up the slide preparation station adjacent to the microscope so that each preparation can be examined immediately. Reference: Kezer and Sessions (1979).

1. Remove a small piece of tissue from 3:1 fixative and submerge it in 45% acetic acid in a small glass dish for at least 2 min (tissue will disintegrate after prolonged exposure).
2. Put a small bit of tissue (e.g., 2 mm^2) in a single drop of 45% acetic acid toward one end of a clean subbed slide.
3. Mince tissue as finely as possible using forceps and scalpel or razorblade. The result should be a cloudy suspension of cells and small clumps of cells. Remove any remaining chunks of tissues, lint, or other solid bits of debris.
4. Cover cell suspension with a clean siliconized coverslip. To avoid bubbles, the coverslip should be lowered gradually by placing one edge down first, in contact with the suspension on the slide, and then slowly lowering the coverslip with forceps.

5. To squash cells, put the slide between layers of absorbant paper, stabilize the coverslip by pushing down firmly with thumb and index finger on the top layer of paper near two edges of the coverslip, and push down very hard with the thumb of the other hand in the center of the coverslip. Slipping of the coverslip, which may ruin the preparation, can sometimes be avoided by tapping gently on the coverslip with a pencil eraser before squashing.

6. The slide can now be examined directly with phase-contrast optics to check for suitable chromosome spreads. A gross phase-contrast effect can be obtained with regular bright field optics by defocusing the condenser lens. It is useful at this stage to record the coordinates of particularly good spreads, and, if working with large chromosomes (e.g., salamanders), to photograph selected examples. Photography is particularly useful at this stage if it is a rare specimen and good chromosome spreads are difficult to find, since subsequent treatment of the slides may destroy or degrade chromosome morphology.

7. The slide can be made permanent with the "dry ice technique" (Conger and Fairchild, 1953) by placing the slide on a block of dry ice for at least 5 min, then quickly prying off the coverslip with the point of a scalpel or razorblade and plunging the slide into 95% ethanol for at least 2 min. The slides are then air dried, and can be stored indefinitely if kept in a sealed slide box with a cotton-stoppered vial of desiccant at 4°C.

Protocol 5: Yeast Method for Mitotic Chromosomes from Small Vertebrates

(Time: >24 hr incubation time, plus ≈3 hr handling)

This technique is based on the stimulation of white blood cell proliferation in bone marrow. For mammals, sufficient bone marrow can be obtained from the long bones of the limbs. For small lizards, bone marrow may also be obtained by removing and crushing the spine (C. Moritz, personal communication). The volumes given are based on tissues obtained from an adult laboratory mouse; they may be reduced or increased for substantially smaller or larger amounts of tissue. Reference: Lee and Elder (1980).

1. Inject animal with active yeast suspension (subcutaneously in dorsal region, or directly into body cavity), 0.5 ml/25 g body weight. One injection followed by a 24-hr incubation period is adequate for subadults and newly caught animals, but two or three consecutive injections at 24-hr intervals may be required for others.

2. After the yeast incubation period, inject the animal with 1 mg/ml colchicine, 0.1 ml/10 g body weight, and incubate for 1 hr (shorter incubation times of 20–40 min will yield less condensed mitotic chromosomes).

3. Kill the animal (e.g., anesthetize with halothane followed by cervical dislocation) and dissect the upper leg bones (femur) and upper arm bones (humerus) and remove as much soft tissue as possible.

4. Cut off both ends of each long bone and use a syringe full of hypotonic KCl inserted in one end to flush out the marrow into a small volume (approximately 3 ml) of

warm hypotonic solution in a 15-ml centrifuge tube. Flick the tube to disperse the cells and, if necessary, add more hypotonic to bring the volume up to 6–12 ml (the solution should be cloudy).

5. Let the cell suspension incubate in the hypotonic solution for 10–15 min at room temperature.
6. Centrifuge the cell suspension (200 g, 5 min) and discard the supernatant.
7. Flick the tube vigorously to loosen the pellet and immediately add approximately 4–6 ml of fresh fixative over the concentrated cell suspension without disturbing it, and let fix for 20 min in ice.
8. Resuspend the cells and let stand for 5 min, then use a pipette to remove any tissue particles that settle to the bottom of the tube.
9. Centrifuge at 200 g for 5 min, and resuspend the pellet in 4–6 ml of 3:1 fixative, and leave on ice for 10 min.
10. Repeat step (9) twice more. At the last change of fixative, resuspend cells in a small volume of fresh fixative (<1 ml) and test cell density by making a slide (via the splash technique, Protocol 6) and examining under the microscope.
11. Cells can now be stored in a vial of fixative in the freezer, or can be used immediately to make slide preparations using the splash technique (Protocol 6).

Protocol 6: Splash Technique for Slide Preparations of Mitotic Chromosomes

(Time: <1 min per slide)

Nearly every lab has a slightly different method for obtaining splash chromosome preparations, indicating that many of the parameters are matters of preference rather than necessity. The following is a generalized protocol that usually works.

1. Take a clean, ice-cold, wet subbed slide, shake it, hold it with one hand at an approximately 30° angle over a trash can or towels, and splash several drops of a fixed cell suspension from a height of 0.5 m or more onto the slide.
2. Gently blow on the slide surface, and dry the slide on a slide warmer or hot plate at 40°C.
3. Check cell density on one or two test slides and adjust the cell concentration if necessary by diluting or spinning down and resuspending the cells in a smaller volume of fixative.

Protocol 7: Mitotic Chromosomes from Peripheral Blood in Vertebrates

(Time: ≈72 hr incubation plus 2–3 hr handling)

This protocol was specifically designed for peripheral blood from humans (obtained via finger puncture), but can be used for a variety of mammalian, avian, and reptilian species. Exact volumes depend on the amount of blood used. Reference: N. Jo (personal communication).

1. Dispense 4–5 ml of culture medium (see Appendix) into sterile culture tubes.
2. Add 1–10 drops of blood into each tube.

3. Incubate tubes at 36–37°C for 72 hr, mixing the contents by inversion at least daily.
4. Add additional antibiotic if cultures become cloudy (contaminated).
5. Thirty minutes before harvesting, add one drop of 0.025% colchicine or 2 drops of 10 μl/ml colcemid to each tube.
6. Centrifuge tubes for 5 min at 200 g.
7. Carefully remove (and discard) supernatant with pipette.
8. Loosen the cell pellet by flicking the tube.
9. Add 37°C hypotonic solution (0.075 M KCl) to produce a dilute cell suspension. Gently resuspend pellet, using aspiration with a pasteur pipette if necessary. Let sit for 5–20 min at room temperature.
10. Now follow steps 6–11 in Protocol 5.

Protocol 8: Mitotic Chromosomes from Fibroblast Cultures (Reptiles)

(Time: >2 days incubation, plus ≈3 hr handling)

This technique (from Yonenaga-Yassuda et al., 1988) could probably be used for any vertebrate, with appropriate modifications in culture media, incubation times, etc.

1. Sterilize hind legs by successive treatment with 70% ethanol, ether, and merfene.
2. Remove muscles aseptically and place in a small sterile bottle with 5 ml of L15 medium with 5% fetal calf serum (FCS) and 50 mg/ml gentamycin, for 24 hr at room temperature.
3. Transfer muscles to a petri dish, cut into small fragments, and culture in 2 ml of Dulbecco's medium with 20% FCS and 50 mg/ml neomycin in a culture flask at 30°C (cultures should be gassed with air plus 5% CO_2 when the phenol red in the medium indicates a rise in pH).
4. When confluent sheets of cells are seen (>24 hr), add 0.02 ml of 0.16% colchicine, and incubate for 1 hr.
5. Harvest cells by detaching them with 0.125% trypsin in 0.02% EDTA.
6. Prepare cells for splash preparations (Protocol 6) using a hypotonic treatment time of 20–30 min.

Protocol 9: Mitotic Chromosomes from Insect Embryos

(Time: ≈8 hr)

This technique is modified from Zhan et al. (1984) for orthopterans.

1. Place eggs separately in a petri dish containing filter paper soaked with Mark's M-20 insect culture medium (Gibco) with 7.5% fetal calf serum (FCS) and 5.0 mg/ml actinomycin D, and incubate at 37°C for 4 hr.
3. Transfer eggs to another petri dish containing fresh M-20 medium with 0.16 mg/ml colcemid and incubate for 1 hr at 37°C.
4. Dissect embryos out of eggs in plain M-20 medium.
5. Transfer intact embryos to a centrifuge tube containing 0.075 M KCl hypotonic (≈0.5 ml/embryo) for 30 min at 37°C.

6. Dissociate embryos by gentle pipetting, and the centrifuge 5 min at 200 g.
7. Discard supernatant and resuspend cells in a large volume of fresh 3:1 metha-nol:acetic acid fixative and fix for 20 min at room temperature.
8. Centrifuge (as above), decant supernatant, and resuspend in fresh fixative; repeat once, resuspending the final cell pellet in approximately 1 ml of fixative.
9. Use final cell suspension to make splash and/or squash preparations.

Protocol 10: Polytene Chromosomes from Dipteran Salivary Glands

(Time: 5–10 min/preparation)

This technique works best with large, healthy, well-fed larvae. Third instar *Drosophila* larvae are usually found actively crawling up the sides of the culture jar. The paired *Drosophila* salivary glands are clear or slightly opaque, somewhat zucchini-shaped, and have pieces of glistening yellowish fat body attached. The glands can be seen clearly by using understage lighting on a dissecting microscope. Good polytene chromosomes for in situ hybridization should be flat and gray with no refractivity, and the banding pattern should be clearly recognizable (Macgregor and Varley, 1983; Pardue, 1986).

1. Remove a large third instar larva from the culture jar.
2. Place larva in 45% acetic acid or in insect Ringer's on a clean slide.
3. Use needles and/or watchmaker's forceps to pinch off the anterior end of the larva just behind the head segment. Pull the body and head apart until paired salivary glands emerge from the anterior opening (if they don't emerge immediately, discard, and select a fresh larva).
4. Tease off as much fat body as possible without damaging glands.
5. Transfer one or two glands to a small drop of 45% acetic acid near one end of a clean subbed slide and fix for 1–2 min.
6. Cover with a clean siliconized coverslip (see step 4, Protocol 4), and tap lightly on the coverslip, directly over the glands with a pointed probe to help spread the chromo-some arms. Monitor the spreading with a phase microscope.
7. When the chromosomes appear well spread, make slides permanent using the squash and dry ice techniques (Protocol 4).

Protocol 11: Lampbrush Chromosomes

(Time: Part A, 15 min; Part B, >1 hr; Part C, several hr; Part D, ≈1 hr)

This technique works for salamanders, and can be easily modified for frogs and reptiles. Generally, medium sized yolky oocytes (0.5–1.0 mm diameter) yield the best lampbrush chromosome preparations; larger, more mature oocytes usually have con-densed, featureless chromosomes. The best dispersion medium varies among taxa (J. Kezer, personal communication; Macgregor and Varley, 1983; Callan, 1986).

Part A. Preparing Ovaries

1. Anesthetize animal (e.g., in 0.1–0.2% MS222).
2. Remove one or both ovaries through an incision in the ventral body wall.
3. Transfer ovaries immediately to a dry, clean embryological watch glass. Ovaries can be stored at 4°C for 2–3 days if watchglass is sealed with parafilm (do not expose ovaries to Ringer's solution).

Part B. Isolation of Nucleus

1. Place a small piece of ovary into "5:1" isolation solution (5 parts 0.1 M KCl:1 part 0.1 M NaCl) in a clean watchglass.
2. Using watchmaker's forceps, tear open ovary and remove an oocyte. Grasp the oocyte with two forceps and pull laterally to break open the oocyte. The yolky contents will spill out.
3. Locate the translucent nucleus ($\approx$0.3–0.4 mm in diameter in salamanders; 0.1 mm in lizards), and suck in and out of a small-bore, flame-polished pasteur pipette several times to remove the adherent yolk (the nucleus is sturdy and can be "bounced" off the bottom of the dish to dislodge adherent bits of yolk).

Part C. Dispersal of Chromosomes

1. Transfer the cleaned nucleus to a dispersion chamber (bored circular plastic disc with a paraffin-attached coverslip on the bottom) completely filled with dispersion medium.
2. Using a black background under a dissecting microscope, grasp the nuclear membrane at the top of the nucleus with one pair of forceps, then take hold with a second pair very near the first, and pull the two forceps apart with a slightly downward motion (nuclear contents should emerge as a gelatinous mass completely separated from the membrane, which should remain attached to one or both forceps). Note: abandon the preparation immediately if the nuclear contents begin to extrude spontaneously through a small hole in the membrane; such preparations will yield only fragmented chromosomes.
3. Cover the preparation with a coverslip. To avoid the disruptive effects of surface tension, the coverslip must be dropped so that the surface of the coverslip is parallel to the surface of the slide.
4. It takes several hours for chromosomes to settle onto the floor of the chamber.

Part D. To Make Permanent Preparations

1. Place the dispersion chamber into a centrifuge tube fitted with an epoxy plug (Figure 6).
2. Centrifuge, using a swinging bucket rotor, in a precooled table-top centrifuge that allows speed to be gradually increased over a 3-min period to 2000–3000 g. If the centrifuge is not refrigerated, it can be precooled by placing dry ice in the chamber for approximately 30 min before use. Centrifuge at 2000–3000 g for at least 15 min.
3. Remove the chambers from the centrifuge tubes, immerse in dispersion medium,

and use a razorblade to remove the coverglass on which the chromosomes now rest. Gently swish the coverslip around in the medium to wash away any remaining nucleoplasm.

4. Fix the preparation in 70% ethanol for 5 min.
5. Remove the preparation to fresh 70% ethanol for at least 15 min, then dehydrate in 95% ethanol (2 × 10 min) and air dry. The preparations are now ready for in situ hybridization, but may be stored desiccated at 4°C until needed.

Protocol 12: C-Banding

(Time: 1 day pretreatment, plus ≈1.5 hr)

This method works for mitotic and meiotic chromosomes of most organisms (including plants, insects, urodeles, anurans, birds, and mammals; Schmid et al., 1979).

1. Bake permanent, unstained, air-dried slides for 1 day in a 60°C oven.
2. Place slides in coplin jar with prewarmed (30°C), saturated barium hydroxide for 5 min at 30°C.
3. Rinse very briefly in 0.1 N HCl, followed by a thorough rinse in distilled water (e.g., fill and empty the coplin jar six times).
4. Place slides in coplin jar with prewarmed 2× SSC (Appendix) for 1 hr at 60°C.
5. Rinse in distilled water (2 min).
6. Stain slides in 8.0% Giemsa in phosphate buffer (Appendix), pH 6.8, for 5 min; load slides into a coplin jar and add 50 ml buffer, then add 4 ml Giemsa and quickly pipette up and down until thoroughly mixed.
7. Rinse out Giemsa stain by flooding the coplin jar with distilled water or fresh buffer to avoid contamination of slides with metallic film that forms on the surface of Giemsa staining solution.
8. Air dry the slides and mount with a coverslip in a xylene-based mounting medium (e.g., Depex or Permount).

Protocol 13: Q-Banding

(Time: ≈15 min/preparation)

This protocol is from Benn and Perle (1986).

1. Place slides in 0.5 mg/ml quinacrine dihydrochloride stain for 10 min at room temperature.
2. Rinse briefly in distilled water to remove excess stain.
3. Soak in McIlvaine's buffer (Appendix) for 1 min.
4. Mount in a few drops of buffer or in aquamount using a thin glass coverslip.
5. Examine and photograph immediately with fluorescent optics using a filter combination approriate for fluorescein dyes (e.g., Zeiss filter set No. 9, BP 450–490 nm, FT 510 nm, and LP 520 nm); the fluorescent image fades quickly.

Protocol 14: G-Banding

(Time: ≈7 min/slide plus 1 hr drying time)
   This protocol is from Benn and Perle (1986).

1. Incubate air dried slides for 20–40 sec in 0.005% trypsin in PBS solution (Appendix); optimal time varies widely for different preparations.
2. Rinse in three changes of ice-cold PBS (dip consecutively in each coplin jar).
3. Stain for 5 min in 5% Giemsa solution in phosphate buffer.
4. Remove Giemsa by flooding under a gentle stream of water.
5. Air dry slides, and cover with a xylene-based mounting medium.

Protocol 15: Fluorochrome Banding with Chromomycin A3

(Time: ≈30 min)
   This stain produces "reverse" banding (relative to G-banding) in mammals, and stains NORs in salamanders, fish, and some plants (Hack and Lawce, 1980; Sessions and Kezer, 1987).

1. Place air-dried slides in a moist chamber and flood each preparation with at least 50 µl of 5 µg/ml chromomycin A3 in chromomycin buffer (see Appendix), cover with a coverslip, and stain for 20 min in the dark at room temperature.
2. Rinse off the chromomycin with distilled water, and place slides (no more than three at a time) in coplin jar with methyl green counterstaining solution (2-ml stock solution in 50-ml phosphate buffer, pH 6.8) for 6 min.
3. Rinse in distilled water.
4. Air dry, and mount in 100% glycerine or aquamount, and examine under UV fluorescence optics using an appropriate filter combination (see Q-banding, Protocol 13).

Protocol 16: AgNO₃-Banding for NORs

(Time: ≈1 min/slide)
   This fast, easy, and reliable technique was published by Hsu (1981), and seems to work for all organisms. Use aged (at least 1 day), air-dried slides.

1. Mix 2 parts of 50% (w/v in H_2O) silver nitrate solution and 1 part developer (Appendix) in a glass vial (allowing at least 150 µl for each preparation), and mix thoroughly.
2. Add 3 drops to each preparation and quickly add a coverslip.
3. Incubate at 90°C for 30–60 sec (or until staining solution has turned muddy yellowish brown).
4. Rinse off coverslip with distilled water (using a squirt bottle, or rinse in a beaker of water).
5. Air dry slides, and mount in oil or permanent mounting medium.

Protocol 17: Differential Replication Banding with BrdU

(Time: several days incubation, plus 1 full working day)

This technique can be used to obtain complex chromosomal banding patterns in organisms in which more conventional banding methods do not work (Dutrillaux, 1975; Benn and Perle, 1986).

1. Set up tissue culture cells.
2. Five to seven hours before addition of colcemid, add 0.01 M bromodeoxyuridine (BrdU) and 0.01 M deoxycytidine to make final concentration of 10^{-4} M each.
3. One hour before harvest add colcemid to final concentration of 0.1%.
4. Harvest and make slides via the splash technique (see above).
5. Soak air dried slides in PBS for 5 min at room temperature.
6. Stain in 0.5 μg/ml Hoechst 33258 for 10 min at room temperature.
7. Mount in McIlvaine's buffer (Appendix).
8. Irradiate slides at approximately 5 cm from a 15-W UV light source at 50°C for 15 min or under a 75-W growlamp for 24 hr.
9. Rinse coverslip away with distilled water, and incubate slides for 15 min in 2× SSC at 65°C.
10. Stain slides in 8% Giemsa in phosphate buffer pH 6.8 for 5–10 min.
11. Air dry slides and mount in a xylene-based medium.

Protocol 18: Modification of BrdU Banding for Salamander Embryos

(Time: 32 hr incubation, 3–5 days slide aging, plus ≈3 hr)

This technique yields complex banding patterns comparable to G-bands in salamanders (Kuro-o et al., 1986).

1. Wash dejellied embryos in several changes of sterile amphibian saline.
2. Transfer embryos to culture dish (35 mm diameter) containing 1.5 ml of 60% Eagle's minimum essential medium (MEM) with 20% FCS, 20% sterile-filtered water, and 400 μg/ml BrdU.
3. Disrupt embryo with a sterile pasteur pipette, and incubate cells in a darkened, humidified incubator under a constant flow of air with 5% CO_2 for 24 hr at 20°C.
4. Add another 1.5 ml of 1.0% colchicine in culture medium and incubate 8 hr at 20°C.
5. Centrifuge cells and medium at 120 g for 7 min.
6. Remove supernatant, and add 10 ml hypotonic solution (amphibian saline diluted 1:7 with distilled water) and incubate for 1 hr at room temperature.
7. Add 0.5 ml fresh 3:1 methanol:acetic acid fixative and fix for 10 min.
8. Centrifuge at 420 g for 5 min, replace supernatant with fresh fixative, and fix for 5 min.
9. Repeat step 8 twice, but resuspend final cell pellet in approximately 1 ml of fixative, and use to make splash and/or squash preparations.
10. Age slides for 3–5 days at room temperature.

11. Stain with 50 µg/ml Hoechst 33258 in calcium- and magnesium-free PBS for 15 min.
12. Rinse briefly in distilled water, mount in PBS, and expose to UV light at a distance of 10 cm for 30 min.
13. Remove coverslips, rinse briefly in distilled water, then incubate in $2\times$ SSC for 30 min at 60°C.
14. Rinse slides in running water, then stain in 3% Giemsa in PBS at pH 6.8 for 4 min.
15. Air dry slides, and mount in a xylene-based mounting medium.

Protocol 19: Labeling Probes via Nick Translation for In Situ Hybridization

(Time: approximately 4–5 hr)

This method uses 1 µg of DNA and produces enough probe for at least 10 slides, with a specific activity of $2–6 \times 10^6$ cpm/µg (Macgregor and Varley, 1983; Pardue, 1985, 1986; Malcolm et al., 1986).

1. Mix 2×10^{-3} µmol each of tritiated precursors (dNTPs) in a microcentrifuge tube.
2. Aliquot 18 µl of the mixture into ethanol-washed microcentrifuge tubes, and quickly freeze-dry under vacuum to prevent radiolysis.
3. Add to a tube containing the dried, tritiated dNTPs 10 µl of nick translation buffer (Appendix), 5 µl of DNA (1 µg), and glass-distilled water to make a total of 94 µl.
4. Incubate the mixture at 15°C for 10 min and then chill the tube in iced water.
5. Add 5 µl (12.5 units) of DNA polymerase I to make a total volume of 99 µl.
6. Add 1 µl of diluted (1 µg/ml) DNase I (1 mg/ml) stock (dilute stock immediately before use).
7. Incubate at 15°C for 1 hr.
8. Slow the reaction by placing the tube on ice, and determine the percentage incorporation of radioactive nucleotides with the following procedure:
   a. Mix 5 µl of the reaction mixture with 995 µl of TCA/BSA in a microcentrifuge tube, and keep on ice for 15 min.
   b. Pass 5 ml of ice-cold 5% TCA through a 2.5-cm-diameter Whatman GF/C glass fiber filter followed by the TCA/BSA reaction mixture.
   c. Wash the filter three times with 5 ml of cold 5% TCA, and dry the filter at 65°C for 20 min.
   d. Measure the radioactivity of the filter in a scintillation counter using a toluene-based scintillation fluid.
   e. To measure total incorporated and unincorporated radioactivity in the reaction mixture, take another 5-µl sample of the reaction mix and put it directly on a clean filter without TCA, dry it, and count it.
   f. The percentage incorporation of radioisotopes into the probe is determined by comparing counts between the two filters. The TCA-treated filter should have 20–60% of the counts obtained from the untreated filter. The DNA should not be used if it shows less than 10% incorporation.

9. Stop the nick translation reaction by adding 100 µl of water-saturated phenol and mixing well with a pasteur pipette.
10. Centrifuge at 5000 g for 5 min.
11. Unincorporated nucleotides can be removed by loading the aqueous supernatant directly onto a Sephadex G-50 column (see Chapter 8) that has been prewashed with distilled water.
12. Elute with distilled water and collect consecutive fractions of 30 drops each. Count 5 µl of each fraction in a scintillation counter (using a tergitol scintillator) and combine the fractions containing the first peak of radioactivity to come off the column.
13. Freeze-dry these combined fractions and redissolve them in 50 µl of distilled water.

Protocol 20: Labeling In Situ Hybridization Probes via Transcription of cRNA Using *E. coli* RNA Polymerase

(Time: 4–5 hr)

This protocol is from Macgregor and Varley (1983).

1. Lyophilize 100 µCi each of tritiated GTP, ATP, CTP, and UTP in a 15-ml siliconized glass centrifuge tube.
2. Add approximately 100 µl of DNA (5–10 µg), 50 µl 0.2% 2-mercaptoethanol, 40 µl distilled water, and 10 µl (5–10 units) *E. coli* RNA polymerase.
3. Cover the tube with parafilm and incubate at 37°C for 60–90 min.
4. Add 0.75 µl of 0.04 M Tris, pH 7.9, and 20 µl DNase I (1 µg/µl), and let stand at 20°C for 15 min.
5. Assess percentage incorporation by spotting two nitrocellulose filters with 10-µl samples each, and dry the filters. Place one filter in cold 5% TCA for 5 min, wash it with 70% ethanol, and dry it. Compare counts in a scintillation counter (incorporation should be 15–40%).
6. Add 200 µl of 5% SDS, and 1 ml of *E. coli* carrier tRNA.
7. Phenol-extract DNA (Chapters 8 or 9) to remove residual proteins.
8. Centrifuge at 10,000 g for 5 min, and load the supernatant onto a Sephadex column (Chapter 8).
9. Collect 25 consecutive fractions of 30 drops each, and put 10 µl of each fraction onto a nitrocellulose filter, dry, and count.
10. Pool the fractions representing peak radioactivity, and incubate at 85°C for 3 min.
11. Filter solution through a 0.45-µm pore nitrocellulose filter, and count 10 µl of the collected volume on a nitrocellulose filter to determine the radioctivity in cpm/ml.
12. The labeled cRNA can now be freeze-dried and dissolved at a higher concentration, if necessary, for use as a probe.

Protocol 21: In Situ Hybridization Procedures for Reiterated Sequences Using a DNA Probe

(Time: Part A, 3.5 hr; Part B, 15 min; Part C, 6–12 hr; Part D, 1.5–2 hr)

Cytological preparations to be hybridized should be air-dried on subbed slides. This method provides enough hybridization reaction mixture for approximately 10 slides, assuming 30 μl with 10^5 counts per slide (Macgregor and Varley, 1983; Pardue, 1986).

Part A. Slide Pretreatment

1. Place slides horizontally in a moist chamber (with black filter paper on the bottom to make it easier to see the cytological material on the slides). For hybridization to a DNA target (but not for an RNA target), put 200 μl of ribonuclease mixture on each preparation, cover with a coverslip (22 × 40 mm), and incubate in a moist chamber at 37°C for 2 hr.
2. Remove coverslips by dipping slides in a large volume of 2× SSC, and wash in 2× SSC in a coplin jar, 3× 10 min.
3. Denature the target DNA by placing the slides in 0.07 M NaOH at 20°C for 3 min.
4. Wash and dehydrate slides in three changes of 70% ethanol and two changes of 95% ethanol, 10 min each, and air dry.

Part B. Preparation of the Hybridization Reaction Mixture

1. Dissolve labeled DNA probe (freeze-dried or ethanol-precipitated) in 30 μl of 0.1 M NaOH in a microcentrifuge tube by flicking with a finger; total radioactivity should be approximately 3×10^6 cpm.
2. Add 150 μl of formamide stock (Appendix) to make a final concentration of 50% and mix well.
3. Add 60 μl of 20× SSC (to make final concentration of 4× SSC) and mix well.
4. Add 30 μl of distilled water, mix well, and cool on ice for 5 min.
5. Add 30 μl of 0.1 M HCl (i.e., enough to exactly titrate the 0.1 M NaOH) and mix well.
6. Keep the hybridization reaction mixture on ice, and use within 10–15 min.

Part C. The Hybridization Reaction

1. Place pretreated, air-dried slides horizontally in moist chambers (again using black paper).
2. Place 30 μl of the hybridization reaction mixture (which has been kept on ice) in the middle of the preparation on each slide.
3. Place a glass coverslip over each preparation, avoiding bubbles and making sure that the entire preparation is covered.
4. Cover the moist chambers and incubate at 37°C for 6–12 hr.

Part D. Washing the Slides

1. Lift each slide from the moist chamber and remove the coverslip by dipping into a large volume of 2× SSC.
2. Place the slide in a coplin jar of fresh 2× SSC at 65°C for 15 min.
3. Wash in 2× SSC, 2 × 10 min, at room temperature.
4. Place slides in a coplin jar of 5% TCA at 5°C for 5 min.
5. Wash in 2× SSC, 2× 10 min at room temperature.
6. Wash in 70 and 95% ethanol, 2× 10 min each.
7. Air dry. The slides are now ready for autoradiography (Protocol 24).

Protocol 22: In Situ Hybridization Using an RNA Probe

(Time: 6–12 hr incubation plus 6–8 hr)

This protocol is based on Macgregor and Varley (1983) and Pardue (1986).

1. Preparation of RNA probe: lyophilized or ethanol-precipitated RNA should be dissolved in 2× SSC or in 4× SSC/50% formamide to provide a total of 2–3×10^6 cpm/ml and 30 µl per slide.
2. Slide pretreatment and hybridization reaction: same as for DNA–DNA hybridization (Protocol 21).
3. Remove coverslips by dipping in large volume of 2× SSC.
4. Place slides in fresh 2× SSC for 15 min at room temperature.
5. Treat each slide with ribonuclease mixture (Appendix) at 37°C for 1 hr.
6. Wash slides in 2× SSC, 2× 15 min.
7. Place slides in 5% TCA at 5°C for 5 min.
8. Wash slides in 2× SSC, 2× 10 min.
9. Wash slides in 70 and 95% ethanol, 2× 10 min each.
10. Air dry the slides; they are now ready for autoradiography (Protocol 24).

Protocol 23: Localization of Single Copy Sequences on Chromosomes

(Time: 8–16 hr incubation plus 5–6 hr)

This protocol is from Harper and Saunders (1984). Use recombinant bacteriophage or plasmid DNA containing single-copy sequences of interest. Probes should be labeled with tritiated dNTPs by nick translation to 20–40×10^6 cpm/µg (Protocol 19).

1. Pretreat slides with RNase (in Protocol 21), rinse in 2× SSC, and dehydrate in ethanol.
2. Dissolve probe in 50% formamide, 2× SCP, 10% dextran sulfate, pH 7.0, along with 500-fold excess sonicated salmon sperm DNA carrier.
3. Denature probe mixture (85°C for 3–15 min, then chill quickly on ice).
4. Apply chilled, denatured probe mixture to slide preparations and cover with a coverslip.
5. Incubate in a moist chamber at 37°C for 8–16 hr.
6. Rinse thoroughly (e.g., 3× 5 min each) in 2× SSC/50% formamide, pH 7.0, and then 2× SSC, pH 7.0, at 39°C, followed by dehydration in ethanol (e.g., 70 and 95%, 2× 10 min each).
7. The slides are now ready for autoradiography (Protocol 24; slides require an exposure time of 5–22 days at 4°C).
8. G-band chromosomes with Wright stain (see Protocol 25).

Protocol 24: Autoradiography for Detection of Radioactive Probes

(Time: Part A, 30 min; Part B, 2–4 hr; Part C, ≈30 min)

Autoradiography requires three main sets of tasks: diluting and aliquoting a new batch of emulsion, coating the slides, and developing the exposed autoradiographs (Macgregor and Varley, 1983).

Part A. Diluting, Aliquoting, and Storing Emulsion

1. Open the package of emulsion (e.g., Kodak NTB2) in the darkroom either in complete dark or under a safelight (e.g., Kodak # 152-1525) and warm the bottle for 30 min in a 45°C waterbath along with a 200-ml flask of distilled water and an empty 500-ml beaker.
2. After the emulsion has melted, pour the entire contents of the bottle very slowly down the side of the prewarmed 500-ml beaker and return beaker to the water bath.
3. Fill the empty plastic emulsion bottle with prewarmed distilled water from the Erlenmeyer flask, mix gently, and pour the contents slowly down the side of the 500-ml beaker.
4. Thoroughly mix the contents of the beaker by swirling gently so as to prevent the formation of bubbles.
5. Dispense the diluted emulsion into scintillation vials, approximately 10 ml per vial. This is enough emulsion to coat approximately 30 slides.
6. Wrap each vial in aluminum foil, place them in a light-proof box, and store at 3–5°C in a refrigerator that is never used for radioisotopes or organic solvents. Stored in this way, the emulsion may be good for up to 5 years.

Part B. Coating and Exposing the Slides

1. Working in complete darkness or under a safelight, place the sealed vial of emulsion and the dipping chamber into the 45°C water bath for 15–30 min (the dipping chamber can be stood in a beaker or diagonally in a coplin jar filled with water, and should be immersed to within 0.5 cm of its top edge).
2. Fill the dipping chamber by slowly pouring the emulsion down its side to avoid bubbles (a small funnel is useful).
3. Dip the slides slowly and smoothly into the chamber, one at a time (taking care not to touch the emulsion with fingers), withdraw and drain briefly against the edge of the chamber, and place in the slide rack to dry.
4. Air dry the slides for at least 2 hr.
5. Store the slides for exposure in light-proof slide boxes tightly sealed with black electric tape. Moisture during the exposure time can cause the latent image to fade, so it is important to place a vial of desiccant into each slide box. The vial of desiccant should be loosely plugged with cotton and can be held in place with a blank microscope slide. The slide boxes should be stored at 4°C for the appropriate exposure time.
6. After the required exposure time, the slides should be warmed to room temperature and developed according to the following procedure (all solutions must be at the same temperature, 15–20°C, to avoid cracking or wrinkling the emulsion).

Part C. Developing the Autoradiographs

1. Gently rock the preparation in freshly mixed developer (e.g., Kodak D-19), 2.5 min at 20°C (a single coplin jar can be used if developing 10 slides or less).

2. Pour out developer and replace with fixer; fix for 5 min at 20°C (lights can come back on after 2 min in fixer).
3. Pour out fixer and rinse slides in distilled water at least five times, 2 min each at 20°C.
4. Air dry the slides.

Protocol 25: Post-autoradiography G-banding with Wright's Stain

(Time: ≈30 min)

This procedure works for mammalian (particularly human) chromosomes (Chandler and Yunis, 1978, cited in Pardue, 1985).

1. Place the slides in a solution of Wright's stain (15 ml in 45 ml of phosphate buffer, pH 6.8) for 8–10 min.
2. Rinse briefly in distilled water.
3. Enhance staining contrast by destaining the slides in 95% ethanol (2 min), chloroform (15 sec), 95% ethanol plus 1% HCl (30 sec), 100% methanol (2 min), and then restaining in Wright's stain (6–8 min); repeat at least once.
4. Rinse in distilled water, air dry, and mount in Permount or other xylene-based mounting medium.

Protocol 26: In Situ Hybridization with Biotin-Labeled DNA Probes

(Time: Part A, 1.5 hr; Part B, 1.5–2 hr; Part C, 24 hr; Part D, 1 hr)

Hybridization signals appear as reddish-brown to black bands depending on strength (Manuelidis et al., 1982; Pinkel et al., 1986).

Part A. Pretreatment of Slides

1. Place slides in 2× SSC at 65°C for 30 min.
2. Wash in 2× SSC, 2 min.
3. Acetylate the slides (reduces nonspecific binding of negatively charged probes):
   a. Rapidly agitate 500 ml of 0.1 M triethanolamine-HCl in a staining dish (use a magnetic stirring bar).
   b. Add 0.625 ml of acetic anhydride.
   c. Turn stirrer off and quickly place slides into solution.
   d. Incubate 10 min.
4. Wash the slides in 2× SSC, 2× 5 min.
5. Dehydrate in 70 and 95% ethanol (2× 5 min each), and air dry.
6. Denature chromosomes in freshly prepared 0.07 N NaOH, 3 min at room temperature.
7. Wash in 2× SSC, 1× 5 min.
8. Dehydrate in 70 and 95% ethanol (2× 5 min each).

Part B. Biotinylation of the DNA Probe (for 0.5 µg DNA or less; quality not important)

1. Mix together in a microcentrifuge tube:
   a. 10× nick-translation buffer (Appendix), 2.5 µl

   b. dATP, dGTP, dCTP, 0.3 mM each
   c. Bio-16-dUTP, 2.0 μl
   d. [^{32}P]dUTP (1 mCi/ml), 1.0 μl
   e. 0.5 μg DNA
   f. DNase I [1:400 dilution of 1 mg/ml in TM10 (10 m M Tris, 10 mM MgCl$_2$, pH 7.5)], 1.5 μl
   g. Distilled water to 25 μl total
2. Add 10 units of DNA polymerase I and incubate at room temperature for 60 min.
3. Stop reaction by adding 25 μl of 50 mM EDTA, pH 8.0.
4. Remove unincorporated nucleotides by separation in a Sephadex G-50 column (Chapter 8) using water or 10 mM Tris-HCl, pH 7.5 as eluants.
5. Check the percentage incorporation by comparing signals of the column and the eluates on a Geiger counter; 5–10% incorporation is sufficient. Oversubstitution (40–50% incorporation) may interfere with the hybridization reaction.
6. Storage: the biotin-labeled DNA probe can be stored at −20°C for long periods of time.

Part C. Hybridization Reaction

1. Add 20 μg of sonicated carrier DNA (e.g., herring sperm DNA) and 5 μl of 3 M sodium acetate.
2. Precipitate the DNA with 2.5 volumes of ethanol, and incubate on dry ice/ethanol for 20 min, or at −20°C overnight.
3. Resuspend the pellet in 75 μl of hybridization buffer (Appendix). For probes less than 1 kb, add 50% dextran sulfate to a final concentration of 10%.
4. Denature the probe by boiling for 3 min, then chill on ice.
5. Add approximately 10 μl of hybridization solution over each preparation and cover with an 18 × 18-mm coverslip (allow no bubbles).
6. Seal the edges of the coverslip with rubber cement to prevent evaporation, and incubate the slides in a moist chamber at 58°C for 12–18 hr.
7. Peal off the rubber cement with forceps and remove coverslips by placing slides vertically in preheated 2× SSC at 53°C (coverslips will float off).
8. Wash in 2× SSC at 53°C, 3× 20 min (do not let slides dry at this point).

Part D. Detection of Signal

1. Wash in PBS, 2× 5 min at room temperature.
2. Wash in PBS with 0.1% Triton X-100 for 2 min.
3. Rinse briefly in PBS (1 min).
4. Place slides in a moist chamber, and cover each preparation with 100 μl of strepta-vidin-biotinylated peroxidase complex.
5. Cover each preparation with a 22 × 40 mm coverslip, and incubate for 30 min at 37°C, or overnight at 4°C.
6. Remove coverslip and wash 2× 5 min in PBS.
7. Wash in PBS with 0.1% Triton X-100 for 2 min.
8. Rinse briefly in PBS.

9. Place slides in a moist chamber, and cover each preparation with approximately 1 ml of freshly prepared DAB staining solution (Appendix) and incubate for 1–10 min at room temperature (monitor staining reaction with a microscope; overstaining results in high background).
10. Stop the reaction by dipping the slides in distilled water and placing them in PBS.
11. Stain slides in Giemsa solution for 20 sec, then flood the Giemsa out of the coplin jar with a gentle stream of water.
12. Air dry the slides, and mount in Permount or other xylene-based mounting medium.
13. Examine the preparations with phase-contrast optics.

Protocol 27: Detection of Hybridized Biotinylated Probes with Fluorescein-Avidin

(Time: ≈1.5 hr)

This protocol is from Pinkel et al. (1986).

1. Incubate slides in blocking solution (Appendix), 30 µl per preparation, for 5 min at room temperature.
2. Remove coverslips and blot briefly to remove excess fluid (but do not let slides dry out).
3. Add fluorescein-avidin (3 µg/ml in blocking solution, 30 µl/preparation), cover with a coverslip, and incubate for 10 min at 37°C.
4. Wash 2 × 5 min in BN buffer (Appendix) at 45°C.
5. Add biotinylated goat anti-avidin antibody, 30 µl/preparation, for 10 min at room temperature.
6. Wash 2 × 5 min in BN buffer at 45°C.
7. Repeat steps 3 and 4, and drain slides.
8. Counterstain with DAPI or propidium iodide (0.25–0.5 µg/ml).
9. Examine using epifluorescence optics with appropriate filters (e.g., Zeiss filter combination 487709 or equivalent).

Protocol 28: In Situ Hybridization with BrdU-Labeled Probes

(Time: Part A, 1 week plus ≈24 hr; Part B, 5 hr)

This method yields a color reaction that is said to be stable for at least 1 year (Frommer et al., 1988).

Part A. Slide Preparation

1. Age slides 1 week.
2. Denature chromosomes with 0.2 N HCl for 20 min at room temperature, or in 70% deionized formamide in 2× SSC (pH 7.0) for 2 min at 70°C.
3. Dehydrate slides and air dry.
4. Denature labeled probe in hybridization mixture (0.5 mg/ml E. coli tRNA and 34% formamide in 5× SSC, pH 7.0) for 3 min at 100°C, then rapidly cool on ice.

5. Place 12 μl of the hybridization mixture on each slide, cover with a 22 × 22-mm coverslip, and seal with rubber cement.
6. Place slides in a moist chamber and incubate for 15–18 hr at 41°C.
7. Remove the coverslips by peeling off the rubber cement and then soaking in 5× SSC.
8. Wash slides extensively (e.g., 3 × 5 min each) at 41°C either in 2× SSC (low stringency washes) or 0.2× SSC (high stringency washes).

Part B. Detection of Hybridized Probe

1. Wash slides 3 × 10 min in PBS or TBS (Appendix), at room temperature.
2. Add 25 μl of primary antibody (anti-BrdU mouse) to each slide, cover with a coverslip, and incubate for 1 hr in a moist chamber at room temperature.
3. Remove coverslip by soaking in TPBS (Appendix).
4. Add 100 μl of secondary antibody without a coverslip, and incubate for 30 min at room temperature.
5. Wash slides in TPBS 3 × 10 min.
6. Repeat steps 4 and 5 with the third and fourth antibodies.
7. After the final TPBS wash, place slides in DAB (diaminobenzidine tetrahydrochloride; 0.2 mg/ml in 40 mM $NaHPO_4$, 40 mM Tris-HCl, pH 7.6) for 1 min.
8. Add hydrogen peroxide to a final concentration of 0.006%.
9. Stop color reaction after 6 min by rinsing in tap water for 5 min.
10. Put slides in TPBS, stain in filtered hematoxylin, rinse in distilled water, and mount in xylene-based mounting medium.

Protocol 29: Immunolocalization of Antigens on Lampbrush Chromosomes

(Time: ≈3 hr)

This protocol is from Lacroix et al. (1985) and Ragghianti et al. (1988).

1. Fix centrifuged lampbrush preparations (see Protocol 11) in 4% paraformaldehyde for 30 min at 4°C.
2. Wash in Tris-NaCl (100 mM NaCl, 10 mM Tris-HCl, pH 7.4), 3 × 5 min.
3. Incubate in inactivated normal rabbit serum diluted 1:20 with Tris-NaCl, 10 min at 4°C.
4. Treat chromosomes with monoclonal antibodies at various dilutions (e.g., 1:250, 1:500, 1:750, 1:1000, 1:2500, 1:5000, 1:10000) in Tris-NaCl for 30–60 min at room temperature.
5. Wash in Tris-NaCl, 3 × 5 min.
6. Incubate slides in a 1:1000 dilution of secondary antibody conjugated with FITC, with 0.1% DAPI, for 30 min at 4°C.
7. Wash in Tris-NaCl, 3 × 5 min.
8. Mount in a 2:1 mixture of glycerol:Tris-NaCl.
9. Examine with phase-contrast and epifluorescence using appropriate filters.
10. Photographs can be made with Ilford FP4 film (phase contrast) or Kodak Ektachrome 1000 or Kodak T-Max 400 film for immunofluorescence.

INTERPRETATION AND TROUBLESHOOTING

Chromosome Bands

Chromosome bands can be scored in terms of their position within and between chromosomes, as well as their relative sizes (Figure 5). The terminology used for the position of bands and other markers differs for different organisms, but has been standardized for dipterans (especially *Drosophila,* Sturtevant and Novitski, 1941) and various species of mammals (Paris Conference, 1971; CSKRN, 1973; ISCN, 1981; Rooney and Czepulkowski, 1986). Banding data are usually presented as a karyotype constructed of chromosomes cut from a photomicrograph. It is helpful to also include an idiogram, presenting relative lengths of chromosomes and positions of bands, especially if the banding pattern is complex.

Chromosome preparation and banding can be capricious, depending on the particular organism and the kind of banding. It is usually easiest to obtain good chromosomes from freshly caught, healthy, well-fed individuals (although there are always exceptions). Colchicine is said to be light sensitive once it is in solution, and it is therefore advisable to make it up fresh just before use, and to keep it refrigerated.

Among available banding techniques, C-banding is perhaps the most fool-proof, reliable method for most organisms, although G-banding works reliably for most species of mammals. Failure to band using the C-banding protocol may be due to poorly aged slides, inferior Giemsa, or the absence of stainable heterochromatin in the chromosomes. It is best, therefore, to make sure that the procedure works on an organism known to have good C-bands before trying it on an untested species. Also, if banding is not produced the first time, it can sometimes be induced by treating the same slides a second time; this is especially important for rare or small organisms from which few preparations are available. Sometimes the same slides may be used for several different banding procedures, using the less stringent methods first (e.g., fluorochrome banding then G-banding then C-banding and/or AgNOR banding). If fluorochrome banding does not work, then either the dye is no good (e.g., it is too old or incorrectly prepared), the wrong excitation filter was used, or the chromosomes are devoid of the kinds of sequences for which the fluorochrome is specific.

In Situ Hybridization

There are many reasons why in situ hybridization may fail to work. The most common problems are either too few (no detectable hybrids) or too many silver grains (high background). Ideally, there should be sufficient grains to unambiguously locate sites of hybridization, but not so many that details of chromosome structure are obscured. For repetitive sequences, hybridized sites are often visibly obvious (Figure 1), but demonstration of single-copy sites usually involves an analysis of silver grain distribution in at least 10 different preparations (Figure 3). Counting the number of grains can yield information on target size or number that can be used to detect amplification or diminution of particular sequences, or duplication events during stages of the cell cycle.

Possible reasons for difficulties with in situ hybridization include (1) probe was not labeled, or was labeled with too low specific activity, or used at too low concentration; (2) presence of contaminating nucleases; (3) target material was not denatured, or was overtreated (loss of DNA); (4) incubation time was too short (or too long), or at the wrong temperature; (5) the washes were too stringent, or not stringent enough; (6) the autoradiographic exposure was too long or too short; (7) slippage or loss of the autoradiographic emulsion during developing, fixing, or washing; and (8) absence of complementary target sequences. Assuming that a suitable labeled probe is available, most of these problems are managed by running appropriate pilot experiments and including replicates with which parameters can be varied. Optimal conditions must be determined empirically for each new probe.

APPENDIX: STOCK SOLUTIONS

AgNOR "developer"

2% gelatin
1% formic acid

Mix gelatin powder in 50 ml distilled water and heat to dissolve. Cool and add formic acid.

Amphibian Ringer's solution

113 mM NaCl
2 mM KCl
0.7 mM $CaCl_2$

Biotinylated goat anti-avidin antibody

5 μg/ml in BN buffer
5.0% goat serum
0.02% NaN_3

Blocking solution for fluorescein-avidin

5.0 % nonfat dry milk
0.02% NaN_3 in BN buffer

BN buffer

0.1 M sodium bicarbonate
0.05% Nonidet P-40

Adjust to pH 8.0.

Chromomycin buffer

0.15 M NaCl
2.5 mM $MgCl_2$
0.03 M KCl
0.01 M Na_2HPO_4

Adjust to pH 7.0.

Culture medium for peripheral blood cells (vertebrates)

Gibco 199 culture medium (cat. No. 320-1153)
1.25 g/ml $NaHCO_3$
10% fetal calf serum
2 mM L-glutamine
100 units/ml penicillin
100 mg/ml streptomycin
1.5% (v/v) phytohemagglutinin

DAB staining solution

0.5 mg/ml diaminobenzidine in PBS
1/100 vol of 1% hydrogen peroxide

Denhardt's solution (1×)

0.02% Bovine serum albumin (BSA)
0.02% Ficoll
0.02% polyvinylpyrrolidone

Dextran sulfate stock (50%)

20 g dextran sulfate

Add distilled water to final volume of 40 ml. Filter through two Whatman no. 1 filter papers (takes several hours under vacuum).

Formamide stock, deionized (>95%)

500 ml formamide
25 g mixed resin, 20–50 mesh [AG 501-X8 (D), Bio-Rad]

Stir for 30 min at room temperature. Filter through a Whatman no. 1 filter. Store in aliquots at −20°C.

Hybridization buffer (for biotin-labeled probes)

50% deionized formamide
0.6 M NaCl
10 mM Tris-HCl, pH 7.5
1 mM EDTA
1× Denhardt's solution
0.5 mg/ml carrier RNA (Sigma, Type IV)
10% dextran sulfate (from 50% stock)

Filter sterilize through 45-μm nitrocellulose filter and store at −20°C.

Lampbrush chromosome isolation medium ("5:1")

5 parts 0.1 M KCl
1 part 0.1 M NaCl

McIlvaine's buffer

Solution A: 0.1 M anhydrous citric acid
Solution B: 0.4 M anhydrous sodium phosphate dibasic
pH 5.6: 92 ml solution A + 50 ml solution B (adjust pH)
pH 7.5: 80 ml solution A + 920 ml solution B (adjust pH)

Methyl green stock solution

0.11 g methyl green
25 ml phosphate buffer, pH 6.8

Nick-translation buffer (10×)

0.5 M Tris-HCl (pH 7.5)
0.1 M MgSO4
1 mM dithiothreitol
500 μg/ml bovine serum albumin

PBS

0.15 M NaCl
0.05 M NaHPO$_4$

Adjust to pH 7.4.

Phosphate buffer, pH 6.8

0.025 M KH$_2$PO$_4$

Titrate to pH 6.8 with 50% NaOH.

Paraformaldehyde (4%)

20 g paraformaldehyde
500 ml PBS

Heat to 65°C, stirring rapidly until completely dissolved.

SCP (1×)

0.12 M NaCl
0.015 M sodium citrate
0.02 M NaPO$_4$

Adjust to pH 6.0, if necessary.

SSC, 20× stock

3.0 M NaCl
0.30 M sodium citrate

Adjust pH to 7.0 with 10 N NaOH.

Subbing solution (for microscope slides)

0.1% gelatin
0.01% chromium potassium sulfate

Dissolve gelatin in hot distilled water, cool and add chromalum. Store at 4°C.

TBS

0.15 M NaCl
20 mM Tris-HCl

Adjust to pH 7.4.

TCA/BSA

50 μl stock BSA (1 mg/ml)
845 μl distilled water
100 μl 100% TCA

TM10

10 mM Tris
10 mM MgCl$_2$

Adjust to pH 7.5.

TPBS

0.15 M NaCl
4 mM NaHPO$_4$
4 mM Tris-HCl

Adjust to pH 7.6.

Tris-NaCl

0.1 M NaCl
10 mM Tris-HCl

Adjust to pH 7.4.

Yeast suspension

2–3 g fresh dry yeast (e.g., Fleischmann's™ "active dry")
5–6 g dextrose
25 ml warm water

Incubate at 40°C until it begins to foam vigorously (30 min).

CHAPTER 7

NUCLEIC ACIDS I: DNA–DNA HYBRIDIZATION

Steven D. Werman, Mark S. Springer, and Roy J. Britten

INTRODUCTION

This chapter focuses on "in solution" hybridization for the quantitative assessment of relatedness of biological species using nuclear DNA. We therefore ignore filter hybridization, which has not yet been shown to be useful for the quantitative evaluation of relationships. In addition to comparing and reviewing different DNA hybridization protocols, we also consider laboratory practice, DNA reassociation kinetics, the significance of genome organization to DNA hybridization data, the interpretation of melting curves, and the application of DNA hybridization data to systematics. Recently, the latter topics have come under close methodological and analytical scrutiny. We believe it is important to give due attention to these issues in a volume on molecular systematics.

Early DNA hybridization/DNA kinetics studies include those of Wetmur and Davidson (1968), Kohne (1970), Kohne and Britten (1971), Bonner et al. (1973), and Britten et al. (1974). These authors provided a sound description and theoretical underpinning for the DNA hybridization technique as well as the underlying reassociation kinetics of DNA. They also introduced several metrics for DNA hybridization data. Although these studies were not primarily concerned with systematics, the distance metrics that they suggested are still in use today.

Large-scale application of the DNA hybridization technique to problems in systematics was pioneered by Charles Sibley and Jon Ahlquist. Their rapid progress was made possible by the construction of an automated thermal elution device, appropriately named the "DNAnalyzer." Although the quantity of data generated by Sibley and Ahlquist is truly impressive, their studies have often been criticized because the individual characters (i.e., nucleotides) remain unidentified. Thus, individual shared derived characters are not isolated and described. Proponents of DNA hybridization have countered that the sheer number of nucleotides under comparison compensates for this lack of individual identification. The mammalian genome, for example, is 3×10^9 nucleotide pairs per haploid set and thus there are about 10,000,000 essentially different fragments when it is sheared to lengths of about 300 nucleotides. The large number of fragments always involved in interspecies comparisons effectively suppresses distance fluctuation due to sampling and the accuracy is determined entirely by the measurement technique. However, it appears to us that the issues of (1) distance data versus discrete character data in phylogenetic reconstruction (see Chapter 11) and (2) the number of nucleotide positions under comparison are largely separate issues. It is therefore appropriate to ask if distance data in general and DNA hybridization data in particular have systematic value. We will attempt to place these issues in perspective.

As mentioned above, there are other points of contention surrounding DNA hybridization. One of these is the choice of an appropriate distance metric, (i.e., ΔT_m versus ΔT_{50H} versus **NPH** versus ΔT_{mode}). We will also comment on these issues.

In spite of these current controversies, DNA hybridization is now a practical technique that is beginning to be broadly used to assess evolutionary relationships. It is primarily applied to single-copy DNA, and selection among sequences is considered to be unimportant to the evolution of the majority of this class of DNA (Britten, 1986). Thus, hybridization is primarily a measure of the neutral drift of the DNA and is not a measure of the events affecting phenotypic change. This decoupling of DNA sequence evolution and phenotypic evolution may have profound consequences for evolutionary morphologists studying processes and rates of phenotypic change. In particular, sequence data and hybridization data may allow the construction of phylogenetic hypotheses that are largely independent of the evolution of morphological characters and/or life history traits. Such phylogenies are sorely needed if we are to understand how phenotypic characters and life history traits evolve.

PRINCIPLES AND COMPARISON OF METHODS

General Principles

DNA hybridization takes advantage of the double-stranded nature of the DNA molecule in which nucleotides on opposing strands are held together by hydrogen bonds. In the case of adenine and thymine, there are two hydrogen bonds. Guanine and cytosine, in turn, are held together by three hydrogen bonds. When double-stranded DNA is heated to 100°C, the hydrogen bonds between complementary base pairs are broken and the opposing strands separate. Subsequent cooling of the solution facilitates reannealing of the complementary strands. Reassociation conditions (e.g., salt concentration, temperature, viscosity, fragment size) determine the amount of base pair mismatch in the hybrid molecules that are permitted to form. At high-stringency reassociation conditions, which are generally achieved by decreasing the salt concentration and/or increasing the temperature, interspecies base pairing will occur only between well-matched sequences. At progressively lower stringency conditions, increased mismatch is tolerated up to a point at which random reassociation occurs.

DNA from two different species can be combined, denatured, and then allowed to reassociate. The double-stranded molecules that form between complementary strands from the two species will contain base pair mismatch because of their evolutionary divergence from a common ancestor. The extent of mismatch determines the temperature at which these hybrid molecules melt when they are placed in a thermal gradient. The depression of melting temperature in a heteroduplex hybrid relative to a homoduplex hybrid then serves as an index of divergence between the DNAs under comparison. The extent of reassociation in homoduplex versus heteroduplex reactions can also be measured, although the factors that influence percentage reassociation are not as easy to disentangle.

Summary of the DNA Hybridization Techniques and Data Analysis

Briefly, double-stranded DNA is isolated and then purified to remove RNA and protein. Long-stranded DNA is then fragmented to short pieces to permit separation of repetitive and single-copy DNA and to reduce viscosity and gel formation. Fractionation of single-copy DNA from repetitive sequences is most easily accomplished using reassociation kinetic techniques developed by Britten et al. (1974). These methods facilitate the construction of C_0t plots (Figure 1), which present the percentage of single-stranded DNA versus the log of C_0t (C_0t = initial concentration of DNA in moles of nucleotides per liter multiplied by time of incubation in seconds). C_0t plots, in turn, allow one to determine a C_0t value (under specific incubation conditions) at which repetitive sequences have reassociated and can be separated from single-stranded single-copy DNA by hydroxyapatite column chromatography (Kohne and Britten, 1971). Fractionated single-copy DNA from one species is then radioactively labeled (**tracer**) and hybridized with unlabeled DNA (**driver**) from the same species (homoduplex reaction) and from different species (heteroduplex reactions). When the hybridization is complete, melting profiles and the extent of reaction are then determined. Melting profiles, in turn, permit the quantification of median and/or modal melting temperatures. Differences in these parameters between homoduplex and heteroduplex curves are then used as the estimates of genetic distance, ΔT_m and ΔT_{mode}. The extent of hybridization for an interspecies heteroduplex measurement may be divided by that for the homoduplex control and multiplied by 100 to obtain a normalized percentage of hybridization (NPH). NPH values are generally considered to

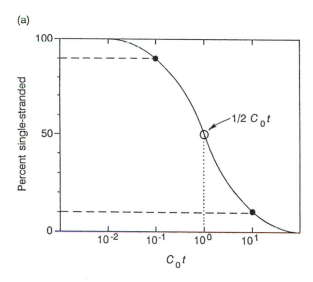

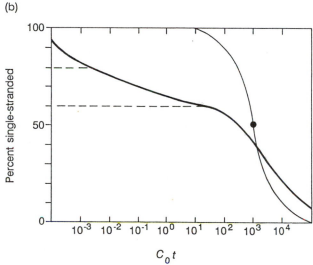

FIGURE 1. (a) Ideal reassociation curve (C_0t curve) for a single class of DNA (i.e., single-copy or a single frequency class of repetitive DNA). The curve tracks the loss of single stranded DNA (determined by $1/1 + kC_0t$; see text) and the formation of duplex DNA over log intervals of C_0t as expressed in [moles of nucleotide/liter] × sec. The "half-C_0t" is the C_0t value (here = 1) at which 50% of the DNA has reassociated. In an ideal reaction, 80% of the DNA reassociates over 2 log intervals, thus the C_0t value at which 90% of the DNA has reassociated is 10 times the half-C_0t. (b) Hypothetical reassociation curve for genomic DNA that includes a mixture of highly repetitive, moderately repetitive, and single-copy components. The individual reassociation curve for the slowest (higher C_0t) component is shown, with the approximate half-C_0t identified by (•). Horizontal dashed lines approximate the percentage of the genome that each class comprises (20% highly repetitive, 20% moderately repetitive, and 60% single copy). Since single-copy DNA is the last component to reassociate ($1/2 \, C_0t = 1,000$), it can be fractionated from the repetitive DNA (over hydroxyapatite) by reassociating the total DNA to a C_0t value of 100. At this point, 90% of the single-copy DNA is single stranded.

be a measure of the fraction of the DNA that has diverged to the point where it will no longer form stable interspecies duplexes under criterion conditions (see below). However, NPH may also be influenced by (1) sequences in the single-copy genome of the tracer species that are deleted in the single-copy genome of the driver species, and (2) kinetic effects (i.e., rates of reassociation decrease as interspecies sequence divergence increases; Bonner et al., 1973).

Finally, ΔT_m and NPH are sometimes incorporated into yet another distance measure, ΔT_{50H}. These different measures of genetic distance can then be used in phylogenetic analysis. Typically, complete matrices in which each taxon has been labeled and compared to all other taxa are used for this purpose. Algorithms for phylogenetic analysis with distance data (see Chapter 11) include phenetic methods (Sneath and Sokal, 1973), best-fit methods (Fitch and Margoliash, 1967; Cavalli-Sforza and Edwards, 1967), minimum length tree methods (Farris, 1972, 1981; Faith, 1985; Saitou and Nei, 1987), and maximum likelihood methods (Felsenstein, 1987).

Properties of Hybridization Data

Genome structure, sequence evolution, population history, and the DNA hybridization technique all affect the content of distance matrices. The task at hand is to understand how these factors affect DNA hybridization data and then to select appropriate tree-construction algorithms. When rates of sequence evolution are not the same in all lineages, for example, UPGMA clustering is an inappropriate tree-construction algorithm if we desire trees that accurately reflect phylogeny. Pairwise tree-construction methods, however, make no assumption about equal rates of change. Partly for this reason, they have become increasingly popular with distance data. On the other hand, pairwise tree construction methods do assume that distance data are additive (at least in expectation). Indeed, an evolutionary interpretation of branch lengths on a best-fit tree requires an underlying additive matrix (Farris, 1981). When distance data are nonadditive in expectation, a simple evolutionary path length interpretation of distances on best-fit trees is confounded. How well do DNA hybridization data fare under the assumption of additivity? Unfortunately, several factors may compromise the additivity of DNA hybridization data (Springer and Krajewski, 1989). These factors include homoplasy (i.e., parallelisms, reversals, multiple hits), sequences that are too divergent to form heteroduplexes, pairing between paralogous sequences, horizontal gene transfer, measurement error, the distribution of rates of sequence change for different sequences, and the history of genetic variation in different lineages. Some of these factors also affect sequence data. On the other hand, violations of additivity do not necessarily preclude the accurate recovery of branching order, even if they cast doubt on the validity of branch lengths (Chapter 11).

In some instances (e.g., homoplasy), there are even remedies for sources of nonadditivity; these remedies make it possible to better approach truly additive data. In other instances, further work will be required to improve the interpretation of DNA hybridization distances. For example, we can use the Jukes and Cantor (1969) model to correct DNA hybridization distances for homoplasy, but this requires that we know the conversion between melting point depression and percentage sequence mismatch. Evidence for a linear relationship between the thermal stability of imperfect hybrids and the extent of sequence divergence shows that about 1°C change in melting temperature corresponds to 1% sequence divergence (Bonner et al., 1973). Other estimates of this relationship are discussed in a following section.

Factors Affecting DNA Hybridization

The kinetics of DNA hybridization are affected by several factors, including genome size, copy number, DNA fragment size, and base composition. Genome size is significant because the rate of reassociation or hybridization is inversely proportional to the number of different sequences in the genome. Since most of the DNA is single-copy (that is, most sequences are different from each other), the rate of hybridization is determined primarily by the genome size or DNA content per haploid set of chromosomes. Genome size varies widely among taxa, ranging for eukaryotes from 10^8 to about 10^{11} nucleotide pairs (Britten and Davidson, 1969). For bacteria the DNA content is much smaller and for viruses it can be less than 10^4 bp. This represents a 10 million-fold range in size, and thus rate of hybridization cannot be ignored and is a central part of experimental design.

Repetitive DNA makes up a significant minority of the genome of all eukaryotes (e.g., Figure 1b) and some prokaryotes and greatly influences the dynamics of hybridization. Repetitive DNA typically shows a large amount of divergence within the genome of an individual and does not usually evolve at the same rate as single-copy DNA. Thus, it is practical and necessary to remove the repeats. Very few data exist for interspecies hybridization of repetitive DNA and it has not been used for effective resolution of systematic issues although the evolution of the repeats themselves is of some interest.

There are two essential problems in the separation of repetitive and single-copy DNA. First, repeats are interspersed throughout the single-copy DNA requiring that the DNA be sheared to small fragments of a few hundred nucleotide pairs (bp). Second, it is not practical to separate low-frequency repeats from single-copy DNA. The more rapid rate of reassociation of repeats, compared to single-copy sequences, has been the only means for separating these two classes (see hydroxyapatite procedure described below). However, separation based on reassociation rate is never absolute, and small numbers of copies of each repeat family will remain in the "single-copy" DNA. This problem has never caused significant uncertainty in the interpretation of hybridization data because the quantity of DNA involved is small. The only situation in which this source of error is likely to be significant is at great evolutionary distances when very small amounts of hybridization are observed.

To fractionate the repetitive and single-copy components, it is practical to reassociate short fragment DNA to the C_0t required for 10% reassociation of single-copy DNA and remove all reassociated fragments (which include most repetitive elements) on hydroxyapatite. DNA concentrations and qualities vary from preparation to preparation and most of the trouble that could arise is suppressed by running the hybridization to completion so that a preparation with 2-fold lower DNA concentration will not much affect the extent of hybridization. The precision will be improved by knowing the required C_0t and accurately controlling the DNA concentrations, purities, and fragment sizes. For fragmented DNA in solution (500-nucleotide-long fragments at 60°C in 0.12 M neutral phosphate buffer) the fraction that remains single stranded (i.e., has no duplexed regions and does not bind to hydroxyapatite) can be simply expressed:

$$S = 1/(1 + kC_0t)$$

where k is the practical reaction rate constant (10^6 divided by the genome size in bp). The C_0t is most easily calculated as $10 \times U \times H \times A$ or $2 \times OD \times H \times A$ where A is an acceleration factor that depends on incubation conditions (= 1.0 at 0.12 M PB at 60°C), U is micrograms per microliter, H is hours, and OD is optical density at 260 nm for a 1-cm path.

For mammalian DNA with 3×10^9 bp a C_0t of about 3000 is required for half reassociation. With 1 µg/µl of DNA under these conditions the C_0t is only 240 per day. To accelerate the process, the DNA concentration and the phosphate buffer concentration can be raised (10 µg/µl is about the practical limit for the former; Britten et al., 1974). These modifications can be used to obtain a C_0t of about 30,000 (or 10 times the half-C_0t) in a day and a half. At this C_0t, the single-stranded fraction is only about 10%, which is sufficient for most applications. Near the end of the reaction an increase of a factor of 10 in C_0t reduces the amount of unreacted DNA by a factor of 10. In practice, DNA degradation may occur and it may not be profitable to use more than a few days of incubation, although **chaotropic agents** may help by reducing the temperature of incubation.

The Criterion and Precision of Reassociation

The precision of the partially matching duplexes that can form during reassociation is determined by the temperature and ionic strength of the incubation buffer. Together they establish the **criterion** (stringency of reassociation) that is usually described as the difference between the T_m of perfect duplexes [about 85°C in most cases for 0.18 M Na$^+$ or 0.12 M phosphate buffer (PB)] in the incubation buffer and the temperature of incubation (Britten et al., 1974: 366). The optimum rate of reassociation occurs at about 25°C below the T_m of the duplexes being formed (Bonner et al., 1973). If the conditions are too stringent (i.e., if the temperature is too high and the salt concentration too low), the NPH is reduced and all of the duplexes have high melting temperatures. The maximum ΔT_m under such conditions is quite small, reducing the resolving power of the method. On the other hand, if the conditions are too relaxed, the rate of reassociation is reduced and dissimilar sequences may form unstable duplexes. Under these circumstances, even very distant sequences form duplexes, possibly to the exclusion of more closely related sequences. It is not known what fraction of duplexes is between nonorthologous sequences, and this may potentially provide misleading information of evolutionary history. However, under more suitable conditions (25–40°C below the T_m) practically all of the duplexes that form are thought to involve orthologous sequences.

Hybridizations in phosphate buffer (PB) are carried out at 60°C in about 0.48 M PB to accelerate the reaction (see above) and thereafter diluted to 0.12 M PB for thermal denaturation. Upon dilution, the effective temperature of incubation is reduced to about 53°C. Consequently, any tracer that elutes below this latter temperature is not due to denaturation of duplexes formed during the hybridization reaction and can therefore be considered an artifact.

Finally, the base composition of duplexes can affect their individual melting temperatures. Since G-C pairs share three hydrogen bonds and A-T pairs share two, DNA double strands that are G-C rich will melt at a slightly higher temperature than an A-T-rich fragment in standard phosphate buffer. This factor tends to increase the width of the melting curve for mixed fragments (i.e., when the tracer is not from a source of cloned fragments) in intra- or interspecies comparisions where the total single-copy fraction is used. The effect of base composition can be eliminated by the use of chaotropic solvents (TEACL, described below) for duplex denaturation.

Comparison of the Primary Methods

Hydroxyapatite   In 0.12 M PB at temperatures from 45 to 60°C, double-stranded DNA binds efficiently to HAP whereas single strands do not. Further, hydroxyapatite (HAP) continues to bind double-stranded DNA until the melting temperature is reached. Thus, the separation of single- and double-stranded DNA is a simple procedure. The practical capacity of HAP for DNA, including divergent sequence duplexes, is about 100 µg/400 mg HAP in buffer, although native DNA may be bound at much higher levels. However, small amounts of duplex DNA are slightly eluted near the melting temperature in 0.12 M PB (Fox et al., 1980b; Martinson, 1973). Thus, it pays to use consistent flow rates and elution volumes for high accuracy.

Many substances, such as 0.3 M NaCl or NaAc, 7 M NaClO$_4$ or 8 M urea, can be present and do not interfere with the binding to HAP. These have little influence on the separation of single- from double-stranded DNA, although they do influence the T_m. However, other substances such as small amounts of CsCl, protein, and some metallic ions interfere with duplex binding. For a detailed account of the HAP procedure see Britten et al. (1974).

The major advantage for the use of HAP for DNA hybridization studies is that it is the most explored technique. It has been investigated extensively at the physicochemical level and applied the most widely to systematic problems. Disadvantages, however, include (1) melting curves that are broader than those obtained with the TEACL method, and (2) the time involved in running individual columns and large numbers of taxa that may require automated procedures.

S1 Nuclease and Precipitation to Assay Melting   Many hybridization experiments (Benveniste and Todaro, 1976; Benveniste, 1985; O'Brien et al., 1985a) have used a straightforward procedure in which, after hybridization, the DNA is treated with the single-strand-specific S1 nuclease and precipitated. S1 nuclease degrades single-stranded, but not double-stranded, DNA. The effective criterion is established by the rigor of the enzyme treatment and these authors have shown that different degrees of S1 digestion can change the criterion significantly. With this procedure the extent of hybridization is never as large as with HAP because the effective kinetics of reassociation are different since all unduplexed regions are digested, and because the unpaired tails of regions containing duplexes bind to HAP but are digested by S1. The result is a kinetic curve such as $S = (1 + kC_0t)^{-0.44}$. Ultimately, there is likely to be steric hindrance and further reduction in rate of reassociation so that the practical extent of completion for high-quality tracer and driver from the same species is likely to be only about 70% S1 resistance. For hybridizations between moderately distant species, the NPH is reduced compared to that observed with HAP. There is apparently a proportionality between NPH and ΔT_m with a slope that depends on the degree of S1 digestion (Benveniste, 1985). Thus, the activity of particular S1 nuclease preparations must be assayed and applied in a consistent fashion.

The Use of Tetraethylammonium Chloride   Tetraethylammonium chloride (TEACL) is a **chaotropic** solvent that essentially eliminates the effect of base composition on hybrid melting temperature at a concentration of 2.4 M (Melchior and Von Hipple, 1973; Hutton and Wetmur, 1973). With the use of TEACL the observed width of a precise duplex melting curve will decrease by a factor of 10, from about 14°C (in 0.12 M PB) to about 1.5°C. The relative technical advantages of TEACL for DNA hybridization are discussed in Powell and Caccone (1989). Additionally, since TEACL interferes with hydroxyapatite chromatography, single-stranded DNA cannot be removed except by digestion with S1 nuclease (see below).

The TEACL procedure has been used effectively to detect intraspecific polymorphism in single-copy sequence divergence (Britten et al., 1978) and evolutionary relationships of closely related *Drosophila* (Caccone et al., 1988a). This procedure, however, is somewhat more complicated and time consuming than the standard PB/HAP system. At present, the method is most useful for closely related species.

TEACL is usually combined with S1 nuclease methods for systematic problems. The advantages of this system include (1) elimination of the need for HAP columns, (2) many samples can be run at a single sitting (depending on the size of the heating block), (3) melting curves are narrow as compared to HAP, and (4) TEACL compensates for A-T, G-C differences. The disadvantages, as compared to HAP procedures, are that (1) S1 nuclease activity must be carefully standardized between assays, (2) the criterion depends on enzyme treatment, and (3) the NPH is more difficult to control.

Melting Curves Combining the Advantages of TEACL and Hydroxyapatite   In 2.4 M TEACL, precisely paired DNA melts at about 62°C regardless of base composition. The width of the melting curve is about 1.5°C. However, this procedure has two disadvantages. First, the criterion of reassociation cannot be set for widely divergent duplexes since such duplexes are digested by S1 almost as fast as single strands. Second, the combination of the HAP technique and TEACL melting characteristics has been restricted by the fact that TEACL and phosphate buffers tend to form precipitates or two-phase solutions under a variety of conditions. However, we have recently observed that in the presence of high concentrations of TEACL the phosphate concentration that permits duplexes to bind but allows single strands to pass through HAP is much reduced.

Although we have not exhaustively varied the concentrations and conditions, a good compromise is 2.0 M TEACL and 0.013 M PB (PT). This solvent is stable from 4 to at least 75°C. Precise duplexes bind HAP very well from room temperature upward and are eluted from HAP only as they are melted by increasing temperature. However, single strands bind below 50°C. Thus, this method is restricted to comparisons of relatively closely related species.

Native long DNA melts in PT at 68°C with a width of about 3°C and sonicated fragments of DNA (500 bp average) melt at 65°C (as determined by elution from HAP in PT). As an example of fractionation, we have used the method to isolate precisely paired repeat duplexes by incubation to C_0t 100 in PT at 60°C and passing the solution over HAP at this temperature. The temperature is raised in steps to 64°C with washes of PT, and then the temperature is dropped and the HAP washed with low concentrations of PB to remove the TEACL. The duplexes are eluted with 0.4 M PB for analysis.

PT is a good solvent for DNA reassociation having a rate acceleration factor of more than 10 (as compared to 0.12 M PB at 60°C) at its optimum temperature of 40°C. The acceleration factor drops to about 1.0 at 65°C or about 3°C below the T_m. It gives narrow accurate melting curves where long native duplexes melt at about 68°C.

A method is now being tested in which DNA is melted in 2.4 M TEACL and then the samples are diluted 40-fold into 0.12 M PB and passed over HAP to separate duplex from single strands (C. Hsiao, personal communication). Those who test this promising method further will have to ascertain that the small concentration of TEACL does not elute some divergent DNA duplexes and perhaps test for the best HAP temperature and the PB concentration for ideal separation of double from single strands.

Automated Melting Assay   In the earliest eukaryotic interspecies DNA hybridizations (Kohne, 1970) HAP elution was carried out with a pump and the temperature was raised with a control on a single column. The accuracy was increased by using two isotopes and an internal reference DNA. Sibley and Ahlquist (1987a and references therein) use an automatic machine to process 25 HAP columns simultaneously. A good compromise might be made between accuracy and efficiency of measurements by avoiding iodination, incorporating the use of an internal standard, and using a large number of columns in an automatic machine. Of course, tracer and driver fragment sizes and concentrations (as well as other critical variables) should be carefully controlled in automated procedures as in manual methods. There are no automated machines as yet available commercially. Most of those in use are based on individual requirements and design.

APPLICATIONS AND LIMITATIONS

Investigations into the kinetics of DNA reassociation form the foundations of DNA hybridization as a tool for questions of systematic and evolutionary relationships. Studies regarding the rate of reassociation of sheared, total genomic single-stranded DNA have provided quantitative estimates of the degree of sequence repetition, length of repeated sequences, and interspersion pattern of these sequences throughout the genome (Britten and Kohne, 1967, 1968; Britten et al., 1974). Highly repeated sequences reassociate far more rapidly than single-copy sequences, and by varying the fragment length of the reassociating DNA it is possible to estimate repeat length and interspersion patterns. Thus, kinetic studies have yielded a wealth of information on genome organization and structure in prokaryotes and eukaryotes, as well as providing for a method useful for the separation of specific sequence classes. Britten and Kohne (1968), as well as Hood et al. (1974: 56), provided explanations of reassociation kinetics and C_0t analysis.

There are at least two other observations derived from kinetic studies that are important to systematic applications of this technique. First, the observed reduction in the thermal stability of reassociated hybrid DNA (ΔT_m) is directly proportional to the sequence difference (in percentage base pair mismatch) between reannealed single strands. Second, this sequence divergence can reduce the rate at which sequences reassociate (Bonner et al., 1973). In other words, as divergence increases between sequences, their reassociation rate slows. This issue is discussed in a later section.

Several studies at the intraspecific level have utilized hybridization techniques to assess interindividual and interpopulational sequence divergence and variation. Britten et al. (1978) determined the magnitude of single-copy sequence polymorphism among individuals of the sea urchin, *Stronglyocentrotus purpuratus,* and, surprisingly, found it to be about 5%. Similarly, divergence estimates have also been made for isogenic (parthenogenetic) strains of *Drosophila mercatorum* (Caccone et al., 1987). Both of these studies employed the use of TEACL to decrease the effective width of the melting curves for more accurate determinations of ΔT_m over standard phosphate buffer conditions. Others have used the latter system to determine intraspecific variation in sea stars (M. J. Smith et al., 1982), herons (Sheldon, 1987), diprotodont marsupials (Springer, 1988), and cave crickets (Caccone and Powell, 1987). The magnitude of intraspecific variation may be important to consider in determining the relationships among closely related species (Chapter 11).

The majority of hybridization studies, as applied to systematics, have involved species and higher taxon relationships, up to family and ordinal level comparisons. The effective limits of resolution depend primarily on the degree of divergence (as related to rate) among taxa under investigation (see below). Powell and Caccone (1989) noted that the smallest interspecific difference accurately resolved in their studies with TEACL was a T_m reduction of 0.27. Interspecific comparisons with a primary focus on phylogenetic relationships among birds and mammals include Sheldon (1987), Sibley and Ahlquist (1987a,b, and references therein), Springer (1988), Springer and Kirsch (1989), and Bledsoe (1987).

As with other molecular techniques used to obtain information useful to phylogenetic reconstruction and systematic relationships, DNA hybridization has limitations. Since many of these details are discussed throughout this chapter, we provide a brief list (arbitrarily ordered) of major points.

1. Direct sequence data are not uncovered, and the data are in the form of distance information.
2. Comparisons are restricted to the single-copy fraction of the genome.
3. Dramatic differences in the size of the single-copy fraction between species pairs could produce errors in reciprocal measurements of NPH, although this has not yet been documented.
4. Large amounts of intraspecific polymorphism can be problematic in the estimation of phylogenetic relationship of closely related species.
5. The upper limit of divergence between species where relationships can be determined by this method is set by the conditions of DNA reassociation. For example, with HAP procedures at standard conditions of incubation it is difficult to estimate relationship with reasonable certainty if the NPH falls below 50% and the ΔT_m is greater than 20°C.
6. DNA hybridization is relatively expensive as compared to other techniques and involves the use of radioisotopes.
7. In many cases, milligram quantities of DNA are required to permit reasonable comparisons with replication. Thus, comparisons among individuals are restricted to organisms from which the required amount of DNA can be extracted. However, the PERT procedure (Protocol 10) requires less DNA, allowing studies to be accomplished with several hundred micrograms.

LABORATORY SETUP

Most of the supplies required for DNA hybridization are those commonly used in other DNA isolation, manipulation, and characterization techniques, including cloning and sequencing (Chapters 8 and 9). General supplies include centrifuge tubes, culture tubes, assorted glassware, ceramic mortar and pestle, pipettes (with microliter to milliliter delivery), filters, razor blades, liquid scintillation vials, etc. Supplies that may be unique to DNA hybridization include capillary tubes (10–100 µl volumes) in which hybridization reactions are carried out and polystyrene disposable chromatography columns with filter disks, 6.5 ml capacity (Figure 2). Equipment needed is shown in Table 1.

PROTOCOLS

1. DNA isolation and purification
2. Preparing sheared drivers from long native DNA
3. Tracer preparation with ^{32}P and ^{3}H
4. Tracer self-reaction and repeat removal
5. Fractionation of single-copy tracer over hydroxyapatite
6. Estimation of tracer fragment length

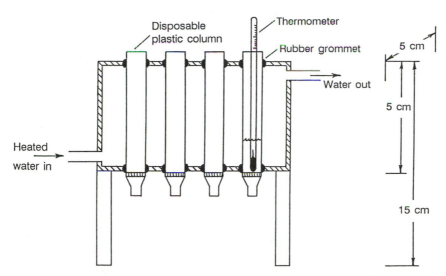

FIGURE 2. Diagrammatic frontal view of a multicolumn apparatus for fractionating tracers which have been self-reacted to remove repeats and/or foldback DNA. The system consists of an acrylic plastic box with legs, through which heated water can be circulated. Holes drilled in the top and bottom allow for the insertion of disposable plastic columns, with filter disks, sealed by rubber grommets. A thermometer placed in the last tube (with water) is used to measure column temperature. The apparatus is hooked up to an externally circulating heated water bath as shown in Figure 3.

7. Preparing tracers by iodination
8. DNA hybridization with hydroxyapatite and phosphate buffer
9. Hydroxyapatite column preparation
10. Phenol emulsion reassociation technique (PERT)
11. Analysis of hybrid thermal stability using the S1 nuclease-TEACL assay

Protocol 1: DNA Isolation and Purification

(Time: 1.5 days)

Genomic DNA used for hybridization studies must be pure and free from contaminants such as glycogen, protein, metallic ions, and other impurities. These contaminants may interfere with DNA reassociation. The following procedures apply to the isolation of DNA from dissected tissues and can include frozen blood samples. This protocol has been adapted from Graham (1978); a rapid method for DNA extraction from blood is given by Jeanpierre (1987). Protocols for small organisms (e.g., fruitflies, etc.) or for small amounts of tissue are described in Chapters 8 and 9.

1. Prechill a small mortar and pestle in dry ice (chill the grinding surface of the pestle only). Grind a small mass of frozen tissue (5–10 g) mixed with an equal quantity of

Table 1. Primary equipment used in DNA hybridization

Equipment	Use
Refractometer (precise to 0.0001 refractive index units)	Determine solution molarity[a]
Balance, analytical	Weighing small samples
Balance, top loading	General weighing
Centrifuge, microtube	Centrifuging small samples
Centrifuge, high-speed re-frigerated, with rotors for 250- or 50-ml bottles	DNA isolation
Electrophoresis apparatus (aga-rose) large and mini submerged horizontal	DNA and tracer sizing
Gamma counter	Detection of ^{125}I[a]
Gradient heating block	Thermal denaturation of DNA[a]
Liquid scintillation counter	Detection of ^{32}P and ^{3}H
Lyophilization apparatus	DNA purification for iodination
Adjustable micropipettes; 20 µl, 100µl, 1000 µl	General solution manipulation
Spectrophotometer	Determining DNA concentration of samples
Ultrasonic cell disruptor (sonica-tor, probe type)	Shearing DNA[a]
UV transilluminator	DNA detection in gels
Waterbaths	
Heated (noncirculating)	Hybridization incubation
Heated (circulating)	Thermal denaturation of DNA[a]
Refrigerated	Nick translation labeling
Water jacket for multiple columns (Figure 2)	Tracer fractionation[a]
Water jacketed glass column with stopcock and sintered glass filter disc	Thermal denaturation of DNA on hydroxyapatite[a]
Vacuum chamber	Drying DNA samples
Vacuum pump	Drying DNA samples

[a] Items specific to DNA hybridization protocols.

    dry ice to a very fine powder. Set aside and let most of the dry ice sublime, but keep the ground tissue frozen. Steps 2–5 should be done as quickly as possible.

2. Rapidly dissolve powder in ice cold SEDTA (Appendix). Use 10–100 volumes SEDTA to tissue depending on DNA content, e.g., sperm requires a larger volume than blood. The resulting solution should be viscous.

3. Rapidly dissolve all lumps of tissue, immediately add 20% SDS to a final concentra-tion of 1%, and stir gently, to avoid shearing the DNA.

4. Add 1/5 volume of 5 M sodium perchlorate solution and mix.

5. Immediately add an equal volume of equilibrated phenol and mix for 30–60 min.

Mix with just enough force to keep the emulsion from separating. Centrifuge at 5000 g for 15 min at 4°C.

6. The phenol phase should be on top because of the density of the perchlorate solution. If the phases do not separate add a small volume of SEDTA and recentrifuge. If you add too much SEDTA, the phenol layer may end up on the bottom. Remove the phenol phase.

7. To the aqueous phase, add an equal volume of 24:1 chloroform:isoamyl alcohol, and mix at room temperature for 30 min. Centrifuge as above and save the aqueous phase (should be on the top). Leave the milky interface. Repeat steps 5–7.

8. The DNA is now ready for spooling. Place the aqueous phase in an acid-washed beaker large enough to hold four times the sample volume. Carefully layer 2 volumes of ice-cold 95% ethanol onto the DNA solution by pouring it down the side. Keep the two solutions from mixing.

9. Take a long acid-washed glass rod and rotate it slowly with a slow mixing action just below the interface of the two solutions. The DNA should wind onto the rod and form a mass large enough so that no more DNA will cling. Remove the glass rod and gently squeeze excess ethanol out of the wound DNA against the side of the beaker. Slice the DNA with a razor blade along the axis of the rod and remove it to a 15-ml sterile tube and repeat the winding until no more DNA sticks to the rod and the two layers are nearly completely mixed.

10. Add 5 ml of TE (or about 0.5 ml/mg DNA) to the DNA and let it swell overnight at 4°C. If the DNA does not dissolve add more TE as appropriate.

11. (Optional, if there is RNA and protein contamination) To the resuspended DNA add 20 µg/ml DNase-free RNase and incubate in a water bath at 37°C for 1 hr. Remove and add 100 µg/ml of proteinase-K solution, 1/10 volume 3.0 M sodium acetate, and 1/100 volume of 25% SDS. Mix and incubate at 60°C for 1 hr. Extract once with phenol, then twice with 24:1, chloroform: isoamyl alcohol. Add 2 volumes of 95% ethanol to precipitate the DNA. Wash once with 70% ethanol and partially dry the DNA pellet under vacuum. Resuspend at 2–3 mg/ml in 0.1 mM EDTA over a few drops of chloroform.

12. In a spectrophotometer, check the optical density (OD) of a dilution of the DNA preparation at 230, 260, and 280 nm. At 260 nm, 50 µg of DNA in 1 ml solution will give an OD of 1.0. The ratio of ODs at 260/280 provides an indication of RNA contamination. The ratio should be close to 1.8. The more RNA present the higher the value. A low ratio of ODs at 260/230 indicates protein contamination; this value should be greater than or equal to 2.3.

13. Electrophorese 500 ng of the DNA solution on a 0.6% neutral agarose gel with DNA markers and stain with ethidium bromide (Chapter 8). The majority of the DNA should migrate as a large band close to the origin.

Protocol 2: Preparing Sheared Drivers from Long Native DNA

(Time: 6 hr)

For interspecies hybridizations both driver and tracer DNA fragments must be approximately 500 bases in length (denatured) to provide for the separation of repetitive and single-copy fractions, since repeats are dispersed throughout the genome at some frequency.

Shearing small samples of DNA is best accomplished using an ultrasonic cell disruptor or sonicator. DNA samples (>20 ml) can be sheared in a motorized tissue homogenizer following Britten et al. (1974) and Hunt et al. (1981). A protocol for sonicating DNA (in solution) is outlined below.

1. To 400 μl sterile water add 50 μl 3.0 M sodium acetate and 100 μg of DNA (= 50 μl at a concentration of 2 mg/ml) in a 2-ml sterile glass screwcap vial. Mix gently and cool on ice for 15 min.
2. Sonicate for 30 sec at 80–90% maximum power with the tip of the sonicator probe just below the surface of the solution. Put on ice for 30 sec and repeat four more times. Place on ice and set up a small chelating resin column (see step 3) to filter out any metallic ions or particles introduced during sonication. Before step 3, it may be desirable to go to step 4 (below) and check the size of the DNA in case additional sonications are required.
3. Clamp a 1-ml pipette tip to a ring stand and push a small piece of sterile glass wool into the tip. Add 0.5 ml of chelating resin (equilibrated to pH 7.0 with 0.3 M sodium acetate) and rinse several times with 1 ml of 0.3 M sodium acetate. As the final rinse of sodium acetate passes through the chelating resin, add the DNA sample and collect it after it passes through to column. Add 250 μl sodium acetate to the column to wash out any DNA and combine with the previously collected sample.
4. Divide sample into 500-μl aliquots in 1.5-ml microcentrifuge tubes and to each add 1 ml of cold 95% ethanol to precipitate the DNA. Spin in a microcentrifuge at high speed for 10 min at 4°C. Wash pellets in 1 ml of 70% ethanol and repeat the spin. Decant the ethanol and partially dry the pellets under vacuum. Resuspend pellets in 20–30 μl 0.1 mM EDTA or in an appropriate volume to obtain a concentration of 5 μg/μl (5 mg/ml).
6. Electrophorese 500 ng of the sonicated DNA on a 2% alkaline agarose minigel (Protocol 6) at 40 V for 2–4 hr with PBR/Hinfl marker (or some other suitable marker for 500-bp fragments). Neutralize gel in 500 mM Tris (pH 7.5) and stain with ethidium bromide to visualize the DNA. If the sheared DNA is much longer than 500 bases in length then it must be sonicated additional times. If the DNA is much shorter than 500 bases, then it cannot be used for driver. Thus, it is best not to overshear the DNA at first.

Protocol 3: Tracer Preparation with ^{32}P or ^{3}H

(Time: 3 hr)

Radioactively labeled tracer DNA can be prepared by standard nick-translation procedures (see also Chapter 8), although iodination has been used extensively in previous systematic applications of DNA hybridization (Sibley and Ahlquist 1987a, and references therein). Iodination procedures (Protocol 7) are somewhat difficult to establish and may require practice to achieve good tracers on a routine basis. An advantage of an iodinated tracer is a long half-life (about 60 days). However, there are also advantages in the use of ^{32}P- or ^{3}H- labeled tracers, which can be counted in a beta counter. ^{32}P has as advantages that it can be counted Cherenkov without the use of scintillation fluid, can be detected by a hand held Geiger counter, and has a high counting efficiency (95%). Tritium does not share these advantages but has a very long half-life and is a lower energy emitter; consequently, tracers made with ^{3}H have an extremely long shelf life. ^{32}P has a half-life of about 14 days; consequently tracers lose their activity and detectability rather quickly. Also, ^{32}P tracers degrade within a few days to a week if labeled to very high specific activity ($>1\times10^{6}$ cpm/µg). Below is a protocol for synthesizing a ^{32}P genomic tracer. ^{3}H can be subsituted or used in combination with ^{32}P, so that a ^{3}H tracer can be easily tracked.

Nick translation of long native genomic DNA is preferred since one has more control in the resulting fragment size by varying the quantity of DNase added to the reaction. If starting with sheared DNA (500 bp), the resulting fragment size will, on average, be considerably smaller, possibly too small to be used as tracer. Additional details regarding nick-translation procedures are presented in Sambrook et al. (1989).

1. In a 1.5-ml microcentrifuge tube add 5 µg of long native DNA to be nick translated. Then add
   5 µl 10× nick-translation buffer (Appendix)
   2 µl each 1 mM dATP, dGTP, dTTP
   1 µl 1 mM dCTP (or less, for higher specific activity)
   5 µl [^{32}P]dCTP (50 µCi at 800 Ci/mM)
   1 µl DNA polymerase
   0.5 µl DNase (10^{6} dilution of a 10 mg/ml stock)
   Sterile water to 50 µl total
   For ^{3}H, use several labeled triphosphates without dilution. Incubate at 12–14°C for 2 hr. Add 1/20 volume 5M EDTA and place on ice. Remove 1 µl and dilute to 500 µl with water in another 1.5 ml tube to check ^{32}P incorporation.
2. To check the amount of radioactive nucleotide incorporated, take 10 µl of the 500 µl sample and dot it onto a Whatman GF/C glass filter disk and set aside. Take another 10 µl and mix it with 5 ml 10% ice-cold trichloroacetic acid (TCA) and 50 µg of sheared salmon sperm DNA. Put on ice for 15 min.
3. Filter the 5 ml of sample plus DNA through a 2.4 cm GF/C glass filter disk and wash five times with 5 ml 10% TCA followed by 2 washes with 95% ethanol. Set the filter aside in a fume hood behind a shield and let it dry; remember both filters are radioactive, as are the wash solutions. Place both filters into separate scintillation vials and add 10 ml of scintillation cocktail (for ^{3}H) or count Cherenkov (no fluid added) for ^{32}P. The unwashed filter represents the total radioactivity added; the washed filter represents the amount of ^{32}P (or ^{3}H) incorporated into the DNA. Incorporation should reach 30–40%.
4. The unincorporated nucleotides must be removed from the nick-translation reaction. This can be accomplished by the "spun column technique" outlined in Chapter 8.

However, we prefer to use the glass powder elution procedure described in Davis et al. (1986:123) which is available commercially as the Geneclean® kit (BIO 101 Inc., P.O. Box 2284, La Jolla, CA 92038-2284). The general procedure is outlined below.

5. To the nick-translation reaction add 10 μl of the glass beads solution (50% slurry in water, = glassmilk of the Geneclean® kit) and gently mix. Add 150 μl sodium iodide solution (Appendix), mix gently and set aside at room temperature for 5 min.

6. Spin at high speed in a microcentrifuge for 5–10 sec. Discard the supernatant (radioactive) and wash the glass pellet three times with 500 μl of the ethanol-Tris wash solution (see ethanol wash in Appendix). Spin 5–10 sec between washes and discard supernatant at each step. Be sure to resuspend the glass pellet with each new wash. On the last wash remove as much of the ethanol solution as possible.

7. Elute the nick-translated DNA from the glass powder by adding 25 μl of TE or 0.48M PB and placing it into a 50°C waterbath for 15 min. Spin down the glass for 30 sec and remove the supernatant to a new tube. The tracer DNA should be free of unincorporated nucleotides and other impurities. It can now be self-reacted to remove repetitive DNA and any "snapback" DNA formed during the nick translation procedure (Protocol 4).

Protocol 4: Tracer Self-Reaction and Repeat Removal

(Time: 1–2 days)

A preliminary hybridization reaction and fractionation must be carried out on the newly labeled tracer. An appropriate C_0t must be chosen to reassociate the repeat DNA while leaving the single-copy component single stranded so as to separate these components using hydroxyapatite chromatography. The C_0t necessary can be calculated by the following formula:

$$\frac{\mu g \ DNA}{sample \ vol \ (\mu l)} \times 10 \times AF \times time \ in \ hours = C_0t$$

AF is the acceleration factor due to an increase in PB concentration over 0.12 M. For 0.48 M PB the AF is 5.6. Usually adjustments to achieve a particular C_0t are made by varying the time required for incubation. For the tracer prepared above (assuming it is primate DNA) the C_0t is

$$\frac{10 \ \mu g}{25 \ \mu l} \times 10 \times 5.6 \times 13.4 \ hr = C_0t \ 300$$

The appropriate C_0t should be determined empirically for each taxon; estimates of C_0t values suitable for repeat removal have been made for *Drosophila* (C_0t 10–50), sea urchin (C_0t 100), and primates (C_0t 300).

1. Aspirate the 25-μl tracer sample into a sterile 50–μl glass capillary tube leaving at least

1 cm air space at each end of the tube. Seal the ends by melting them closed over a gas flame. Label with a piece of waterproof tape and immerse in a boiling water bath for 2 min to denature the DNA.

2. Remove the capillary tube and place it in a large screwcap vial filled with water at 60°C. Submerge this vial into a water bath at 60°C. The vial will protect the capillary tube from damage. Incubate for the required amount of time.

3. Remove the capillary tube from the vial and quickly place on ice. To a sterile 1.5-ml microcentrifuge tube add 75 μl sterile water (which will dilute the reaction mix to 0.12 MPB) and 400 μl of 0.12MPB (NEVER add the hybridization reaction mix directly to pure water, as some PB must be present, to prevent premature denaturation of duplexed fragments). Mix and set aside at room temperature.

4. File a small groove near each end of the capillary tube and break off the sealed tips. Carefully force out the hybridization reaction into the 1.5-ml tube and mix gently by rocking the tube back and forth for 30–60 sec. Fractionate over HAP at 50°C (Protocol 5) and collect the fraction not bound to the HAP; this is the single-copy tracer.

Protocol 5: Fractionation of Single-Copy Tracer over Hydroxyapatite

(Time: 3 hr)

Removal of repeated DNA from the tracer is best accomplished with the use of disposable plastic columns that fit through a clear plastic box that can be filled with circulating water adjusted to 50°C (Figure 3). The high radioactivity of the total tracer may contaminate the columns if used later for interspecies melts.

1. Add 300 mg of dry HAP to a plastic column and rinse three times with 3 ml of sterile water. Blow the water through with a small air pressure hose fitted with a rubber stopper on the end to fit snugly into the top of the column. Wash the HAP twice with 3 ml of 0.12 M PB. Load 1 ml of 0.12 M PB and raise the circulating water temperature to 50°C and blow the buffer through the column.

2. Add the tracer preparation (about 500 μl) and mix the HAP bed with a 5-μl sealed capillary tube to remove trapped air bubbles. Let the column equilibrate to 50°C, blow the tracer through the column, and collect into a 1.5-ml tube. Wash the HAP three times with 1 ml of 0.12M PB, collecting each wash. Then wash the HAP twice with 0.48 M PB to elute the adsorbed double-stranded DNA (mainly repeated sequences) and collect each into a separate tube.

3. Remove 1 μl of each sample (blank, load fraction, 0.12 wash #1, #2, #3, 0.48 wash #1, #2) and place into scintillation vials containing 1 ml of water. Add 10 ml of scintillation fluid and count each vial for 2 min (for ^{3}H, all scintillation solutions must be identical). The load fraction should have the highest cpm/μl and is to be used as the tracer. Remember the tracer is in 0.12 M PB, which should be considered in the calculation of the final hybridization PB concentration (or the tracer may be concentrated with the glass elution method, Protocol 3).

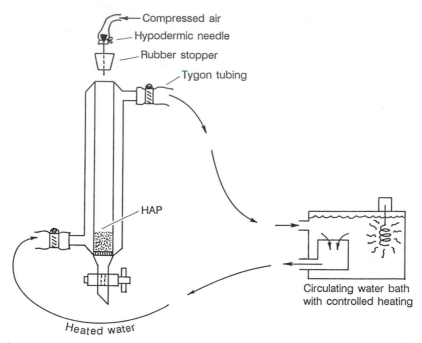

FIGURE 3.   Standard water-jacketed glass column for DNA hybrid melts, attached to a circulating, controlled heating water bath. These columns are not available commercially, but can be made with the aid of a glassblower by modifying standard chromatography glassware. Plastic syringes can also be modified to this configuration, with glass wool replacing the sintered glass disc. The glass column (shown) can be attached in series of up to six, depending on the flow rate of the water bath. Also shown is a rubber stopper/needle setup required to pressurize and blow the samples through the column.

Protocol 6: Estimation of Tracer Fragment Length

(Time: 6 hr)

Tracers below 500 bases in length have the effect of lowering the T_m by (500/tracer length) in degrees C. Thus, a tracer of length 250 will lower the T_m by 2°C. Tracer size must be estimated in the denatured state and this can be accomplished with alkaline agarose gel electrophoresis as described below.

1. Prepare a 2% agarose gel by adding 3 g agarose to 150 ml of gel buffer (50 mM NaCl, 1 mM EDTA) and microwave or boil to dissolve. Pour into an appropriately sized gel mold and let cool.
2. Place the gel into a submerged electrophoresis chamber and add enough alkaline running buffer to cover the gel with 0.5 cm of buffer. Let the gel equilibrate for 1 hr. Remove enough running buffer so that only about 2 or 3 mm of buffer lies on top of the gel.
3. Add 1/20 vol 1M NaOH to 1000–2000 cpm of tracer and incubate at 37°C for 10

min. Add 1/5 vol loading dye (Chapter 8, Appendix), mix, and load onto the gel in the second or third lane. Treat 500 ng of appropriate marker (PBR/*Hinf*I) in the same fashion and load onto the first lane. Run gel at 35 V for 6–10 hr.

4. Following electrophoresis, cut the lanes into strips separating the marker lane from the tracer lane(s). Neutralize marker lane in 0.5 *M* Tris-HCl, pH 7.5 and stain with ethidium bromide. Visualize and photograph (include a ruler) on a UV transilluminator.

5. Cut the tracer lane into 0.5-cm segments, starting from the origin (loading well) and place each piece into a separate scintillation vial in order; add 10 ml of scintillation fluid and count for 5 min each. Graphically compare the position of the modal cpm with the measured marker fragments to estimate the average tracer size. Use this size to adjust the T_m if desired. The primary reason for sizing is to avoid degraded tracers or those that are exceptionally long.

Protocol 7: Preparing Tracers by Iodination

(Time: Part A, 4 days; Part B, 24 hr)

Unlike the nick-translation procedure, iodination of DNA for hybridization experiments in systematics is generally carried out on sheared, single-copy DNA. We will therefore describe a procedure in which single-copy DNA is first isolated and then radiolabeled. This procedure is derived from the general protocols given in Commorford (1971), Davis (1973), Tereba and McCarthy (1973), Orosz and Wetmur (1974), Scherberg and Refetoff (1975), Chan et al. (1976), Anderson and Folk (1976), Prensky (1976), and Sibley and Ahlquist (1981a). The primary result of the iodination reaction is the replacement of a hydrogen atom at the C-5 position of cytidine by an iodine atom. Iodination is much more efficient when DNA is single stranded.

A few words of caution should be mentioned for investigators contemplating the use of radioiodine. The temperature, acidic pH, and presence of an oxidizing agent in the iodination reaction all contribute to the volatization of a fraction of the radioiodine (Prensky, 1976). This danger requires rigid measures of monitoring and protection. The review paper of Prensky (1976) is particularly relevant.

Much larger quantities of DNA are generally labeled with radioiodine. When DNA samples are in short supply, this is a significant consideration. Also, radioiodine must be assayed in a gamma counter.

Part A. Preparation of Samples for Iodination with [125]I

1. Boil 1.0–1.5 mg of sheared, native DNA in 0.48 *M* phosphate buffer for 10 min and incubate at 60°C to a C_0t value at which repeated sequences have reannealed. Dilute the sample to 0.12 *M* phosphate buffer and apply to a hydroxyapatite column at 55°C. Elute the single-stranded, single-copy DNA with 20 ml of 0.12*M* phosphate buffer.

2. Dialyze the single-copy fraction of DNA against deionized water for 48 hr to remove phosphate buffer. Change water frequently.

3. Transfer dialyzed sample to a serum bottle and freeze at $-20°C$. Lyophilize sample for 24 hr until DNA sample appears like cotton.
4. Refrigerate sample until subsequent iodination (not more than 24 hr).
5. Rehydrate lyophilized sample in a small volume (50–100 µl) of 0.2 M NaAc adjusted to pH 7.5 with glacial acetic acid. It is convenient to carry out the rehydration on a piece of parafilm.
6. Transfer the sample to a 1.5-ml microfuge tube. Vortex for 15 sec and centrifuge in a microfuge for 30 sec to remove insoluble debris. Determine the concentration of DNA using a spectrophotometer. One to 2 µl of the sample diluted in 2 ml of water is generally sufficient for this purpose.
7. Combine an aliquot of the sample containing 100 µg of DNA with 0.2 M NaAc (pH 5.7) to bring the total volume to 130 µl in a 1-ml stoppered serum vial. Add 6 µl of 2 mM KI and 11 µl of bromcresol green dye (BGD is a pH indicator). Adjust the pH of the reaction mixture to 4.7–4.8 with 0.2 M NaAc (pH 4.0), using pH color standards. Place sample on ice.

Part B. Preparation of Iodine

Iodinations are conveniently carried out for several samples at once. For sample reactions prepared as above, eight DNA samples can be radiolabeled with 5 mCi of ^{125}I. The following protocol is thus designed for the simultaneous iodination of eight samples with 5 mCi of ^{125}I, although the basic protocol can be adapted to any number of samples by using more or less iodine.

All of the manipulations described below should be carried out under a hood while wearing two pairs of latex gloves.

1. Start with 5 mCi of ^{125}I in a volume of 10 µl. Vent the rubber seal of the iodine container with a 23-gauge needle. Using a 23-gauge needle and a 1-ml syringe, dilute the iodine solution with 340 µl of 0.2 M NaAc and 10 µl of 1 mM KI and allow to equilibrate for 1 hr. Do not remove the rubber top on the iodine container; iodine is highly volatile at this stage.
2. Using a 23-gauge needle and a 1-ml syringe, carefully draw out all of the iodine solution and add 40 µl of isotope (0.625 mCi) to each of the eight samples. Do not remove the rubber stoppers from the serum vials.
3. Add 60 µl of 18 mM thallium chloride (TlCl) to each sample. Again, use a 23-gauge needle and 1-ml syringe to deliver the TlCl through the rubber stopper that caps each sample reaction.
4. Incubate samples at 60°C for 15 min in a temperature block.
5. Place samples on ice for 5 min.
6. Use a 23-gauge needle and 1-ml syringe to add 30 µl of 1.0 M Tris (base) to each sample.
7. Heat samples for 10 min at 60°C.
8. Place samples on ice for 5 min.
9. Transfer samples to dialysis bags and dialyze overnight against a 4 liter solution of 0.4 M NaCl, 0.01 M phosphate buffer, and 0.2 mM EDTA.

10. Transfer samples from dialysis bags to screw top vials using pasteur pipettes.
11. Determine concentration of DNA using a spectrophotometer (see Protocol 1, step 12). Cuvettes committed for this purpose will remain "hot" and should not be used for other laboratory work.
12. Count 1 µl of each sample in a gamma counter.
13. Store iodinated DNA tracers at −20°C until needed for hybridization.

Protocol 8: DNA Hybridization with Hydroxyapatite and Phosphate Buffer

(Time: 2 hr setup; up to 2 weeks incubation)

The reassociation of DNA hybrids and their melting properties in neutral phosphate buffer (PB) has been used extensively to study genome structure and systematic relationships. The description given here is for the simplest manual procedure, which is adequate but less accurate than some automated methods (e.g., Kohne et al., 1972; Britten et al., 1974). In setting up hybridizations in the PB system with properly prepared tracer, careful attention must be given to the following: (1) The concentrations of the PB stock solutions, the hybridization reaction mix, and the PB used to elute single-stranded DNA from the hydroxyapatite column are critical and must be known with accuracy. (2) The reaction volume must be as small as possible (10–100 µl) with a DNA mass:volume ratio of at least one or greater. Long periods of time are required to achieve high C_0t if the volume is large and driver concentration is small (see below). (3) The driver must be from 1000- to 10,000-fold in excess over the tracer mass for total single-copy reactions to prevent significant self-reassociation of the tracer. (4) For a set of interspecific measurements using a particular tracer, three reactions must be set up for thermal fractionation: (a) tracer × driver DNA of the same individual or species; (b) tracer × driver DNA from different species; and (c) tracer × greatly divergent DNA (to control for self reaction of tracer). Hybridizations using ^{32}P or ^{3}H are set up as follows (^{125}I hybridizations require approximately 250,000 cpm of tracer):

1. In a 1.5-ml microcentrifuge tube combine 500–1000 cpm of tracer (mass of tracer can be calculated from its specific activity) with a 1000-fold excess of sheared (500 bp) driver (e.g., for 5 ng tracer add 5 µg or more of driver). Remember the tracer will be in 0.12 M PB and the driver will be in water or 0.1 mM EDTA.
2. Adjust the final PB concentration to 0.60 M with 2.4 M stock (pH 6.8), mix, and draw the solution into a sterile glass capillary tube leaving at least 1 cm of air space at each end. Flame seal the ends and inspect under a dissection microscope to ensure the integrity of the seal.
3. Repeat for homoduplex and control reactions.
4. Mark tubes with water-resistant tape and boil at 100°C for 2 min. Place immediately into a 50-ml screw cap glass tube filled with water at 60°C and submerge in a 60°C waterbath. Incubate for the appropriate length of time to achieve the desired C_0t. This can be calculated as follows:

$$C_0t = (\mu g \text{ driver}/\mu l \text{ reaction volume}) \times 10 \times AF \times \text{hr of incubation}$$

For PB concentrations over 0.12 accelerate the reaction: e.g., 0.48 M = 5.6 times faster and 0.60 M = 6.5 times faster (see Britten et al., 1974).

5. Following the incubation, remove the capillary tube and break off each end by first filing a small groove and snapping it manually. Immediately dilute to 0.12 M PB. This is accomplished by calculating the volume of water necessary to dilute the 0.60 M reaction mix to 0.12 M and adding this volume of water to 500 µl 0.12 M PB. The hybridization solution can be dissolved directly into this solution. Do not dilute the reaction mix into pure water, as denaturation may take place prematurely. PB must be present.

6. If the hybridizations are not to be fractionated immediately, they should be taken out of the 60°C bath and placed directly into a dry ice–ethanol bath and quick frozen. They then can be stored at −20°C for a few days. Slow freezing is disastrous as everything binds to HAP.

7. Once the hybridizations have been diluted to 0.12 M PB they can be loaded on a column for thermal denaturation (Protocols 9–10).

Protocol 9: Hydroxyapatite Column Preparation

(Time: 3 hr)

1. Rinse the column (Figure 3) twice with distilled water (all solutions must be blown through the column under air pressure) and load 400 mg dry hydroxyapatite (HAP). Rinse the HAP three times with 3 ml water followed by two rinses with 3 ml 0.12 M PB. Load each column with 3 ml 0.12 M PB and increase the circulating water temperature to 50°C. Blow this solution through the column into individual scintillation vials and use them as the "blanks" for background counts.

2. Load the hybridization solution (now in about 600 µl of 0.12 M PB) onto the HAP bed and add 2.4 ml of 0.12 M PB. Gently stir the HAP with a glass rod to remove trapped air and record the temperature of the column. Remove the thermometer and blow the PB through the column into a new vial. Add 3 ml of 0.12 M PB to the column, let it equilibrate to 50°C, and blow through and collect. Repeat the 0.12 M PB wash twice more at this temperature.

3. Raise the column temperature at 3–5°C intervals to 100°C, collecting a 3 ml 0.12 M PB wash at each interval. Be sure to gently mix the HAP at each interval and allow the temperature to remain constant for 3–4 min. Check and record the temperature before blowing the wash through.

4. When all the fractions are collected, add 10 ml of scintillation fluid and count each vial twice for 5 min, or longer (10 min) if counts are low (<50 cpm).

5. The first fraction collected is the blank. The load fraction and the three 50°C washes are used to determine the proportion of unreacted single-stranded tracer. The remaining fractions at the temperatures above 50°C are used to determine the T_m (see Interpretation and Troubleshooting).

Protocol 10: Phenol Emulsion Reassociation Technique (PERT)

(Time: 2 hr, plus a 1–5 day incubation)

The PERT system (Kohne et al., 1977) is a method that achieves high C_0t value at room temperature by using a phenol emulsion phase to accelerate the reassociation reaction as much as 10,000-fold. We have found this to be a high stringency of reassociation system and it is essentially unexplored for studies of relationships. Advantages of this system include high C_0t in a short period of time (half C_0t of *Drosophila* scnDNA is about 10 min), the elimination of a low temperature foot on resulting melting curves (due to degraded tracer, etc.), and the reduction in the amount of driver necessary (optimal acceleration at about 5 μg). However, the system is limited in that the stringency criteria probably cannot be manipulated and the NPH falls more quickly with increasing evolutionary distance than with the standard PB system. The systematic value of this technique is yet to be determined, but the protocol is included here to promote further study.

1. Combine 5 μg of sheared driver and 500–1000 cpm tracer in a 1.5-ml microcentrifuge tube and adjust the PB concentration to 0.48 M using 2.4 M stock. Add 0.48 M PB to a final volume of 1 ml, mix, and place in 100°C waterbath for 3–5 min to denature the DNA.
2. Cool the mixture to room temperature and add 100 μl equilibrated phenol and vortex. Shake continuously to keep the phases from separating for one to several days. Mixing can be accomplished by attaching the tubes to a "wristaction" flask shaker, which shakes the tubes in a vigorous up-and-down motion.
3. Remove from shaking and add 400 μl ether and vortex to extract the phenol. Remove the ether and discard.
4. In a sterile 15-ml tube combine 3 ml water, 1 ml 0.12 M PB, and add exactly 1 ml of the hybridization solution. This will yield 5 ml of a 0.12 M PB solution that can be fractionated over HAP as in Protocols 8 and 9, except for the following: instead of adding 0.12 M PB to the loaded sample to bring it to 3 ml, simply add the 5 ml to the column, mix, and blow it through the column, dividing the sample equally into two scintillation vials. Wash three times with 0.12 M PB at 50°C and continue as in Protocol 9, step 3.

Protocol 11: Analysis of Hybrid Thermal Stability Using the S1 Nuclease-TEACL Assay

(Time: 3 hr setup, incubation up to 2 weeks, 6 hr fractionation)

This procedure requires the heating of a number of samples simultaneously in a temperature gradient from 45 to 65°C. This is accomplished with the use of an aluminum block with heat exchangers at each end that can be connected to two separate temperature-controlled circulating water baths. A series of 11-mm-diameter holes drilled into the block in a staggered pattern can accommodate 1.5-ml microcentrifuge tubes after the holes have been partially filled with water. The block must be well insulated with styrofoam (six sides) and the resulting temperature gradient along the block is linear and should be accurate within 0.1°C (Britten et al., 1978). This should be checked with a precision thermometer moved from hole to hole.

1. Tracer and driver DNA must be in 0.1 M EDTA. Tracer should be at least 200 cpm/μl and sheared driver must be at a concentration of 5 μg/μl. Add about 8000–10,000 cpm tracer to 300–400 μg driver (e.g., 30 μl tracer to 70 μl driver). To this mixture add an equal volume of 3.0 M TEACL to reach a final concentration of 1.5 M TEACL. Seal in a 200-μl capillary tube as described above (in the PB system), and incubate in a boiling water bath for 2 min. Due to the large bore of the capillary tube, it is useful to break a smaller diameter tube (10–20 μl) and insert a small 0.5 cm-piece into the ends of the 200-μl tube before flaming the ends. This will ensure a good seal.

2. After denaturation, place the tube in a 45°C water bath for a sufficient length of time to achieve an appropriate C_0t. The acceleration factor for incubations in 1.5 M TEACL is four times that of 0.12 M PB. Thus, C_0t can be calculated using the formula: $C_0t = [(\mu g/\mu l) \times 10 \times 4 \times \text{hr incubation}]$.

3. Following the reassociation incubation, cool the sample to room temperature, and add an equal volume of 2× S1 nuclease buffer to decrease the pH to 4.4 and the TEACL concentration to 0.75 M. Add S1 nuclease (1–2 μl of a 5 unit/μl stock) for 95% single-strand digestion (appropriate incubation times may have to be determined empirically). Vortex and incubate at 37°C for 10 min. Remove and vortex again to remove lumps, and incubate at 37°C for 50 min.

4. Chill the sample on ice and add 1/10 volume of 0.30 M EDTA to stop the reaction. Remove 10–20 μl of the mixture and purify with the glass beads elution procedure (Protocol 3) or the spun column technique (Chapter 8) and determine the duplex fragment size by electrophoresis through an alkaline agarose gel electrophoresis, as outlined in Protocol 7.

5. To remove digested single-stranded DNA from the remainder of the sample and bring it to 2.4 M TEACL, load the sample on to a 3-ml Sephadex G-100 column previously equilibrated with 2.4 M TEACL and wash through with 10–15 ml of 2.4 M TEACL. Collect 20 to 30 drop fractions and count Cherenkov in a scintillation counter to identify the exclusion peak (maximum concentration of duplex DNA that may be in more than one of the fractions). Bring the peak fraction to 1.8 ml with 2.4 M TEACL, and determine the precise concentration by means of a refractometer.

6. Remove 100 μl aliquots from the 1.8-ml sample and place into 16 microcentrifuge tubes. Place 14 of these in the heating block (with a little water in each hole). Keep one of the remaining samples unheated and heat the other to 70°C for 30 min, prior to S1 digestion. Heat the samples in the heating block for 30 min in the thermal gradient described above. These samples will determine the melting curve. To determine the total amount of radioactivity in each sample, count 100 μl of the 1.8-ml sample in liquid scintillation fluid.

7. After heating, place all the samples on ice and add 100 μl of sterile water, 200 μl of 2× S1 nuclease buffer, and 10 μl of S1 enzyme solution (Appendix); use enough S1 nuclease to digest 99% of single strands. Vortex and incubate at 37°C for 1 hr. Put on ice and add 1/10 volume 0.5 M EDTA (= 40.5 μl) to each tube to stop the reaction. At this point the remaining duplex DNA must again be separated from the digestion products (both of which are radioactive).

8. Prepare 16 individual (3 ml) Sephadex G-100 columns that have been equilibrated with 1.0 mM EDTA. Load the 16 samples and elute with 3 ml 1.0 mM EDTA and collect the excluded fraction. Mix with an appropriate volume of scintillation fluid (10 ml) and count for 5–10 min per vial. This excluded fraction is the portion of DNA remaining as undigested duplex. The melting curve can be constructed by plotting the increasing fraction of the sample digested by the S1 nuclease versus increasing temperature. The cpm of duplex in the unheated sample divided by the total cpm of the predigested 100 μl sample ($\times$ 100) provides the extent of reassociation in percent.

9. As an alternative to the final Sephadex G-100 fractionation, the DNA in duplex can be precipitated with cetyltrimethylammonium bromide (CETAB) and counted separately from the digested single-stranded DNA remaining in the supernatant (Hereford and Robash, 1977; Hall et al., 1980). Add 150 μg of calf thymus DNA and 1/2 volume of 9% CETAB and NaAc to a final concentration of 0.1 M. Spin in a microcentrifuge for 10 min and resuspend the pellet in a small volume of 1.0 mM EDTA; count this sample separately from the supernatant.

INTERPRETATION AND TROUBLESHOOTING

Calculation of Melting Curves From Raw Counts: An Example

Several different measures are possible for estimating evolutionary distance from interspecific DNA hybridizations and melting curves: ΔT_{mode}, ΔT_m, percentage hybridization (ΔNPH), and ΔT_{50H} which combines ΔT_m and ΔNPH into a single number.

The T_m is the interpolated temperature at which 50% of the hybrids that were formed remain in duplexes. T_m can be easily determined by eyeballing an integral melting curve, or nonlinear least-squares regression methods can be used to increase accuracy. The T_{50H} is an estimate of the temperature at which 50% of the DNA remains in duplexes; this measure differs from T_m if all DNA fragments do not form duplexes. The T_{mode} depends on the determination of the interpolated temperature at which maximum amount of hybrids melt (the peak in a differential plot of a melting curve).

The following example (Tables 2–4) is based on a PERT, interspecies tracer–driver reassociation, with a homoduplex control. The curves are derived from actual data and are of good quality. Poor curves, influenced by a variety of factors, are illustrated and described at the end of this section.

The tracer has reacted to the extent of 87% with driver from which the tracer was originally made (homoduplex control); this degree of hybridization is not uncommon for PERT reactions. Thermal denaturation of the duplex DNA is as described in Protocol 9 where 15 fractions (vials) are collected and counted, with washes at 5°C intervals (narrower intervals would be preferable). The fractions and the cpm tracer in each fraction based on 5 min counting of each in a liquid scintillation counter (beta) are presented in Table 2 (a gamma counter is needed for ^{125}I).

Table 2. Raw counts data for homoduplex hybridization for the calculation of NPR and an intergral melting curve

Temperature (°C)	Vial	cpm eluted	cpm − blank	
50	1. Blank (3 ml PB wash)	12.4	0.0	
50	2. Load fraction	41.8	29.4	Unbound
50	3. 3 ml PB wash	59.2	46.8	fraction
50	4. 3 ml PB wash	20.8	8.4	
50	5. 3 ml PB wash	12.0	0.0	
55	6. 3 ml PB wash	14.2	1.8	Bound
60	7. 3 ml PB wash	17.6	5.2	fraction
65	8. 3 ml PB wash	16.8	4.4	
70	9. 3 ml PB wash	25.6	13.2	
75	10. 3 ml PB wash	44.4	32.0	
80	11. 3 ml PB wash	95.4	83.4	
85	12. 3 ml PB wash	216.2	203.8	
90	13. 3 ml PB wash	200.2	187.8	
95	14. 3 ml PB wash	34.2	21.8	
100	15. 3 ml PB wash	12.6	0.2	
Total			638.4	

Adding the blank-corrected cpm from #2 to #15 gives the total cpm (= 638.4). The proportion of tracer that did not bind to the HAP on loading is the sum of fractions 2 through 5 divided by the total cpm × 100 (84.6/638.4 × 100 = 13.25% unbound). The bound fraction, or the extent of reaction, is the percentage tracer that formed stable hybrids during the reassociation and is represented by the sum of the counts from vials 6 through 15 divided by the total, and expressed as a percentage [(553.8/638.4) × 100 = 86.74%].

Table 3. Counting data for calculating an integral melting curve normalized to 100% reactivity

Temperature (°C)	cpm − blank	eluted (%)
55	1.8 (vial 6)	1.8/553.8 = 0.33
60	7.0 (vials 6-7)	7.0/553.8 = 1.27
65	11.4 (vials 6-8)	11.4/553.8 = 2.06
70	24.6 (vials 6-9)	24.6/553.8 = 4.45
75	56.6 (vials 6-10)	56.6/553.8 = 10.2
80	140.0 (vials 6-11)	140.0/553.8 = 25.3
85	343.8 (vials 6-12)	343.8/553.8 = 62.1
90	531.6 (vials 6-13)	531.6/553.8 = 96.0
95	553.4 (vials 6-14)	553.4/553.8 = 99.96
100	553.8 (vials 6-15)	553.8/553.8 = 100.00

Table 4. Calculation of heteroduplex melting curve data[a]

	A	B	C	D	E	Elution temperature (°C)
1.	15.7	0.0				
2.	46.4	30.7				
3.	129.6	113.9				
4.	26.0	10.3				
5.	13.4	0.0		0.00	12.18	50
6.	17.0	1.3	1.3	0.27	13.05	55
7.	20.4	4.7	6.0	1.25	13.89	60
8.	25.4	9.7	15.7	3.26	15.65	65
9.	53.0	37.3	53.0	11.02	22.41	70
10.	97.8	82.1	135.1	28.90	37.30	75
11.	150.4	134.7	269.8	56.10	61.71	80
12.	155.0	139.3	409.1	85.07	86.98	85
13.	82.8	67.1	476.2	99.02	99.14	90
14.	20.4	4.7	480.9	100.00	100.00	95
15.	11.8	0.0	480.9	100.00	100.00	100

[a]Data are from raw counts normalized to 100 % reaction (column D) and normalized with respect to the homoduplex control (column E). Column A, raw counts, B, raw counts minus blank, C, cumulative sum of counts, D, % of total bound (= 480.9) versus temperatures, and E, counts normalized to homoduplex control.

To obtain the data to draw an integral melting curve normalized to 100% reactivity, the cpm from vials 6–15 are added sequentially and each total is divided by the total cpm bound and multiplied by 100 to give the total percentage eluted at each temperature (Table 3). The curve (Figure 4) is generated by plotting percentage eluted (y axis) versus temperature (x axis). The T_m of this curve is the point on the temperature scale where the curve intersects 50% eluted.

In a heteroduplex DNA hybridization reaction involving the tracer (in the first example) and a driver from a different species, the curve (normalized to 100%) can be determined and the T_m identified as in the example above. However, for T_{50H} estimates the homoduplex curve is normalized to 100%, as above, but the heteroduplex curve is normalized to the homoduplex melt. For example, if the homoduplex melt reacted 86.7% and the heteroduplex comparison reacted 75.6%, then the homoduplex curve would be normalized to 100% and the heteroduplex curve would be normalized to the former, resulting in an NPH (normalized percentage hybridization) of 87.2%. The procedure for normalizing is illustrated in Table 4. For this melt unbound cpm = 154.9, bound cpm = 480.9, and total cpm = 635.8.

The T_m (Figure 4) can be derived from column D (Table 4). To estimate T_{50H}, these data and the resulting curve must be normalized to the homoduplex melt. This can be accomplished by dividing the heteroduplex percentage reaction (percent bound) by the homoduplex percentage reaction: 75.6/86.7 = 0.8719, which gives heteroduplex NPH

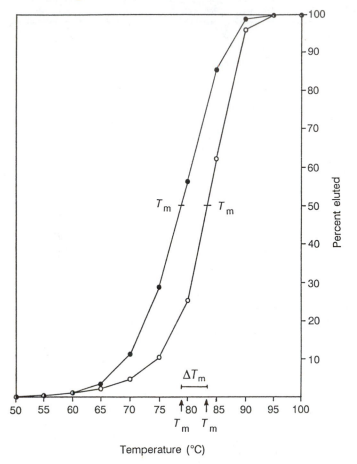

FIGURE 4.   Homoduplex (○) and heteroduplex (•) melting curves illustrating T_m and ΔT_m. Both curves are normalized to 100% (see text for explanation).

of 87.19. The data from column D are transformed by multiplying the heteroduplex values by NPH/100 (0.8719) and adding the y intercept of the curve (12.81).

Utilizing the information in the far right column (E) of Table 4, a curve can be drawn that is normalized to the homoduplex melt (Figure 5). The value of T_{50H} can be obtained by inspection where both curves intersect the 50% eluted line. For the homoduplex curve in this example, $T_{50H} = T_m$. The difference between the homoduplex T_m and the heteroduplex T_m or the difference between the T_{50H} values is the numerical value used in distance calculations for phylogenetic analysis (Chapter 11, see also Felsenstein 1981a, 1987).

Problematic Melting Curves

Precision and accuracy in the execution of hybridization methods are paramount for the consistent generation of high-quality melting curves and the related distance in-

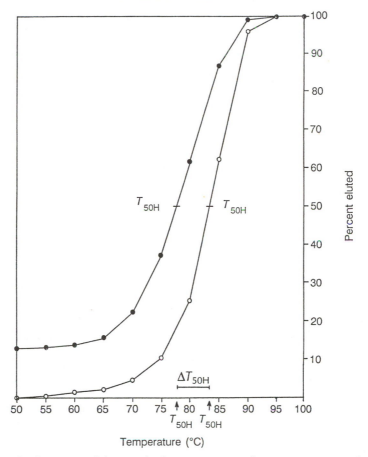

FIGURE 5.   Homoduplex (○) and heteroduplex (•) curves illustrating T_{50H} and ΔT_{50H}. The homoduplex curve is normalized to 100% whereas the heteroduplex curve is normalized relative to the homoduplex curve (see text for explanation).

formation. Precision in replication can be enhanced with the use of automated methods that can run many columns simultaneously. Accuracy can be improved by close attention to technical details.

There are several sources of error that can result in poor melting curves and lead to difficulties in interpretation. Consequently, poor melting curves, and the resulting ambiguous distance information, ultimately result in weak hypotheses of phylogenetic relationship. Many of these potential errors are manifested by (1) the presence of a low temperature melting component, or a "foot" near or at the criterion temperature, and (2) the broadening of a melting curve over a wider temperature range (Figure 6). In certain instances, these observations are real and not due to experimental error; the significance of these phenomena is currently under debate (Britten, 1990). Below is a

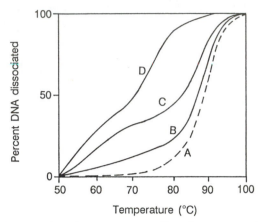

FIGURE 6. Idealized integral melting curve profiles that may indicate potential problems of tracers used in hybridization reactions based on melting characteristics of homoduplex controls. Curve A represents the expected curve for a homoduplex reaction, with good tracer and driver. It is characterized by a steep slope near the median duplex melting temperature (T_m here = 85°C), with very little DNA eluted below 70°C. Curve B may result if the tracer includes a significant amount of repeated DNA. This usually results in a slightly depressed T_m and broad lower temperature "foot". The profile of C indicates contamination with short, degraded tracer fragments, which usually leads to a steady elution of fragments at low temperatures with only a moderate reduction in the T_m. Curve D illustrates that the tracer is dominated by short, degraded fragments, resulting in an excessive low temperature component and a dramatic depression of the expected T_m.

list of possible sources of contamination in the hybridization reaction that may give rise to poor curves, alone or in combination.

Repeated Elements in the Single-Copy DNA Used to Make Tracers   Repeated elements often have divergent members that may lead to extensive paralogous rather than orthologous reassociation. This can contribute to unstable, low-temperature melting duplexes. Although it is unlikely that all repeated elements can be removed (i.e., low-frequency repeats), it is important to fractionate and remove repeated elements by obtaining a C_0t value where at least 10% of the single-copy DNA has reassociated.

Short or Degraded Tracer Fragments   Short tracer fragments (<25 bases) usually melt at low temperatures, or simply broaden the resulting curve if there exists a distribution of small lengths. Also, if the entire tracer sample is composed of short fragments the T_m will be greatly reduced. A solution would be to use fresh tracer that has been properly sized and is free of possible DNase contamination and unincorporated nucleotides.

Interindividual Sequence Variation   Individual heterozygosity, coupled with the use of different sources of tracer and driver in intraspecies melts, can lead to broad melting curves. This can also be the case if the driver or tracer is derived from a mixed source (Britten et al., 1978). Good quality control reactions may provide insight into the contribution of these factors to questionable melts.

Incomplete Reactions   If hybridization reactions are not carried out to the appropriate C_0t to permit reassociation of at least 90% of the single-copy DNA, low-temperature melting, hybrid duplexes may result. Depression of reassociation rates with sequence divergence must be considered in allowing for reactions to approach termination, without compromising tracer stability due to autoradiochemical degradation.

Inadequate DNA Purification   DNA used for driver and tracer preparations must be free of protein, carbohydrate, metallic ion, and RNA contamination. The presence of any of these substances can interfere with reassociation or HAP binding. Occasionally proteins bound to DNA or contaminating hemoglobin (from samples derived from whole blood) are difficult to remove and repeated phenol extractions and proteinase treatments are required. Glycogen and RNA are usually eliminated by spooling DNA during isolation, whereas metallic ions are excluded by passage of the DNA through a chelating resin column (see Protocols 1 and 2).

Characteristics of Distance Estimates Derived from Raw Melting Curve Data

The precise shape of melting curves depends on how the single-copy DNA evolves; this is not a simple issue because there are differences between different systematic groups. Among insects, it has been clearly shown that there is a wide range of rates of subsitution among different regions of the genome of an individual lineage (Caccone et al., 1987). This has also been shown to be true for sea urchins (Grula et al., 1982), although they have less heterogeneity. High rate variability may also be typical of mammals (based on the fraction of interspecies duplexes digested by S1 nuclease; Benveniste, 1985), but further investigation is needed. Figure 7 shows freehand curves for closely and more distantly related species, with the T_{mode}, T_m, and T_{50H} indicated. Figure 7a shows how these three measures change with distance between the species, and Figure 7b shows an example (Hall et al., 1980) of greater distances for sea urchin DNA.

The range of rates of DNA change, distributed among different sequences in the genome, determines the shape of melting curves at different interspecies distances. The slowly changing DNA shows little T_m reduction and maintains the higher temperature part of the melting curve as distance increases. The rapidly changing fraction makes the foot of the curves (at lower temperature) change rapidly and leads to incomplete hybridization at small distances, depending on the conditions of hybridization. The T_{mode}

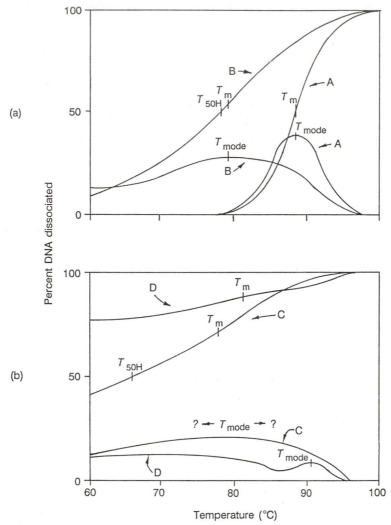

FIGURE 7. (a) Melting curves of heteroduplexes at different distances, showing T_m, T_{50H}, and T_{mode}. Curve A is for intraspecies hybrids and curve B is for fairly closely related species. The lower curves are smoothed representations of the amount of DNA dissociated at each temperature (differential melting curve). (b) Curves C and D show the results to be expected at moderate and great distances. For curve C the T_{50H} is clearly the preferred measure. The T_{mode} cannot be determined due to the flat melting curve. For very great distance, as in curve D, no established measure is in use and future work is required. The T_m is meaningless; the T_{mode} depends on details of the distribution of conserved DNA sequences and, in the case shown, is almost the same as for native DNA. The best measure for this comparison is perhaps the extent of hybridization.

is not appreciably affected by the rapidly changing DNA sequences. In contrast, the T_m is affected by both the rapidly and slowly changing sequences for closely related species. At moderate distances the most divergent sequences fail to hybridize and the rapidly changing sequences then principally affect the extent of hybridization rather than the T_m. If the distribution of rates was well known, the shape of the melting curves would be evident. In that circumstance, the relationships between T_m, T_{mode}, NPH, and T_{50H} would also be apparent and each of the others could be calculated if one were measured. However, it will be a while before that much information is available for other taxonomic groups, although Sheldon and Bledsoe (1989) have investigated the relationship between these distance measures for bird data and Kirsch et al. (1989) have presented data for marsupials. Below, we list some characteristics of each of these measures.

T_{MODE} ΔT_{mode} has been advocated as the distance measure of choice for phylogenetic applications by Sarich et al. (1989). Like ΔT_m, ΔT_{mode} is moderately precise and does not suffer from the high standard error that NPH and ΔT_{50H} do (Bledsoe, 1987; Kirsch et al., 1990). This is clearly an advantage of the former two measures for the range over which they can be measured. Another putative advantage of ΔT_{mode} is that the criterion conditions may not form a boundary at which differences become compressed; the mode may simply shift below criterion (with increasing interspecies divergence) and cease to be a property of melting curves. Thus, ΔT_{mode} may index comparable suites of sequences over its range rather than the increasingly smaller subsets indexed by ΔT_m; this would render ΔT_{mode} values more additive than ΔT_m values. An exception to this pattern is noted in Hall et al. (1980), where slowly evolving components of the genome in distant interspecies comparisons may retain a T_{mode}.

T_{mode} measures a peak in the distribution of diverging DNA. If the distribution of rates were Gaussian, the variance were small, and if the technique were good enough to determine the peak accurately, then T_{mode} would be the measure of choice. However, melting curves spread with increasing divergence and it becomes increasingly difficult to determine T_{mode} accurately (Figure 7b); false modes may result from the scatter of individual measurements if the curve is broad and flat. Also, different algorithms for T_{mode} determination (e.g., modified Fermi–Dirac curve fitting, parabolic curve fitting, graphic methods; see Sheldon and Bledsoe, 1989) do not always give the same answer.

It is worth noting that much of the width of HAP melting curves is due to the range of base composition present in the DNA and the failure of complete washing at each temperature. It has never been possible to draw conclusions about the range of rates of divergence from HAP measurements for this reason. If the distribution of rates of divergence were broad and uniform, and if there were no variance in base composition, no mode could exist. In reality, however, spectrophotometric and hydroxyapatite melting curves for native DNA are sigmoidal owing to the effects of base composition, and in the case of hydroxyapatite, incomplete washing. Thus, one always observes a mode and it is not yet possible to know how such artifacts affect the apparent mode in a particular measurement. It may be the case, however, that the mode indexes that component of the genome that has the modal base composition as well as the modal rate of divergence.

NPH   The normalized percentage hybridization (NPH) falls for heteroduplex comparisons compared to homoduplex controls, even if the heteroduplex comparison involves closely related species. If there is a rapidly evolving class of DNA, NPH is a measure of that class. However, it is not yet known how much of the reduction in NPH is due to kinetic effects; rates of reassociation differ in homoduplex versus heteroduplex reactions and the amount of driver DNA available for hybridization may be used up before the divergent tracer–driver sequences can hybridize. Clearly, there is a need for measurements of this kinetic effect.

The rate of reassociation for interspecies hybrid formation is retarded by a factor of two per 10°C divergence in T_m (Bonner et al., 1973). Late in the incubation there is less driver DNA available to complete the slower formation of more divergent tracer–driver duplexes. One published measurement (Galau et al., 1976) suggests a large kinetic effect for a *Xenopus* interspecies hybridization, but it has not been confirmed. Recent measurements with *Drosophila* DNA (Werman et al., unpublished) and primate DNA (Bonner et al., 1980; Britten and Stout, unpublished) suggest that the kinetic effect is small for modest (6%) divergence. The underlying process is complex, since the duplexes of driver DNA form between randomly terminated fragments and single-stranded regions remain available at their ends. Thus, the concentration of the single-stranded part of the driver falls at a rate of $(1 + KC_0t)^{-0.44}$, so that there is some single-stranded driver available very late in the incubation (Britten and Davidson, 1985). The kinetic effect can therefore be reduced by incubation to high C_0t, as is usually done. Since there are no satisfactory measurements, anyone planning to use NPH for major phylogenetic work is advised to make some determinations of the kinetic effect under the conditions of incubation. The obvious method is to rehybridize the nonhybridizing fraction in a standard incubation.

In many measurements (Sibley and Ahlquist, 1981a, 1983; Kirsch et al., 1989; Powell and Caccone, 1990) NPH is not accurately determined. However, in other work it apparently has been determined more reproducibly (Hall et al., 1980; Benveniste, 1985). The technical problems that are due to limited C_0t and to variations in length and concentration of tracer and driver preparations are undoubtedly solvable, so we may look forward to more precise determinations of NPH. At large evolutionary distances where a majority of the DNA can no longer form interspecies duplexes at the criterion temperature, NPH is the primary available measure of interspecies relationships. T_{mode} and T_m, in turn, are not useful at such high levels of divergence.

T_M For fairly closely related species the T_m is a good measure of the amount of the DNA that hybridizes. The T_m falls steadily with increasing divergence until it reaches about halfway between the criterion temperature and the T_m of precise duplexes. At greater divergences, the amount of DNA that hybridizes continues to fall while the T_m changes very little. There may be some additional decrease in T_m if there is a well defined peak in the distribution of rates of DNA sequence change since the peak may move down toward the criterion. Thus, there is a linear relationship between T_m reduction and divergence over short, evolutionary time. At increasing divergence the curve begins to level off. This causes compression of the estimates of greater divergence.

T_{50H} The T_{50H} measure was devised by Kohne et al. (1972) to correct for the reduction in normalized percentage hybridization (NPH) that occurs even for closely related species and to remedy the compression of ΔT_m values that is forced by criterion conditions. The method of calculation is shown in Figure 5 and Table 4. As discussed, T_{50H} is a measure of the median sequence divergence between species (Hall et al., 1980). Obviously, if median sequence divergence could be accurately estimated, the result would be independent of the criterion used in a particular measurement. It was shown by M. J. Smith et al. (1982) that a good compensation for different temperatures of incubation can be achieved by using T_{50H}. The compensation for different criteria even extends to S1 methods. The S1 nuclease digests the more divergent duplexes and the undigested DNA is the better paired fraction, so the observed reduction in T_m with S1 is less than with HAP. However, the NPH is also less and as a result the T_{50H} observed with S1 is about the same as with hydroxyapatite. Also, T_{50H} gives a more linear relationship with sequence divergence than do the other measures of distance (Britten, 1986).

ΔT_{50H}, however, has its own limitations. Some of the initial reduction is due to kinetic effects (see NPH) and this could exaggerate the actual amount of divergence. The T_{50H} continues to fall more or less linearly with increasing divergence until the NPH falls below 50%, at which point it becomes difficult to determine and requires extrapolation beyond the observed melting curve. Estimates of T_{50H} obtained using extrapolation must be regarded as more unreliable than those that are obtained directly from melting curves. Another problem with T_{50H} results from the error in determining NPH. The effect of this error in measuring NPH has much less of an impact on T_{50H} when the slope of the melting curve is very steep (as in close relationships measured with 2.4 M TEACL).

One approach to remedy the kinetic problems of ΔT_{50H} is to calculate the expected decrease in NPH (for a heterduplex reaction) based solely on kinetics and add this to the observed NPH value before calculating ΔT_{50H}. However, data are not yet available to make such a correction and such measurements would be valuable. A practical suggestion is to carry out hybridization reactions to high C_0t values in an attempt to minimize the NPH differences caused by kinetic effects.

T_{25H} An alternative for very distant species would be to use the T_{25H} or the temperature at which 25% of the hybridizable tracer remains in duplexes, but this has not yet been tested. In the case of melting curves for two distant species of sea urchin studied by Hall et al. (1980) and Angerer et al. (1976), only about 20% of the DNA hybridized. The reduction in T_{mode} and T_m, in turn, was only a few degrees, since the DNA that hybridized was dominated by a high melting temperature component. The 20% NPH is a good measure for this case but a T_{25H} estimate for the distant species would be easier to combine with other data for more closely related species.

Hybridization Data in Phylogenetic Reconstruction

Phylogenetic reconstruction is discussed in Chapter 11, so the discussion here is restricted to factors that specifically apply to DNA hybridization. If DNA evolutionary changes were additive, so that Buneman's (1971) four-point metric was satisfied and all base pair changes were accurately indexed over all pairwise comparisons, then reconstruction of phylogeny would be a trivial operation. Optimality criteria such as those developed by Fitch and Margoliash (1967) and Cavalli-Sforza and Edwards (1967) provide unambiguous criteria for choosing among competing topologies when distances are additive: the correct topology will exhibit a perfect fit to the matrix of distances. Springer and Krajewski (1989) have proved a Perfect-Fit Theorem to substantiate this argument. An unambiguous outgroup taxon then allows one to root the topology and specify net amounts of shared derived and uniquely derived change on that topology. Furthermore, such an analysis produces a topology that is equivalent to the topology that one would obtain using the individual characters and a parsimony algorithm (see Chapter 11).

In reality, however, several factors may compromise the additivity of DNA hybridization data and destroy the precise correspondence between trees derived from distance data and trees derived from parsimony analysis of individual characters. Some of these factors result from processes of DNA evolution and also influence sequence data; others are peculiar to different measures of genetic distance derived from DNA hybridization data. At close distances, for example, NPH is somewhat inaccurate and it is preferrable to use ΔT_{mode} or ΔT_m. At larger distances, ΔT_{mode} is subject to error in its determination and ΔT_m exhibits a saturation, making ΔT_{50H} the method of choice. At even larger distances ΔT_{50H} cannot be determined and NPH remains perhaps the only meaningful measure.

Sources of Nonadditivity and Error   Below we consider sources of error relevant to DNA–DNA hybridization data: homoplasy, uneven distribution of rates of change, measurement error, paralogous sequences, differences in genome size, and intraspecific variation.

Homoplasy.   Homoplasy (i.e., reversals and parallelisms) causes observed sequence differences to underrepresent actual amounts of sequence divergence. As a consequence both DNA sequence data and DNA hybridization data are nonadditive in expectation, because they do not index all base pair changes that have taken place. Furthermore, the accumulation of homoplastic changes is nonlinear and becomes progressively more important for increasingly divergent sequences. Indeed, the accumulation of homoplasy in DNA sequences has been studied extensively and several mathematical models have been developed to describe the effect of accumulated homoplasy on sequence divergence. One of the simplest models is based on a Poisson process and is given as follows:

$$T = -(3/4)\ln[1 - (4/3)\, D]$$

where D is the observed fraction of sequences that are different for any pairwise comparison and T is the expected sum (expressed as a fraction) of homoplastic changes plus observed differences (Jukes and Cantor, 1969). Since this model makes unrealistic assumptions about DNA sequence evolution, more sophisticated models have been developed to account for biased codon usage, synonymous versus nonsynonymous substitutions, position-dependent differences in substitution probabilities, and base-dependent differences in substitution probabilities (Fitch, 1971a, 1976a, 1986; Kimura, 1980, 1981; Golding, 1983; Tajima and Nei, 1984; Li et al., 1985; Gillespie, 1986b; Nei and Gojobori, 1986; Nei, 1987). [Shoemaker and Fitch (1989), however, argue that all of these models are too conservative since not all nucleotide positions are replaceable.] Unfortunately, these models require actual DNA sequences and cannot be used in conjunction with DNA hybridization data. Even so, for observed sequence differences up to 50% all of these models provide estimates of T that are in excellent agreement with the Jukes and Cantor model; discrepancies become important only at larger distances. Since DNA hybridization distances are generally much less than 50% divergence, the Jukes and Cantor model is therefore appropriate, albeit slightly conservative.

It should also be noted that the expected amount of homoplasy is deterministic but that stochastic influences lead to variance around this expectation. For DNA sequence data, the stochastic component is much more important than for DNA hybridization data. Indeed, the variance component associated with the expected amount of homoplasy is trivial when the entire single-copy genome is under comparison (Nei, 1987). This is an advantage of DNA hybridization data over DNA sequence data. Mitigating against this putative advantage is the increased measurement error associated with DNA hybridization. Felsenstein (1987), for example, has suggested that measurement error reduces the resolving power of DNA hybridization data to 4472 bases, although this estimate did not take into account the increased precision that results from additional replicates for the same interspecies comparison, nor the increased precision that can now be achieved using the S1 TEACL technique (Powell and Caccone, 1990).

We also need to investigate the consequences of homoplasy for phylogenetic reconstruction if a correction for homoplasy is not employed. Most importantly, branch lengths on resulting topologies will be too short and the relative proportionality of branch lengths will be distorted. Thus, homoplasy cannot be ignored if we are interested in the relative timing of branching events. The sequence of branching events on a topology is much less affected by homoplasy, however, and, in most instances, corrections for homoplasy do not affect the sequence of branching events (Springer and Kirsch, 1989; Springer and Krajewski, 1989). This results from the deterministic fashion in which homoplasy accumulates (i.e., homoplasy is a function of divergence) when a large number of nucleotides are under comparison, as is the case for DNA hybridization data. In contrast, no one has ever proposed (or documented) that homoplasy among morphological characters is such a predictable, deterministic function of divergence. When homoplasy is not a function of divergence, or when the variance associated with this function is extremely large, homoplasy is much more of an obstacle to phylogenetic reconstruction.

Application of the Jukes and Cantor correction requires that we know the conversion between delta values and percentage mismatch. Empirical estimates from the literature range from 0.7 to 2.0% base pair mismatch per 1°C depression in ΔT_m (Bautz and Bautz, 1964; Laird et al., 1969; Kohne, 1970; Britten et al., 1974; Hutton and Wetmur, 1973; Caccone et al., 1988b). The conversion most often used is that 1% sequence mismatch corresponds to 1°C of T_m depression, which is partly a matter of convenience and standardization. The recent estimate of 1.7% sequence divergence per degree of T_m depression (Caccone et al., 1988b) may well be correct for the ribosomal sequences studied, since these sequences have conserved regions and clustered substitutions. It presumaby does not apply to typical single-copy DNA since most of this DNA is noncoding and might be expected to exhibit a more random distribution of substitutions. Further work will be required, however, to determine the mean value of the conversion between percentage sequence divergence and melting temperature depression for a population of sequences (i.e., the single-copy genome) that undoubtedly exhibits great variation in the clustering of substitutions.

The Distribution of Rates of Sequence Change.   A desirable property of any DNA hybridization distance measure is that it represents the mean amount of sequence divergence between equivalent portions of all genomes under comparison. However, once a rapidly evolving fraction of the DNA has diverged such that its T_m is less than the temperature of its reassociation, the average or mean divergence can no longer be measured. The median can be measured out to about 50% NPH and perhaps estimated further as mentioned above. It seems likely that median distances would then need to be corrected only for homoplasy to generate reliable estimates of additive genetic distance, but this has not yet been shown. The shape of the distribution of rates of DNA sequence change (see Springer and Krajewski, 1989) is also important. If this distribution were Gaussian, for example, ΔT_{mode} would provide a reliable estimate of mean sequence divergence until it went off scale. If kinetic effects were accounted for, and if deletions were an unimportant source of NPH reduction (as they very well may be; see Meyerowitz and Martin, 1984), then ΔT_{50H} should also converge on the same value as ΔT_{mode}. In reality, we do not know the distribution of rates of change for the suite of sequences in the single-copy genome, but it is most likely to differ among taxonomic groups. Thus, it is unclear if modal or median values of sequence divergence provide better estimates of mean sequence divergence. We hope that this issue can be evaluated quantitatively in the future.

Measurement Error. Measurement error is potentially the single biggest problem with DNA hybridization distances. Springer and Krajewski (1989) discuss such error in the context of imprecision and inaccuracy, where precision refers to the repeatability of replicate measurements and accuracy refers to the reliability of a measurement as an estimate of some quantity. Reciprocity is also a useful concept for dealing with matrices of DNA hybridization distances. Sarich and Cronin (1976) defined the percentage nonreciprocity for a pairwise comparison as [(distance AB − distance BA/(distance AB + distance BA)] × 100. Average percentage nonreciprocity for a distance matrix is then the mean value of this parameter over all pairwise comparisons.

Precision of DNA hybridization measurements is generally indexed as the standard error or standard deviation of replicate measurements. Sibley et al. (1987) reported an average standard deviation of 0.35 degrees for ΔT_m measurements. Krajewski (1989), in turn, reported an average standard deviation of 0.48 degrees for ΔT_m measurements of cranes. Furthermore, Sibley et al. (1987) found that the standard deviation for ΔT_m values increases as a function of sample size up to $n = 5$ and then remains stable. Also, standard deviation does not depend on the magnitude of ΔT_m values (Sibley et al., 1987; Springer et al., 1990; Krajewski, 1989). Finally, average percentage nonreciprocities for matrices of ΔT_m values generally fall between 3 and 10% (Sheldon, 1987; Springer and Kirsch, 1989; Springer et al., 1990).

Similarly, ΔT_{mode} values are moderately precise (Kirsch et al., 1989; Bledsoe, 1987). NPH and ΔT_{50H} values, on the other hand, exhibit more scatter for replicate measurements (Sheldon, 1987; Krajewski, 1989; Kirsch et al., 1989). This measurement error may obscure branching patterns revealed by ΔT_m and ΔT_{mode} matrices (see Kirsch et al., 1989). An alternative strategy is to use a regression equation to convert ΔT_m values into ΔT_{50H} values. This approach is much less sensitive to the effects of measurement error, yet it allows one to reduce the effects of compression that plague ΔT_m values and obtain better estimates of branch lengths on output topologies. If our intent is to use DNA hybridization distances to estimate the timing of branching events, this issue cannot be overlooked. Catzeflis et al. (1987) and Springer et al. (1990) have developed exponential regressions of ΔT_{50H} on ΔT_m for DNA hybridization data on rodents and marsupials, respectively; additional equations would have to be developed for other groups. A major disadvantage of this approach is that it may prove intractable for some taxonomic groups, e.g., a consistent relationship between ΔT_m and ΔT_{50H} may not hold for all taxa under study because of differences in genome size or variation in the amount of rapidly evolving DNA. A second disadvantage is that NPH and ΔT_{50H} are most useful when ΔT_m can no longer provide resolution, but a regression equation should be used only over a range where ΔT_m and ΔT_{50H} are both monotonically increasing functions of sequence divergence.

In contrast to imprecision, inaccuracy is often caused by systematic biases that affect a whole suite of measurements. One such bias deserves mention. Most workers who have used ^{125}I tracers are familiar with a compression of ΔT_m values associated with specific tracers (Springer and Kirsch, 1989). Short tracer fragments are probably the culprit. Compression can increase the average percentage nonreciprocity in a distance matrix. The effects of compression, however, can be reduced through the use of an algorithm developed by Springer and Kirsch (1989), which, in turn, is a modification of an earlier algorithm developed by Sarich and Cronin (1976) for immunological distances. For an uncorrected matrix of ΔT_m values given in Springer and Kirsch (1989), the average percentage nonreciprocity was 3.12%. After several iterations of the correction algorithm, this value was reduced to 1.05%.

Both imprecision and inaccuracy affect the internal inconsistency of distance data and reduce the fit between observed distances and distances on an output topology. This internal inconsistency casts doubt on the validity of branching arrangements when clades are united by short branch lengths (see also Chapter 11).

Differences in Paralogous Sequences and Genome Size.   Sequences whose differences are a consequence of independent evolutionary change arising after speciation are referred to as orthologous sequences (Fitch, 1976a). In contrast, paralogous sequences evolve in parallel in a single line of descent subsequent to their origin through gene duplication (see Chapter 1). A salient point is that cross-matched paralogous sequences from two different species may contain differences that predate speciation.

Fox and Schmid (1980) and Sarich et al. (1989) have argued that such cross-matched hybrids may be present in significant numbers when the conditions of reassociation are too relaxed, and that these paralogous hybrids form a low-melting temperature component characteristic of many melting curves. Furthermore, they argue that this low-melting temperature component seriously compromises the phylogenetic value of certain hybridization distances, such as ΔT_m and ΔT_{50H}. However, there are no measurements in any species that precisely quantify the amount of such low-copy number elements in the single-copy fraction of the genome. Significant quantities of hybridizing paralogous sequences may be present only under relaxed reassociation, although with many species the situation may be different for polyploid plant genomes.

In addition, for iodinated tracers, which constitute most of the melting curves to which Sarich et al. (1989) refer, short fragment size is often a contributing cause (if not the most important cause) of the low-melting temperature component. Furthermore, if paralogous sequences are shown to exist, it is easy enough to calculate a T_m for a higher temperature component and discard the lower melting temperature component. Finally, the effect of paralogous hybrids on ΔT_m or ΔT_{50H} values is probably to decrease their absolute magnitude without having much impact on their relative magnitude or on the branching patterns that are derived from matrices of such values (Springer and Krajewski, 1989).

Differences in genome size may also affect certain DNA hybridization metrics, at least in theory. Consider a case in which differences in genome size reside entirely in the single-copy fraction of the genome: one species has all of the genes that are found in the single-copy genome of the second species as well as its own unique set of genes. When the DNA from the species with the larger single-copy genome is labeled, part of that DNA will be incapable of reacting with driver DNA from the second species simply because there is none. This will result in a decreased NPH value for the heteroduplex reaction, although ΔT_m and ΔT_{mode} may or may not be affected. When DNA from the second species is labeled, however, an NPH reduction will not occur because the driver species has all of the genes that are present in the single-copy genome of the tracer species. Thus, we expect nonreciprocity in NPH values. This, in turn, will affect ΔT_{50H} values.

As a second example of the effect of genome size, consider a case in which the size and constituency of the single-copy genome is the same in each of two species, but for which the repeated DNA fractions differ considerably. If repeats constitute 80% of the genome in one species and only 10% of the genome in the second species, and if this remains unassessed, then effective concentrations of single-copy driver DNA (which react with single-copy tracer DNA) will differ when the total concentration of DNA (repeats plus single-copy) is the same. Theoretically, these differences could have a profound impact on NPH values and lead to nonreciprocity, although there is not yet any empirical evidence that bears on this question. In any event, some knowledge of genome size is probably important for studies that seek to understand nonreciprocity among NPH and ΔT_{50H} values.

Intraspecific Variation   Intraspecific polymorphisms can obscure relationships, particularly among closely related species (Chapters 2, 11, and 12). Differences in the melting profiles of conspecific individuals are generally small in most vertebrate species that have been investigated. Sheldon (1987), for example, provides evidence that the mean intraspecific ΔT_m values were only 0.28°C in herons. Similarly, Springer (1988) reported a mean intraspecific ΔT_m value of 0.36°C for diprotodontian marsupials. In contrast, high levels of intraspecific variation have been discovered in several invertebrate taxa, including sea urchins, cave crickets, and fruit flies (Britten et al., 1978; Caccone et al., 1987; Caccone and Powell, 1987). In some instances, intraspecific ΔT_m values are as high as 5°C for individuals from the same population. If population bottlenecks are also important in the evolutionary history of such lineages, then phylogenies derived from distance matrices could prove positively misleading (Roberts et al., 1985), especially with respect to branch lengths and the timing of divergence events. While it is generally not possible to know when and where bottlenecks occurred, it is possible (and advisable) to assess the magnitude of intraspecific variation versus the magnitude of interspecific distances.

APPENDIX: STOCK SOLUTIONS

Alkaline electrophoresis tray buffer (10X)

300 mM NaOH
10 mM EDTA

Alkaline electrophoresis gel buffer (10X)

500 mM NaCl
10 mM EDTA

DNase I

10 mg/ml in 150 mM NaCl, 50% glycerol

EDTA 0.5 M

18.6 g EDTA dihydrate
60 ml water

Add 10 N NaOH to pH 8.0. Add water to 100 ml.

Ethanol wash

50 ml 95% ethanol
50 ml buffer:
   20 mM Tris, pH 7.5
   1 mM EDTA
   100 mM NaCl

Store at −20°C.

Ethidium bromide solution

Add 200 mg to 20 ml water. Store in a light-proof container at 4°C.

Nick-translation buffer (10X)

500 mM Tris, pH 7.5

100 mM magnesium chloride
1 mM DTT
500 µg/ml BSA

Nucleotides (dNTPs)

Resuspend in water to 10 mM concentration. Adjust pH to 7.0 with 50 mM Tris by spotting small samples on pH paper, aliquot into small volumes and store at −20°C.

Phenol, equilibrated to pH 7.4

500 gm bottle phenol (solid)

Add 100 ml 2 M Tris, pH 7.4, 100 ml water, heat slowly to 37°C, mix layers, and let stand. Remove aqueous layer, add equal volume 1 M Tris, pH 7.4, mix and let stand, remove aqueous layer. Repeat until Tris remains at pH 7.5. Add 500 mg 8-hydroxyquinoline. Store at 4°C under 1 M Tris pH 7.5.

Phosphate buffer (PB) 2.4 M

500 ml 2.4 M sodium phosphate monobasic
500 ml 2.4 M sodium phosphate dibasic
(pH should be 6.8)

PB 0.48 M

100 ml 2.4 M stock
400 ml water

Check refractive index of solution against that of water. The difference should be exactly 0.0098 (e.g., if water = 1.3320, PB at 0.48 M = 1.3418); adjust up or down with 2.4 M PB and water, respectively.

PB 0.12 *M*

125 ml 0.48 *M* PB (checked with
   refractometer)
375 ml water

RNase (DNase free)

Resuspend RNase in TE at 10 mg/ml. Heat to
70°C for 15 min, cool and store at −20°C.

Salmon sperm DNA

Resuspend at 5 mg/ml in 0.1 m*M* EDTA, and
force through a 23-gauge needle several
times to shear or sonicate 4X at 80%
maximum power. Store over chloroform.

SEDTA

0.1 *M* NaCl
50 m*M* EDTA

Adjust pH to 8.0.

S1 nuclease buffer (2X)

1.2 ml 5 *M* NaCl
664 μl 100 m*M* zinc sulfate
400 μl 3 *M* sodium acetate, pH 4.5

Add water to 4 ml.

S1 nuclease solution

Resuspended S1 nuclease in 1X S1 buffer at
a concentration of 5 units/μl.

Sodium acetate 3.0 *M*

20.4 g sodium acetate, trihydrate

Add water to 50 ml; adjust pH to 7.5 with
acetic acid.

Sodium hydroxide 10 *N*

200 g NaOH
400 ml water

Dissolve, add water to 500 ml.

Sodium iodide solution (or use NaI of Geneclean® kit)

91.0 g sodium iodide
1.5 g sodium sulfite

Add water to 100 ml, filter through
Whatman # 1 filter paper. Add 0.5 g sodium
sulfite. Store at 4°C in a light proof bottle.

TE

10 m*M* Tris, pH 7.5
0.1 m*M* EDTA

TEACL (tetraethylammonium chloride)

Dissolve at a concentration of about 300
g/liter. Vacuum distillate to about 3.0 to 3.2
M. Pass twice over activated charcoal and
filter through a 0.45 μm Millipore filter;
adjust pH to 7.0 with tetraethylammonium
hydroxide. Dilute to about 80% and check
refractive index (R.I.):

R.I. (difference versus water @ 1.3520)
   of 2.5 *M* = 1.4065 (0.0745)
        2.4 *M* = 1.4032 (0.0712)
        2.3 *M* = 1.3999 (0.0679)

Note: the molarity of TEACL changes with
prolonged storage, so it must be carefully
checked before each use.

Trichloroacetic acid (TCA) 10%

500 g solid TCA

Add 227 ml water to make a 100% solution.
Dilute to 10% for working stock solution and
refrigerate.

Tris-HCl, 2.0 *M*, pH 7.5 and 8.5

850 ml water
242.2 g Trizma base
20 ml HCl (conc.)

Adjust pH to 7.5 or 8.5; add water to 1 liter.

Tris-acetate, neutral gel and tray buffer (10X)

850 ml water
48.4 g Tris
27.22 g sodium acetate trihydrate
3.8 g EDTA

Adjust pH to 7.0 with acetic acid; add water
to 1 liter.

CHAPTER 8

NUCLEIC ACIDS II: RESTRICTION SITE ANALYSIS

Thomas E. Dowling, Craig Moritz, and Jeffrey D. Palmer

PRINCIPLES AND COMPARISON OF METHODS

General Principles

In the past few years there has been a remarkable increase in the application of DNA analysis to problems in population genetics and systematics. Analysis of DNA has several significant advantages: (1) the genotype rather than the phenotype is assayed, (2) one or more sequences appropriate to a problem can be selected on the basis of evolutionary rate or mode of inheritance, (3) the methods are, for the most part, general to any type of DNA, and (4) DNA can be prepared from small amounts of tissue and is relatively stable. The last attribute means that genetic information on rare or endangered species can be obtained without destructive sampling (e.g., Plante et al., 1987) and it is somtimes possible to analyze suitably preserved tissues from extinct taxa (e.g., Higuchi et al., 1987; Pääbo, 1989).

There are several approaches for assaying DNA variation. DNA–DNA hybridization provides an estimate of the amount of sequence divergence between genomes (Chapter 7) but cannot provide discrete character data. Direct comparison of sequences offers extremely high resolution and yields character data that can be converted to estimates of sequence divergence if so desired (Chapter 11). An alternative assay for sequence variation involves comparison of the number and size of fragments produced by digestion of the DNA with restriction endonucleases. These variations are **restriction fragment length polymorphisms (RFLPs)**. In contrast to DNA–DNA hybridization, RFLP analysis is more laborious and expensive, but provides information on the nature, as well as the extent, of differences between two DNA sequences. RFLP analysis is (at present) simpler and cheaper than direct sequence comparisons, but offers less information on the evolution of the sequence itself. However, because of its relative ease of use, it is currently possible to analyze more loci per individual by RFLP analysis than by sequencing. A combination of direct sequencing and RFLP analysis can provide high-resolution information with high efficiency (e.g., Cann and Wilson, 1983; Cann et al., 1984).

Restriction endonucleases (REs) are enzymes that cut DNA at a constant position within a specific recognition sequence, typically 4–6 base pairs (bp) long (e.g., Tables 1 and 2). Over 400 REs have been isolated from bacteria and characterized (Roberts, 1984). Each cleaves DNA at a characteristic recognition sequence, although REs isolated from different types of bacteria may have the same recognition sequence (these are called isoschizomers). The natural function of REs is to protect bacteria from foreign DNA. Methylation of the bacterial DNA at the sequence cleaved by the RE prevents the RE from cleaving. Foreign DNA lacks this methylation and is therefore cleaved (Lewin, 1987). The sequences recognized by REs are symmetrical in that the sequence is the same when read from the 5' end of each strand. Depending on the location of cleavage within this sequence, REs can produce ends with a 5' overhang, a 3' overhang, or no overhang (Table 1), and the symmetry ensures that the ends can be rejoined (see Chapter 9).

The specificity of cleavage by REs means that complete digestion of a particular sequence of DNA will yield a reproducible array of fragments. Changes in the number and/or size of fragments can be brought about by sequence rearrangements, the addition or deletion of DNA, or base substitution within cleavage sites (Upholt, 1977; Figure 1). The first two types of mutation will affect the fragment patterns of all REs that cleave in the region. An inversion that spans a cleavage site reduces the size of one fragment and increases the size of another by the same amount. A tandem direct duplication that includes a cleavage site results in an additional fragment equal to the size of the duplication. The effects of an inverted duplication are more complex, particularly where there is high-frequency recombination between repeats (see Palmer, 1986a). The gain or loss of

Table 1. Examples of recognition sequences and types of end produced

Enzyme	Recognition Sequence	End	Type
EcoRI	5' -G'AATT C- 3' 3' -C TTAA,G- 5'	-G -CTTAA	5' Overhang
RsaI	5' -GT'AC- 3' 3' -CA,TG- 5'	-GT -CA	Blunt
SacII	5' -CC GC'GG- 3' 3' -GG,CG CC- 5'	-CCGC -GG	3' Overhang

Table 2. Properties of commonly used restriction endonucleases[a]

Enzyme	Buffers			Site	Number of Sites	
	L	M	H		cpDNA[b]	mtDNA[c]
AluI	+	+++	+++	AG'CT	341	64
ApaI	+++	+++	++	GGGCC'C	13	5
ApaLI	+++	++	+	G'TGCAC	8	1
AseI[d]	+	+++	+++	AT'TAAT	114	10
AvaI	+++	+++	+++	C'YCGUG	70	3
BamHI[e]	+	++	+++	G'GATCC	40	1
BanI	+++	+++	++	G'GYUCC	32	8
BanII	+++	+++	+++	GUGCY'C	75	15
BclI[d,f]	+	+++	+++	T'GATCA	54	4
BglI[g]	+	+++	+++	GCN$_5$'GGC	9	2
BglII	++	+++	+++	A'GATCT	60	0
BstBI[h]	+++	+++	++	TT'CGAA	94	7
BstEII[e,i]	+	++	+++	G'GTNACC	12	2
BstNI[e,i]	++	++	+++	CC'(A/T)GG	128	4
BstUI[i]	+++	+++	++	CG'CG	89	6
BstXI[e,j]	++	+++	+++	CCAN$_6$'TGG	26	4
Bsu36I	+	++	+++	CCT'NAGG	15	1
ClaI	+++	+++	+++	AT'CGAT	59	1
DdeI[e]	++	+++	+++	C'TNAG	309	72
DraI	++	+++	+++	TTT'AAA	64	4
EcoO109I	+++	+++	+++	UG'GNCCY	67	14
EcoNI	+++	+++	+++	CCT'N$_8$AGG	18	5
EcoRI	*	+++	+++	G'AATTC	97	3
EcoRV[e]	+	+	+++	GAT'ATC	36	3
HaeII	+++	+++	+++	UGCGC'Y	25	7
HaeIII	+++	+++	+++	GG'CC	196	50
HincII	++	+++	+++	GTY'UAC	57	12
HindIII	*++	+++	+++	A'AGCTT	33	3
HinfI	++	+++	+++	G'ANTC	718	36
HinPI	+++	+++	+++	G'CGC	89	17
HphI	+++	+++	+++	GGTGAN$_8$'	128	55
KpnI	+++	+	+	GGTAC'C	15	3
MboI	++	+++	+++	'GATC	623	22
MluI	++	+++	+++	A'CGCGT	9	0
MspI	+++	+++	+++	C'CGG	214	23
NciI[k]	+++	+++	++	CC'(C/G)GG	113	8
NcoI[e]	+	++	+++	C'CATGG	NA	4
NdeI[e]	+	+	++	CA'TATG	NA	3
NheI	+++	+++	+++	G'CTAGC	7	1
NsiI[e]	++	++	++	ATGCA'T	43	3
PstI	+++	+++	+++	CTGCA'G	14	2
PvuII	+++	+++	+++	CAG'CTG	12	1
RsaI	+++	+++	+++	GT'AC	286	35
SacI	+++	++	+	GAGCT'C	21	2

(Continued)

Table 2. (Continued)

Enzyme	Buffers			Site	Number of Sites	
	L	M	H		cpDNA[b]	mtDNA[c]
SacII	+++	+	+	CCGC'GG	7	2
SalI[e]	+	+	++	G'TCGAC	11	0
ScrFI	++	+++	+++	CC'NGG	239	22
SmaI	+	+	+	CCC'GGG	16	0
SpeI	++	+++	+++	A'CTAGT	28	9
SspI	++	+++	+++	AAT'ATT	137	11
StuI	+++	+++	+++	AGG'CCT	16	13
StyI	+	++	+++	C'C(A/T)(A/T)GG	100	22
TaqI[i]	+++	+++	+++	T'CGA	639	29
XbaI	+	+++	+++	T'CTAGA	49	5
XhoI[e]	++	+++	+++	C'TCGAG	24	1

[a]The number of +'s under buffers indicates the range of salt conditions under which the REs are active: + <10%; ++, 10-30%; +++, >30% activity (modified from the 1989 New England Biolabs catalog). The buffer most commonly used in single-enzyme digests (L = 0 mM NaCl, M = 50 mM NaCl, H = 100 mM NaCl) is underlined. For the recognition sequence, the location of cuts is indicated by the prime ('), and the bases filled in by end-labeling are underlined. Blunt ends and 3' overhangs depend on the exonuclease activity of the polymerase to provide a suitable primer for labeling; therefore, the base left of the center of the recognition sequence is indicated. To ensure good labeling, it may be prudent to use the two or three bases left of the center point. "NA" indicates that the information was not available, "*" indicates star activity (see supplier's catalog for details).
[b]Number of tobacco (Shinozaki et al., 1986) cpDNA fragments produced (= number of cleavage sites minus one inverted repeat segment, plus one).
[c]Number of human (Anderson et al., 1981) mtDNA fragments produced (= number of cleavage sites).
[d]optimum salt concentration = 150 mM KCl.
[e]optimum salt concentration = 150 mM NaCl.
[f]optimum temperature = 50°C.
[g]optimum salt concentration = 75 mM NaCl.
[h]optimum temperature = 65°C.
[i]optimum temperature = 60°C.
[j]optimum temperature = 55°C.
[k]optimum salt concentration = 25 mM NaCl.

DNA between two cleavage sites increases or decreases the size of that fragment by the corresponding amount. Base substitution can result in the gain or loss of cleavage sites. This leads to three fragment changes: e.g., for a site gain, one fragment is lost and two, summing to the size of the lost fragment, are gained. These effects are usually specific to a single RE, although REs with overlapping recognition sequences (e.g., *Bam*HI and *Mbo*I, Table 2) may show coincident changes in fragment patterns.

The fragments produced by digestion of DNA with REs are sorted according to their size by gel electrophoresis. At neutral pH, the sugar-phosphate backbone of the DNA is negatively charged and causes the molecule to migrate through an electric field. The media used, agarose and polyacrylamide, form a dense matrix through which smaller fragments can move more easily than larger fragments. In practice, the distance migrated is proportional to the logarithm of the molecular weight. Fragments of known size are run on each gel to act as an internal standard against which the sizes of other fragments are estimated by interpolation from a calibration curve (Figure 1).

The basic steps in RFLP analysis are (1) selecting the DNA sequence(s) to be analyzed, (2) preparing the DNA, (3) cleaving the DNA with selected REs, (4) sorting the fragments by gel electrophoresis, (5) visualizing the sorted fragments, and (6) analyzing the results (Figure 2). Following a

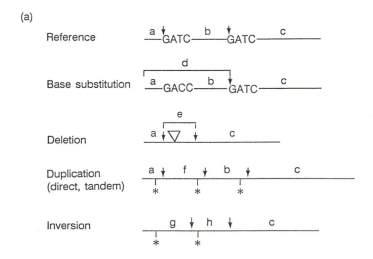

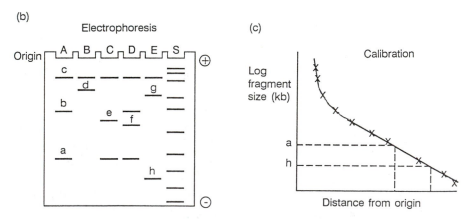

FIGURE 1. The effect of different kinds of sequence change on RFLPs. (a) DNA fragments (a–h) are generated by RE digestion, (b) electrophoretically separated by size, and (c) their sizes determined using a calibration curve based on a sample with fragments of known size run on each gel (lane S = size standard). Vertical arrows indicate cleavage sites and asterisks indicate the boundaries of rearrangements.

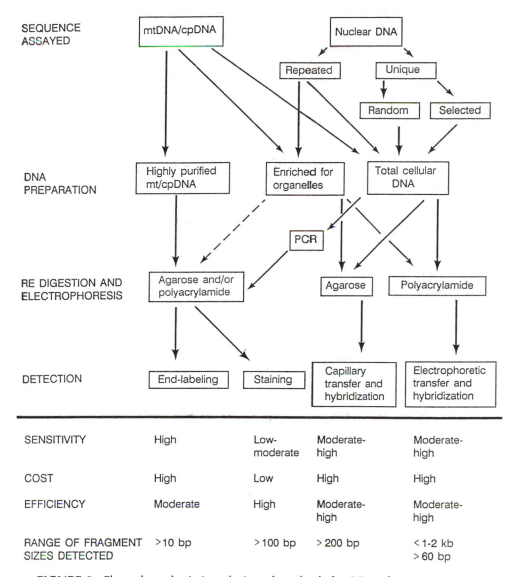

FIGURE 2. Flow chart depicting choice of methods for RE analysis (see text).

discussion of some of the general assumptions of RFLP analysis, we discuss the choices that must be made in steps 1 to 5, first in the context of techniques, and then in terms of different hierarchical levels of application. The analysis of results, step 6, is covered under the section on gel interpretation and in Chapters 10 and 11.

Assumptions

Causes of Gain/Loss of Cleavage Sites A basic assumption of RFLP analysis is that changes in fragment patterns are due to one of the types of mutation outlined above. However, because the activity of most REs is sensitive to methylation, variation in the state of methylation can, in principle, mimic the gain/loss of cleavage sites. Methylation-induced artifacts can be detected by comparing the fragment patterns produced by iso-schizomers that differ in sensitivity to methylation. For example, Groot and Kroon (1979) treated mitochondrial DNA (mtDNA) with *Msp*I (methylation insensitive) and *Hpa*II (methylation sensitive), both of which recognize CCGG. The same fragment pattern was observed, indicating that methylation did not affect cleavage at these sequences. By contrasting the enzymes *Sal*I and *Hinc*II, Wilson et al. (1984) suggested that extreme variation for *Sal*I sites in nuclear ribosomal DNA (rDNA) of primates was due to different patterns of methylation. Another approach is to compare the fragment patterns produced by the crude DNA and the same sequence cloned in a methylation-deficient strain of *Escherichia coli* (e.g., Castora et al., 1980). It is not clear how frequent these artifacts are. It appears that methylation is not a problem for RFLP analysis of mtDNA and cpDNA (Palmer, 1985a), but can result in hypervariation and apparent homoplasy of specific sites in nuclear sequences (Wilson et al., 1984; Jorgensen and Cluster, 1988).

Causes of Changes in Fragment Migration   A second general assumption is that changes in the mobility of DNA fragments reflect differences in molecular weight. However, the migration of DNA in acrylamide (but not agarose) gels is also affected by conformation. Singh et al. (1987) demonstrated that changes in mobility of some fragments of human DNA were due to base subsitions affecting the degree of bending of DNA. Such changes will affect the fragment patterns produced by all REs that cleave on either side of the conformation mutation and can be misinterpreted as addition/deletion mutations.

Identity of Shared Fragments or Cleavage Sites   A third assumption made in many studies is that if two samples share a particular sized fragment, they must also share flanking cleavage sites. This assumption is generally true for sequences from closely related individuals and perhaps even most intra-specific comparisons (depending on the rate of evolution of the DNA sequence in question). However, the likelihood of convergence, that there are two samples having fragments of the same size but produced by different cleavage sites, increases as the sequences become more different. Upholt (1977) suggests that simple fragment comparisons should be restricted to sequences that differ by less than 15%. If it is assumed that all differences in fragment patterns stem from the gain and loss of cleavage sites (i.e., no structural variation), then the proportion of shared fragments (F) can be used to estimate the amount of sequence divergence (p) (see Upholt, 1977; Nei and Li, 1979). However, the relationship between F and p is curvilinear, so that small errors at low values of F result in large errors in the estimation of p. For this reason, Kessler and Avise (1985b) suggest that simple fragment comparisons should not be used where less than 25% of fragments is shared. To that we would add that this approach should not be used where there is structural variation.

Comparison of mapped cleavage sites eliminates problems of convergent fragment lengths, although the sites themselves may be convergent. This approach extends the validity of RFLP analysis to sequences that differ by up to 25% (Upholt, 1977; but see Templeton, 1983a; Nei and Tajima, 1985); beyond that point there is a reasonable probability that shared cleavage sites may be convergent. The probability of convergent site losses is far greater than that of convergent site gains because a site loss is caused by any point mutation within a cleavage site, whereas a site gain requires a specific base substitution at a particular base pair (Templeton, 1983b; DeBry and Slade, 1985; Li, 1986). These inequalities should be considered in using restriction site data for phylogenetic analysis (Chapter 11). Mapping is also essential to identify and localize sequence rearrangements (e.g., Palmer et al., 1985; Jansen and Palmer, 1987a), duplications (e.g., Moritz and Brown, 1986, 1987), and minor length variants (e.g., Cann and Wilson, 1983; Densmore et al., 1985; Palmer et al., 1985). Misinterpreting such changes as gain/loss of cleavage sites leads to gross errors in subsequent analyses.

Comparison of the Primary Methods

Choice of Sequence   Selecting the sequence(s) to be analyzed is probably the most important decision to be made in designing an RFLP study (see also Chapter 2). Sequences should display sufficient variation to enable population genetic or phylogenetic analysis, but not so much that there is substantial homoplasy of fragment lengths or cleavage sites. In broad terms, choices are made according to evolutionary rate and mode of inheritance (Table 3). Non-coding sequences typically have high evolutionary rates (Li et al., 1985a) as do most regions of animal mtDNA (Brown et al., 1979; reviewed by W. M. Brown, 1985; Table 3). In contrast cpDNA and mtDNA from plants and fungi have a relatively slow rate of base substitution and the substitution rate of plant mtDNA is slower still (Table 3). Some genes, notably nuclear rDNAs (Hillis and Davis, 1986; Jorgensen and Cluster, 1988), encompass both highly variable and conserved regions, providing information across a broad phylogenetic spectrum. The inheritance of mtDNA and cpDNA is usually uniparental and effectively haploid (reviewed by Birky, 1983). This has a marked effect on the population genetics, most notably a 4-fold reduction in N_e when males and females are equally frequent (Birky et al., 1983), and a high rate of turnover within populations (Avise et al., 1984, 1988).

Table 3. Evolutionary properties of different genomes and lineages[a]

Genome	Lineage	Inheritance	Relative Mutation Rate or Amount of Variation		
			Point mutations	Size range (kb)	Rearrangements
mtDNA	Animals	Maternal	High	14–26	Very rare
mtDNA	Plants	Maternal	Very low	200–2500	Very frequent
mtDNA	Fungi	All[b]	Low	20–200	Frequent
cpDNA	Plants	All[b]	Low	120–217	Rare
nDNA	Animals	Biparental	Variable	$1–1000 \times 10^5$	Frequent
nDNA	Plants	Biparental	Variable	$1–1000 \times 10^5$	Frequent
nDNA	Fungi	Biparental	Not known	$0.1–10 \times 10^5$	Frequent

[a]Data summarized from Cavalier-Smith (1985b), Palmer (1985a,b), Moritz et al. (1987), Wolfe et al. (1987, 1989), Palmer and Herbon (1988), and Neale and Sederoff (1988).
[b]All refers to maternal, paternal, and biparental modes of inheritance.

The rate of sequence rearrangement may also be a factor in selecting a sequence for analysis (Table 3). Rearrangements, unless carefully characterized, complicate fragment comparisons and can lead to gross overestimates of sequence diversity. However, once identified, the rearrangements are themselves potential sources of phylogenetic information (see below). Most cpDNA and animal mtDNA sequences are stable in this regard, although there are exceptions (e.g., Moritz and Brown, 1987; Palmer et al., 1987, 1988a). In contrast, plant mtDNAs (Palmer and Herbon, 1988) and, to a lesser extent, fungal mtDNAs (Bruns and Palmer, 1989) are in general unsuitable for RFLP analysis, owing to their high rate of structural change (length mutations and rearrangements), which makes it difficult to align homologous sequences for comparison of RE sites. Although these rearrangements can potentially serve as phylogenetic characters, the frequency with which they occur in plant mtDNAs severely limits their utility, because later mutations tend to obscure earlier ones and also because the amount of work required for proper analysis is not usually rewarded at those taxonomic levels at which they occur with sufficient rarity. For these reasons, this chapter will be limited to the "simple" (i.e., fairly free of structural mutations) genomes of animal mitochondria and plant chloroplasts, as well as to specific regions of nuclear genomes that have demonstrated evolutionary and systematic utility.

Transposable elements are essentially unknown in plant and animal organelle genomes, although they are common in nuclear genomes. The insertion/excision of transposable elements has been shown to modify fragment patterns for nuclear sequences (e.g., Aquadro et al., 1986) and may therefore complicate RFLP analyses. Lawrence et al. (1989) used the location of transposons to estimate relationships among closely related strains of *E. coli,* but noted that transposons moved too frequently for such characters to be of use among more divergent isolates of *E. coli.*

Repeated Nuclear Sequences.   Nuclear sequences (nDNAs) may be present in high, moderate, or low copy number and may be dispersed or arranged in tandem arrays. Tandemly repeated sequences have the advantage that they are relatively simple to assay because of their high copy number. In contrast, members of some dispersed repeat families are operationally equivalent to single-copy sequences in terms of their detection by transfer hybridization. This can also be an advantage as use of dispersed repeats as probes enables variation to be assayed at a large number of loci simultaneously (e.g., hypervariable sequences, see below).

Repeated sequences are prone to **concerted evolution.** This refers to the tendency for copies of such sequences to become homogeneous, first among gene copies within genomes and then among individuals within populations (Zimmer et al., 1980; Coen et al., 1982; reviewed by Dover et al., 1982; Arnheim, 1983; Dover and Tautz, 1986), and is thought to result from **unequal crossing-over** and/or **gene conversion.** Concerted evolution has important implications for the use of repeated sequences in population genetic or phylogenetic studies. The homogenizing process among genomes, if significant, makes such sequences inappropriate for estimating fundamental population genetic parameters such as rates of gene flow, effective population size, and deviations from Hardy–Weinberg equilibrium (see Chapter 10). However, if the aim is simply to identify discrete genetic units (demes), then the reduction of interpopulation variation predicted to follow from concerted evolution may actually increase resolving power (Hoelzal and Dover, 1987). The same effect could simplify sampling for phylogenetic studies: if the rate of interindividual homogenization is much greater than the rate of speciation, then it should be possible to identify synapomorphies with relatively small samples (Hillis and Davis, 1988; see also Chapter 1). Concerted evolution also introduces the potential for horizontal transfer of variants, leading to discrepancies between the gene tree and the organismal phylogeny. For example, it is conceivable that biased gene conversion could result in the rapid spread of a variant introduced into a gene pool by hybridization.

The degree and pattern of concerted evolution of repeated genes within species appear to be variable. rDNA length variants frequently have a nonrandom distribution among chromosomes (Appels and Dvořák, 1982; Arnheim et al., 1982; Williams et al., 1987; Separack et al., 1988) indicating that within-chromosome homogenization is more efficient than that among chromosomes. Several studies suggest that among-chromosome homogenization is more effective in primates than in mice, possibly because only the former have nucleoli including rDNA cistrons on nonhomologous chromosomes (Arnheim et al., 1982; Arnheim, 1983). The effectiveness of concerted evolution may also vary with the chromosomal location of the genes. For example, Williams et al. (1987) reported less variation among X chromosome rDNA repeats than among those on the Y chromosome of *Drosophila melanogaster.*

Evidence for biased gene conversion of rDNA variants has come from studies of hybrid grasshoppers (Arnold et al., 1988) and barley. In hybrid barley, Saghai-Maroof et al. (1984) found that one rDNA locus conformed to Mendelian inheritance whereas another had a significant excess of homozygotes, possibly due to biased gene conversion. Biased gene conversion (Arnold et al., 1988) may be partly responsible for marked asymmetry of the distribution of rDNA markers across a hybrid zone of *Caledia captiva* defined by chromosomes, repeated sequences, and allozymes (Shaw et al., 1988).

The above observations do not argue against the use of repeated nuclear sequences in population biology and systematics. Such sequences, rDNA in particular, have been widely employed, with the results often being congruent with other studies (see below). However, there is need for caution in interpreting differences among individuals of the same, or closely related, species.

Unique Nuclear Sequences.   Unique and low copy-number sequences are technically more difficult to visualize from preparations of total cellular DNA. However, these sequences can be applied to a broad range of population genetic and phylogenetic problems. An important choice here is whether to assay randomly selected sequences (e.g., Quinn and White, 1987a,b; Figueroa et al., 1987) or genes of known function (e.g., Aquadro et al., 1986; Nadeau et al., 1988). Depending on the size of the cloned probe and its location relative to noncoding sequences, both approaches are likely to reveal extensive polymorphism. The latter choice has the advantage that surveys of variation will also add to our understanding of the evolution and function of those sequences. Mutation rates, and thus number of alleles per locus and heterozygosity, vary widely, from the hypervariable sequences used in DNA fingerprinting (see below) to highly conserved coding sequences (e.g., histones, heat shock proteins).

Organellar Genomes.   mtDNA and cpDNA are technically simple to assay because they are present in high copy-number and can be separated from the nuclear genome. This has resulted in their widespread application in molecular systematics (see below). Particular use has been made of their maternal inheritance and tendency to become homogeneous within populations (reviewed by Wilson et al., 1985; Avise et al., 1987; Moritz et al., 1987; Palmer, 1986b, 1987; Palmer et al., 1988b). However, as with any sequence, the recovered phylogeny is of the molecules and not necessarily the organisms. When closely related species are examined, discrepancies can arise because of introgression or sorting of ancestral polymorphisms. In animals, introgression is usually limited to populations near a hybrid zone, whereas in plants, hybridization and introgression are more general phenomena. Sorting of ancestral polymorphisms among taxa is expected to cause the appearance of polyphyly at first, progressing through paraphyly to monophyly as variants arise and become fixed (Neigel and Avise, 1986; Pamilo and Nei, 1988). These potential problems can, to some extent, be alleviated by sampling multiple populations within each species (see also Chapter 2) and by comparing phylogenies derived from organellar and nuclear sequences.

Method of DNA Preparation   The optimal method of DNA preparation depends on the type of tissue available and the sequences to be assayed (Figure 2). Preparations of total cellular DNA can be used for analysis of any sequences. If only organellar and highly repeated nuclear sequences are to be assayed, detection of the former may be facilitated by enriching the organellar fraction by a round of differential centrifugation (Lansman et al., 1981; Palmer et al., 1985). Some simple and reliable protocols for DNA extraction are given below (see protocols) and in Chapter 9.

If only organellar sequences are to be assayed, or if they are to be assayed separately, then additional purification may be desirable. For large amounts of tissue (e.g., >1.0 g vertebrate organ tissue, >10 g wet weight of plant tissue), enriched organellar DNA can be extracted from organelles isolated via sucrose gradients of varying densities (Lansman et al., 1981; Palmer, 1986a). This is the approach most often used to purify cpDNA. However, for animal mtDNA, the level of nuclear contamination in such preparations is usually still too high for visualization of the DNA via end-labeling. Maximum purity can be achieved using neutral CsCl equilibrium gradients with intercalating dyes such as ethidium bromide or propidium iodide. Invertebrate and fungal mtDNAs and some algal cpDNAs (but not vertebrate mtDNAs or plant cpDNAs) typically have a strong bias toward A and T (W. M. Brown, 1985; Palmer, 1985b) and can be separated from unbiased or GC-rich DNA in neutral CsCl gradients, especially with the addition of dyes, such as bisbenzimide (Hoescht 33258), that preferentially bind AT-rich regions (Hudspeth et al., 1980; Gargouri, 1989). Care must also be taken to avoid copurification with nuclear density satellites (Arnason and Widegren, 1984).

The time and expense involved in obtaining highly purified organellar DNAs are considerable, but are justified when fragments are to be detected by staining or end-labeling (e.g., animal mtDNA). Numerous alternatives to purification via CsCl gradients have been proposed (e.g., Chapman and

Powers, 1984; Powell and Zuniga, 1983; Palva and Palva, 1985; Jones et al., 1988) that are generally cheaper and faster and seem adequate for particular tissues or organisms and for detection methods of low sensitivity. However, they are not as generally applicable as CsCl banding, and any nuclear contamination can lead to misinterpretation of fragment patterns (see Interpretation and Troubleshooting). If such alternatives are to be used, we recommend that mtDNA be extracted via CsCl gradients from a subset of samples to act as nDNA-free controls. For analysis of cpDNA and mtDNA RFLPs via transfer hybridization, total cellular DNA is quite adequate and may have advantages in some situations (see under Applications).

If two or more highly conserved regions have already been identified near the boundaries of the sequence to be assayed, then the polymerase chain reaction (PCR) can be used to amplify that sequence from even minute amounts of total DNA (Saiki et al., 1988; White et al., 1989; Innis et al., 1990; Chapter 9). This process involves cycles of denaturation, annealing oligonucleotide primers complementary to the conserved sequences, and primer extension by polymerase, resulting in an exponential increase in the concentration of the sequence bounded by the primers. The target sequence can then be assayed with restriction enzymes or can be sequenced directly. PCR permits direct analysis of any defined DNA sequence and is particularly useful for assaying variation in organellar (Wrischnik et al., 1987; Kocher et al., 1989; Pääbo, 1989) and single-copy nuclear genes (Saiki et al., 1985; Wong et al., 1987; reviewed in White et al., 1989). This method is already finding widespread application in population genetic and phylogenetic studies and is likely to become widely used.

Restriction Enzymes   Depending on the application, REs are selected on the basis of how many cuts they are likely to make, the range of salt conditions under which they are active (broad for double-digest mapping), the types of ends produced (5′ overhang or blunt for end-labeling, see below), and cost (Table 2). In general, REs that cleave at 4-bp sites will cut more often than those that cleave at 6-bp sequences. The base content of the recognition sequences is also important: REs with GC-rich recognition sequences will make few cuts in sequences that have low G + C content. Sensitivity to methylation may also be relevant where a mixed sample of nDNA and organellar DNA is analyzed. Palmer (1986a) lists methylation-sensitive REs that cut plant nDNA rarely, but cpDNA sufficiently, to permit RFLP analysis. Analysis of large sequences such as entire cpDNAs (~150 kb) typically employ 6-bp cutting REs that produce fragments that mostly range from 1 to 5 kb, although REs that produce more, smaller fragments can be used to compare closely related sequences. Closely related animal mtDNAs can be compared using 4-bp recognizing REs (e.g., Brown, 1980), but beyond ~2% sequence divergence it becomes too difficult to identify individual gains or losses of cleavage sites and mtDNAs are best analyzed by mapping cleavage sites for 5-bp or 6-bp recognizing REs. For animal mtDNA, these approaches are most powerful in conjunction with end-labeling and electrophoresis through both agarose and polyacrylamide gels (see below).

Electrophoresis of Fragments   Fragments may be sorted according to size by electrophoresis through agarose gels, polyacrylamide gels, or both. The range over which the size of fragments can be accurately estimated varies with gel concentration and buffer. Agarose gels are commonly used at between 0.6 and 2.0%, using TAE or TBE buffers (Appendix), allowing accurate estimation of fragment size over the range from 300 bp to 20 kb. TAE provides better separation of large fragments, but poorer resolution of small fragments. Polyacrylamide gels are typically composed of between 3.5 and 6.0% acrylamide, providing for analysis of fragments ranging from 10 bp to 1 kb.

   Agarose gels are simpler to prepare and have the widest range of applications. Gels may be run horizontally or vertically. Horizontal gels are easier to prepare but are thicker than vertical gels. This is an advantage when large amounts of DNA are to be loaded per lane (e.g., staining with intercalating dyes, hybridization methods) as the DNA concentration at the gel interface should not exceed 1 μg/mm^2. However, the thinner (1-mm) vertical gels are easier to dry onto filter paper for autoradiography of end-labeled fragments. Polyacrylamide gels are potentially hazardous, as unpolymerized acrylamide is a potent neurotoxin. However, these are the only gels that allow very small fragments (<0.2 kb) to be detected with high resolution and are used in conjunction with high-sensitivity techniques, e.g., end-labeling (Brown, 1980) and some hybridization methods (Kreitman and Aguade, 1986). They are run vertically using a TBE buffer and long plates (see Protocols).

   For some applications (e.g., end-labeling of animal mtDNA), it is advantageous to run each sample on both types of gel. Use of both agarose and polyacrylamide gels produces extremely accurate cleavage site maps and ensures that all fragments are visualized. This often clarifies the arrangement of closely spaced fragments in the overlapping region (~1.5–0.4 kb for 1.2% agarose and 3.5% polyacrylamide). It can also reveal conformation-induced changes in fragment mobility, as these are restricted to the polyacrylamide gels (Singh et al., 1987).

Methods of Detection   DNA fragments can be detected by direct staining, end-labeling, or hybridization. Direct staining, usually with ethidium bromide (EB), is the simplest and cheapest, but least sensitive method and can be applied only to purified sequences (Figure 2, e.g., cpDNA, mtDNA, PCR amplified nDNA sequences). The dye binds to the DNA, so that staining intensity is proportional to DNA concentration, and, thus, fragment size. The minimum amount of DNA in a band detectable by this method is about 2 ng. Thus, small fragments can be detected only if a large amount of DNA is loaded: for example, 100 ng of a 10-kb sequence must be loaded to detect fragments of 200 bp. Silver staining of DNA is reported to be more sensitive, allowing detection of 10–100 pg amounts of DNA (Guillemette and Lewis, 1983) and has been applied to analysis of mtDNA RFLPs (Tegelstrom, 1986). Although the intensity of staining is still proportional to fragment size, this is a substantial improvement over staining with ethidium bromide.

End-labeling involves adding ^{32}P-tagged nucleotides (dNTPs) to the ends produced by cleavage with REs and, again, can be applied effectively only to highly purified sequences. Because each fragment has the same number of ends, intensity is independent of fragment size (i.e., a 10-bp fragment should be as intense as a 10-kb fragment). This method is also highly sensitive: end-labeled fragments of any size can be visualized from 1–5 ng of digested DNA. With end-labeling, it is preferable to use REs that produce 5′ overhangs or blunt ends (Table 2). The large (Klenow) fragment of *E. coli* polymerase has both polymerase and 3′ exonuclease functions. The polymerase will add radioactive nucleotides using the 5′ overhang as a template. The 3′ exonuclease can convert blunt ends or 3′ overhangs to 5′overhangs but is relatively inefficient. Ends can be labeled with any [^{32}P]- or [^{35}S]dNTP so long as it occurs among the bases to be filled (Table 2). However, if several different REs are being used, it is most efficient to use all four [^{32}P]- or [^{35}S]dNTPs. This makes end-labeling a relatively expensive approach.

Transfer hybridization (Southern, 1975) has two basic steps. First, a membrane-bound replica of the gel is made by treating the gel with base to denature the electrophoresed fragments and transferring the single-stranded fragments to a nylon or nitrocellulose membrane. Second, a labeled single-stranded DNA probe is allowed to hybridize with complementary membrane-bound sequences. The probe is usually labeled with radioactive (^{32}P or ^{35}S) dNTPs by nick translation (Rigby et al., 1977) or random priming (Feinberg and Vogelstein, 1983), although nonradioactive methods have also been developed (e.g., the biotin-streptavidin method, Leary et al., 1983). The labeled strands will hybridize to complementary membrane-bound single-stranded sequences, allowing them to be detected by autoradiography (radioactive probes) or staining (biotin probes). The amount of base pair mismatch permitted (**stringency**) can be controlled by varying temperature, salt, and formamide concentration (Sambrook et al., 1989). This approach has many advantages. Any sequence for which there is a probe can be analyzed from heterogeneous (total cellular) DNA. Thus, hybridization is the only practical approach in which the sequence cannot be readily purified. The method is highly sensitive, allowing detection of picogram (pg) quantities of a fragment. It is also an efficient approach to assaying multiple sequences (e.g., Shaw et al., 1988; Baker et al., 1989; Sites and Davis, 1989). Multiple (>20) probes can be sequentially applied to membrane-bound DNA by dissociating the probe and target strands under conditions in which the latter remain attached to the filter.

Transfer hybridization does have some disadvantages. Using standard methods it is difficult to detect fragments smaller than 250 bp, making the technique less able to detect the gain or loss of closely spaced cleavage sites. However, Kreitman and Aguade (1986) used electrophoretic transfer of digested DNA from denaturing polyacrylamide gels, combined with high sensitivity hybridization conditions (Church and Gilbert, 1984), to detect fragments smaller than 100 bp. By digesting the DNAs with 4-bp REs and using these methods, they detected 50 variants of a 2.7-kb sequence for alcohol dehydrogenase among 87 isochromosomal lines of *Drosophila melanogaster* taken from two localities (see also Simmons et al., 1989).

A second potential pitfall of hybridization is the danger of detecting fragments other than those from the sequence to be assayed. In most cases, this problem can be minimized by hybridizing at the highest possible stringency. For organelle sequences, this problem can be due to the not infrequent movement of sequences to other organelles or to the nucleus (Gellisen et al., 1983; Jacobs et al., 1983; Timmis and Scott, 1984; Palmer 1985a). However, in practice, these paralogies do not usually interfere with hybridization assays of cpDNA and animal mtDNA because of the high copy number of the organellar sequences relative to nDNA copies and, in plant leaves, the abundance of cpDNA relative to mtDNA (but see Quinn and White, 1987b).

Applications of transfer hybridization to assaying sequence variation are restricted by the availability of suitable probes. These must have sufficient sequence similarity to the target DNA to form a stable hybrid at moderate to high stringency. Optimally, the probes should come from the same species, but this usually requires cloning and is more demanding technically (see Chapter 9). Alternatively, if the sequence contains some highly conserved regions, probes prepared from other species can be used. The use of such heterologous probes is exemplified by analyses of cpDNA and rDNA sequences (see below). Although the use of heterologous probes is technically simpler, caution is needed in interpreting results as fragments wholly included within a rapidly evolving region may not be detected (Hillis and Davis, 1988; Jorgensen and Cluster, 1988; Williams et al., 1988).

In summary, the choice of detection method must consider cost, effort, the amount and type of DNA that can be prepared, and the sensitivity desired (Figure 2). It should be noted that the methods described above are not exclusive. For example, RFLP variation can be studied in detail using end-labeling of a highly purified sequence for a representative subset of taxa, following which broader scale surveys could use hybridization methods. This approach maximizes both accuracy and efficiency.

APPLICATIONS AND LIMITATIONS

Population-Level Comparisons

RFLPs of cpDNA, animal mtDNA, and unique and repeated nuclear sequences have provided useful genetic markers for the analysis of population-level variation. Applications include estimating the extent of variation within and among populations, levels of gene flow, effective population size, patterns of historical biogeography, and analyses of parentage and relatedness.

Chloroplast DNA   The chloroplast genome is conservative in most respects (Table 3; reviewed by Palmer 1985a,b; Wolfe et al., 1987). Although the range of cpDNA sizes among land plants is large (120–217 kb), most of this variation is due to a few exceptional genomes and length mutations are typically short (<1 kb) and of restricted occurrence. The order and arrangement of chloroplast genes are quite invariant. Most land-plant cpDNAs have identical organization and most variants stem from one or a few simple inversions. The rate of nucleotide substitutions in cpDNA appears to be less than that of animal and plant nDNAs, and much less than for animal mtDNAs (Table 3). Perhaps the most variable feature of cpDNA is its mode of inheritance, which may be strictly maternal (most angiosperms), biparental, or paternal (conifers; reviewed in Sears, 1980; Neale and Sederoff, 1988). However, even with biparental inheritance, transmission is essentially clonal as recombination has not been observed in land plants.

The slow rate of change in cpDNA sequence and structure (Table 3) is reflected in the low levels of within- and between-population variation apparent from most of the studies to date (i.e., Banks and Birky, 1985; Wagner et al., 1987; Neale et al., 1988). However, two recent studies of flowering plant species have found substantial levels of within-population variation (Soltis et al., 1989a,b). In a large number of studies principally aimed at clarifying interspecific relationships, cpDNA polymorphisms were either absent or rare among conspecific populations (reviewed in Palmer, 1987; Crawford, 1989). The ultimate utility of cpDNA as a population marker remains unclear, and is likely to vary from species to species, depending on extrinsic and intrinsic factors (i.e., the number of restriction sites and average length of restriction fragments surveyed, the age of the species and its populations, and their rate of cpDNA evolution). Recent data suggest that point mutation rates in cpDNA may vary several fold among closely related taxa (Palmer et al., 1988b), whereas rates of rearrangements and length mutations have been shown to be highly variable, principally due to differences in the amount of short dispersed and tandem repeats, respectively (Palmer et al., 1985, 1987).

Animal Mitochondrial DNA   Considerable attention has been given to the analysis of RE site variation in animal mtDNA within and among populations (reviewed by Avise and Lansman, 1983; Brown, 1983; Wilson et al., 1985; Avise, 1986; Birley and Croft, 1986; Avise et al., 1987; Moritz et al., 1987). Animal mtDNA is easily isolated, evolves relatively rapidly, and is maternally inherited without recombination, making it a valuable marker for the study of variation within and among populations. Much of the variation in animal mtDNA is due to base substitution, with transitions greatly outnumbering transversions. Recent analyses of mtDNA from phylogenetically diverse animals have shown that length variation is also common with differences occurring within (**heteroplasmy**) as well as between individuals (reviewed in Moritz et al., 1987).

Analyses of levels of RE site variation have demonstrated that mtDNA varies within populations (Brown, 1980; Avise and Lansman, 1983; Cann et al., 1984) and may be useful as a marker for tracing maternal lineages. Variants detected at this level typically have low levels (<1%) of sequence divergence (reviewed by Avise et al., 1987) and are best detected by digestion with 4-bp REs that sample a larger fraction of the genome. Despite the low levels of sequence divergence within populations, the levels of variation generally found within populations allow for discrimination of virtually every individual sampled (Brown, 1980; Cann et al., 1984; Avise et al., 1989; A. A. Echelle et al., 1989). Kessler and Avise (1985a) attempted to trace maternal lineages within a population of cotton rats using mtDNA RFLPs. They found significant spatial heterogeneity in the distribution of variants, but raised two important cautions regarding interpretation of the results. First, immigration of individuals with genotypes previously found within the study area will provide false indications of relationship. Therefore, such studies are best performed on isolated demes. Second, the segregation of heteroplasmic restriction site (Hauswirth and Laipis, 1985) and length (Solignac et al., 1984; Rand and Harrison, 1986b) variants within lineages may also obscure matrilineal relationships.

Levels of differentiation among populations tend to be significantly higher than those within populations (Dowling and Brown, 1989; reviewed in Avise et al., 1987; Moritz et al., 1987). Because of this pattern, mtDNA can be used to estimate phylogenies of populations, and thus to investigate patterns of historical biogeography (e.g., Bermingham and Avise, 1986; Bowen and Avise, 1989; reviewed by Avise et al., 1987). This phylogenetic approach provides a qualitative assessment of genetic population structure within species. A more quantitative approach, using fixation indices (e.g., Takahata and Palumbi, 1985) or alternative statistics (see Chapter 10), is needed where the variation within populations is substantial relative to that among populations. Using F statistics, DeSalle et al. (1986, 1987b) found significant heterogeneity for mtDNA among populations of *Drosophila* that were homogeneous for allozymes.

The unique characteristics of mtDNA also result in some disadvantages. The lack of recombination makes mtDNA comparable to a single allozyme locus with many alleles. Consequently, estimates of gene diversity obtained from mtDNA are expected to exhibit larger standard errors than comparable estimates made using a large number of nuclear loci. The extremely rapid evolution of some mtDNA genes can result in convergence, confounding phylogenetic relationships even within some species (Aquadro and Greenberg, 1983; Lansman et al., 1983; Dowling and Brown, 1989). This problem seems to be particularly severe for length variants that result from variation in copy number of tandemly repeated sequences (reviewed by Moritz et al., 1987). In such a situation, the number of character states is discrete and finite and, with a high mutation rate, the likelihood of convergence is high. In general, length variants should be used as markers for the analysis of population subdivision only when there is evidence for their stable inheritance, and even then with caution.

Nuclear Sequences   Sequences in the nuclear genome provide an effectively inexhaustible supply of genetic markers if they can be accessed. Attention has so far focused on single-copy sequences, particularly those that are hypervariable, and repeated sequences such as rDNA cistrons. These vary widely in the form and rate of mutation, which has important implications for how they are used.

Hypervariable (VNTR) Sequences.   The recent discovery of hypervariable sequences and their use in "DNA fingerprinting" may well revolutionize the analysis of population-level variation, particularly the assessment of parentage, and will therefore contribute to studies of sexual selection, mating behavior, and population ecology (Jeffreys et al., 1985a,b; Burke and Bruford, 1987; Jeffreys and Morton, 1987; Wetton et al., 1987; Dallas, 1988; Burke et al., 1989; reviewed by Burke, 1989). Hypervariable sequences are each present in low copy number and generally include a series of tandem repeats of a short minisatellite sequence that is itself dispersed throughout the genome (Jeffreys et al., 1985a, 1987). The hypervariable sequences so far described are quite heterogeneous, coming from several unrelated minisatellite families (Jeffreys et al., 1985a, b, 1987; Nakamura et al., 1987). Some hypervariable loci contain a sequence similar to one found in the phage M13 (Vassart et al., 1987; Georges et al., 1988; Dallas, 1988;

Rogstad et al., 1988), and Epplen (1988) described hypervariable loci revealed by hybridization to a (GA[T/C]A)^n oligonucleotide. Mutation rates at these loci vary widely, and may be as high as 10^{-3} per gamete per generation, with numbers of alleles and heterozygosity approaching the number of individuals and one, respectively (Jeffreys et al., 1988). This extensive variation is mainly due to differences in copy number of the tandem repeats, probably generated by a combination of unequal crossing over and replication slippage (Jeffreys et al., 1985a).

Hypervariable sequences are analyzed by digestion with REs (those that do not cleave within the tandem repeats) and transfer hybridization, using minisatellite sequences, an entire hypervariable sequence, or synthetic oligonucleotides as probes. There are two distinct strategies. One option is to use the minisatellite probes to simultaneously reveal variation at a large number of hypervariable loci (e.g., Jeffreys et al., 1985b). The result is complex multifragment patterns that are usually unique to an individual and are extremely powerful for testing parentage, where the putative parents can be lined up next to the individual in question. A major advantage of the method is that many of the probes can be applied across a wide spectrum of plants and animals. The alternative approach is to assay hypervariable loci one at a time using synthetic oligonucleotides as probes (e.g., Nakamura et al., 1987). This approach is more laborious, but has the major advantage that alleles can be assigned to specific loci and genotypes identified. A possible alternative to the above two methods was reported by Boerwinkle et al. (1989), who used PCR amplificiation to assay variation in the hypervariable human apolipoprotein gene. This information can then be used for extremely high-resolution analyses of genetic population structure as well as testing parentage.

Although comparison of the multifragment patterns generated by the minisatellite probes remains the method of choice for testing parentage, there are several technical and statistical difficulties with using this method for broader problems, such as estimating the relatedness of randomly sampled individuals (see Lynch, 1988). These include (1) assigning specific fragments to a particular locus and thus identifying alleles and determining genotypes, (2) potential comigration of nonhomologous fragments (convergence), inflating the variance of the estimate of similarity, (3) obtaining a complete picture of the variation because of the inability to reliably detect small fragments (see also Jeffreys et al., 1987), and (4) correlations among loci due to linkage (Lynch, 1988). Use of high stringency probes to assay variation at single loci overcomes the first three difficulties, but applications are currently limited by availability of suitable probes.

Other Single-Locus Sequences.   Variation within and between populations has been examined for unique sequences of unknown function (e.g., Quinn and White, 1987b; Figueroa et al., 1987) and for genes of known function [e.g., alcohol dehydrogenase (ADH), Aquadro et al., 1986; Kreitman and Aguade, l986; Simmons et al., 1989; H-2 polymorphism, Nadeau et al., 1988; and globins, Flint et al., 1986] by RFLP analysis. These studies have typically revealed a wealth of polymorphism and have provided information on the evolutionary history of the populations, on the action of selection and drift on the sequences concerned, or both. For example, Quinn and colleagues selected random unique sequence probes from a genomic library of the lesser snow goose *Anser caerulescens* (Quinn and White, 1987b) and, using these, demonstrated multiple maternity and paternity within broods (Quinn et al., 1987). Each of the 17 probes selected revealed useful polymorphism and analysis of selected broods confirmed the identification of genotypes and established the Mendelian inheritance of the markers. This approach circumvents many of the pitfalls associated with DNA fingerprinting (see above) and has considerable potential for the analysis of natural populations.

Repeated Nuclear Genes.   Because of their greater ease of use, multicopy nuclear genes, such as those encoding subunits of ribosomal RNA (reviewed in Long and Dawid, 1980; Jorgensen and Cluster, 1988), have received more attention than single-copy sequences. In eukaryotes, the rDNA repeat unit typically consists of three coding regions (18 S, 28 S, and 5.8 S), internal and external transcribed spacers, and an external nontranscribed spacer region, and is typically associated with the nucleolar organizer regions (NORs). Plants have a relatively large number of rDNA copies, with most species examined possessing >1000 copies. Except for some amphibians, animals usually have fewer copies of the rDNA cistron.

Studies of rDNA restriction site variation in plants and animals indicate that coding sequences are conserved, whereas the spacer regions are variable (Long and Dawid, 1980; Appels and Honeycutt, 1987; Hillis and Davis, 1986, 1987; Jorgensen and Cluster, 1988). Variation exists both within and among individuals and populations and is usually the result of variation in copy number of short, repeated sequences found in the nontranscribed spacer (NTS) region (Arnheim, 1983; Saghai-Maroof et al., 1984; Williams et al., 1985, 1987; Flavell et al., 1986; Schaal et al., 1987; Suzuki et al., 1987), although restriction site variation has also been documented within and among individuals (Arnheim, 1983; Wilson et al., 1984; Schafer and Kunz, 1985). At this time, rDNA has seen limited use for analysis of population level variation. These variants may reveal population subdivision (e.g., Learn and Schaal, 1987; Templeton et al., 1990). However, as noted above (see under Choice of Sequence) any concerted evolution among chromosomes invalidates use of this variation to quantify the level of gene flow.

Analysis of Hybrid Zones

RFLPs have been invaluable for the analysis of hybridization between distinct forms (e.g., Shaw et al., 1988; Baker et al., 1989; Dowling et al., 1989; Rand and Harrison, 1989; Sites and Davis, 1989). The sequences most frequently utilized up to this point have been rDNA and mtDNA, mostly due to their ease of application. mtDNA is particularly useful for these types of studies, with the strict maternal inheritance providing a means for identifying the maternal form involved in the production of hybrids and the assessment of directionality of hybridization and introgression. Y-chromosome-specific sequences can provide a similar haploid marker for tracing the male contribution, although their use has been limited (VanlerBerghe et al., 1986). It is important to remember, however, that haploid markers such as these require the use of some other character for identification of hybrids and determination of the extent of hybridization. Allozymes (Chapter 4) and rDNA have proven to be excellent markers for the identification of hybrids. For markers such as rDNA, it is important to keep in mind the effect concerted evolution could have on the distribution of variants within individuals and populations and the estimation of deviations of observed numbers of hybrids relative to those expected.

Species-Level Comparisons

As with other levels of comparison, the ideal is to find characters that vary among, but not within, the groups (species) being studied. Further, the differences among groups should not be so large that convergences and parallelisms obscure the true phylogeny. The choice of sequence for analysis is critical to achieving this balance. Regions with rapid evolutionary rates and moderate to low intraspecies polymorphism are most appropriate for analyzing relationships of closely related species, whereas those with slower evolutionary rates may provide useful characters for studying relatively ancient divergences. It should be restated that the phylogenies produced are of the molecules, and may differ from the organismal phylogeny for various reasons including introgression, gene conversion, and sorting of polymorphism.

Animal Mitochondrial DNA   The application of mtDNA RFLPs to phylogenetic analysis of congeneric species has been reviewed extensively (Brown, 1983; Avise and Lansman, 1983; Wilson et al., 1985; Birley and Croft, 1986; Avise, 1986; Moritz et al., 1987). In general, the approach has proved useful for resolving relationships of closely related species. Phylogenetic analysis of mtDNA restriction sites has also identified the bisexual species that acted as the maternal parent of hybrid-parthenogenetic species (Brown and Wright, 1979; reviewed in Moritz et al., 1989b).

The main problems encountered in such studies stem from sorting of polymorphism where recently separated species are being compared, and from high levels of noise (homoplasy) where distantly related species are examined. Using simulation studies, Neigel and Avise (1986) showed that sequences from recently separated monophyletic sister taxa appear poly-phyletic initially, then appear paraphyletic, and then monophyletic as the original polymorphic lineages are terminated and replaced by variants unique (i.e., apomorphic) to each taxon. The simulations indicated that, for a haploid marker such as mtDNA, this process may take $4N$ generations, where N is the effective population size. However, the time frame is also likely to be affected by the amount and distribution of polymorphism within each species, the geographic mode of speciation, and the demo-graphic history of the two species (see also Avise et al., 1984, 1988). This problem is not restricted to recently separated taxa. Theoretical studies (Pamilo and Nei, 1988) indicate that if an ancestral taxon was highly polymorphic and multiple speciation events occurred over a short time relative to effective population size, then the probability of obtaining the correct topology from a single sequence is low. This has undoubtedly contributed to the debate over the phylogeny of higher hominoids as de-duced from mtDNA and other sequences (reviewed by Holmquist et al., 1988). There seems to be no obvious solution to the problem of polymorph-ism. However, it does stress the need for adequate geographic sampling for phylogenetic analyses. If there is a strong geographic component to the intraspecific polymorphism, inadequate sampling may lead to erroneous phylogenies.

Homoplasy can be a substantial problem where distantly related taxa are compared (cf. Carr et al., 1987). This is particularly so where compar-isons are restricted to fragment sizes rather than mapped cleavage sites (e.g., Honeycutt et al., 1988). The upper limit to useful RFLP comparisons of mtDNA presumably is set by constraints on sequence evolution. Sequence comparisons indicate that primate mtDNAs reach a plateau of sequence divergence at about 25% (Brown et al., 1982). Surprisingly, *Drosophila* mtDNA sequences seem to plateau at only 8% divergence (DeSalle et al., 1987a). Once these levels are reached, further base substitutions are con-centrated at positions that have already changed, which is likely to increase homoplasy among RFLPs.

Where homoplasy does appear to be obscuring relationships, it may be possible to improve the signal-to-noise ratio by restricting comparisons to a slowly evolving region. Obviously, this approach can be used only where variable cleavage sites have been mapped in relation to gene order (which can also vary; see below).

Chloroplast DNA   Nucleotide sequence divergence values for cpDNAs of congeneric species typically range up to 2.0% (see references in Palmer,

1987; Palmer et al., 1988b; Crawford, 1989). Given a typical genome size of 150 kb, sampling with 10–20 REs that cleave from 20 to 100 times each will allow coverage of 1–5 kb of sequence, which is usually adequate to produce a highly resolved phylogeny. Such phylogenies have thus far been remarkably untroubled by problems of homoplasy (0–5%) and have contributed to a better understanding of a host of phylogenetic problems, including the identification of crop plant origins from wild species, identification of the maternal and paternal ancestry of a number of hybrid and polyploid species, detection of unsuspected cases of introgression, and identification of the progenitor genus of a putatively monotypic, morphologically isolated genus (reviewed in Palmer, 1987; Palmer et al., 1988b; Crawford, 1989).

Early studies of cpDNA restriction site variation within a genus were accomplished by direct inspection of restriction fragment patterns of purified cpDNA. However, most current efforts use a transfer hybridization approach in which cloned cpDNA fragments are hybridized sequentially to filter blots containing digests of genomic DNA (Protocols 7 and 8). Although more laborious, this approach has two main advantages. The use of total DNA as compared to cpDNA has major advantages with respect to yield (therefore much less starting material is required) as well as extraction flexibility and adaptability (see Palmer et al., 1988b for fuller discussion). By probing with cloned portions of the chloroplast genome, the complexity of the fragment patterns is greatly reduced, allowing a more critical analysis of fragment differences in terms of discrete mutations and often permitting the direct mapping of restriction fragments and sites. Fortunately, many complete clone banks are readily available for a wide range of land plant cpDNAs (reviewed in Palmer, 1986a; Palmer et al., 1988b).

Nuclear DNA   Some single-copy nDNA sequences have been compared among species by RFLP analysis (e.g., ADH among *Drosophila;* Langley et al., 1981; Bishop and Hunt, 1988), but the data are too few for particular advantages and limitations to be identified. RFLP analysis of multigene families is exemplified by studies of globin variation among primates (Zimmer et al., 1980; Barrie et al., 1981). Analysis of multigene familes requires particular care when using heterologous probes as low stringency hybridization is likely to detect variation in duplicate copies as well as the target sequence. It then becomes important to distinguish between variation in orthologous (shared by descent) and paralogous (duplicate) copies for phylogenetic analysis. Even if this distinction can be made (e.g., by relative intensities of hybridization, Barrie et al., 1981), gene conversion among members of a multigene family (e.g., Slightom et al., 1987) could still cause the gene tree to differ from the species tree.

Of repeated genes, the rDNA cistrons have been used most widely in interspecific comparisons (e.g., Coen et al., 1982; Wilson et al., 1984; Arnold et al., 1987a). The variation revealed in these studies was typically, though not exclusively, in the transcribed or nontranscribed spacers and was due to length mutations or to the gain/loss of cleavage sites. The phylogenetic information obtained from these studies has typically been consistent with previous studies. However, Arnold et al. (1987b) found that divergence of a highly repeated sequence was inconsistent with other evidence on the relationships among subspecies of *Caledia captiva,* and attributed the discrepancy to historical introgression.

Higher-Level Systematics

In contrast to sequence data (see Chapter 9), there have been relatively few applications of the RFLP approach to higher-level systematics. Investigations at this level have used both changes in cleavage sites and gross structural rearrangements as characters for phylogenetic analysis.

Animal Mitochondrial DNA Although sequence evolution of animal mtDNA is typically rapid, certain aspects are highly conserved. These include gene order, genetic code, and the secondary structure of tRNA and rRNA sequences (reviewed in Attardi, 1985; Wolstenholme et al., 1985; Moritz et al., 1987). The order of mtDNA genes varies among phyla, with the position of tRNA genes more variable than other coding sequences (reviewed in Moritz et al., 1987). There are some indications of minor variations (e.g., tRNA transpositions or conversions, Dubin et al., 1986; Cantatore et al., 1987; Haucke and Gellissen, 1988) within classes or phyla, making it imperative to further investigate within-group diversity before applying gene order as a tool for estimating relationships among phyla.

Aside from structural changes, some coding sequences (reviewed in W. M. Brown, 1985) may be conservative enough to provide characters useful for phylogenetic analysis among genera. However, because of the small size of many of these sequences, they will have to be characterized with 4-bp REs (Kreitman and Aguade, 1986) or sequenced (Chapter 9) to make a substantial contribution.

Chloroplast DNA Comparative restriction site mapping of cpDNA (Protocol 8) can be successfully employed at the highest levels within many families of flowering plants (e.g., Jansen and Palmer, 1988). In general, this approach cannot be used above the family level, or across diverse families such as the Fabaceae and Onagraceae. This restriction is due to excessive DNA divergence, both in sequence and in structure.

Two other approaches for extracting phylogenetic information from cpDNA at higher levels are DNA sequencing and rearrangement analysis. The general practice of DNA sequencing is discussed in detail in Chapter 9 and its application to cpDNA, in particular the chloroplast *rbc*L gene, is discussed by Palmer et al. (1988b). Rearrangements that are useful as higher level characters are major events such as (1) inversions, (2) deletions/insertions of introns, (3) partial or complete deletions/insertions of genes, and (4) deletion of a segment of the large inverted repeat found in most chloroplast genomes. Major rearrangements are quite rare relative to nucleotide substitutions and therefore cannot be expected to produce by themselves a fully resolved phylogenetic tree. However, their very rarity and lack of homoplasy relative to substitutions make each rearrangement a character that should be weighted more heavily than a single substitution.

Gene and intron losses/gains are detected by a simple presence/absence test based on filter hybridization experiments (e.g., Figure 3). Inversions and losses of the large inverted repeat are detected by hybridization assays that analyze linkage relationships between two or more probe fragments from the ends of the inversion or deleted repeat segment (Palmer et al., 1988b; for examples, see Jansen and Palmer, 1987a,b; Lavin et al., 1990).

Nuclear DNA   Given the vast complexity of the nuclear genome, many nuclear genes should be useful for inferring higher order phylogenetic relationships. The best example of the broad utility of nuclear genes is the rDNA repeat unit, which has intervening sequences that vary within and between populations (see above) and coding sequences that are so highly conserved as to be useful for comparisons among kingdoms (Pace et al., 1986; reviewed in Chapter 9). However, despite the potential of the approach, there has as yet been virtually no use made of nDNA RFLPs to investigate relationships among genera or higher taxa other than the studies of ribosomal (Hillis and Davis, 1987) and globin genes (Barrie et al., 1981) discussed above.

LABORATORY SETUP

Major equipment items needed for RFLP analysis are included in Table 1 of Chapter 9. The most expensive of these is an ultraspeed centrifuge and appropriate rotors. Although it is possible to carry out many of these analyses without it, an ultraspeed centrifuge greatly extends the types of analyses that can be performed. Other essential items include an autoclave (or access to one), a fume hood, and a source of high-purity water. Single-distilled or deionized water can be used for rinsing glassware and making up electrophoresis buffers, but solutions used for preparing or manipulating DNA require even greater purity, i.e., double-distilled or distilled–deionized water. Standard laboratory items that are used include glassware, e.g., various sizes of beakers, graduated cylinders, pipettes, Erlenmeyer flasks, side-arm flasks, and bottles. High-strength, acid/solvent resistant centrifuge

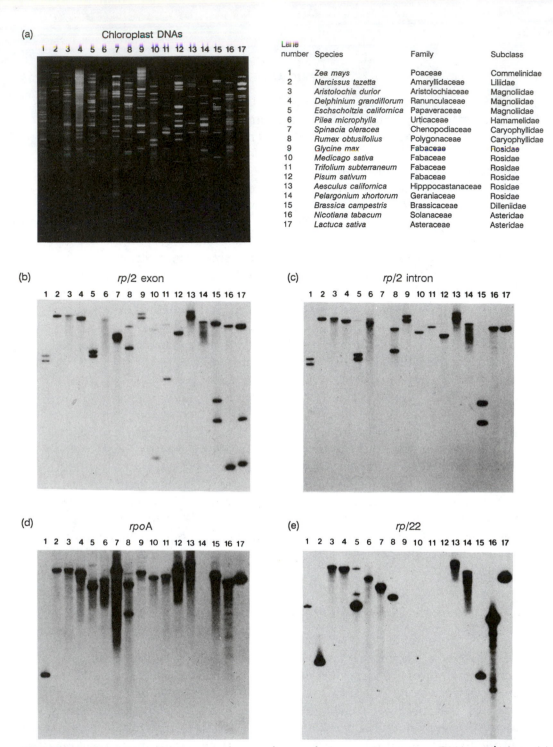

(a) Chloroplast DNAs

Lane number	Species	Family	Subclass
1	*Zea mays*	Poaceae	Commelinidae
2	*Narcissus tazetta*	Amaryllidaceae	Liliidae
3	*Aristolochia durior*	Aristolochiaceae	Magnoliidae
4	*Delphinium grandiflorum*	Ranunculaceae	Magnoliidae
5	*Eschscholtzia californica*	Papaveraceae	Magnoliidae
6	*Pilea microphylla*	Urticaceae	Hamamelidae
7	*Spinacia oleracea*	Chenopodiaceae	Caryophyllidae
8	*Rumex obtusifolius*	Polygonaceae	Caryophyllidae
9	*Glycine max*	Fabaceae	Rosidae
10	*Medicago sativa*	Fabaceae	Rosidae
11	*Trifolium subterraneum*	Fabaceae	Rosidae
12	*Pisum sativum*	Fabaceae	Rosidae
13	*Aesculus californica*	Hipppocastanaceae	Rosidae
14	*Pelargonium xhortorum*	Geraniaceae	Rosidae
15	*Brassica campestris*	Brassicaceae	Dilleniidae
16	*Nicotiana tabacum*	Solanaceae	Asteridae
17	*Lactuca sativa*	Asteraceae	Asteridae

(b) *rpl*2 exon

(c) *rpl*2 intron

(d) *rpo*A

(e) *rpl*22

FIGURE 3. Detection of intron and gene losses during angiosperm cpDNA evolution. (a) Electrophoresis in a 0.9% agarose gel of cpDNA fragments produced by digestion with *Eco*RI (lanes 1 and 11), *Sac* I–*Pvu*II (lanes 2–6, 8, 9, 13, and 14), *Sac*I– *Pst* I (lanes 7, 12, and 15), and *Hind*III (lanes 10, 16, and 17). Filter replicas of the gel were made by bidirectional blotting and were then hybridized sequentially with the gene probes indicated in panels b–e. (b) Hybridization with a 772-kp fragment internal to and containing 90% of the coding region of the *rpl*2 gene from spinach. (c) Hybridization with a 545 fragment internal to and containing 82% of the intron of the *rpl*2 gene from tobacco. (d) Hybridization with a 1040-bp fragment containing 96% of the coding region of the *rpl*A gene from spinach and 78 bp of 5′ noncoding sequence. (e) Hybridization with a 209-bp fragment internal to and containing 45% of the coding region of the *rpl*22 gene from tobacco. *Reprinted from Palmer et al. (1988b).*

tubes are needed for many applications. Expendable supplies include pipette tips, pasteur pipettes, and microcentrifuge tubes.

Reagents should generally be of reagent grade or better, although there are some exceptions (see below). In particular, chemicals used in the preparation and manipulation of DNA must be of high quality, as must the media used for electrophoresis. Commonly used reagents include Tris-base, sodium chloride, ethylenediaminetetraacetic acid (EDTA, disodium, dihydrate), sucrose, sodium dodecyl sulfate (ultrapure), cesium chloride (technical grade), propidium iodide, light mineral oil, hydrochloric acid, sodium hydroxide, isopropyl or isobutyl alcohol, proteinase (e.g., proteinase K or "pronase"), phenol (ultrapure), chloroform, isoamyl alcohol, ethidium bromide, RNase A, DNase I, DNA polymerase (Klenow fragment), DNA polymerase (Kornberg enzyme), restriction enzymes (see Table 2), bovine serum albumin (crude for addition to hybridization solutions, ultrapure for other applications), ethanol, sodium acetate, 2-mercaptoethanol (BME), sorbitol, hexadecyltrimethylammonium bromide (CTAB), ammonium acetate, potassium chloride, dithiothreitol, agarose (ultrapure), acrylamide (ultrapure), bisacrylamide (ultrapure), ammonium persulfate (ultrapure), NNN'N'-tetramethylethylenediamine (TEMED), boric acid, and sodium acetate.

PROTOCOLS

1. Isolation of animal mtDNA using CsCl-PI gradients
2. Isolation of cpDNA using sucrose step and CsCl-EB gradients
3. Digestion of DNA with restriction endonucleases
4. Agarose and polyacrylamide electrophoresis
5. Staining with ethidium bromide
6. ^{32}P end-labeling
7. Transfer hybridization
8. Construction of cleavage maps

DNA Isolation

The optimal method of DNA isolation depends on the type of sequence to be assayed, the level of resolution desired, and the type and condition of tissues (Figure 2). Protocols for isolating total cellular DNA from plants and animals are given in Chapter 9 and a method for large-scale DNA isolation is given in Chapter 7. Here, we concentrate on methods for purifying organellar DNAs. These protocols employ CsCl gradients in conjunction with the intercalating dyes, propidium iodide (PI) or ethidium bromide (EB). mtDNA is obtained by lysing an enriched mitochondrial preparation and is separated from contaminating nDNA on the basis of conformation. Supercoiled mtDNA molecules bind less of the lower density dye than does linear DNA. The supercoiled molecules therefore have higher density in the presence of dye and band below the linear nuclear (and damaged mitochondrial) DNA in the appropriate CsCl gradient (Smith et al., 1971). Unlike mtDNA, cpDNA is large enough that effectively all the molecules are damaged such that they band with linear (nuclear) DNA; hence, CsCl gradients are used here to purify DNA only relative to other kinds of molecules (i.e., proteins). Techniques for isolation of fungal and plant mtDNA are given in Hauswirth et al. (1987).

Protocol 1: Isolation of Animal mtDNA Using CsCl-PI Gradients

(Time: Part A, 2–3 hr; Part B, 20–36 hr; Part C, 15–30 min; Part D, 4 hr; Part E, 20 hr; Part F, 24 hr. Total: 2–4 days)

The steps in this protocol reduce to two basic operations: (1) the preparation of a mitochondrially enriched fraction from a cell or tissue homogenate by differential centrifugation, and (2) the further purification of mtDNA from nDNA using CsCl-PI gradients. When using tissues unusually rich in mitochondria (e.g., amphibian or fish oocytes, avian cardiac muscle) the DNA obtained from repeated differential centrifugation (repeat steps 3–6, Part A) may be adequately purified for mtDNA for most applications. However, in most cases it will be necessary to proceed to the CsCl-PI gradient. A single gradient (Part B) will usually suffice, although to obtain maximum purity (e.g., for use as a probe), a velocitization step (Part D) and a second CsCl-PI gradient (Part E) are added. The velocitization differs from equilibrium gradients in that molecules are pelleted at a rate proportional to their size and largely independent of conformation. Thus, intact mtDNA (and large nDNA fragments) can be purified away from smaller DNA fragments, RNA, and proteins. Velocitization can also be used to clean up partially degraded samples and to remove any contaminating DNases.

The yield and purity of mtDNA is highly dependent on tissue type and condition. Fresh tissues generally provide good yields ($\cong$ 1 μg/g of tissue) so that amounts adequate for >50 digests (using end-labeling) can be obtained from relatively small amounts of tissue (<250 mg). The best source for animal mtDNA is unfertilized eggs; heart, liver, kidneys, gonads, and brain also provide good yields. For crustaceans, heart and pleopod muscle provide the best yields, but should be gently homogenized. Likewise, adductor muscle of bivalves and flight muscle of insects are adequate; all striated muscle should be homogenized and centrifuged in high volumes. mtDNA has been successfully isolated from the white blood cells obtained from 200–250 ml of whole mammalian blood (W. M. Brown, personal communication) and 5–10 ml of whole reptile blood (L. D. Densmore, personal communication). Frozen tissues typically yield about half the amount of mtDNA relative to fresh material; this may be due to rupturing of mitochondrial membranes, exposing the mtDNA to cytosolic DNases and reducing the efficiency of enrichment. Tissues should be removed from the freshly killed specimen and snap frozen (<−70°C). Alternatively, if mtDNA is to be prepared within a few days, it may be preferable to store tissues at 4°C in STES buffer (Appendix). However, storage of tissue in this buffer softens tissue considerably, making membranes more susceptible to breakage. This may be a function of the high concentration of EDTA, since Avise and co-workers (Lansman et al., 1981; Ball et al., 1988) report good results using their buffer, which contains less EDTA. Therefore, this strategy should be tested for the different combinations of tissue, species, and buffers. Yields from ethanol-preserved tissues are poor, possibly because of damage to the mitochondrial membranes (S. Palumbi, personal communication).

Part A. Preparation of Crude mtDNA

1. Sacrifice or, if frozen, partially thaw animals and remove tissues. If using only cells (i.e., blood), pellet and begin at step 7.
2. Homogenize thoroughly in cold STES buffer (Appendix: 12 ml total/g tissue, 12 ml minimum). The concentration of EDTA may be adjusted, depending on levels of DNase activity. EDTA inhibits DNases by chelating divalent cations required for their function. A good starting concentration is 100 mM EDTA. Isolations of mtDNA from organisms with high levels of DNase activity (i.e., mollusks) have been more successful using 200 mM EDTA in their grinding buffer, whereas initial studies of teiid lizards and terrestrial mammals worked well with 1 mM EDTA. It is important to note that increasing EDTA concentration decreases the stability of membranes. High EDTA concentrations limit the loss of mtDNA due to degradation, but mtDNA is lost due to membrane breakage and inability to recover the molecules from the supernatant. Therefore, it may be necessary (particularly when working with small amounts of tissue) to determine empirically which EDTA concentration provides the best yields.
3. Centrifuge homogenate for 5 min at 1200 g, 4°C, to pellet nuclei and large cellular debris. This pellet may be saved for nuclear DNA extraction. Repeat this step, when using large samples (>1 g) until pellet is the same size in two consecutive spins.
4. Transfer the supernatant to a 50-ml polypropylene or polyallomer screw cap centrifuge tube. Centrifuge at 23,000 g, 4°C, for 20 min to pellet mitochondria and other remaining cellular debris. Decant supernatant and drain pellet.
5. (Optional: For large amounts of tissue) Purify mitochondrial fraction on a 1.0 M/1.5 M sucrose step gradient as follows:
   a. Resuspend pellet in 20 ml 0.25 M sucrose (in ThE; see Appendix).
   b. Make the sucrose gradient by underlayering 10 ml of 1 M sucrose (in ThE) with 8 ml of 1.5 M sucrose.
   c. Carefully overlayer the sample onto the gradient.
   d. Centrifuge at 25,000 rpm (81,000 g), 4°C, for 1 hr (no brake) in a Beckman SW28 rotor (or equivalent).
   e. After centrifugation, aspirate off the top of the gradient and carefully remove the mitochondrial fraction (appears as a band at the 1.0–1.5 M interface; Figure 4).
   f. Resuspend mitochondrial fraction in three volumes of ThE and centrifuge at 23,000 g to pellet.
6. Resuspend the pellet (from step 4 or 5f) in 1.0 ml ThE at room temperature and mix vigorously. If the pellet volume is greater than 0.3 ml, resuspend in 4 volumes of ThE.

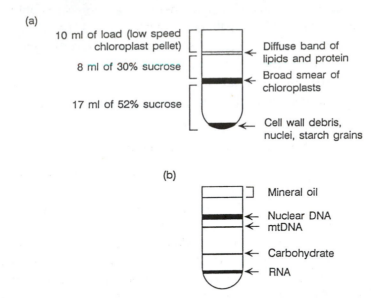

FIGURE 4. Purification of chloroplasts and DNA by gradient centrifugation. (a) Sucrose step gradient purification of chloroplasts. The sucrose step gradient purification of mitochondria appears the same, with the mitochondria banding in the same position as depicted for the chloroplasts (1.0–1.5 *M* sucrose interface). (b) CsCl-propidium iodide density gradient purification of mtDNA, prepared as a step gradient (1.40–1.70 g/ml). For the alternative method (1.55 g/ml), the RNA will pellet at the bottom of the tube.

7. Add 0.125 ml (1/8 resuspended volume) 20% SDS (w/v in H_2O) to lyse membranes, mix gently, and leave at room temperature for at least 10 min.
8. Add 0.188 ml (1/6 volume) CsCl-saturated water to precipitate nuclear DNA–SDS–CsCl, mix gently, and place on ice for at least 15 min. Larger samples (>1 g tissue) may require longer incubation times (i.e., overnight) to complete precipitation. (This mixture can be stored at 4°C overnight or longer at this point.)
9. Centrifuge at 17,000 g, 4°C, for 10 min. Transfer supernatant to an ultracentrifuge tube (if continuing) or a 5-ml culture tube (for storage). If the liquid is extremely viscous, force the solution through a 30-gauge needle six times to shear any nuclear DNA. (Do not shear the DNA if you plan on saving nuclear DNA.) This solution can be stored at −20°C after adding CsCl (step 1, Part B).

Part B. Ultracentrifugation of CsCl-PI Gradient

If the sample volume is less than 1.5 ml, mtDNA can be isolated in a preparative ultracentrifuge (e.g., Beckman SW60Ti rotor or equivalent) using the rapid step-gradient method (steps 1–4). If smaller gradients are run (e.g., in a Beckman TLS-55 rotor), or if the sample volume is greater than 1.5 ml, follow steps 5–6.

NOTE: In this protocol, the intercalating dye is propidium iodide (PI) instead of ethidium bromide (EB). More PI can intercalate into a closed-circular molecule, allowing for visualization of smaller amounts of mtDNA. EB may also be used for isolation of mtDNA; however, higher dye concentrations are required for the same result. These dyes are mutagenic, so wear gloves whenever using them.

1. Measure the sample volume and add the appropriate amount of solid CsCl to adjust the sample density to 1.40 g/ml (Table 4), i.e., 0.53 g of CsCl/ml of sample plus 0.12 g to account for the dye volume to be added (0.23 ml of 2 mg/ml PI in TE). Samples may be stored for months at $-20°C$ by adding only the CsCl. The PI is added just prior to ultracentrifuging.
2. Add 0.23 ml of 2 mg/ml PI stock (in TE). Check the density of each sample by (a) repeatedly weighing 1 ml of the solution, (b) accurately measuring the sample volume and weighing the sample, or (c) using a refractometer. Adjust to 1.40 g/ml by addition of distilled–deionized water (if too heavy) or solid CsCl (if too light).
3. Place samples in ultracentrifuge tubes and check the volume of each. To form a step gradient, carefully underlayer the sample with 1.33 ml of 1.70 g/ml solution (Appendix) per ml of sample (see Table 4). Overlayer the step gradient with mineral oil to within 1–3 mm of the top (there should be at least 2 mm oil). Balance paired tubes (i.e., those to be in buckets directly opposite one another) to within $\pm$ 0.02 g.
4. Put tubes into rotor buckets, carefully hook buckets onto rotor, and place rotor on drive shaft. For a Beckman SW60Ti rotor or equivalent, set run parameters to temperature = 21°C, maximum temperature = 35°C (if adjustable), speed = 36,000 rpm (140,000 g), and running time = 20–24 hr. The running time depends on the amount of DNA in the sample; the more DNA, the longer it takes for the sample to attain equilibrium. Larger samples may require more than 24 hr to reach equilibrium. Now go to step 8 (see below).
5. For small volume gradients (total <2.5 ml, to be run in a Beckman TLS-55 rotor or equivalent) or for large initial volumes (>1.5 ml of sample, to be run in a Beckman SW60Ti rotor or equivalent), measure volume of supernatant from Part A and adjust density to 1.52–1.57 g/ml by adding the amount of CsCl indicated in Table 5 (this includes the volume of PI to be added later).

Table 4. Approximate amounts of PI and CsCl to adjust sample densities to 1.40 g/ml[a]

$V_{initial}$ (ml)	PI (ml)	CsCl (g)
1.0	0.23	0.65
1.1	0.23	0.71
1.2	0.23	0.76
1.3	0.23	0.81
1.4	0.23	0.86
1.5	0.23	0.92

[a]See step 1 under Part B of Protocol 1 (Isolation of animal mtDNA). $V_{initial}$ is the volume of sample prior to the addition of PI and CsCl.

Table 5. Approximate amounts of PI and CsCl to adjust sample densities to 1.55 g/ml[a]

$V_{initial}$ (ml)	PI (ml)	CsCl (g)
1.0	0.21	0.93
1.1	0.23	1.01
1.2	0.25	1.11
1.3	0.27	1.20
1.4	0.29	1.29
1.5	0.31	1.39
1.6	0.33	1.48
1.7	0.35	1.57
1.8	0.37	1.66
1.9	0.39	1.76
2.0	0.41	1.85
2.1	0.43	1.94
2.2	0.46	2.04
2.3	0.48	2.13
2.4	0.50	2.22
2.5	0.52	2.32

[a]See Step 5 under Part B of Protocol 1 (Isolation of animal mtDNA). $V_{initial}$ is the volume of sample prior to the addition of PI and CsCl.

6. Just prior to centrifugation, add the amount of 2 mg/ml PI needed to bring final concentration to 350 mg/ml (Table 5) and mix. Measure the density of the solution. The final density should be 1.52–1.57 g/ml. If necessary, adjust by adding distilled–deionized water or solid CsCl.

7. Place samples in tubes and fill to within 1–3 mm of the top with light mineral oil. Balance tubes to within ±0.02 g of each other. Run parameters for a Beckman TLS-55 rotor (or equivalent) are 50,000 rpm (140,000 g), 21°C, and >20 hr. For the larger Beckman SW60Ti rotor (or equivalent), parameters are as in step 4, except that minimum run time is 36 hr.

8. To end the run, push "stop" with brake on and remove tubes from buckets.

Part C. Recovery of DNA

In room light, the nuclear DNA (actually all linear and relaxed circular DNA, i.e., including damaged mtDNA) should be visible as an intense red band; the band containing mtDNA, which is from 2 to 6 mm below the nuclear DNA band, will probably not be visible (Figure 4). Bands of carbohydrate are white to light pink in room light and may be present below the mtDNA band. RNA is found at or near the bottom of the gradient.

1. Wear safety glasses and gloves. Using a long-wave (305 nm) UV light source, locate the mtDNA band. If the mtDNA band is not visible, the area 2–6 mm below the main band should be collected.

2. Puncture the tube bottom with an 18- to 21-gauge syringe needle with a wire inserted in it (apparatus in Figure 5). Use a thin wire (e.g., guitar string) to dislodge (by pushing up) the small plastic plug that may clog the needle, then remove the wire from the needle. The flow can be regulated by placing a gloved finger over the top of the tube.

3. Collect the mtDNA fraction in a 1.5 ml microcentrifuge tube. If the mtDNA is to be further purified (Parts D and E) and the mtDNA bands are faint (or invisible), include the first drop of nuclear DNA as a reference point for further gradients. Otherwise, avoid contaminating the mtDNA fraction with any DNA from the top band. The top band DNA can also be collected and is usually adequate for transfer hybridization analysis of nuclear sequences. Proceed to extraction and dialysis (Part F), unless further purification (Parts D and E) is desired.

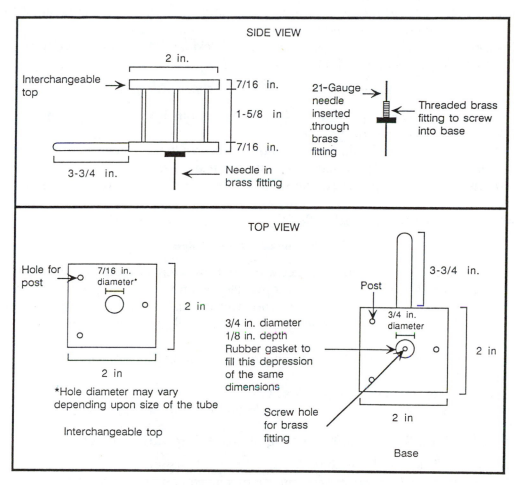

FIGURE 5. Apparatus for bottom puncture of ultracentrifuge tubes.

Part D. Velocity Centrifugation on a Step Gradient

1. Measure the volume of the sample collected from the equilibrium gradient and add an equal volume of TE (at least 2/3 of the sample volume) and mix. Addition of TE reduces the density of the sample below 1.40 g/ml, allowing it to be layered over the step gradient. The combined volume of sample and TE should be less than 1 ml.
2. The sample is overlayered onto a step gradient consisting of two layers, 0.7 ml of 1.70 g/ml solution (Appendix) and a quantity of 1.40 g/ml solution (Appendix) determined by the volume of the diluted sample. The amount of 1.40 g/ml solution is calculated using the following formula:

$$\text{volume (ml) of 1.40 g/ml solution} = 3.8 - \text{volume of diluted sample}$$
$$- \text{0.7 ml 1.70 g/ml solution}$$

3. Add the correct amount of 1.40 g/ml solution to an ultracentrifuge tube. Using a pasteur pipette, underlayer this with 0.7 ml of 1.70 g/ml solution.
4. Carefully layer the diluted sample on top of the gradient, add light mineral oil to within 1–3 mm of the top and balance tubes to within 0.02 g.
5. Put the tubes into rotor buckets and place rotor (Beckman SW60Ti or equivalent) into ultracentrifuge. Centrifuge at 45,000 rpm, 21°C for 3.5 hr, with no brake.

Part E. Sample Recovery and Final Equilibrium Gradient

1. Puncture tubes as in step 2, Part C. Collect the bottom 1.4 ml of the step gradient into a 1.5-ml microcentrifuge tube.
2. Put 1 ml of 1.55 g/ml solution (Appendix) into an ultracentrifuge tube, add the sample, and mix. Add light mineral oil and balance tubes as above.
3. Use the same centrifugation conditions as in step 7, Part B, with run time reduced to 18–20 hr.
4. Recover sample as described in Part C.

Part F. Extraction of Dye and Dialysis

1. To remove PI from a sample, extract with isopropyl alcohol (saturated with CsCl-saturated water—top layer is the isopropyl alcohol) and spin briefly in microcentrifuge. The saturated alcohol forms the top layer (pinkish from the dye) and is discarded after each extraction. Repeat this process until the sample (lower layer) is clear.
2. Place samples into 8-mm dialysis tubing (for preparation, see Appendix) and tie or clip tightly.
3. Dialyze against two changes of 2 liters 0.5× TE, for 24 hr.
4. Remove and store purified mtDNA (should be in 0.2–0.5 ml) at −20°C.

Protocol 2: Isolation of cpDNA Using Sucrose Step and CsCl-EB Gradients

(Time: Part A, 3 hr; Part B, 6–18 hr)

This method involves two steps, purification of intact and broken chloroplasts using a sucrose step gradient, and purification of the cpDNA released from the organelles, together with any contaminating nDNA and mtDNA, using a CsCl gradient with the intercalating dye ethidium bromide (EB). Although the sucrose gradient procedure does not give cpDNA as absolutely pure as the DNase I procedure of Kolodner and Temari (1987), it is much more applicable to a wide range of plants for which it is difficult or impossible to prepare intact, DNase I-resistant chloroplasts, or for which tissue quantities are limiting. For details and modifications of this procedure, and for discussion of alternative procedures for purifying cpDNA, see Palmer (1986a) and Palmer et al. (1988b).

Part A. Isolation of Chloroplasts and Lysis

1. Use young, unexpanded green leaves if at all possible since they will have smaller cells than older fully expanded leaves, and hence will yield more DNA. If practical, prior to extraction, place plants in the dark for 1–4 days to reduce chloroplast starch levels. This is usually not essential.
2. Cut leaves into small pieces, 2–10 cm^2 in surface area. Wash cut leaves in tap water (only if visibly dirty or buggy).
3. Place 10–100 g of cut leaves in 50–400 ml of ice-cold cpDNA isolation buffer (Appendix).
4. Homogenize in a blender for three to five 5-sec bursts at high speed.
5. Filter through four layers of cheesecloth (with squeezing).
6. Centrifuge filtrate at 1000 g for 15 min at 4°C.
7. Resuspend the pellet from 10–50 g of starting material in 5–8 ml of ice-cold wash buffer (Appendix) using a soft paint brush and vigorous swirling.
8. Load the resuspended pellet onto a step gradient consisting of 17 ml of 52% sucrose overlayed with 8 ml of 30% sucrose, both in 50 mM Tris-HCl, pH 8.0, 25 mM EDTA. The overlay should be added with sufficent mixing to create a diffuse interface and thereby prevent trapping of nuclear material in the band of chloroplasts that form at the 30–52% interface.
9. Centrifuge the step gradients at 25,000 rpm (81,000 g) for 30–60 min at 4°C in a SW-27 (Beckman) or AH-627 (Sorvall) rotor.
10. Remove chloroplast band from the 30–52% interface (Figure 4) using a wide-bore pipette, dilute with 3–10 volumes wash buffer, and spin at 1500 g for 15 min at 4°C.
11. Resuspend chloroplast pellet in 1–2 ml wash buffer (or 15 ml if to be further purified).
12. Add 1/20 volume of a 20 mg/ml solution of self-digested (2 hr at 37°C) proteinase and incubate for 2 min at room temperature.
13. Gently add one-fifth volume of lysis buffer (Appendix). Slowly invert tube several times over a period of 10–15 min at room temperature, then make the CsCl gradient (Part B, below).
14. A cpDNA-enriched "total" DNA preparation can be prepared by resuspending the pellet of the sucrose gradient in 1.5 ml wash buffer, lysing (steps 12 and 13), clearing spin (10 min, 1750 g), and CsCl banding (see below).

Part B. CsCl-EB Purification of cpDNA

This method is described for cpDNA, but is applicable to any crude DNA preparation. A smaller volume, more rapid protocol is described by Weeks et al. (1986).

1. Bring the DNA sample (e.g., chloroplast lysate, resuspended isopropanol pellet from a total DNA CTAB extraction, Chapter 9) to a volume of roughly 3 ml. Add 3.35 g of freshly powdered CsCl and dissolve by gentle mixing. Add EB to a final concentration of 200 μg/ml and distilled H_2O to bring to a final volume of 4.45 ml and a final density of 1.55 g/ml.
2. Centrifuge for 4–16 hr at 220,000–290,000 g at 20°C in a vertical rotor (e.g., Sorvall TV-865, Beckman 65Vti).
3. Remove any scum (this will be considerable in the case of a directly banded chloroplast lysate) from the top of the gradient using a 1-ml pipette tip with the end cut off. Use a second 1-ml pipette tip with end cut off obliquely to remove the visible band of DNA. This should be removed in as small a volume as possible, i.e., 0.5–1.0 ml.
4. If the DNA fraction is visibly dirty after the first gradient (as is often the case with direct banding of chloroplast lysates), it can be banded a second time. Simply bring the DNA/CsCl fraction to a volume of 4.45 ml by adding a premixed 1.55 g/ml density solution of CsCl with 100 μg/ml EB and TE, and repeat steps 2 and 3.
5. Remove EB by three extractions with isopropanol (uppermost layer) as described in Protocol 1, Part F. 6. There are two ways to remove the CsCl. Either dialyze (Protocol 1, Part F) or ethanol precipitate as described below:
   a. Remove the aqueous layer from the third isopropanol extraction and add two volumes of H_2O to dilute the CsCl. Mix gently and add six volumes of ice-cold ethanol to precipitate DNA. Place at −20°C for 30 min to overnight. Do not place at −80°C or the CsCl will precipitate.
   b. Centrifuge at >1750 g for 10 min to collect the DNA precipitate.
   c. Wash pellet with 70% ethanol. Spin at >1750 g for 2 min to collect the DNA.
   d. Resuspend pellet in 0.1–0.5 ml of TE.
7. Store the DNA at 4°C for short-term use and at −20°C for long-term use.

Protocol 3: Digestion of DNA with Restriction Endonucleases

(Time: Part A, 2–6 hr; Part B, 2–6 hr)

The activity of REs varies with temperature, pH, and salt (Na^+, K^+, Mg^{2+}) concentration. However, it is usually possible to achieve acceptable levels of activity using a small range of buffers that differs in the final concentration of Na^+ (low [L], 0 mM; medium [M], 50 mM; high [H], 100 mM; Table 2, Appendix, and see manufacturer's instructions). REs vary widely in stability: those that denature rapidly are best used at relatively high concentration, whereas stable REs can be used at lower concentrations (1–2 units/sample) for extended periods (e.g., Crouse and Amorese, 1986). Because many REs are heat sensitive, they should be stored at −20°C, preferably not in a frost-free freezer, and removed for as short a period as possible. The enzymes are stored in 50% glycerol to prevent denaturation by freezing. The glycerol can affect RE activity if present at greater than 5% of the final reaction mixture. Thus, the volume of RE added to a reaction should always be less than 10% of the total.

Part A. Digestion of Single Samples

1. For each sample, the final reaction volume should be 5–30 µl. For a single digest, add the following to a sterile microcentrifuge tube:
   a. Appropriate 10× buffer stock (Table 2 and Appendix) is added at 1/10 final volume.
   b. Water (sterile, deionized, distilled) is added to dilute the reaction mixture to the calculated final volume (see below).
   c. DNA according to amount required: 1–5 ng for end-labeling, 0.1–10 µg for staining or transfer hybridization, depending on the sequence assayed and the size of fragment to be detected. The volume depends on concentration (e.g., mtDNA purified according to Protocol 1 can usually be used at 1-10 µl per digest for end-labeling).
   d. One to two units of the appropriate RE. More units per microgram of DNA may be needed for large amounts (>1 µg) of DNA or for heat-labile REs.

   Example:
   1 µl 10× buffer stock (1/10 final volume)
   5 µl DNA sample (depends on DNA concentration)
   1–2 units of RE (volume varies with RE concentration)
   H_2O to final volume of 10 µl

2. Mix well, and incubate at 37°C (or higher temperature as recommended by suppliers, Table 2). Digestion of purified mtDNA or cpDNA is usually complete in 1–3 hr, although some samples take longer or may require adding a second aliquot of enzyme after a few hours for complete digestion. Digestions of large amounts of total cellular DNA with expensive but long-lived REs are typically left overnight.
3. Remove from the incubator, spin briefly in microcentrifuge (not necessary for large reaction volumes), and place on ice or store in the freezer until needed (indefinitely if desired).

Part B. Multiple Samples and Double Digests

If multiple samples will be digested with the same RE, prepare a "digest-mix" and then add an aliquot (e.g., 3 µl) to each sample. For double digests involving REs with compatible salt requirements (Table 2), an aliquot of each RE is added to the DNA sample, although for REs that are inhibited by high salt concentrations, the sample volume should first be increased by adding an equal volume of TE.

Example: To digest 14 DNA samples of mtDNA with volumes per digest varying from 3 to 7 µl:

1. Bring all samples to the same volume by adding TE (e.g., up to 7 µl).
2. Prepare a digestion mix sufficient for 14 samples with some allowance for pipetting error [e.g., $(14 \times 3) + 3 = 45$ µl total]. The amount of 10× buffer stock must include the volume of DNA as well as the other ingredients of the digestion mix. In this example, each tube will contain 7 µl of DNA and 3 µl of digest mix. Allowing for error, there is a total of 15×10 µl = 150 µl. Thus, the mix should contain:
   15 µl 10× buffer stock
   15–30 units RE (add last)
   H_2O up to 45 µl
3. Mix thoroughly and aliquot 3 µl of the digest mix to each sample, mix, and incubate as above.

Protocol 4: Agarose and Polyacrylamide Electrophoresis

(Time: Part A, 2 hr preparation, 2–18 hr electrophoresis; Part B, 2 hr plus exposure time)

The fragments produced by digestion are separated according to size by electrophoresis through agarose or polyacrylamide gels. For analysis of double digests (e.g., for restriction site mapping) or analysis of single digests that produce small fragments (e.g., 4-bp REs) by end-labeling, each sample should be run on both types of gel to accurately resolve fragments over a wide size range (e.g., 10 kb to 20 bp). Most other applications just use agarose gels (see under Electrophoresis of Fragments, above).

Part A. Gel Preparation and Electrophoresis

Agarose gels.   Agarose gels can be run horizontally or vertically. Horizontal gels are used for most applications, e.g., staining and transfer hybridization; vertical gels are easier to dry and offer better resolution for autoradiography of end-labeled fragments. Agarose of the high-purity grade necessary for electrophoresis of DNA is expensive so that the gel is usually kept as small and thin as the application allows. Minigels (e.g., 50 × 100 mm) are often used for checking DNA samples or clones (Chapter 9). Larger gels (see molds in Figures 6, 7, and 8) are used for RFLP analysis.

The range of fragment sizes that can be accurately measured varies with the concentration (w/v) of the agarose. Fragments as small as 200–300 bp can be visualized using 2.5% gels, while fragments as large as 30 kb can be resolved in 0.6% gels. Agarose cannot be poured easily at concentrations of greater than 2.5%, although some special preparations (e.g., Nu-Sieve, FMC Corp.) can be used at much higher concentrations (at least 4%) for detecting smaller fragments.

The steps involved are to prepare a mold, prepare the agarose, pour the gel, insert the well-forming comb, and remove the comb after the agarose has set.

1. Preparation of the gel mold depends on the type of unit. Horizontal units have preformed molds that are taped on opposite sides to contain the agarose solution and have combs inserted to form the wells (Figures 6 and 7). For large molds, tight taping across the top prevents warping. For vertical units (Figure 8), two polished glass plates (one notched) are separated by spacers and clamped together. The bottom is sealed

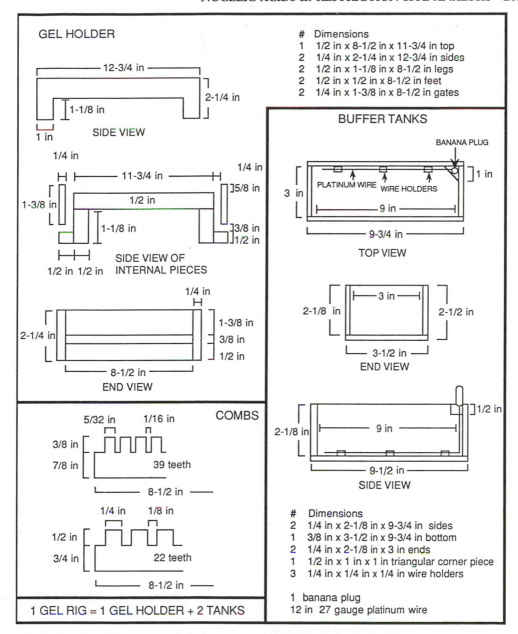

FIGURE 6. Plans for a nonsubmarine type, horizontal agarose gel electrophoresis unit with agarose wicks (1 unit = 1 gel holder plus 2 tanks). The design of two types of gel combs is also shown. Gel rig plans are based on the published plan of McDonell et al. (1977), with modification by M. Murray, W. Thompson, R. Jorgensen, and J. Palmer. Figure courtesy of Nanette Mussy and Jim Manhart.

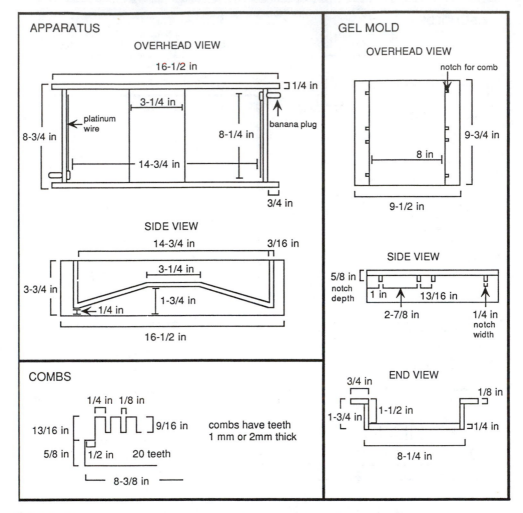

FIGURE 7. Plans for a submarine horizontal gel rig. Gel mold is made of ultraviolet transparent acrylic plastic for use in EB staining.

using tape or by pouring an agarose plug while the mold unit is held vertically in a stand with a central well.

2. Mix agarose, 10× stock of gel buffer (usually TBE or TAE, see Appendix), and distilled water. For example, to make 200 ml of a 1% gel, combine 2 g of agarose, 20 ml of 10× buffer, and 180 ml of H_2O. Mix the ingredients thoroughly in a flask and boil vigorously with intermittent swirling. If using a microwave, add a Teflon-coated stir bar to avoid superheating. The preparation is ready when all of the particles have gone into solution. When cooking agarose (especially in a microwave oven), loss of water due to evaporation can be significant. Check the final volume, add water to replace what has boiled away, and reheat briefly to ensure that the agarose is well

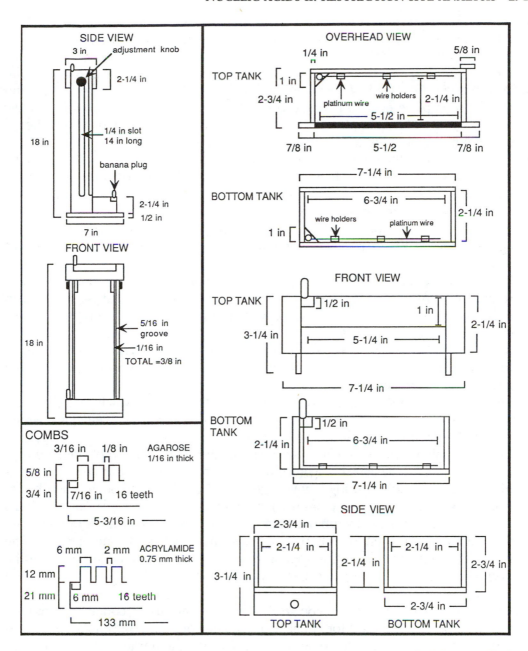

FIGURE 8. Plans for an adjustable vertical gel rig. Glass plates are 3.2-mm double strength glass, 16.5 × 19 and 16.5 × 44.5 cm; in sets of two, where one has a notch in the top that is 1.9 cm deep × 14 cm wide (centered). Spacers for the agarose gel (small gel) are 2.0 mm thick. Spacers for the polyacrylamide gel (large gel) are 0.75 mm thick. Combs routinely have 16 wells for both gels. The blackened area in the overhead view of the top tank depicts the notch in the front tank.

mixed and dissolved. Molten agarose may be stored for several days at 70°C (or allowed to set at room temperature), or after sufficient cooling (when the flask is no longer too hot to handle; ≈50–55°C), can be poured into the vertical or horizontal mold. Pouring agarose that is too hot will crack the plates or warp the plexiglass mold.

3. Pour the slightly cooled agarose into the level mold. For horizontal gels, the comb should be in place prior to pouring. For vertical gels, insert the comb immediately after pouring and fix it in place by clamping the comb to the back plate. Let the gel set until it is cool to the touch and opaque.

4. Carefully remove the comb to prevent tearing of the wells or the teeth separating them. For vertical gels, squirting a small amount of buffer between the gel and the comb oftens helps. Remove the tape from the mold, place the gel in the rig and submerge in buffer to prevent the gel from desiccating. For vertical gels, squirt molten agarose between the plate and rig before clamping together to provide a good seal against buffer leakage. The gel is now ready to use or may be kept as is for at least 1 day.

5. Prior to electrophoresis, wells should be tested by preloading with dilute (i.e., 1×) running dye (Appendix) and electrophoresis. In the case of vertical gels, thin layers of agarose need to be removed from the wells manually (Hamilton syringes work well for this) and by gentle rinsing.

6. Connect the electrical leads to the gel apparatus. DNA migrates to the anodal (positive) pole, therefore the wells should be closest to the cathodal pole (for vertical gels, anode at the bottom, cathode at the top).

7. Add 1/5 volume of loading solution (Appendix) to the sample (which should already be end-labeled if necessary). A size standard (e.g., HindIII or AvaI/BglII digested lambda DNA) must be included on each gel.

8. Using a Hamilton syringe or adjustable micropipetter, load each sample into the well, splitting samples between the agarose and acrylamide gels if both are used. Fragments are best resolved using low voltages (1.0–1.5 V/cm), although much higher voltages (≈10 V/cm) are sometimes used for rapid running of minigels. Full length (20 cm) agarose gels are usually run overnight. Electrophoresis is typically stopped when the dye front (equivalent to ≈500 bp in a 1% gel) has reached the end of the gel. The gel mold is removed from the apparatus and the gel is treated to visualize the fragments (e.g., see Staining and Gel Drying below).

Polyacrylamide gels.   Polyacrylamide gels are prepared at varying concentrations (typically 3.5–6.0%) and are used for visualizing small fragments (<1000 bp). Unlike agarose, polyacrylamide gels are run only vertically (Figure 8), and transfer from polyacrylamide gels to a hybridization filter must be done electrophoretically (Church and Gilbert, 1984; Kreitman and Aguade, 1986).

CAUTION: Polyacrylamide is a cumulative neurotoxin and must be handled with extreme care. Always use gloves, and also wear a face mask when handling the powder.

1. Wash plates (one notched as for agarose gels) with ethanol. If the gel consistently sticks to both plates, apply silane (Appendix) to the top (notched) plate. Place spacers

between the plates and clamp tightly in place. Tape the bottom of the plate to complete the mold.

2. Wearing gloves, mix the appropiate amounts of bisacrylamide, acrylamide, buffer, and distilled water (Appendix) in a flask.

3. Add 10% ammonium persulfate and TEMED to the mixture, mix by swirling, and immediately pour between glass plates. While pouring, make sure that no large bubbles form, as these will interfere with migration of fragments. When the mold is full, lay flat on a raised surface. Insert comb approximately 1–2 cm into the gel (depending on sample volume to be loaded) and fix by clamping the two plates over the comb with large binder clips. This minimizes the amount of polymerized acrylamide in the wells.

4. After 1 hr, carefully remove the comb from the gel. If teeth of gel break, they usually may be put back in place and still provide a good barrier between lanes.

5. Place the gel in the apparatus (Figure 8) as for vertical agarose gels. Fill the buffer tanks to prevent desiccation. The gel may be stored as is overnight or used immediately.

6. The remaining steps are as described for agarose gels. Polyacrylamide gels can be run in a minimum of 4 hr or overnight. However, application of strong current (typically >300 V) can severely distort the migration front due to differential heating. When the dye front has migrated the appropriate distance (28 cm for a 40-cm 3.5% gel), the gel mold is removed from the apparatus. The gel is then treated accordingly (see below).

Part B. Gel Drying and Autoradiography

When using gels to separate end-labeled fragments, it is best to dry the gels to a piece of chromatography paper (Whatman 3MM) before autoradiography. Dried gels are easier to handle and the fragment patterns much sharper.

CAUTION: For ^{32}P end-labeled fragments, the solution in the bottom tank of the gel rig contains the unincorporated nucleotides. Therefore, this solution is highly radioactive, requiring caution in handling and proper disposal.

1. Remove the gel mold from the apparatus. Remove a side spacer and carefully split the top (notched) plate away from the bottom with a spatula. For polyacrylamide gels, the gel will sometimes stick to both plates. The gel can be removed from either plate by gently squirting with water as the plates are separated.

2. Gently rinse the exposed side of the gel with water to remove excess nucleotides and reduce background contamination. For agarose gels only, excess water should be removed by gentle blotting with an absorbent wipe. Drain by tilting, allowing the water to run off.

3. Remove the gel from the glass plate by adhesion to the filter paper.

4. Rinse and blot the opposite side of the gel as previously described (step 2) and place a second piece of filter paper the same size as the first beneath the gel and filter paper. Cover the gel with plastic wrap, trim the plastic wrap and filter paper to the size of the gel, and place in the gel dryer. Apply vacuum and turn on heat: 1.5-

mm-thick vertical gels usually dry in 30–45 min; thicker, horizontal gels take considerably longer. The gel is now fixed to the top piece of filter paper.

5. Remove plastic wrap and extra filter paper and dispose as radioactive waste. Load the dried gels and film into an autoradiograph cassette. The number of intensifying screens to be used is determined by monitoring the gel with a geiger counter (for ^{32}P labeling) and past experience. At $-70°C$, intensifying screens enhance the intensity of the image (including the background contamination) by a factor of 4 (one screen) to 10 (two screens). However, the use of two screens reduces the crispness of image. If one intensifying screen is used, the orientation is intensifying screen (shiny side up), film, and dried gel (gel side toward film). If two screens are used, the orientation is intensifying screen (shiny side up), film, intensifying screen (shiny side down), and gels (gel side facing the film).

6. After exposure for the appropriate length of time [depending on the amount of DNA labeled, efficiency of the labeling reaction, age and type of nucleotide used (^{32}P or ^{35}S), and the number of intensifying screens], the autoradiograph is developed, fixed, and allowed to dry.

Visualization of RFLPS.   Fragments may be visualized by staining (Protocol 5) end-labeling (Protocol 6), or transfer hybridization (Protocol 7). The relative merits of these alternatives were discussed above, under Methods of Detection.

Protocol 5: Staining with Ethidium Bromide

(Time: 30 min)

Fragments may be visualized simply using UV fluorescing dyes such as ethidium bromide (EB, a powerful carcinogen) that bind to the DNA molecule. This is used to observe RFLPs where large amounts of purified sequence are available and there is no need to detect very small fragments. Staining is also an important step in the transfer hybridization method (Protocol 7). The method below is used to stain gels after electrophoresis. Alternatively, EB can be included in the gel mix or added to the electrophoresis buffer.

1. Trim the gel (e.g., at slots and 4 cm below the bromophenol blue) and place on a acrylic plastic sheet.
2. Stain gel in 500 ml distilled H_2O with 0.5 µg/ml EB for 10–20 min. Shake gently. Pour off EB solution and rinse for 1 min in distilled H_2O.
3. Shake gel in second rinse of distilled H_2O for 5–30 min.
4. Photograph gel (using a Polaroid camera) with a plastic ruler next to the size marker. If the photograph is to be enlarged or published, save the negative (wash with water, then sodium sulfite, then water).

Protocol 6: ^{32}P End-Labeling

(Time: 45 min)

End-labeling of the DNA fragments produced by RE cleavage with radioactive nucleotides ([^{32}P]- or [^{35}S]dNTPs) can detect minute amounts of DNA, enabling more digests per sample. [α-^{32}P] dNTPs are used most frequently because their high energy emission results in relatively short exposure times for autoradiography. The alternative, ^{35}S, has a longer half-life (half-life of 60 days compared to 14 for ^{32}P) and produces crisper images, but requires much longer exposure times and is more difficult to detect in the laboratory. Where several different REs, each with its own type of end, are used, it is simplest to use all four radiolabeled dNTPs for end-labeling.

The reaction uses the large (Klenow) fragment of DNA polymerase I that has 5' →3' polymerase and 3' exonuclease functions (see under Methods of Detection). The polymerase function is far more active than the 3' exonuclease. Labeling generally is carried out at room temperature or at 4°C. However, fragments with blunt ends or 3'overhangs (Table 1) are best labeled at 37°C, maximizing the exonuclease activity. Under these conditions randomly sheared fragments may also be labeled, increasing background. This can be reduced by adding only the first nucleotide to be inserted (e.g., for *Rsa*I digests, just add [^{32}P]dTTP).

1. Prepare a labeling mix to be added to each sample. This consists of 10× label buffer (Appendix), radioactive dNTPs, the large (Klenow) fragment of DNA polymerase I, and distilled water. The amount of 10× label buffer added must take into account the volume of the digests as well as the labeling mix itself. For example, if 5 µl of label mix is to be added to each of 16 tubes (e.g., 15 digested DNA samples and a size standard) that already contain 10 µl of digest, the total volume, including an aliquot for pipetting error, is 17 × 15 = 255 µl. For this example, this mix would include:

   25.5 µl 10× label buffer
   5 units (>0.25 U/sample) Klenow polymerase
   2 µl of 800 Ci/mM [^{32}P]dNTPs (at 0.5 µl = 5 µCi each)
   H$_2$O up to 85 µl (i.e., 17 aliquots at 5 µl)

2. Add 5 µl of label mix to each sample and leave at appropriate temperature (see above) for 20–30 min.
3. Add 1/5 volume of loading dye to each sample. This can be mixed by vortexing or by gentle aspiration in the Hamilton syringe during loading.
4. Load samples into wells, splitting each between agarose and acrylamide gels.

Protocol 7: Transfer Hybridization

(Time: Part A, 5 hr–overnight; Part B, 2 hr–overnight; Part C, 6–24 hr; Part D, 2 hr plus exposure time)

This method consists of five basic steps that follow digestion and electrophoresis on agarose gels: (1) transfer of the DNA from the gel onto a filter, (2) prehybridization of the filter, (3) labeling the probe DNA, making it single-stranded, and hybridization, (4) washing the filter, and (5) autoradiography. Significant variations include transfer by vacuum instead of capillary action ("vacuum-blotting"), transfer under alkaline conditions (Reed and Mann, 1985), production of radioactive probes by random priming instead of nick translation (Feinberg and Vogelstein, 1983), and modifications of the stringency of hybridization and washing (see Hames and Higgins, 1986). For some lower sensitivity applications (e.g., excluding dextran sulfate from hybridization mix) or some types of membranes (e.g., noncharged membranes), the prehybridization step can be omitted without a substantial increase in background.

Part A. Transfer of DNA to the Membrane

The electrophoresed fragments are made single-stranded by alkaline treatment and transferred in the same orientation from the gel to a membrane to which they are bound.

1. After electrophoresis, stain the gel with EB and photograph with a ruler next to the size marker to allow fragment sizes to be determined from final autoradiograph. Trim the gel to minimum size, slicing at the origin and 1–2 mm from the outside DNA lanes. Cut the bottom at the 150–200 bp position for gels with frequent cutting REs (based on previous experience) and at 500–600 bp for rare-cutting REs. For large-scale survey work, the sizes of gels are calculated so that two (or sometimes three or four) fit precisely onto a single piece of film: e.g., two 20 × 12.5-cm-trimmed gels will result in membranes that can be exposed together on a standard size (20 × 25 cm) piece of film.
2. If the membrane is to be cut into smaller pieces after transfer, mark between lanes with India ink to indicate positions of cuts. Mark a corner to orient gel and membrane.
3. Shake gel gently in 0.25 M HCl until the bromophenol blue turns yellow. The exact time varies depending on the thickness and percentage of the gel: 5–10 min usually gives the best results. The acid depurinates the DNA, breaking large fragments into smaller pieces for more efficient transfer. Prolonged exposure will result in excessive depurination, producing small fragments that pass through the filter or hybridize poorly with the probe DNA. Insufficient exposure may result in incomplete transfer.
4. Pour off acid solution and rinse gel with distilled H_2O for 1 min.
5. Shake gel in 0.4 M NaOH until the dye becomes blue again (>20 min; can be much longer).
6. Pour off the NaOH and rinse gel with distilled H_2O for 1 min.
7. Neutralize the gel by shaking in 3 M NaCl/0.5 M Tris-HCl, pH 7.5 for 30 min.
8. For double-sided transfer, cut out two pieces of hybridization membrane to the exact size of the gel or, at the most, 5 mm larger in each dimension. We prefer nylon membranes over nitrocellulose as these can be rehybridized many times and are tougher. Membrane can be purchased in long (30-m) rolls that are approximately two gel-widths wide to minimize wastage. Mark filters carefully on the edge with appropriate data (e.g., date, experiment number, etc.).
9. Cut out four pieces of robust filter paper (e.g., Whatman 3MM) to the same size as the gel.

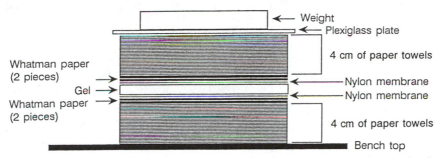

FIGURE 9. Setup of transfer according to the two-sided, dry-blot method.

10. Soak membranes in distilled H_2O for >20 min; then soak membranes and filter paper in 20× SSC for 1–10 min prior to setting up the transfer.
11. Make a symmetrical "gel-blot sandwich" (Figure 9) as follows:

    4-cm stack of paper towels, trimmed to size of membrane and flat
    2 pieces of Whatman paper
    1 piece of nylon membrane
    gel (do not slide it around)
    1 piece of nylon membrane
    2 pieces of filter paper
    4-cm stack of paper towels trimmed to fit
    plexiglass or glass plate
    weight

    Any bubbles or creases will cause uneven transfer and should be rolled out with a pasteur pipette or other suitable round object. Allow transfer to proceed for 3 hr to overnight.
12. Disassemble the "gel sandwich," taking care to mark the filters at any spots where they are to be cut. Also mark the filter to define orientation relative to the gel.
13. Shake membrane in 2× SSC for 10 min.
14. Air dry for 30–120 min on filter paper. Some types of membrane may require further drying in a vacuum oven for 30 min at 80°C to irreversibly bind the single-stranded DNA fragments to the membrane.
15. Trim off any extra filter around outside. Remember, filters must be appropriately sized for autoradiography. Slice filters into smaller strips as necessary. Store new filter at room temperature or 4°C until needed for hybridization.

Part B. Prehybridization of the Filter

1. Wet filters in 500 ml 2× SSC for 5 min.
2. If appropriate, remove probe from the previous hybridization by shaking the filter in 500 ml of boiling 0.1× SSC for 3 × 5 min. New membranes should be washed in

500 ml of 0.1× SSC, 0.5% SDS at 65°C for 1 hr to minimize background on subsequent hybridizations.

3. Remove excess solution from membranes and place in a plastic bag (carefully heat-sealed around the margins leaving 2 cm at the top) or in a plastic tub.

4. Prepare hybridization solution (4× SSC, 1% SDS, 0.5% nonfat dry milk), first adding the dry milk powder to distilled H_2O to nearly full volume and mixing gently, then SDS (from 20% stock solution, Millipore-filtered) and finally the SSC (20× stock solution, also Millipore-filtered). Cover flask and heat for 2 hr in a 65°C H_2O bath. One or two filters in a plastic bag can be prehybridized (and hybridized) in 10 ml of solution. For large-scale experiments, 100 ml of solution will suffice for approximately 20 filters (each 12.5 × 20 cm) hybridized together in a plastic tub or in plastic bags, either individually or in smaller groups of filters.

5. For tubs, add the hybridization solution and cover with a lid. For bags, add the hybridization solution, carefully remove any bubbles, and heat seal 2–3 cm from the edge of the filter. This will leave room for additional sealing after adding the probe. Place all bags together in a single plastic tub with lid.

6. Shake gently for 2 hr to overnight at 65°C.

Part C. Labeling of Probe and Hybridization

1. Prepare 10 µl of nick-translation buffer cocktail per reaction:

    3.0 µl 10× nick-translation NT buffer (Appendix)
    0.5 µl DNA Polymerase I (10 U/µl)
    1.0 µl each of 5 mM dTTP, dATP, dGTP
    1.0 µl [^{32}P]dCTP (i.e., 20 µCi of 3000 Ci/mmol stock)
    1.0 µl DNase I (0.1 µg/ml stock)
    3.5 µl H_2O

    More of the [^{32}P]dCTP can be added to achieve greater incorporation if necessary. Keep mixture on ice. Do not mix cocktail vigorously as this may denature the enzymes.

2. Add 10 µl of nick-translation cocktail to each tube containing template DNA in a volume of 20 µl (>50 ng, typically 2 µl of plasmid DNA + 18 µl distilled H_2O).

3. Incubate reactions for 2 hr in a 15°C water bath.

4. Prepare spun columns for removal of unincorporated nucleotide (for alternative method see Chapter 7, Protocol 3):
    a. Heat Sephadex G-50 [hydrated in STE (Appendix) and stored at 4°C] in 65°C waterbath for 15 min.
    b. Pack a small wad of glass wool into the bottom of a 1-ml syringe.
    c. Fill each syringe with Sephadex G-50 and let drip dry in used 15-ml polypropylene tubes.
    d. Add additional Sephadex G-50 and spin down briefly at 1750 g in a benchtop centrifuge. Repeat, if necessary, until the packed volume is 0.9 ml.

   e. Fill columns with STE and centrifuge at 1750 g for 2 min. Repeat once. Do not allow the columns to run dry.

5. Stop reactions by adding 75 μl of STE and mix well.
6. Place the spun columns in new, labeled 15-ml polypropylene tubes.
7. Load each column with reaction mixture.
8. Centrifuge at 1750 g for 4 min and discard columns as radioactive waste.
9. Add 600 μl TE to effluent and denature by boiling for 10 min.
10. For hybridization in bags, inject probe into bag using a 1-ml syringe. Heat seal the bag twice just above the filter to minimize total bag volume. Mix bag contents well. For buckets, dump probe directly in bucket with filters and mix well.
11. Shake gently overnight at 65°C.

Part D. Washing and Autoradiography

1. Prepare 4 liters of filter wash buffer (2× SSC, 0.5% SDS). Add water first (3.5 liters), then 20× SSC (400 ml and mix), then 1.0% SDS (100 ml).
2. If using a bag, cut off top of bag and remove the hybridization solution using a disposable 10-ml pipette. This can be stored for further use or discarded as radioactive waste. Slide the filter out and place it in a plastic bucket with 500 ml of room temperature wash buffer. Discard the bag into the dry radioactive waste container. If using a bucket, remove each filter, one by one, into bucket with wash buffer and store or discard the radioactive solution as above.
3. Shake filters for 5 min and discard the wash solution as radioactive waste. Repeat room-temperature wash once more, discarding wash solution into radioactive waste.
4. Wash filters two to three times for 30–40 min at 65°C. Discard wash solution into radioactive waste.
5. Remove the filters from the final wash and remove as much excess liquid as possible, but do not to let the filters dry out as this makes it difficult or impossible to strip hybridized probe for subsequent rehybridization. Wrap filters in plastic wrap.
6. In a darkroom, load each filter into a cassette with film as described previously. The film can be marked to identify each filter. As for end-labeled gels, the exposure times and number of screens used are determined by the strength of the emission signal detected on manual scanning of the wrapped filter with a geiger counter.
7. Remove the cassettes from the freezer. Allow >15 min for them to defrost before developing. There is a danger of cracking and damaging the cassettes if they are opened for development when still partially frozen.
8. Develop and fix the film.

Protocol 8: Construction of Cleavage Maps

(Time: variable)

A cleavage map can be constructed by determining the order of restriction sites for each RE and their location relative to restriction sites for other REs. This information greatly extends the phylogenetic applications of RFLP analysis and is essential to localize length mutations. Three methods are commonly employed: two of these methods, partial digests and double digests, are particularly appropriate for small genomes or sequences (e.g., animal mtDNA); the third approach, sequential hybridization, is useful for much larger genomes such as cpDNA.

Incomplete digestion of DNA results in a set of larger fragments, each one equal to the sum of two or more adjacent fragments. The order of fragments can therefore be determined by comparing the sites of partially versus completely digested fragments (Figure 10). This method is particularly useful where there are several closely spaced cleavage sites for an RE not separated by sites for any other REs. The general approach is to label the DNA at one end only and then to generate a series of partial digests, either by varying digestion time or by serial dilution (Danna, 1980; Ausubel, 1989). Multiple digestion experiments compare the fragment patterns of REs used alone and in combination. This enables the cleavage sites of the different REs to be located relative to one another (Figure 10). This approach works well for sequences of up to 30 kb (e.g., animal mtDNA and nuclear rDNA) and for REs that make relatively few cuts. The third approach, sequential hybridization (Figure 3), is particularly useful for large (>30 kb) sequences such as entire cpDNAs (Palmer, 1982, 1986a), although it can also be used to map cleavage sites in smaller sequences (e.g., Sites and Davis, 1989). The basic strategy is to sequentially hybridize a series of radioactive fragments ("probes"), that together make up the entire sequence, to fragments produced by single and double digests after the latter have been transferred to nylon membranes. Adjacent fragments will hybridize to the same probe whereas physically separated fragments will not. By hybridizing to both single and double digests, it is possible to obtain information on both the order of fragments produced by each RE and the relative location of cleavage sites for different REs (see Palmer, 1982). Methods for double digesting and sequential hybridization are given below.

Part A. Double Digestion Experiments

1. Determine which samples need to be mapped and for which REs. To start, the fragment pattern for each RE is compared across representative samples (e.g., one per locality; Protocols 3–7). Not all samples need to be mapped for all sites: a reference sample can be fully mapped and other samples that share multifragment patterns or that have easily interpreted site losses can be compared with no further digests. However, REs with single sites and those with inferred site gains must be characterized further.

2. Perform double digests (Protocol 3) for a single reference sample to map sites relative to each other. Our strategy is to begin with all pairwise combinations of three to four REs that cleave only a few sites, i.e., up to three and that cut in at least two buffers (Table 2). These "core" REs are then used in double digests with the other REs, building up a crude map. Actually determining the relative location of cleavage sites from the double digests requires considerable patience; alternative solutions must be tried out and followed through for each combination. Finally the relative positions of closely spaced sites are tested by selected double digests. The accuracy of the crude

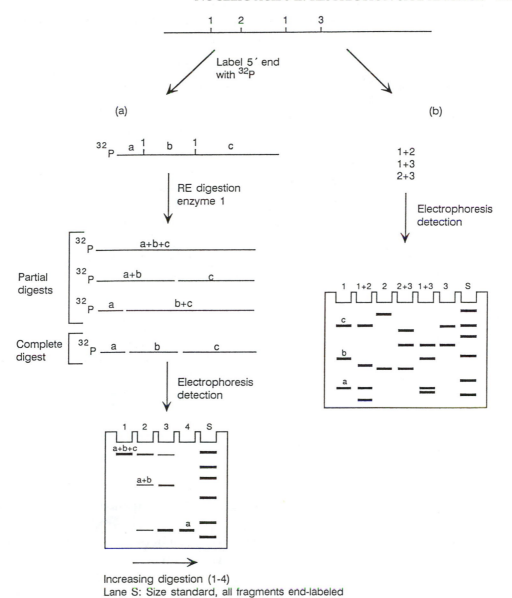

FIGURE 10. Illustration of methods for mapping cleavage sites: (a) partial digestion and (b) double digestion.

map is tested in this final phase and also by comparison to site losses or partial digests observed in the single digest comparisons.

3. After construction of the first map, the strategy for mapping the other samples depends on the extent of divergence. For similar sequences, any REs that cut once should be critically compared using side-by-side double digests. Site gains relative to the reference sample can usually be placed with a single double digest. Samples that

show considerable sequence divergence (>5%) should have their sites mapped independently to guard against site convergence.

Part B. Sequential Hybridization

The approach used to survey for and map RE variation for large, complicated target sequences (i.e., cpDNA) is as follows:

1. Digest each DNA sample (total cellular or purified cpDNA) with 10–20 REs that each cleaves the genome 20–100 times (Protocol 3).
2. Separate digests of different DNA samples, each produced with the same RE, by electrophoresis through 1.0–2.0% agarose gels (Protocol 4). The loading dye can be run 10 cm, which allows the gel to fit onto a 12.5-cm membrane.
3. Make two filter replicas of each gel by bidirectional blotting onto durable nylon membranes (Protocol 7).
4. Hybridize the two identical sets of filters with two different probes in plastic buckets. After autoradiography, strip the probes and rehybridize with other clones until the entire target sequence has been covered. For cpDNA, a total of 6–40 probes, requiring 3–20 rounds of hybridization, is usually used. The lower the taxonomic level and expected level of variation, the fewer and larger the probes that can be used, although one can compensate for these factors by using enzymes that cut at a high frequency. If the entire genome is to be covered with only a few hybridizations several clones can be pooled in a single nick-translation and hybridization reaction.
5. Interpret autoradiograms and make restriction maps (see Interpretation and Troubleshooting).
6. Identify character-state changes and perform phylogenetic analysis (Chapter 11).

A typical survey for rearrangements in a genome such as cpDNA ($\approx$ 150 kb) can be performed as follows:

1. Digest each DNA sample with between one and four REs that cut the genome 30–70 times.
2. Place all digests for a given DNA sample in adjacent lanes of a 1.0–1.5% agarose gel and electrophorese until tracking dye has moved 5–10 cm.
3. Make two filter replicas of each gel by bidirectional blotting onto durable nylon membranes.
4. Hybridize the two identical sets of filters with two different probes in plastic buckets. After autoradiography, strip the probes and rehybridize with other clones. Clones may range in size from fairly large ones (5–15 kb) as illustrated in Jansen and Palmer (1987b) to small, gene-specific ones of a few hundred basepairs (see Palmer et al., 1988b).
5. If the initial survey was designed principally to find rearrangements, then a secondary survey may have to be performed to sample more intensively among taxa related to those that have rearrangements.
6. In some cases, more detailed molecular characterization of a particular rearrange-

ment, such as fine-structure mapping or sequencing of its endpoints, may be needed before the event can be used with confidence as a phylogenetic character.

INTERPRETATION AND TROUBLESHOOTING

General Aspects

The general aim of studies utilizing RFLPs is to obtain unambiguous data on the gain and loss of cleavage sites. This information can be used to characterize the DNAs themselves (i.e., length changes, rearrangements), provide insight into evolutionary processes at the population and species level (i.e., population structure, hybridization), or formulate hypotheses of phylogenetic relationships. Several different approaches may be used, depending on the choice of sequence and levels of divergence. These approaches are illustrated below, using as examples the analysis of animal mtDNA (primarily by [^{32}P]dNTP end-labeling) and cpDNA (primarily by transfer hybridization). The concepts illustrated in these discussions are meant to be examples, with these methods applicable to other sequences (i.e., large nDNA sequences) and methods of analysis and detection.

There are fundamentally two different approaches for analyzing RFLP variation: comparing fragments or comparing sites. Fragment comparisons suffer from several drawbacks. If RE sites were randomly distributed, all enzymes with the same number of nucleotides in their recognition sequence would produce many fragments of approximately the same size, resulting in a high proportion of comigrating fragments. Sites are not randomly distributed; nevertheless, many enzymes produce comigrating fragments, visible by difference in intensity from other fragments produced by the same digest. Under the best conditions (end-labeling), comigrating fragments complicate the determination of fragment homology among genotypes. In the worst case (EB staining, transfer hybridization), it can be difficult to identify comigrating fragments, let alone assign homology. Because of these problems and difficulties presented by length variation (discussed previously), we would argue that the fragment method should be restricted to very closely related sequences, and then used with caution. The site comparison approach involves the ordering, by either direct or indirect means, of restriction fragments and sites. This allows the interpretation of fragment pattern differences in terms of individual mutations that affect the presence and position of restriction sites. For animal mtDNA and cpDNA the great majority of mutations identified by mapping are restriction site mutations, i.e., the loss or gain of restriction sites, which are generally caused by single nucleotide substitutions within the 4- or 6-bp site surveyed. The extra time taken to produce cleavage site maps (either by direct mapping or inference) is well worth the investment.

The particular approach used for restriction mapping and identification of mutations will depend on the size of the genome or genome region being studied, the amount of variation expected, and the number and size of restriction fragments generated. In the case of very limited variation, most site mutations can be inferred directly by inspection of the raw pattern of fragments (either viewed as a stained gel or autoradiogram). In this situation a limited amount of mapping information is gained by the recognition that the pair of fragments unique to one fragment pattern and summing in size to a third larger fragment unique to a second fragment pattern must lie adjacent in the first genome. There are two weaknesses to this approach. First, it is not possible to discriminate between small length mutations and site mutations near the end of a fragment, although this problem can be alleviated to a large extent by using polyacrylamide gels to visualize small fragments. Second, for small molecules such as animal mtDNA, it can be difficult to differentiate samples with no sites for a particular enzyme from those with one site. This is not a problem for linear genomes because the number of sites is one less than the number of fragments, and samples that are uncut will exhibit only a single fragment. Uncut circular DNA occurs in three conformations (Hauswirth et al., 1987), form I (supercoiled), form II (nicked, circular), and form III (linear), exhibiting the following gradient of mobilities (slowest to fastest): I>III>II. All three forms may be visible in mtDNA preparations (depending on the quality) when EB staining or hybridizing with a radioactive probe; however, only forms II and III are visible in end-labeled preparations, with II generally less intense than III. In addition, circular genomes with one site will exhibit a single fragment of the same length as another sample with a single site in a different position. The only way to alleviate these problems for circular DNAs is to directly map the sites in question. In practice, all samples showing nicked circles and/or linear molecules for a particular RE must be critically examined using an appropriate double digest.

When variation is greater, such that site changes cannot be directly inferred from fragment changes (usually more than two site changes), the sites must be mapped directly. In the case of small genomes such as animal mtDNA, or small genomic regions such as the 10–20 kb typically sampled using a nuclear single-copy gene probe, mapping is generally performed by the double-digestion approach (Protocol 8). Larger genomes (i.e., cpDNA) can be mapped using sequential hybridization (see below).

Levels of variation also determine the choice of REs (4-bp versus 6-bp recognition sequence) to be used, each with its own particular set of difficulties. When sequence divergence in animal mtDNA is relatively high (>2–4%), it is usually possible to obtain enough characters from 6-bp REs. These typically produce fewer, larger fragments with sites that are easily mapped. The difficulty is related to fragment size: large fragments have larger errors of measurement than smaller fragments. Thus, maps and genome size estimates generated from larger fragments (especially >10 kb) are less accurate than those from small fragments. The use of polyacrylamide gels for mapping increases significantly the accuracy of mapping because small fragments produced in double digests can be visualized. When sequence divergence in animal mtDNA is low (<2%), it may not be possible to generate enough characters with 6-bp REs to address the question at hand. In such instances, 4-bp REs, which tend to produce more fragments, are likely to provide more sensitivity, based on the larger number of characters examined. Unfortunately, 4-bp REs tend to produce complicated fragment patterns (i.e., many fragments with comigration), making it difficult to infer site changes. The combination of 4- and 6-bp REs provides a powerful approach to analyzing variation across a wide range of divergence levels.

Length Variation

Length variation can be tentatively identified by the absence of fragments predicted by a site-gain or site-loss hypothesis and by a strong correlation of effects among REs. If the aim is to examine site gains and losses, suspected length variation should be characterized by mapping to remove confounding effects from the analysis.

Several different types of length variation are commonly observed (e.g., in animal mtDNA, reviewed by Moritz et al., 1987). Minor length variation can result from changes in the copy number of small (e.g., <100 bp) tandemly repeated sequences or even in the number of nucleotides in a string of the same base [e.g., poly(C) tracts]. These changes are most obvious where the variable region is contained within a relatively small fragment and under these conditions the variation should be correlated across REs. However, for some REs the variation may not be obvious if the variation resides in a large (e.g., >10 kb) fragment. Also, if the RE has a site within the repeated sequence, the variation will be reflected by differences in intensity of a fragment of the repeat size, rather than differences in length (e.g., Densmore et al., 1985). This type of variation occurs at high frequency and is often heteroplasmic.

Larger scale length variation may be due to insertions or deletions and, again, these should be obvious with all REs. This type of variation can be illustrated by a direct tandem duplication (Figure 11) and a deletion that occurs in approximately half of the molecules (i.e., heteroplasmic; Figure 12). In each case, the nature of the modification is highly predictable given a cleavage map for a closely related sequence lacking the length mutation. More complicated changes that require analysis by hybridization as well as mapping (see below) include duplicative transpositions and insertions of exogenous material.

Partial Digestion and Heteroplasmy

The presence of substoichiometric bands and a total size of fragments exceeding the sequence size beyond acceptable error suggests either incomplete digestion or length heteroplasmy. Such fragments should not be ignored or confused with RFLPs due to changes in restriction sites. Length heteroplasmy should be strongly correlated across REs (see above), although this may be complicated if the length variation is dispersed [e.g., in *Hyla (Pseudacris) crucifer* mtDNA, Moritz et al., 1987]. Incomplete digestion can be

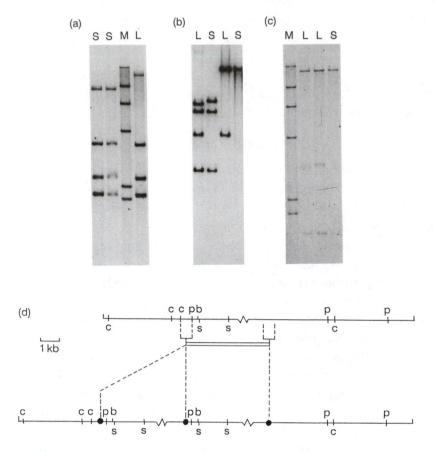

FIGURE 11. Effects of a 4.8-kb direct tandem duplication on fragment patterns. Three types of changes are apparent in comparing the standard length (S) and long (L) genomes: (a) one fragment becomes larger if there is no cleavage site for the RE within the duplication (e.g., *BclI*); (b) one *additional fragment of the same size as the duplication is present if the duplication includes a cleavage site (e.g., PvuII and BamHI)*; and (c) there are multiple additional fragments that sum to the size of the duplication and all but one of which comigrate with previously existing fragments. (d) The location of cleavage sites for *BclI* (C), *PvuII* (P), *BamHI* (B), and *SacII* (S) in relation to the 4.8-kb duplication (see Moritz and Brown, 1986, for further details). The M lane is the size marker; lambda DNA digested with *HindIII*.

caused by technical artifacts (see below) such that not all molecules are cleaved at all sites, or by true heteroplasmy for a restriction site that results in large fragments equal to the sum of two or more smaller fragments present in complete digests. A crude but simple test is to repeat the digest with a new batch of the RE or to digest with an excess of enzyme for extended time periods. Typically, partial digestion produces a variety of larger fragments, whereas site heteroplasmy results in only one larger fragment. However, there may also be considerable heterogeneity among sites in their cleavage rate.

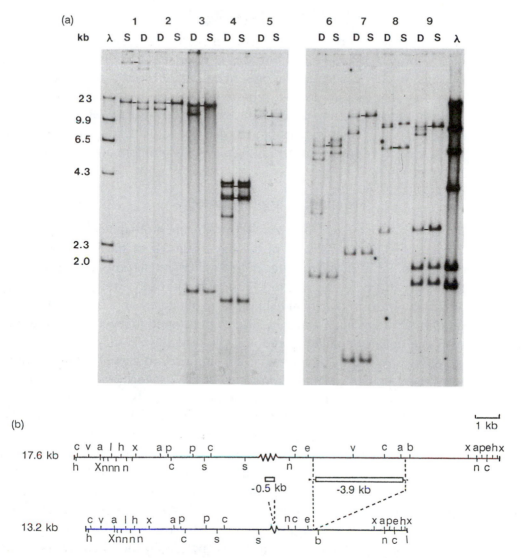

FIGURE 12. Effects of a heteroplasmic 3.9-kb deletion in *Cnemidophorus* mtDNA. (a) Comparison of end-labeled fragments of the standard 17.6-kb mtDNA (S) and the heteroplasmic 17.6/13.2-kb sample (D). The two types of genome in the heteroplasmic sample are present in approximately equal quantities. The first pair of lanes is partial digests showing two size classes of relaxed circles and linear molecules in the D sample. The other eight pairs of lanes are digests with: (2) *Bam*HI, (3) *Sac*II, (4) *Bcl*I, (5) *Eco*RV, (6) *Nci*I, (7) *Xba*I, (8) *Eco*RI, and (9) *Ava*I. The bars indicate fragments from which 3.9 kb was deleted in the 13.2-kb genome. The size marker is lambda DNA digested with *Hind*III. (b) Map of the location of the deletions (3.9 and 0.5 kb) in relation to the S genome cleavage map. For REs that have no site within the deletion one fragment simply gets smaller (e.g., lanes 2, 3, 7). For REs that cut within the deleted region, two fragments are missing and have been replaced by a larger fragment equal to their sum minus the deletion (e.g., lane 5, 10.7 and 6.7 replaced by 13 kb; lane 9, 10.0, and 3.0 replaced by 8.6 kb). Abbreviations: a, *Ava*I; b, *Bam*HI; c, *Bcl*I; e, *Eco*RI; h, *Hind*III; l, *Sal*I; n, *Nci*I; p, *Pvu*II; s, *Sac*II; v, *Eco*RV; x, *Xba*I. The sawtooth region indicates a set of small tandem repeats that has been reduced in copy number in the deletion genome.

A more sophisticated way to discriminate between partial digestion and site hetero-plasmy is to clone from the sample and test for the presence of both restriction types among the clones. Bands suspected of representing partial digests should always be measured to ensure that they do indeed represent the sum of two or more smaller fragments. All too often interesting phenomena such as duplications or deletions are missed because additional bands were assumed to be due to partial digestion. Also, the combinations of fragments present in partial digests is useful information for mapping cleavage sites (Protocol 8).

Mapping Large Sequences

Larger sequences, such as cpDNA (typically 150 kb), are more difficult to map using the methods described above because of the large number of fragments produced (20–100 per enzyme, Table 2); yet, the large number of fragments provides the potential for easily analyzing many more sites. The most common approach to restriction mapping of cpDNA variation employs sequential hybridization of filters with a battery of cloned cpDNA fragments. By adjusting the average size of the cloned fragment relative to the average size of the fragments being scored and the amount of variation expected, a series of autoradiograms is generated of sufficient simplicity to allow the critical inter-pretation of fragment pattern differences in terms of site or length mutations, and, in many cases, the construction of more or less complete maps. As an illustration of this approach, Figure 13 shows the hybridization of a cloned 10.6-kb fragment from the 151-kb chloroplast genome of *Lactuca sativa* to a hybridization filter containing either total DNA or purified cpDNA from each of eight genera in the family (Asteraceae), to which *Lactuca* belongs. Note that even though the enzyme used here, *Bst*XI, cuts cpDNA relatively infrequently (22–30 times) compared to many 6-bp cutters, the complexity of fragment pattern differences apparent in the purified cpDNAs precludes analysis by direct inspection of whole genome patterns (Figure 13a). However, by probing with a fragment representing 7% of the genome, a subtle and readily interpretable pattern is produced (Figure 13b).

As a starting point, consider the simple two-banded pattern of DNA 4 in Figure 13b: relative to this pattern, DNAs 1, 3, 5, and 7 have lost a 14.2-kb fragment and gained fragments of 9.0 and 5.2 kb. The simplest explanation of these differences is that these four DNAs have gained an extra *Bst* XI site located 5.2 kb from one end of the 14.2-kb fragment in DNA 4. Relative to DNAs 1, 3, 5, and 7, DNAs 6 and 8 have lost the 9.0-kb fragment and gained fragments of 6.3 and 2.7 kb. Again, the simple inference is the gain of an extra *Bst*XI site in the latter two DNAs.

These observations establish the existence of two restriction site mutations of po-tential phylogenetic significance. The actual mapping of the sites involved is not a requirement for their use in phylogenetics, but can be accomplished within the frame-work of the overall hybridization analysis (see below) and can be important in cases of potential confusion between site and length mutations (see next paragraph). In the

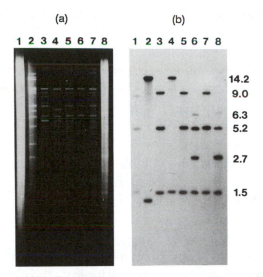

FIGURE 13. Hybridization analysis of cpDNA restriction site variation. (a) Photograph of EB-stained 1.0% agarose gel containing *Bst*XI digests of total DNAs (lanes 1 and 8) and cpDNAs of varying purity (lanes 2–7) of species from eight genera in the Asteraceae. (b) Autoradiogram showing hybridization to the gel at left with a ^{32}P-labeled, nick-translated plasmid containing a 10.6-kb *Sac*I fragment cloned from the cpDNA of *Lactuca sativa*. Numbers at right indicate fragment sizes in kb. Figure courtesy of Bob Jansen.

example shown in Figure 13, the order of the mutated fragments could be established in two ways. First, in cases in which the probes used are identical or nearly so to the DNAs on the filter, one can read the intensity of hybridization signals in terms of the approximate amount of overlap of probe- and filter-bound fragments. Since the 10.6-kb probe hybridizes, in the simplest pattern, to two fragments whose sizes significantly exceed the probe length, the largest of these two, the 14.2-kb fragment, must extend past the end of the 10.6-kb probe by at least 5 kb. So too must one of the two smaller fragments (9.0, 5.2 kb) produced by site gain within the 14.2-kb fragment. Given their sizes and relative hybridization intensities, the only possible interpretation is that the 5.2-kb fragment is internal to the 10.6-kb probe and that the 9.0-kb one extends across its end. Similarly, in DNAs 6 and 8, the 6.3-kb fragment, which hybridizes only weakly with the 10.6-kb probe, must overlap the probe only slightly and the 2.7-kb fragment must lie internal to it. This logic, therefore, established *Bst*XI fragment orders of 1.5–5.2–9.0 in DNAs 1, 3, 5, and 7 and 1.5–5.2–2.7–6.3 in DNAs 6 and 8. The second way to establish these fragment orders is by using as hybridization probes fragments adjacent to the 10.6-kb fragment. One of these two fragments should hybridize to the mutated fragments, of 9.0 and 6.3 kb, that span the junction between it and the 10.6-kb probe.

DNA 2 differs from all other DNAs in Figure 13 in lacking the 1.5 kb *Bst*XI fragment and featuring instead a fragment of 1.3 kb. In the absence of any other information one cannot distinguish between two alternative explanations for this fragment difference: (1) a deletion of 0.2 kb in this region in DNA 2 relative to all others, and (2) a site mutation occurring 0.2 kb from one end of the 1.5-kb fragment, with the additional 0.2-kb fragment in DNA 2 having gone undetected, probably being run off the end of the gel. Length mutations can be recognized by aligning the fragment maps constructed for each enzyme and observing correlated size changes overlapping the variable fragment in question (see above).

The general form of the site-mapping analysis used in cpDNA studies consists of three steps: (1) construct a reasonably complete map of the cpDNA of one of the taxa under study (the reference genome) for each enzyme used. Logical choices for the reference genome could be the one used as the source of cloned hybridization probes or one that has been completely sequenced (e.g., tobacco; Shinozaki et al., 1986) and for which computer-generated maps can be made easily. If such genomes do not fall within the study group then the choice of reference genome should be limited to taxa for which purified cpDNA is available, as this will facilitate the mapping effort. Alignment of each enzyme map for the reference genome is greatly aided by including on each enzyme gel or set of gels a double digest of the reference genome with the enzyme specific to that gel and an enzyme used in common in all the double digests. Draw the maps on two sets of sheets. On one sheet draw all of the aligned maps for the reference genome, one on top of the other. Use this for step 3. Draw the reference map for a single-enzyme on a second sheet. Include below this one-line map the aligned map of the clones used as hybridization probes. Use this set of maps for step 2. (2) Group the autoradiograms by enzyme and by order of probe. Using the single-enzyme map sheets, draw the map for each taxon on a separate line above the reference map. The completeness of the mapping information needed will depend on the amount of variation detected. For divergent genomes, it may be necessary to write each site and fragment size for the whole genome; for similar genomes simply writing the variable sites and fragments may suffice; for genomes of intermediate variability writing all the sites and just the variable sizes may be enough. This step will identify all clear-cut site mutations and all cases of ambiguity regarding length mutation/site mutations near ends of fragments. (3) Regroup the autoradiograms according to probe fragment. Analyze these together with the unified reference genome map to resolve length mutation/site mutation ambiguities as described in the preceding paragraph.

Troubleshooting

Problems encountered in analyses described in this chapter typically occur at any of three stages: (1) during DNA isolation and storage, (2) during digestion and electrophoresis, and (3) during transfer hybridization. Alternatively, inherent properties of the sequence studied could present problems. Below is a list of problems we have encountered and some solutions.

Part 1. DNA isolation and storage

DNA degradation in vivo and/or in vitro

Problem: in vivo: old tissue, senescent
Remedy: use fresh tissue

Problem: improper storage of tissues (slow freezing, freeze–thaw)
Remedy: store properly

Problem: breakage during isolation
Remedy: extract gently

Problem: breakage after isolation (freeze-thaw, nuclease or bacterial contamination, particularly in dialysis)
Remedy: be certain dialysis tubing has been treated appropriately (Appendix); store DNA clean and frozen. Test storage solution (TE) for nuclease activity. When the sample is degraded, use frequent-cutting enzymes to reduce the average size of fragments compared, thereby minimizing degradation effects. If the sample has been contaminated, clean by: (1) velocity gradient centrifugation protocol 1, part D), dye and salt removal (Protocol 1, Part F), and concentration by isobutanol extraction; (2) phenol:chloroform extraction (Chapter 9), or (3) commercially available wash solutions.

Part 2. Digestion and Electrophoresis

a. DNA degradation during RE digestion (Figure 14)

Problem: endonuclease or exonuclease contamination
Remedy: titrate enzymes properly; switch to cleaner enzymes/suppliers; clean sample as described above

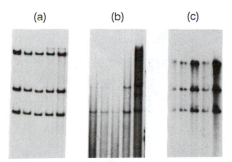

FIGURE 14. The effect of nuclease contamination on samples that have been digested with an RE and end-labeled. (a) *Bcl*I digests of uncontaminated *Lacerta* mtDNA. (b) *Bcl*I digests of nuclease contaminated *Lacerta* mtDNA. Note that degradation is most severe for the larger fragments and produces a smear of randomly degraded DNA. The sample on the right is also contaminated with nuclear DNA, which is obvious as relatively high-molecular-weight DNA. (c) The same mtDNAs digested with *Bcl*I after contaminating nucleases have been removed by velocitization (Protocol 1, Part D). The largest fragment is now clear, indicating negligible degradation due to nucleases. The poor resolution of fragments in some lanes is due to insufficient care in clearing out the wells prior to loading the samples.

b. Partial digestion (Figure 15)

   Remedy: use more enzyme; 2-step digestion; switch to better enzyme/supplier; clean up DNA (as above); redialyze to remove excess salt

c. Electrophoretic artifacts

   Problem: retardation due to excess DNA
   Remedy: use less DNA, use purified or semipurified organellar DNA

   Problem: smiling bands
   Remedy: remove any bubbles from wells, increase run times

   Problem: fuzzy bands
   Remedy: reduced buffer capacity of running buffer—make new buffer

   Problem: missing small bands
   Remedy: do not run too far or use a combination of agarose and polyacrylamide gels

   Problem: missing large bands
   Remedy: too much BSA in digests

   Problem: nonspecific background in gels loaded with end-labeled samples (particularly agarose gels)
   Remedy: use higher grade agarose, making certain it is completely dissolved; make sure plates and apparatus are clean; rinse gels before drying down

Part 3. Transfer Hybridization

a. Poor transfers

   Problem: bubbles and spots of no transfer

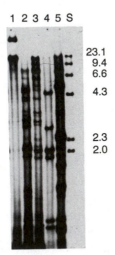

FIGURE 15. Minnow (*Notropis cornutus*) mtDNAs digested with *Ava*I and end-labeled with ^{32}P, demonstrating various states of digestion. The sample in lane 1 is uncut, showing relaxed circular and linear molecules (upper and lower fragments, respectively). Samples in lanes 2, 3, and 5 show varying degrees of partial digestion, while the sample in lane 4 is digested to completion. S, lambda DNA digested with *Hind*III.

Remedy: treat filter carefully—do not touch with bare hands; roll out bubbles in setting up blot

Problem: bottom filter weaker (and blurrier) than top

Remedy: avoid excess weight in blotting; use bottom filters (which may be weaker due to less DNA transferred) with high copy number probes; hybridize with total mtDNA or cpDNA to assess bad spots

Problem: double images

Remedy: don't slide gel or filters around while setting up the blot

b. Hybridization problems

Problem: nonspecific background

Remedy: don't let filters dry in wrapping, washing, and exposing; improper pre-hybridization (especially if large, bubble-shaped blotches), strip and repeat

Problem: hybridization to contaminating vector DNA

Remedy: use isolated inserts; avoid vector contamination of DNAs

Problem: inability to strip completely previously hybridized probe

Remedy: let filters decay; use low copy number and more divergent probes first and high copy number and conserved probes last

Problem: weak bands on autoradiogram (new filter)

Remedy: be sure transfer was complete (stain gel with EB following transfer); be sure probe is labeled

Problem: weaker signals with time and reuse

Remedy: don't wash filters excessively; use low copy number and more divergent probes first and high copy number and conserved probes last

Problem: unequal hybridization efficiency to different fragments of the same digest

Remedy: probe produced by random-priming (Feinberg and Vogelstein, 1983) not labeling randomly, use nick-translated probe; heterologous probe not binding equally well to all fragments, use one more similar to the target DNA or reduce the stringency of wash conditions

Part 4. Intrinsic Biological Problems

a. Heteroplasmy

Remedy: characterize and correct for the phenomenon in comparisons

b. Cross hybridization to another genome or to repeated sequences within the genome (Figure 16)

Remedy: switch probes; use subportions of the probe lacking the cross-hybridizing region; make sure probe is free of contaminating DNA

c. Distinguishing point mutations from small length mutations

Remedy: use both agarose and polyacrylamide gels to visualize all possible fragments

d. Comigrating, nonidentical bands

Remedy: establish homology by constructing restriction maps

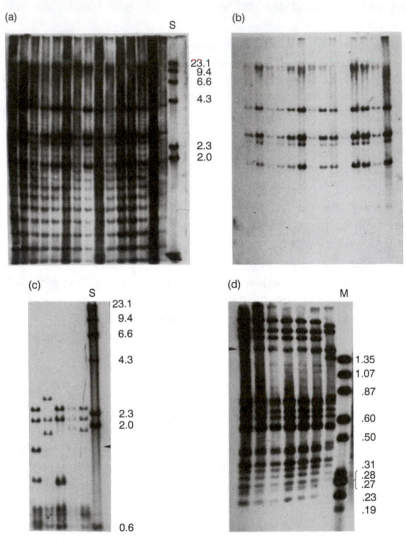

FIGURE 16. The effect of contaminating nuclear DNA in hybridization probes. Contamination can be obvious, as portrayed in panels a and b, or subtle, as demonstrated in panels c and d. (a) *Tursiops truncatus* total DNA samples digested with *Hin*PI and hybridized with *Tursiops* mtDNA contaminated with nuclear DNA. S, lambda DNA digested with *Hind*III. (b) Same filter as (a), hybridized with a *Tursiops* mtDNA sample lacking nuclear DNA contamination. (c) *Tursiops truncatus* mtDNAs digested with *Mbo*I and end-labeled with [32]P. Note the faint band (indicated by the arrow) of nuclear origin (identified as nuclear by EB staining of total DNA digests—gel not shown). (d) *Tursiops truncatus* total DNA samples digested with *Mbo*I and hybridized with *Tursiops* mtDNA, contaminated with nuclear DNA. The arrow identifies a fragment of nuclear origin (see panel c). M, ϕX174RF DNA digested with *Hae*III.

APPENDIX: STOCK SOLUTIONS*

CsCl stock solutions (1):

	Density (gm/ml)		
	1.40	**1.55**	**1.70**
CsCl gm	53.3	73.3	93.3
10×TE ml	10.0	10.0	10.0
H$_2$O ml	71.7	66.7	61.7
PI (2 mg/ml in TE; ml)	5.0	5.0	5.0
Total	100	100	100

Dialysis tubing (1, 2; modified from Maniatis et al., 1982)

1. Cut the tubing (MWCO = 12,000–14,000, diameter = 6.4 mm) to a convenient length.
2. Boil vigorously for 10 min in 1 liter of 2% sodium bicarbonate, 10 mM EDTA, pH 8.0.
3. Rinse thoroughly with distilled water.
4. Boil (as above in step 2) for 10 min in 10 mM EDTA, pH 8.0.
5. Allow to cool and store at 4°C, being certain that there is enough liquid to keep tubing submerged at all times.

End-labeling buffer (6)

see Restriction endonuclease buffer L

Isolation buffer (2)

0.35 M sorbitol
50 mM Tris-HCl, pH 8.0
5 mM EDTA
1.0% BSA, 0.1% BME

Loading dye (4)

40% sucrose (w/v)
0.25% bromophenol blue in 5× TBE

Lysis buffer (2)

5% sodium sarcosinate (w/v)
50 mM Tris-HCl, pH 8.0
25 mM EDTA

Nick translation buffer (7)

50 mM Tris-HCl, pH 7.2
10 mM MgSO$_4$
1 mM DTT

Polyacrylamide gel, 4.0% (4)

(38:2 acrylamide:bis stock solution = 95 g acrylamide, 5 g bisacrylamide; dissolve in 125 ml of distilled water and bring volume up to 250 ml using distilled water. Store at 4°C in a dark bottle).
7.5 ml 38:2 acrylamide:bis solution
59.6 ml distilled water
7.5 ml 10× TBE
0.5 ml 10% ammonium persulfate
75 µl TEMED

*Protocol numbers are given in parentheses.

Restriction endonuclease buffers (3)

(each recipe is for 1 ml total)

Buffer L, 10× stock

1M KCl: 60 µl
2M Tris: 50 µl
1M MgCl$_2$: 100 µl
2mg/ml autoclaved gelatin:
    100 µl
14M BME: 5 µl
sterile distilled water: 685 µl
(also used for end-labeling with Klenow polymerase)

final concentrations used:

6 mM KCl
10 mM Tris
10 mM MgCl$_2$
0.02 mg/ml

7 mM BME

Buffer M, 10× stock

3M NaCl: 166 µl
2M Tris: 35 µl
1M MgCl$_2$: 70 µl
2mg/ml autoclaved gelatin:
    100 µl
14M BME: 5 µl
sterile distilled water: 624 µl

final concentrations used:

50 mM NaCl
7 mM Tris
7 mM MgCl$_2$
0.02 mg/ml gelatin

7 mM BME

Buffer H, 10× stock

3M NaCl: 332 µl
2M Tris: 35 µl
1M MgCl$_2$: 70 µl
2 mg/ml autoclaved gelatin:
    100 µl
14M BME: 5 µl
sterile distilled water: 458 µl

final concentrations used:

100 mM NaCl
7 mM Tris
7 mM MgCl$_2$
0.2 mg/ml

7 mM BME

Buffer 150, 10× stock

3 M NaCl: 500 µl
2 M Tris: 35 µl
1 M MgCl$_2$: 70 µl
2 mg/ml autoclaved gelatin:
    100 µl
14 M BME: 5 µl
sterile distilled water: 390 µl

final concentrations used:

150 mM NaCl
7 mM Tris
7 mM MgCl$_2$
0.02 mg/ml

7 mM BME

STES (1, 2)

Mix 1 part 1.5 M sucrose in ThE with 5 parts 10 mM Tris, 100 mM EDTA, 10 m M NaCl, pH 7.5

TAE, 10× buffer (4)

4 M Tris
0.05 M sodium acetate
0.01 M disodium EDTA, pH 8.2 (if necessary, adjust pH using glacial acetic acid)

TBE, 10× buffer (4)

0.89 M Tris
0.89 M Boric acid
0.11 M disodium EDTA, pH 8.3 (if necessary, adjust pH using hydrochloric acid)

ThE (1)

10 mM Tris
100 mM EDTA, pH 7.5

TE (general)

10 mM Tris
1 mM EDTA, pH 7.5

Wash buffer (2)

0.35 M sorbitol
50 mM Tris-HCl, pH 8.0
25 mM EDTA

CHAPTER 9

NUCLEIC ACIDS III: SEQUENCING

David M. Hillis, Allan Larson, Scott K. Davis, and Elizabeth A. Zimmer

PRINCIPLES AND COMPARISON OF METHODS

General Principles

Advances in sequencing technology over the past decade have been phenomenal. Although nucleic acid sequencing is a comparatively new approach for systematics (as it is for all of biology), the power of the technique has ensured that DNA sequencing has become one of the most utilized of the molecular approaches for inferring phylogenetic history. The primary attractions of nucleic acid sequencing include the facts that the characters (nucleotides) are the basic units of information encoded in organisms (which presents some advantages in analysis; see Chapter 11), and that the potential sizes of informative data sets are immense. Some species contain more than 10^{11} nucleotide pairs per haploid genome, although the number of independent characters that could be used in phylogenetic analysis is considerably lower. To use nucleotide sequence positions in phylogenetic studies, orthologous sequences must be aligned. The number and size of orthologous sequences that can be aligned will differ depending on the level of comparison, but for most studies systematically informative variation is essentially inexhaustible.

Nucleic acid sequence data are regularly compiled in several data bases; GenBank (under contract with the U.S. National Institutes of Health) is perhaps the best known and most widely used (another well known data base is compiled by the European Molecular Biology Laboratory). These compilations represent, in effect, the largest comparative data sets ever collected. At the time of writing, nearly 10^8 base pairs have been compiled in GenBank, and the data set is doubling every 1 to 2 years. With increased numbers of laboratories collecting nucleotide sequence data, and with continued advances in sequencing technology, this rate is likely to increase. Although only a tiny fraction of these data is collected for systematic

318

studies, many of the data can be used for this purpose. Unfortunately, most of the available data have been collected from a very few species; over half of the sequences are from fewer than 10 species of commercial or medical importance to humans (Hillis, 1987). Thus, although sequence data collected for other purposes serve as a useful starting point for some systematic studies, systematists must usually collect comparative data for relevant species of interest.

Although there are a number of strategies for obtaining sequence data for use in systematics, all methods have four basic steps. First, a particular target sequence must be identified that contains an appropriate amount of variation across species or individuals for the problem that is to be addressed (this step is discussed below, under Applications and Limitations; also see Chapter 8). Second, large numbers of copies of the target sequence must be isolated and purified from each individual to be examined. Third, the purified DNA or RNA must be sequenced. Finally, homologous sequences must be aligned (alignment is discussed in Interpretation and Troubleshooting and in Chapter 11). The various methods differ primarily in how the nucleic acid is isolated: "direct" methods involve either directly amplifying the target DNA or isolating abundant RNA transcripts; cloning methods involve the preparation and isolation of viral and/or bacterial vectors that contain copies of the sequence of interest. Each of the methods has distinct advantages and disadvantages and is particularly appropriate under certain conditions. In the next few sections, we outline the differences of the most widely used methods and discuss the relative merits of each approach.

Isolating Target Sequences   In molecular systematics, three methods are commonly used to isolate nucleic acids for sequencing (Figure 1). Cloning using recombinant DNA technology is the most widely used approach in molecular biology, although systematists have often been reluctant to use this strategy because of the time and effort involved. Alternatively, target DNA sequences can be amplified from whole genomic DNA and sequenced directly, or RNA transcripts of the genes of interest can be isolated and sequenced. Which of these methods is used in a particular study will depend on the nature of the target sequence, the accuracy desired, the resources available, the time and energy one is willing to invest, and the desire of the investigator to make his or her material available to others for verification and extension of the findings.

Cloning.   Whole genomic DNA can be used to construct **genomic libraries,** which contain virtually all of an organism's DNA cloned in pieces into a viral host (usually one of several derivatives of the **lambda (λ)** bacteriophage). The library often contains 10^8 or more copies of packaged viral

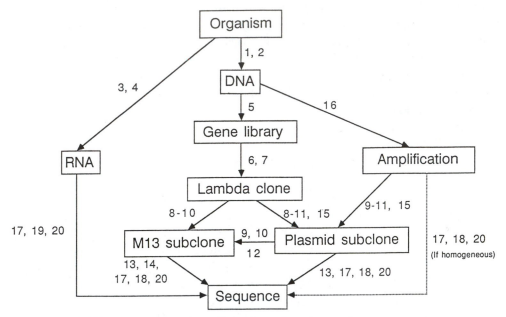

FIGURE 1   Sequencing flow chart. Numbers refer to protocols.

DNA, each with a fragment of the original organism's genome. To use the library, an investigator must find the viruses that contain the gene or region of interest and grow additional copies of these viruses (Figure 1). Although DNA can be isolated and sequenced at this point, it is usually desirable to subclone the target sequence into a sequencing vector first (usually a bacterial **plasmid** or the virus **M13** and its derivatives). Plasmid clones can be stored indefinitely in a freezer and grown in quantity whenever desired, and the target DNA can be easily isolated and sequenced. Thus, the sequence is available for verification and extension of the findings of the sequencing analysis.

Bacteriophage λ is one of the most extensively used cloning vectors for initial cloning steps, such as construction of genomic and subgenomic libraries (for reviews, see Frischauf, 1987; Sambrook et al., 1989). Numerous λ derivatives have been constructed, each of which has advantages for certain cloning applications. This bacteriophage is a double-stranded DNA virus of approximately 50 kilobases (kb) in length, with single-stranded complementary ends that allow the λ DNA to circularize after entering a bacterial host. In the host bacterium, the λ DNA is replicated by one of two pathways (**lytic** and **lysogenic cycles**). In lytic growth, many copies of the bacteriophage DNA are produced via rolling circle replication, and are then packaged into a protein coat that consists of a head (that contains the DNA) and a tail. The host bacterium that contains these mature viruses is then lysed, and the progeny phage are released. In a petri dish, this lysis can be visualized against a background of a bacterial lawn as a **plaque** (a clear spot in which the bacteria have been lysed).

Modifications of λ bacteriophage for cloning have deletions of a central, nonessential (for lytic growth) portion of the genome, into which foreign DNA may be inserted. They also have been selected to have only one or two restriction sites for a given restriction enzyme, which are the target sites for cloning. In vectors with two cloning sites (known as replacement vectors), a fragment of the bacteriophage DNA is replaced with the foreign DNA; in vectors with a single site (known as insertion vectors), the foreign DNA is inserted into the bacteriophage. Numerous vectors have been constructed that contain a diversity of cloning sites and accommodate a relatively large range of foreign fragments. Many of these are commercially available as predigested phage arms, so that one need only to digest the target DNA with an appropriate restriction enzyme, ligate the DNA into the cloning vector, and package the resulting recombinant λ DNA with commercial packaging extracts to produce subgenomic libraries (they are subgenomic libraries because restriction fragments of inappropriate sizes will not be represented). The choice of cloning vector will depend on the desired target site and on the size range of inserts to be cloned. λ vectors are generally limited to inserted fragments of less than 23 kb, and many vectors can contain only much smaller fragments (typically in the range of 2–15 kb). If larger fragments must be cloned, then the libraries should be constructed in cosmids, which are specialized cloning vectors designed to accommodate fragments up to 45 kb in length (see DiLella and Woo, 1987, or Sambrook et al., 1989 for more information).

To use the gene library, one must screen the recombinant λ (or other) clones to find the particular gene or DNA region of interest. Numerous methods have been developed for screening gene libraries (see Berger and Kimmel, 1987; Ausubel, 1989). Most of these methods involve either hybridizing with a nucleic acid probe (see Protocol 7) or immunoscreening for expressed proteins of interest. Hybridization is the more general procedure, although it is not as efficient for screening for protein-coding genes that can be expressed in vivo.

If the library is composed of random fragments of average size I from a genome of size G, the number of independent clones (N) that must be screened to isolate a single-copy fragment of interest with a probability of P can be calculated by the formula

$$N = \ln(1-P)/\ln[1-(I/G)]$$

As a rough approximation, one must screen approximately five times the number of base pairs of DNA that are in the genome [e.g., $(I \times N)/G = 5$] to have a 99% chance of locating a specific single-copy gene (Seed et al.,

1982). However, the total number of clones screened can be much smaller for sequences that are present in high copy number (e.g., rRNA genes, mtDNA, and cpDNA). Screening efficiency can be increased greatly by cleaving the genomic DNA with appropriate restriction enzymes (rather than random shearing) and selecting λ vectors that accept restriction fragments only in the size range of the desired target sequence. Of course, for this approach, a restriction map must be obtained before the library can be constructed (see Chapter 8).

Once the appropriate DNA fragment has been cloned and isolated in a λ vector, it can be subcloned into a plasmid or the virus M13. Subcloning is accomplished by cleaving the target sequence from the λ with the same restriction enzyme with which it was originally cloned, and then ligating the ends of the DNA with plasmid or M13 DNA that has been cleaved with a restriction enzyme to produce compatible ends (see Protocol 9 and Chapter 8). The subcloning vector is then introduced into a bacterial host in a process known as transformation (see Protocols 10–12). This allows the cloned fragment to be grown in quantity, easily isolated, and sequenced. DNA amplified in vitro (see below) can also be cloned in this manner to purify heterogeneous amplification products and produce a stable clone. Until recently, M13 was the most widely used sequencing vector, because it allowed single-stranded sequencing which provided superior autoradiographs compared to double-stranded sequencing. However, new double-stranded sequencing protocols have greatly improved sequencing of plasmid clones, and the greater ease with which plasmids are grown, manipulated, and stored is a strong point in their favor. Also, some recently developed plasmids have single-stranded forms that can be used for single-stranded sequencing.

In Vitro Amplification.   Direct sequencing from complex genomic DNA has been made possible with the development of the **polymerase chain reaction (PCR)** technique (Kleppe et al., 1971; Mullis and Faloona, 1987; Saiki et al., 1988; Ochman et al., 1988; Innis et al., 1990). Starting with virtually any amount of DNA, it is possible to amplify a target sequence up to microgram quantities, which (if the specimen was homozygous at the target locus and the amplified fragment is relatively short) can be sequenced directly (Figures 1 and 2). The basic procedure involves in vitro construction of oligonucleotide primers that are complementary to opposite strands of the target sequence and separated by up to a few kilobases. The target DNA is denatured into its respective strands by heating, and the primers are annealed to their complementary sequences. DNA polymerase from the thermophilic bacterium *Thermus aquaticus* (*Taq* polymerase) is used to replicate the complementary strands of the target sequence starting from the two primer sites. The complementary strands are then allowed to

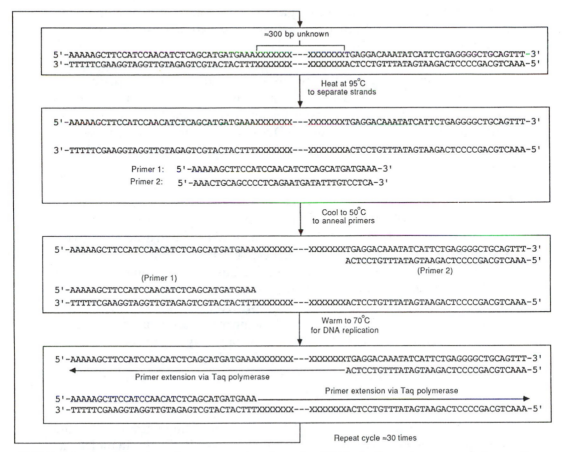

FIGURE 2 Amplification of a conserved region of mtDNA via the polymerase chain reaction. The primers shown amplify a region of the mitochondrial cytochrome *b* gene in vertebrates and some invertebrates (Kocher et al., 1989).

reassociate by cooling (Figure 2). Because *Taq* polymerase is not denatured during the heating step, the process can be repeated many times without adding new polymerase. During each cycle, the target DNA is replicated by a factor of two. Thus, within 30 cycles, one can amplify a single strand of DNA up to 2^{30} times. Because each cycle takes only about 4 min to complete (see Protocol 16), enough DNA can be generated for multiple sequencing and/or cloning experiments in less than a day. Automated thermal cycling machines have been developed that allow an investigator to add the appropriate reagents and genomic DNA samples to a series of tubes and then return in a few hours to amplified DNA ready for sequencing.

Double-stranded DNA produced by PCR amplification can be sequenced directly (DuBose and Hartl, 1990) or by generating single-stranded DNA from the amplification product. Single-stranded DNA is generated by asymmetric reamplification using an excess of one of the primers (Gyllensten and Erlich, 1988), by treatment with an exonuclease (Higuchi and Ochman, 1989), or by use of biotinylated primers (Mitchell and Merrill, 1989). The only requirement for PCR is that the sequences of the regions flanking the target sequence are known so that primers to these regions can be constructed (however, a method called inverted PCR can be used to amplify outside of a known region; Ochman et al., 1988). Using this methodology, it is possible to sequence DNA isolated from a wide variety of sources, including preserved museum specimens and subfossil tissue (Pääbo et al., 1988; White et al., 1989).

RNA isolation.   The third approach to isolating target sequences is to isolate the transcribed RNA and sequence it using reverse transcriptase (Hamlyn et al., 1978; Protocol 19). This method has been used extensively in sequencing the ribosomal RNA genes; this is possible because rRNA accounts for a large fraction of the total cellular RNA. Some regions of rRNA sequences are conserved throughout most living organisms, and several "universal primers" have been constructed that are complementary to these regions (Lane et al., 1985). These primers can be used to directly sequence several regions of rRNA from virtually any organism. This technique has had a major impact on systematic studies of prokaryotes (e.g., G. E. Fox et al., 1980) and has been applied throughout metazoan and plant groups as well (e.g., Field et al., 1988).

Nucleic Acid Sequencing   Although protein sequencing became a routine (albeit costly and labor-intensive) method for the study of protein molecular evolution by the late 1950s, it has been only in the last decade that nucleic acid sequencing has had practical applications in molecular systematics. Until the mid-1970s, only stretches of DNA 15–20 base pairs in length had been sequenced. Breakthroughs in nucleic acid sequencing were published almost simultaneously by Maxam and Gilbert (1977) and Sanger et al. (1977). These two procedures are outlined in Figures 3 and 4.

Maxam–Gilbert (chemical) sequencing.   Maxam–Gilbert, or chemical, DNA sequencing relies on the use of base-specific modification and cleavage reactions (Figure 3). In the original versions of this method (Maxam and Gilbert, 1977, 1980), a DNA fragment was electrophoretically separated into its two complementary strands, each of which was then end-labeled with ^{32}P. The two complementary strands were each sequenced, and the results were compared to check for errors. Development of sequencing vectors especially designed for use in Maxam–Gilbert sequencing (Eckert, 1987) greatly facilitated the procedure; these vectors allow selective end-labeling, so that either strand can be sequenced without separation into

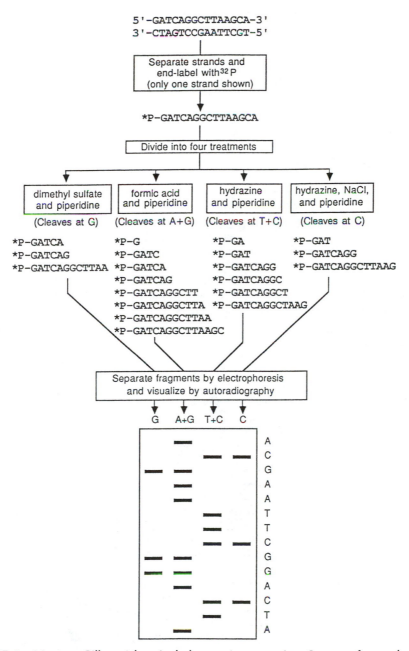

FIGURE 3   Maxam–Gilbert (chemical cleavage) sequencing. See text for explanation.

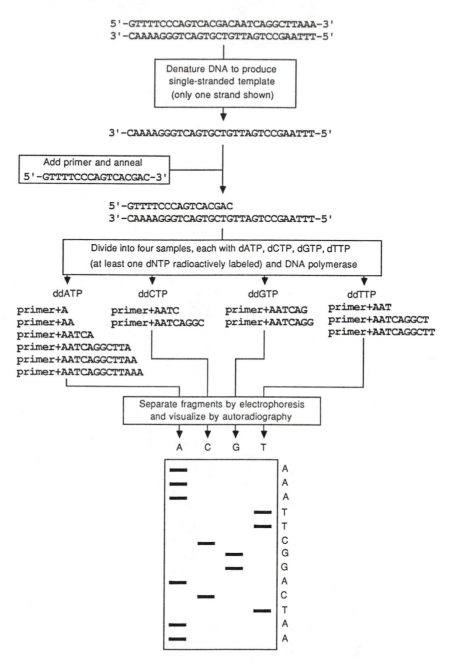

FIGURE 4   Sanger (enzymatic) sequencing. See text for explanation.

single-stranded fragments. Sequencing is accomplished by dividing the target DNA into four subsamples and then treating the subsamples with a series of base-specific chemical reagents that partially cleave the DNA (Figure 3). The first sample is treated with dimethyl sulfate, which methylates a few percent of the guanines in the sequence, and piperidine, which displaces the methylated guanines and thereby cleaves the DNA at these sites. The second sample is treated with formic acid, which protonates a few percent of purine ring nitrogens, and piperidine, which then displaces the affected purines (adenosine and guanine). The third sample is treated with hydrazine, which removes cytosine and thymine from the DNA and leaves ribosylurea. The DNA is then cleaved at these sites with piperidine. The fourth sample is treated like the third, except that the hydrazine treatment is conducted in the presence of NaCl, which suppresses the reaction of thymine with hydrazine (so the DNA is cleaved only at cytosines). In all of these subsamples, chemical cleavage is carried out under conditions in which only a few of the respective nucleotides in any given fragment are affected. However, because the cleavage is random and the population of DNA fragments examined is large, some fragments will be cleaved at each of the nucleotide positions (Figure 3). The radioactively labeled fragments from the four subsamples are then electrophoretically separated by size on a denaturing polyacrylamide gel and visualized by exposing the dried gel to X-ray film to produce an autoradiograph. The sequence of the DNA sample can then be read directly from the autoradiograph (Figure 3).

Sanger Dideoxy Sequencing.   Sanger sequencing, or controlled interruption of enzymatic DNA replication, uses dideoxynucleotide analogs in primer-directed DNA extension to produce discrete DNA fragments (Figure 4). The double-stranded DNA is first denatured to produce single-stranded DNA (or single-stranded DNA is isolated from single-stranded vectors). Next, a short segment of DNA (typically 15–18 bp) known to be complementary to a segment on the target DNA (or in the adjacent sequencing vector) is annealed to the target sample; this short fragment is known as a primer. The sample is then divided into four subsamples, to each of which is added the four deoxynucleotides (i.e., dATP, dCTP, dGTP, and dTTP, at least one of which is radioactively labeled) and DNA polymerase. In addition, one of four dideoxynucleotides (ddNTP) is added to each of the tubes (ddATP, ddCTP, ddGTP, or ddTTP, respectively). The primer has a free 3'OH group on its deoxyribose, to which additional nucleotides can be attached. The DNA sequence is extended by the DNA polymerase using the target DNA as a template (Figure 4). On some strands in the sequencing reaction, a given ddNTP will be incorporated in the growing strand, at which point the polymerization is terminated because the ddNTP lacks a 3'OH group. The radioactive fragments in the four subsamples are then separated by denaturing polyacrylamide gel electrophoresis and visualized by autoradiography, as with Maxam–Gilbert sequencing. The fragments in each subsample will terminate with the corresponding ddNTP (which is complementary to the dNTP on the template sequence), and the sequence of the target DNA can be read directly from the autoradiograph (Figure 4).

Automated Sequencing.   There are a number of types of automated DNA sequencing, but most use Sanger sequencing with fluorescently labeled (rather than radioactively labeled) DNA fragments. These fragments are detected during electrophoresis with the use of a tunable laser. The laser is stationary with respect to the electrophoresis apparatus, and fragments are recorded as they pass a single point. The process is "automatic" in that one does not visually inspect an autoradiograph and manually record the results; instead, the sequence is recorded directly into a computer or onto paper.

Assumptions

In all broad-scale comparative studies, certain implicit assumptions are made by the investigators. In the case of DNA sequencing, these include assumptions about biochemical methodology and assumptions about genome and organismal evolution. It is important to realize the limitations on the interpretation of sequence data that arise from such assumptions.

At the biochemical level, the homogeneity of input DNA and the fidelity of DNA replication are issues of primary importance. Contamination of template DNA, prepared either in vivo via cloning or in vitro via PCR, or polymorphism in the sequence based on interallelic variation or pooling of individual samples, will lead to uninterpretable or incorrect sequences. Contamination is a potentially serious issue with PCR, because even a single strand of foreign DNA can be amplified and thereby confound results. High fidelity of DNA replication is important both in the amplification of the input DNA (done either in vivo or in vitro) and in the various methods that constitute controlled interruption of enzymatic replication. When sequencing amplified DNA, one assumes that the sequence determined is the original one isolated from the genome and not a variant produced during DNA replication. During chemical and enzymatic sequencing, all bases, independent of identity or context (nearest neighbor sequence), ideally should be equally susceptible to cleavage or incorporation/interruption, respectively.

With respect to computer analyses, both alignment and phylogenetic inference steps involve implicit assumptions and certain subjective decisions. Alignment algorithms, which presently are effective only for pairs of sequences, are usually designed to maximize percentage sequence similarity and minimize the number of insertion/deletion events (see Interpretation and Troubleshooting and Chapter 11). Thus, base substitution mutations are assumed to be more frequent during evolution and are penalized less severely by the alignment algorithms. Furthermore, alternative alignments may be equally parsimonious, and the extent of reciprocal illumination between alignment and phylogeny reconstruction procedures has not been standardized. Choice of methods for phylogenetic inference and character weighting strategies also rely to varying degrees on subjective criteria (see Chapter 11).

When the sequences utilized are part of a gene family, it is also important to use a typical or consensus repeat sequence and it is desirable to determine in advance that the rate of concerted evolution (gene family homogenization) of the repeats in that family significantly exceeds that of speciation for the groups of organisms compared (Hillis and Davis, 1988). A preliminary restriction endonuclease cleavage analysis (Chapter 8) can produce an estimate of the homogeneity of the gene family to be compared and this may dictate the appropriate sampling strategy (Chapter 2). Finally, assumptions of level of homology are critical to comparative sequence studies; for all phylogenetic studies except study of gene phylogenies, it is critical that orthologous (rather than paralogous) genes are being compared (see Chapter 1).

Comparison of the Primary Techniques

Cloning versus Direct Sequencing   Development of the PCR method is having a major impact on systematics, because it is much faster and easier to amplify DNA using PCR than to clone. The disadvantages are that the sequences of the flanking regions usually must be known, the individual usually must be homozygous for the sequence of interest, only relatively short fragments can be sequenced directly, the cost of *Taq* polymerase is quite high, and the product may not be available to others for verification and/or extension of the work as is true for a clone. Because of the need for known flanking regions, the technique has been used mainly for regions for which complete and closely related sequences are available (e.g., Wrischnik et al., 1987) or regions that are flanked by highly conserved sequences (e.g., Kocher et al., 1989). Recently, however, methods have been developed for sequencing outside (rather than inside) of known flanking regions (Ochman et al., 1988; Loh et al., 1989). Most of the disadvantages of PCR (except for the requirement of known flanking regions and the high cost) can be overcome by combining DNA amplification and cloning strategies; cloning from the amplified DNA products is considerably faster than cloning from whole genomic DNA (see Protocol 16). However, sequencing directly from PCR amplification products reduces problems associated with errors made by *Taq* polymerase, because errors in all but the earliest rounds of amplification will appear as ambiguities. Length polymorphisms of any size in the target DNA cause the greatest difficulties for PCR, because two offset sequences produce unreadable sequence gels. In heterozygous individuals, both alleles will be amplified unless the primers are specific for a particular allele (which is virtually never the case in systematic studies). In these situations, cloning from the amplification product is often the best solution. A second method for further purifying amplification products involves using two internested sets of primers. After the initial amplification round has been completed using the two external primers, the DNA is reamplified with a set of primers internal to the first. This helps purify the amplification product and reduces ambiguity (Mullis and Faloona, 1987), but usually cannot correct for intron or allelic polymorphism.

Sequences from preserved specimens (Kocher et al., 1989) and subfossil tissues (Pääbo et al., 1988) have been obtained using the PCR procedure. Because most nucleic acids from such sources are usually largely degraded, PCR is the only approach to sequencing that is generally applicable under these conditions.

The other direct sequencing technique, RNA sequencing, is also faster and easier than cloning and sequencing. Direct RNA sequencing is used primarily to obtain sequence information for nuclear-encoded ribosomal RNA for phylogeny reconstruction and rapid identification of microorganisms. Direct RNA sequencing has the disadvantages of having only a single strand available (so that two-strand verification is not possible), requiring fresh (or in some cases frozen) tissues, providing access only to transcribed regions, and greater difficulties in regions of strong secondary structure. The regions that are accessible to direct RNA sequencing (primarily the nuclear ribosomal genes) are also accessible to PCR; thus, sequencing the amplified products of PCR is usually the preferred direct sequencing method because of the greater accuracy and the possibility of using poorer quality tissue samples.

Maxam–Gilbert versus Sanger Sequencing   Although both of these methods have been used extensively for determining DNA sequences, we emphasize Sanger sequencing in this chapter for several reasons. Modifications of Sanger sequencing can be used to sequence both DNA and RNA, whereas Maxam–Gilbert sequencing is applicable only to DNA. Furthermore, Maxam–Gilbert sequencing requires a prior knowledge of the restriction map of the target sequence, because it is necessary to cleave the DNA into manageable size pieces for sequencing. Sanger sequencing requires no prior knowledge of the sequence, because primers can be complementary to the cloning vectors or, in the case of PCR, amplification primers can also be used as sequencing primers. Finally, it is possible to read more sequence information per gel with Sanger than with Maxam–Gilbert sequencing, because of better band resolution. However, Sanger sequencing can present problems in sequencing regions with strong DNA secondary structure, which is not a problem with Maxam–Gilbert sequencing. Occasionally, a section of DNA will be highly resistant to Sanger sequencing and only can be sequenced using chemical degradation. In addition, Maxam–Gilbert sequencing is easily adaptable to massive sequencing efforts (see below).

Other Methods of DNA Sequencing Several other modifications to DNA sequencing have been developed, each of which has specialized applications. Genomic sequencing (Church and Gilbert, 1984) can be used to assess sequence information directly from genomic DNA. In this procedure, fragments from completely restricted whole genomic DNA are partially cleaved using chemical degradation. These fragments are then separated on an acrylamide gel and electrophoretically transferred to a nylon membrane. The fragments on the membrane are then hybridized to a series of specific probes so that DNA sequences can be visualized. The technique thereby eliminates cloning or in vitro amplification steps. Another advantage to this approach is that information on DNA methylation is preserved, whereas this information is lost during cloning and amplification steps. However, this procedure is not generally applicable to most systematic studies, because each visualized sequence requires a separate probe and an appropriately located restriction site. In addition, the low copy number of each target sequence results in a weak signal on the hybridized blot (Church and Kieffer-Higgins, 1988). A related technique known as multiplex sequencing was described by Church and Kieffer-Higgins (1988) and holds great promise for massive sequencing of entire prokaryotic genomes (or similarly large sequencing efforts). Although several variations are possible, the key to multiplex sequencing is the combined sequencing of numerous clones on a single gel, each of which is incorporated in a distinct vector. The DNA is then transferred to a nylon membrane as in genomic sequencing, and the sequences of the individual clones are visualized by successive hybridization to vector-specific probes. The advantage of the technique is the reduction of repetition of many of the sequencing steps by a factor of 20 or more. The technique will undoubtedly be extremely useful for major sequencing efforts, although it is unlikely to be incorporated in most smaller scale systematic studies because of the need for cloning into a large number of distinct vectors.

Manual versus Automated Sequencing Several machines are now available that eliminate the steps of autoradiography and gel reading. Additional technological improvements in automation are inevitable, especially given the recent interest in sequencing the entire human genome (Roberts, 1989). The currently available automated sequencing machines increase the speed at which data can be collected in a given laboratory, but are extremely expensive. In addition, the error rate of the machines is somewhat higher than manual reading by an individual who is experienced in gel interpretation, especially if the machines are allowed to "interpret" long sequences. For most laboratories, the current price is too high and the benefits are too low to justify the use of automated sequencing. However, this situation is likely to change rapidly, so recent progress in automation should be monitored closely.

APPLICATIONS AND LIMITATIONS

There are three major applications of comparative nucleic acid sequencing in systematic studies: (1) construction of molecular phylogenies to evaluate the evolutionary diversification of particular genes or gene families and their RNA or protein products, (2) the tracing of organismal genealogies within species, and (3) construction of species phylogenies to evaluate macroevolutionary patterns and processes. The systematic characters used in sequence comparisons are nucleotide positions in aligned orthologous sequences. These characters may also be useful at the intraspecific level for studies of individual relatedness, geographic variation, and hybridization.

All applications must begin by matching the level of variability of the molecule to be studied with a set of systematic comparisons representing the appropriate evolutionary time scale. However, it must be cautioned that ephemeral lineages that existed relatively far into the past are unlikely to be unambiguously recovered by any molecular (or other) method. The evolutionary variability of any molecule is a balance between mutational input and the constraints of structure and function. The relative constancy of these interactions across taxonomic groups for some molecular sequences produces the useful result that the expected rate of evolution of these sequences can be predicted within broad confidence intervals (see Chapter 12 and Wilson et al., 1987). Therefore, specific sequences can be recommended for various applications with reasonable expectation that they will provide the desired resolution. Molecular sequences will be readily accessible for systematic study, however, only if sequence information is already available for homologous sequences from related organisms.

Gene Phylogenies

For the investigation of gene phylogenies, the sequence to be studied is chosen for its own intrinsic interest, and the appropriate methods for analysis depend on (1) whether the DNA sequence is located in a nuclear or organellar genome, (2) the number of copies of the DNA sequence present per cell, (3) whether the DNA sequence is transcribed and, if so, (4) which tissues contain the largest relative abundance of the transcript. If the sequence is present in an organellar genome, fractionation of the cellular components is a useful first step for sequence analysis, although this can be avoided if sequence information is available for homologous regions from a sufficiently closely related organism. In the latter case, cloned or synthetic DNA can be used to select the sequence of interest from a clone library made from a sample of total cellular DNA (Sambrook et al., 1989); alternatively, synthetic DNA can be used to amplify the sequence of interest from a preparation of total cellular DNA using the polymerase chain reaction (see In Vitro Amplification, above). Sequences located in organellar genomes are present in sufficiently high frequency in cellular DNA preparations that cloning and amplification are relatively routine.

For nuclear DNA, cloning and amplification are routine for sequences that are extensively repeated relative to the total size of the nuclear genome. The genes encoding ribosomal RNA, for example, are in this category. If the DNA sequence is transcribed, direct sequencing of the RNA transcript is possible where the frequency of the transcript in total cellular RNA (or in the isolated polyadenylated or nonpolyadenylated RNA fractions) is sufficient to permit direct sequencing. Weisman et al. (1986) show that this latter approach is feasible even for a nonrepetitive gene whose transcript is abundant only in a certain tissue at a certain stage of the life cycle, and where prior knowledge of homologous sequences is fragmentary. However, the sequence evolution of alleles at most loci must be studied by first cloning the gene and its flanking regions (e.g., Ponath et al., 1989a,b).

Intraspecific Diversity

Most studies of organism genealogies and other intraspecific applications concern sequences that are easily accessible, either because of their high repetition in the genome or abundance in the cytoplasmic RNA. Organism genealogies are traced most effectively using molecular sequences that do not undergo genetic recombination in the process of transmission. Mitochondrial DNA (mtDNA) has been the most effective molecule to use for tracing genealogies within animal species because it demonstrates a strictly maternal pattern of inheritance and is easily accessible for analysis (Wilson et al., 1985; Chapter 8). However, the control regions of mtDNA contain "hotspots" for base substitution that may change so rapidly as to produce considerable homoplasy (Aquadro and Greenberg, 1983; Greenberg et al., 1983). mtDNA is present in sufficiently high frequency in preparations of total cellular DNA that both molecular cloning and amplification by PCR are feasible. For species having XY sex-chromosome heteromorphisms, tracing of paternal lineages is feasible using Y-chromosome specific sequences, although this is less routine than use of mtDNA.

Although few studies have yet tapped the potential of PCR, this methodology opens up many possibilities for studies of intraspecific variation at the species level. Studies of individual relatedness, geographic variation, and hybridization are all possible. However, there is a trade-off between the detailed information at one or a few loci in a sequencing study and the less detailed information at many more loci in studies of allozymes (Chapter 4) and restriction site variation (Chapter 8). Sequence information can also provide heretofore unavailable details about the molecular processes that are responsible for many evolutionary phenomena. For instance, several distinct hypotheses have been proposed for the origin of rare alleles of enzymes in hybrid zones, including intracistronic crossing over, hybrid dysgenesis (the release of previously controlled transposable elements), and differential selection in hybrid zones (see Chapter 4). Sequencing of the rare and parental alleles would resolve this debate. In addition, sequencing could offer new insights into the causes and effects of concerted evolution (see Chapter 8).

Interspecific Diversity

The nuclear and mitochondrial genes encoding ribosomal RNA have been particularly important for inferring species phylogenies because they are easily accessible and collectively demonstrate a wide range of evolutionary rates and therefore a wide potential for phylogenetic resolution across a large time scale. Phylogenetic sequence comparisons have concentrated on the coding portions of the ribosomal genes and their RNA products. Nuclear-encoded ribosomal RNA demonstrates an unusual pattern of evolution featuring the interspersion of relatively rapidly evolving sequences with highly conserved ones that represent some of the most highly conserved macromolecular sequences known (Gerbi, 1985). The most highly conserved sequences have been useful for investigating the oldest divergences in the history of life (G. E. Fox et al., 1980; Küntzel and Köchel, 1981; Spencer et al., 1984; Hasegawa et al., 1985; Lane et al., 1985; Olsen, 1987; Field et al., 1988; Cedergren et al., 1988; Lake, 1988; Ghiselin, 1988). Comparisons of mitochondrial and chloroplast ribosomal genes with prokaryote ribosomal genes have helped resolve relationships among these eukaryote organelles and their prokaryotic relatives (Yang et al., 1985; Turner et al., 1989).

Sequence studies of the more variable regions of the rRNA genes, including the rapidly evolving sequences known as "divergent domains" or "expansion segments," have proven useful for investigating evolutionary divergences that occurred during the past 500 million years or so. For instance, phylogenetic problems examined to date using relatively variable sequences within the rRNA genes include such diverse questions as the relationships of coelacanths to tetrapods (Hillis et al., 1990), amphibian phylogeny (Larson and Wilson, 1989; Hillis, 1990), seed plant relationships (Zimmer et al., 1989), tribal relationships within the plant family Poaceae (Hamby and Zimmer, 1988), protist phylogeny (Elwood et al.,

1985), and relationships among amniotes (Hillis and Dixon, 1989). To date, relatively few sequencing studies have examined the transcribed and nontranscribed spacer regions of rDNA arrays, but restriction site analyses suggest that these regions will be useful for reconstruction of phylogenies over at least the past 50 million years (Hillis and Davis, 1986). The nuclear-encoded ribosomal RNA sequences can be studied by sequencing cloned copies of the ribosomal genes (for example, Ware et al., 1983; Hassouna et al., 1984; Elwood et al., 1985; Hillis and Dixon, 1989), or by sequencing the ribosomal RNA directly (Hamlyn et al., 1978; Youvan and Hearst, 1979; Qu et al., 1983; Hamby et al., 1988; Larson and Wilson, 1989). The latter is facilitated by the fact that the majority of the RNA in any cell is nuclear-encoded ribosomal RNA. For study of the rDNA spacer regions, cloning is usually required (Hillis and Dixon, 1989). Sequence comparisons of selected regions of animal mtDNA (including the mtDNA ribosomal genes) are useful for inferring phylogenetic relationships of species whose divergences are more recent than those accessible using nuclear-encoded ribosomal RNA (e.g., Brown et al., 1982; Higuchi et al., 1984; Hixson and Brown, 1986; Hayasaka et al., 1988; Holmquist et al., 1988). The methodology for obtaining this sequence information is identical to that described for the use of mtDNA sequences for tracing intraspecific genealogies.

Many other loci have been cloned and sequenced from several species and used for phylogeny reconstruction. Loci for which considerable comparative sequence information is available include (in addition to the rRNA genes and mtDNA genome discussed above) the globin loci in primates (e.g., Koop et al., 1986; Goodman et al., 1987; Miyamoto et al., 1987; Holmquist et al., 1988), the immunoglobulin genes in rodents (e.g., Ponath et al., 1989a,b), and the alcohol dehydrogenase loci in fruit flies (e.g., Bodmer and Ashburner, 1984; Schaeffer and Aquadro, 1987; Bishop and Hunt, 1988). The number of loci examined in a comparative manner and the number of taxa for which such information is available should increase dramatically over the next several years.

Many comparative plant DNA sequencing studies have utilized the chloroplast genome. Complete chloroplast DNA (cpDNA) sequences have been obtained for two species of plants, *Nicotiana tabacum* and *Marchantia polymorpha* (Shinozaki et al., 1986; Ohyama et al., 1986). However, broad-scale comparisons of cpDNA sequences have been conducted primarily for *rbcL*, the gene that encodes the large subunit of ribulose-1,5-biphosphate carboxylase (Aldrich et al., 1986a,b; Ritland and Clegg, 1987; Zurawski and Clegg, 1987). These initial studies have focused on domesticated species and only recently have been extended to include groups that will allow resolution of higher order relationships among seed plants or green algae (Turner et al., 1989; Morden and Golden, 1989).

Summary

From the preceding discussion, it should be clear that nucleic acid sequencing can be used to study virtually any systematic problem, from studies of intraspecific variation to the phylogeny of life. However, this does not mean that sequencing is necessarily the best approach to any problem, because it is often not a cost- or time-effective method for obtaining relevant data (see Chapter 12). Because of the time and expense involved with collecting appropriate sequence data, other techniques are best used for many systematic applications. This is especially true of studies that require examination of many individuals, such as most studies of intraspecific variation (e.g., geographic variation, reproductive modes, and heterozygosity estimates) and most studies of closely related species (hybridization, cryptic species, and recent phylogeny). However, for problems of phylogeny reconstruction over relatively ancient spans of time (greater than 50 million years), no other molecular information is as likely to be informative as appropriate nucleotide sequence data.

LABORATORY SETUP

Although all nucleic acid sequencing work requires a relatively sophisticated laboratory, needs will vary depending on the method(s) chosen for isolating the target sequence. In general, cloning requires somewhat more equipment (and experience) than PCR amplification or direct RNA sequencing. Table 1 provides a general idea of the requirements of a typical sequencing laboratory.

Beyond the equipment listed in Table 1, the supply needs for a sequencing laboratory are similar to those described in Chapter 8 for restriction enzyme analysis. Cloning work requires a few additional supplies, such as disposable petri dishes, culture tubes, and wire loops for spreading bacterial colonies. Oligonucleotide primers for sequencing are commercially available for the various cloning vectors, or specific primers can be made to order by many companies and centralized institutional facilities. If an oligonucleotide synthesizer is available, primers can be designed and constructed with little time or effort (see Barnes, 1987).

In addition to the materials described for restriction site analysis in Chapter 8, cloning and sequencing require a few specialized enzymes, antibiotics, and other chemicals. Sanger dideoxy sequencing requires one of several DNA polymerases, or reverse transcriptase for RNA sequencing (see Protocols 18 and 19). PCR requires *Taq* polymerase, which is also an effective polymerase for DNA sequencing. DNA ligase is needed in cloning to ligate compatible fragments (see Protocol 9). Plasmid subcloning requires

Table 1 Equipment needed for a sequencing laboratory[a]

Equipment	Use (protocols in parentheses)
Agarose gel apparatus, small	Quick check of DNA fragments*
Agarose gel apparatus, large	Separation of DNA fragments*
Autoclave	Sterilization*
Balance, analytical	Weighing small samples*
Balance, top-loading	General weighing*
Camera, Polaroid	Gel photography*
Centrifuge, microtube	Centrifuging small samples*
Centrifuge, high-speed refrigerated	Centrifuging large samples
Rotors:	
for 250-ml bottles	Spinning down cells (6, 14)
for 50-ml tubes	DNA, cell isolation (1–4, 6–14
Centrifuge, ultra-speed	Isolation of cellular components
Rotors:	
swinging bucket, ≈36-ml tubes	mtDNA, cpDNA isolation (Chapter 8)
vertical or fixed angle, 2–5-ml tubes	Plasmid, mtDNA, and cpDNA isolation (14, Chapter 8)
Centrifuge, vacuum	Drying DNA/RNA samples (N.E.)
Computer, with hard disk	Sequence storage and analysis*
Darkroom, with developing tanks	Developing autoradiographs (7, 10, 21)
Distilled water source	Purified water*
Film cassettes, autoradiography	Autoradiography (7, 10, 21)
Freezer, −80°C	Tissue and sample storage*
Freezer, −20°C, not frost-free	Sample and enzyme storage*
Fume hood	Use of caustic reagents*
Geiger counter	Radiation detection and safety (7, 18–21)
Gel dryer	Drying gels for autoradiography (21)
Gel reader	Reading sequence gels (N.E.)
Goggles, UV protective	Safety*
Heating blocks, ambient–100°C (≈3)	Heating reactions*
Ice machine	Cooling samples or reactions*
Incubator, 37°C	Bacterial incubation (6–15)
Incubator, 55–65°C, with rocker	DNA hybridization (7; N.E.)
Laminar flow hood	Sterile work area (6–15; N.E.)
Lucite screens	Radiation protection (7, 18–21)
Micropipetters (set from 1 µl–1 ml)	Pipetting small volumes*
Microwave oven	Heating liquids*
pH meter	Adjusting pH of solutions*
Plastic bag sealer	Filter hybridization (7)
Power supply, 2,000–5,000 V	Acrylamide gel electrophoresis (20)
Power supply, 250–500 V	Agarose gel electrophoresis*
Refrigerators, 4°C	Cold storage*
Sequencing gel apparatus	DNA or RNA sequencing (20)
Shaker, orbital	Gel fixation (21)
Shaker, heated water bath	Bacterial growth, hybridization (6–15)
Spectrophotometer	DNA/RNA quantification, purity assessment, cell concentration (1–14)
Stirring hot plate	Mixing solutions*

Table 1 (*Continued*)

Equipment	Use (protocols in parentheses)
Thermal cycler	PCR amplification (16; N.E.)
Timer	Timing protocol steps*
Tissue homogenizer	Tissue grinding (1–4)
UV light box, long wave	Visualizing and photographing gels*
Vacuum oven (or UV oven)	Crosslinking DNA to membranes (7)
Vacuum pump	For vacuum oven and centrifuge
Vortexer	Mixing samples*
Water baths, ambient to 65°C (≈3)	Constant temperature of samples*
Water bath, cooling	Preparing radioactive probes (7; N.E.)

[a]Not all items listed are needed for some of the applications described in this chapter. Therefore, the protocols for which each item is needed are listed; items that are essential for many protocols are marked with an asterisk. Items that are nonessential but facilitate a procedure are marked N.E.

the use of various antibiotics and substrates (to ensure the presence of plasmids and for screening recombinant from nonrecombinant plasmids; see Protocol 11). Vector DNA and host bacterial strains are available commercially or through exchange with other laboratories. The remaining reagents are the same as described for restriction site analysis (Chapter 8).

PROTOCOLS
All of the stock solutions given in the following protocols are listed in the appendix to this chapter.

1. DNA isolation from animals, protists, and prokaryotes
2. DNA isolation from plants, fungi, and algae
3. Isolation of RNA from animals
4. Isolation of RNA from plants
5. Preparation of partial gene libraries in λ bacteriophage vectors
6. Growing bacteriophage
7. Screening bacteriophage libraries
8. Miniprep isolation of λ DNA
9. Subcloning into plasmids or M13
10. Preparation of frozen competent cells for transformation
11. Transformation of *E. coli* with plasmid DNA
12. Transformation of M13 bacteriophage DNA
13. Isolation of plasmid DNA
14. Miniprep isolation of M13 DNA
15. Preparing permanent frozen stocks of plasmid clones
16. DNA amplification by the polymerase chain reaction
17. Preparing a sequencing gel

Protocol 1: DNA Isolation from Animals, Protists, and Prokaryotes

(Time: day 1, ≈3 hr; day 2: ≈30 min)

There are many protocols for isolation of high-molecular-weight DNA; the following protocol is useful for isolating DNA from small tissue samples and will produce more than enough DNA for all applications in this chapter. This procedure works well for many multicellular animals, but it may not work for crustaceans and some fish, among other organisms; unicellular organisms may be processed by starting at step 2. For any organism, if the final DNA is degraded, it may be necessary to skip step 6. This will greatly reduce yields, but will almost always produce high-molecular-weight DNA. Protocol 2 will also work well on some animal tissues. For vertebrates, muscle tissue produces the highest quality DNA, although liver tissue usually produces the greatest yields. Vertebrate blood is also a good source of high-quality DNA; for all vertebrates except mammals, a few drops of blood can be diluted directly into STE in step 2 (much greater quantities of mammalian blood are needed, because mammalian erythrocytes are enucleate and the DNA must be isolated from the leukocytes). Extraction of DNA from many species of mollusks (especially gastropods) has proven difficult; it is usually necessary to experiment with several different tissues (the gonads work well in many species). Most insects can be processed whole after removal of the digestive tract. If DNA is to be isolated from ethanol-preserved samples, the tissue should be lyophilized first. See Chapter 8 for protocols for isolating mtDNA and cpDNA and Chapter 7 for a large-scale protocol for isolating nuclear DNA.

1. If the tissue sample cannot be disrupted easily, grind the sample to a fine powder in liquid nitrogen with a mortar and pestle. (Note: Be very careful while powdering tissue as the mortar and pestle can shatter due to the extreme cold.)
2. Place ≈100 mg of the tissue into 500 µl of STE (Appendix) in a 1.5-ml microcentrifuge tube.
3. Add 25 µl of a 10 mg/ml stock of proteinase K in STE and mix well. If the tissue was not powdered, grind the tissue with a pellet pestle or similar instrument.
4. Add 25 µl of 20% SDS.
5. Mix again, taking the time to break up any remaining clumps of tissue.
6. Incubate 2 hr at 55°C. Mix occasionally during the incubation to keep the tissue suspended.
7. Add an equal volume of PCI (Appendix), mix gently but thoroughly and incubate at room temperature for 5 min. If the phases separate, gently mix again.
8. Centrifuge for 5 min at 7000 g.
9. Carefully remove the aqueous layer with a micropipette and a wide-bore tip and

transfer to a clean tube. The aqueous layer is usually the top layer, although high salt concentrations can cause inversion of the phases. Be careful not to disturb the cellular debris on the interface.

10. Reextract the aqueous phase with PCI (repeat steps 7–9).
11. Add an equal volume of CI (Appendix), mix gently, and incubate at room temperature for 5 min. Remix once a minute to prevent the phases from separating.
12. Centrifuge for 3 min at 7000 g.
13. Carefully remove the upper (aqueous) layer with a micropipette and a wide bore tip and transfer to a clean tube. Be careful not to disturb the interface.
14. Reextract the aqueous phase with CI (repeat steps 11–13).
15. Add one-tenth volume (about 45 µl) of 2 M NaCl or 3 M NaAc and 1 ml of cold (−20°C) absolute ethanol to precipitate the DNA.
16. Incubate on ice for 10–20 min.
17. Spin down the DNA precipitate. If large wisps of DNA are visible, centrifuge for 20 sec at 7000 g. If DNA is not clearly visible, centrifuge for 1–2 min at 7000 g. Overpelleting the DNA will make resuspension difficult.
18. Decant the ethanol and dry the pellet in a vacuum centrifuge until the ethanol has just evaporated. Overdrying the pellet will make resuspension difficult.
19. Resuspend the pellet in 250 µl of 1× TE (this may require up to 24 hr). Check concentration and purity of sample in a spectrophotometer by taking readings at 260 and 280 nm. An optical density of 1 at 260 nm corresponds to a double-stranded DNA concentration of approximately 50 µg/ml. The ratio of the readings at 260 nm/280 nm should be approximately 1.8; lower readings indicate contamination with protein and/or phenol.

Protocol 2: DNA Isolation from Plants, Fungi, and Algae

(Time: day 1, ≈2 hr; day 2, ≈30 min)

Many methods have been developed for isolation of high molecular-weight-DNA from plants (Zimmer et al., 1981; Saghai-Moroof et al., 1984; Rogers and Bendich, 1985; Doyle and Dickson, 1987). These methods differ primarily with respect to their requirements for input material (fresh, frozen, or lyophilized; gram or hundreds of gram quantities) and the use or nonuse of ultracentrifugation steps. The protocol given below is relatively simple and is useful for the preparation of the small samples of DNA needed for applications in this chapter. See Chapter 8 for protocols for isolating cpDNA.

1. Grind leaf or flower tissue to a fine powder in liquid nitrogen with a mortar and pestle. (Note: Be very careful while powdering tissue as the mortar and pestle can shatter due to the extreme cold).
2. Add 2-mercaptoethanol (BME) to 2× CTAB extraction buffer (Appendix) to a final concentration of 0.2%. Heat the CTAB plus BME solution in a 60°C water bath for 5 min.
3. Aliquot 500 µl of the above buffer into a 1.5-ml microcentrifuge tube. Add ≈100 mg fine nitrogen-powdered tissue and place in a 60°C water bath for 45 min.

4. Add 500 µl of CI. Close the tube and extract by gently inverting the tube. Extract for 10 min.
5. Centrifuge for 5 min at 7000 *g* in a microcentrifuge.
6. Transfer the upper (aqueous) phase to a fresh tube using a wide-bore pipette tip and reextract with CI. Centrifuge as above and transfer the aqueous phase to a fresh tube.
7. Add 1 ml of absolute ethanol and allow the DNA to precipitate at −20°C at least 30 min. (Note: Precipitating overnight substantially increases the yield.)
8. Centrifuge for 1–5 min at 7000 *g* to pellet the DNA.
9. Decant the ethanol and briefly dry the pellet in a vacuum centrifuge.
10. Redissolve the pellet in 100 µl of 1× TE. Add 10 µl of 3 *M* NaAc and 2.5 volumes of 95% ethanol. Precipitate at −20°C for 30 min.
11. Centrifuge for 5 min at 7000 *g*. Decant the ethanol and add 1 ml of 70% ethanol to wash the pellet. Recentrifuge for 2 min at 7000 *g*. Two 70% ethanol washes may be necessary to remove traces of CTAB or chloroform.
12. Dry the pellet in a vacuum centrifuge until all visible traces of ethanol are gone. Do not overdry the pellet. Redissolve in 200 µl of 1× TE. Determine concentration and purity of the sample as in Protocol 1, step 19.

Protocol 3: Isolation of RNA from Animals

(Time: day 1, ≈2 hr; day 2, ≈3 hr; day 3, ≈1 hr)

RNA is much less stable than DNA, and the tissues must be as fresh as possible. All glassware used for RNA work must be baked at 250°C for at least 4 hr to remove RNase. Water should be treated with 0.1% diethylpyrocarbonate (see Appendix). All plasticware should be sterile. It is a good idea to use a separate set of glassware and other reusable supplies exclusively for RNA work to avoid contamination with RNase.

The following protocol is useful for isolating total cellular RNA from vertebrates and many other multicellular animals for direct sequencing of rRNA. Protocol 4 is preferred for isolation of RNA from plants and algae, as well as most insects.

First Day

1. Place frozen phenol in a 60°C waterbath to melt.
2. Weigh the frozen tissue and place in a mortar. Cover with liquid nitrogen.
3. Measure 5 ml of guanidinium isothiocyanate solution (Appendix; with 1/100 volume 2-mercaptoethanol added) per gram of tissue (from step 2). Add to a 50-ml centrifuge tube and set aside.
4. Grind the tissue to a fine powder with a pestle. (Liquid nitrogen may have to be added several times as it evaporates.)
5. Pour the guanidinium isothiocyanate/2-mercaptoethanol solution into the mortar and stir until the mixture freezes.
6. Place the mortar in a 60°C waterbath until the mixture melts. Stir, then pour the mixture into the centrifuge tube and place it in a beaker of water in the 60°C waterbath.

7. Draw the mixture into a syringe (10 ml volume, fitted with an 18-gauge needle) and forcibly eject it into the centrifuge tube. Repeat until the viscosity of the mixture is reduced.

8. Add phenol (5 ml/g of tissue), preheated to 60°C, and continue to pass the emulsion through the syringe.

9. Add 5 ml of ATE (Appendix) per gram of tissue.

10. Add 5 ml PCI/g of tissue and shake vigorously for 10–15 min while maintaining the temperature at 60°C.

11. Cool on ice and centrifuge for 10 min at 4°C using a swinging bucket rotor at moderate speed ($\approx$3000 g).

12. Recover aqueous (top) phase (use siliconized Pasteur pipette) into a new 50-ml centrifuge tube.

13. Reextract with an equal volume of PCI at 60°C.

14. Repeat steps 11 and 12.

15. Reextract twice with CI, centrifuge, and recover (at room temperature).

16. Add 2–2.5 volumes absolute ethanol, then store at −20°C overnight.

Second Day

17. Centrifuge for 20 min at 4°C using a swinging bucket rotor at moderate speed ($\approx$3000 g).

18. Pour liquid into a waste container, and dry the pellet in a vacuum centrifuge.

19. Dissolve the pellet in the original starting volume (step 3) of STE plus 0.2% SDS.

20. Add 20 µl proteinase K (10 mg/ml in STE) per ml starting volume. Incubate for 1–2 hr at 37°C.

21. Heat to 60°C. Add 1/2 volume (from step 19) of phenol heated to 60°C and mix. Add 1/2 volume (from step 19) of CI; mix for 10 min at 60°C.

22. Cool on ice and centrifuge at 4°C for 10 min ($\approx$3000 g).

23. Recover aqueous (top) phase (with siliconized Pasteur pipette) into a new 50-ml centrifuge tube.

24. Repeat steps 21–23.

25. Extract twice with CI at room temperature (as in steps 21–23, except for temperature)

26. Add 2–2.5 volumes of absolute ethanol and store at −20°C overnight.

27. Repeat steps 17 and 18.

28. Resuspend the pellet in DEP-treated, distilled water (<1 ml).

29. Take optical density readings on a 1/50 dilution (10 µl sample in 490 µl DEP-dH$_2$O) at 260 and 280 nm in a spectrophotometer. An optical density of 1 at 260 nm corresponds to $\approx$40 µg/ml for RNA. Pure samples of RNA have a ratio of optical density readings at 260 nm/280 nm of $\approx$2.0; lower readings indicate contamination by protein and/or phenol.

30. Separate the sample equally into two microcentrifuge tubes. For long-term storage, precipitate one sample with 2–2.5 volumes absolute ethanol and store at −80°C.

31. For immediate use, add dithiothreitol (DTT) and RNasin to the second tube as follows: 1 µl of 2.5 M DTT per 500 µl of sample (or 10 µl of 0.25 M DTT per 500

μl). Spin, vortex, and spin again. Add 12.5 μl RNasin per 500 μl of sample. Spin, vortex, and spin. Store at −80°C.

Protocol 4: Isolation of RNA from Plants

(Time: day 1, 8–10 hr; day 2, ≈ 4 hr)

This technique is the most effective for isolation of RNA from plants and algae, as well as from many insects. It is a modification of the procedure of Hall et al. (1978) (for further information, see Hamby et al., 1988).

1. Place 5–10 g of liquid-nitrogen-powdered tissue in a 50-ml polypropylene tube, add 25 ml hot (90–95°C) borate buffer (Appendix), and homogenize the sample in three 10-sec bursts.
2. Filter the extract through sterile cheesecloth into a fresh tube. Add 0.3 ml of 10 mg/ml proteinase K solution. Incubate for 1 hr at 37°C. Add 1 ml of 2 M KCl to the tube and chill on ice for 5–10 min.
3. Centrifuge at 16,000 g in a swinging bucket rotor for 10 min at 4°C. Filter the supernatant through a double layer of laboratory wipes into a 30-ml glass centrifuge tube. Add 1/4 volume of 10 M LiCl. Freeze the sample on dry ice for 30 min and then keep at 4°C for 2–4 hr.
4. Centrifuge at 13,000 g in a swinging bucket rotor for 15 min at 4°C. Pour off the supernatant immediately as the RNA pellet will be loose.
5. Wash and resuspend the pellet with 5 ml of cold 2 M LiCl. Centrifuge as in step 4.
6. Resuspend the pellet in at least 2 ml of 2 M potassium acetate, pH 5.5. This pellet often requires extensive vortexing and some warming to redissolve. Add 2.5 volumes of ice-cold ethanol and store at least 4 hr at −20°C.
7. Centrifuge at 12,000 g in a swinging bucket rotor for 15 min at 4°C.
8. Air dry the pellet and dissolve in 5 ml STE. (Remove aliquots of 20–50 μl here and at steps 9 and 10 to assay RNA integrity on agarose minigels.) Add 5 ml of PCI and extract the sample with thorough mixing for 2–5 min. Let stand on ice for 10 min.
9. Repeat step 7. Remove the top layer and put in a fresh glass centrifuge tube (remove minigel aliquot) and then add 1 ml of 4 M ammonium acetate and 10 ml ice-cold absolute ethanol. Mix well and store 4 hr to overnight at −20°C.
10. Repeat step 7. Dry the pellet in a vacuum centrifuge. Dissolve in 1–2 ml of 1× TE (use more TE with larger pellets). Determine the RNA concentration and purity as in Protocol 3, step 29.

Protocol 5: Preparation of Partial Gene Libraries in λ Bacteriophage Vectors

(Time: day 1 [steps 1–6], ≈6–10 hr; day 2 [steps 7–9], ≈2–3 hr)

The following protocol presents an example of λ cloning that is typical for many commercially available λ cloning vectors. One can also grow and purify λ DNA and make packaging extracts rather than using commercial preparations (see Berger and Kimmel, 1987, for details). In step 1, the DNA may be digested to completion with a particular endonuclease that is known to flank the region of interest. If this information is not known, it is usually preferable to partially digest the DNA with an endonuclease that has a short recognition sequence (e.g., *Mbo*I) to generate fragments of the desired size.

1. Digest 1 µg of target DNA with the desired cloning enzyme (e.g., *Eco*RI).
2. Ethanol-precipitate the restriction digest by the addition of 1/10 volume of 2 M NaCl and 2.5 volumes of absolute ethanol. (Note: 3 M NaAc can be used in place of 2 M NaCl, but small traces of NaAc seem to be more detrimental to ligation efficiency.)
3. Incubate two or more hr at $-20°C$, then centrifuge at 7000 g for 20 min. Decant the ethanol and dry the pellet in a vacuum centrifuge.
4. Resuspend the pellet in 10 µl of water.
5. Assay 5 µl of the digested DNA on a minigel with standard lanes containing 0.1 and 0.5 µg of DNA. This will verify that the restriction digest and subsequent ethanol precipitation were successful.
6. Add an equal molar ratio of target DNA to λ phage arms. For instance, if the approximate average target cloning size is 8 kb and the λ arms total 40 kb, then add 0.2 µg of the digested DNA ($\approx$ 2 µl) to 1 µg of λ phage arms and bring the total volume to 3.5 µl. Next add 0.5 µl of 10× ligation buffer, 0.5 µl of T4 DNA ligase (2 Weiss units), and 0.5 µl of 10 mM ATP, pH 7.5. Mix the ligation reaction thoroughly and incubate for 1 hr at room temperature, then overnight at 4°C.
7. Package the ligation using a commercial packaging extract, following the manufacturer's instructions. (This step varies slightly with various packaging extracts, but usually involves simply adding the ligation mixture directly to a freeze-thaw lysate and a sonic extract and incubating at room temperature for a few hours.)
8. Dilute the packaged phage with 0.5–1 ml of PDB (Appendix). The resulting gene library should contain 10^6–10^9 recombinant phage (depending on the efficiency of the packaging extract used and the quality of the DNA ligation) with inserts from 1 to 23 kb (depending on the cloning vector used).
9. Plate serial dilutions (1 µl/0.1 µl/0.01 µl) of the gene library to determine the titer and recombination efficiency (see Protocol 6).

Protocol 6: Growing Bacteriophage

(Time: day 1, $\approx$10 min; day 2, $\approx$1 hr [plus incubation time])

Recombinant λ bacteriophage are grown by adding aliquots or serial dilutions of the phage library to appropriate host bacteria, and then plating the bacteria and selecting the resulting plaques. For titering libraries, it is usually desirable to plate several 10-fold serial dilutions of the stock to accurately determine the concentration. If relatively few recombinant phage are obtained, or if larger quantities of the library are desired, the library can be amplified (see Berger and Kimmel, 1987). However, it should be cautioned that some recombinant bacteriophage will replicate much faster than others (because of the size of the insert), and that the amplified library will therefore overrepresent some clones and underrepresent others. Therefore, it is usually best not to amplify the library unless absolutely necessary.

For growing λ bacteriophage, strains of bacteria are selected that do not allow recombination among the phage (recA⁻ strains); these strains are typically supplied with the phage arms and the recA⁻ phenotype can usually be maintained by antibiotic selection. Systems for detection of recombinant versus reconstituted λ bacteriophage also vary with different host strains; some systems use color selection by IPTG/X-Gal (see Berger and Kimmel, 1987) and others use bacteria that allow only recombinant λ growth. The basic protocol for growing bacteriophage is given below; variations may be required for particular bacterial host strains.

1. Pick a single colony of the host strain from a plate that contains the antibiotic that allows selection for the recA⁻ phenotype, and add to L-broth (Appendix) plus 0.2% maltose plus 10 mM MgSO₄ using sterile technique (250 ml of L-broth is enough for most applications). Grow overnight with vigorous shaking (≈300 rpm) at 37°C.
2. Centrifuge in sterile tubes at 1000 g for 10 min to pellet the cells.
3. Resuspend the cells in one-half of the original volume of sterile 10 mM MgSO₄.
4. Remove L-broth plates from 4°C and warm them in an incubator (37°C).
5. Mix 200 µl of cells for a 100-mm plate or 450 µl of cells for a 150-mm plate with the phage stock in a sterile culture tube. Incubate at 37°C for 15 min with gentle (≈100 rpm) shaking.
6. While the cells plus phage are incubating, melt L-broth top agarose in a microwave oven and allow it to cool to 48°C. Hold at 48°C in water bath. (Top agarose is preferable to top agar, because the former will not stick to filter lifts as readily.)
7. After the infection is complete, add 3 ml (100 mm plate) or 7 ml (150 mm plate) of 48°C top agarose to the culture tube, vortex gently, and pour over the surface of the plate. Tilt the plate to spread the agarose evenly. Grow 6 hr to overnight in a 37°C incubator until plaques are approximately 1 mm in diameter.

Protocol 7: Screening Bacteriophage Libraries

(Time: step 1, see Protocol 6; step 2, ≈2 hr; steps 3–7, ≈2.5 hr; step 8, ≈2 days; steps 9–10, see Protocol 6; steps 11–12, ≈4 days. Total time: ≈1 week)

To find the particular gene or DNA region of interest, one must screen the gene library by plating the phage at an appropriate density (typically 2000–50,000 plaques per plate), transferring the phage DNA to a binding membrane (filter lift), and hybridizing the filter lift with an homologous probe. This procedure is relatively easy if the gene is present in high copy number (e.g., the rRNA genes, heterochromatic repeats, or mtDNA fragments) and is flanked by appropriate restriction sites for the library that has been constructed. Single-copy genes require screening of many more plaques (often as many as 10^6); this may require plating on larger plates than in the protocol below or use of a λ strain that accepts larger fragments.

This protocol is among the simplest for identifying clones of interest, although numerous other techniques are more applicable in particular situations. For a review of the various methods, see Berger and Kimmel (1987) or Ausubel (1989).

1. Plate out the phage at a density where the plaques cover the majority of the plate, but do not overlap significantly. Square plates are preferable to round plates, as square filter lifts save film during autoradiography. For a 100 mm square plate, approximately 2000–10,000 plaques can be screened efficiently. Incubate plates for ≈8 hr at 37°C.
2. Cool the plates several hours at 4°C to harden the top agarose.
3. Carefully lay a nylon or nitrocellulose filter onto the surface of the plate and wait about 2 min for it to absorb moisture (and phage DNA) from the plate. No bubbles should be trapped under the nylon or areas of the plate will not transfer well. While waiting, stick a hypodermic needle containing waterproof ink through the filter into the plate in three to five places. This should mark both the filter and the plate with ink dots for later realignment.
4. Carefully peel the nylon filter off and place it into denaturing solution for ≈2 min. Meanwhile, lay a second filter on the plate and repeat the process, this time waiting ≈4 min before removing the filter. Mark the second filter in the same spots as the first with the waterproof ink. Place the second filter into denaturing solution (Appendix) for 2 min.
5. Transfer the nylon filters to neutralizing solution (Appendix) for 5 min.
6. Transfer the nylon filters to 2× SSC (Appendix) for 30 sec.
7. Air-dry the filters and bake for 2 hr at 80°C in a vacuum oven (or 30 sec in a UV oven).
8. Hybridize to the desired probe sequence (see Chapter 8 for details of this procedure). See Figure 5 for results of filter-lift hybridization. Positive plaques should appear on autoradiographs from both filter lifts; dark marks on the autoradiograph of only one filter are false positives.
9. For primary screening, align the plate and the resulting autoradiograph using the ink marks. Use the wide end of a sterile Pasteur pipette to "plug" the plate at the region containing a positive plaque. Place the agar plug into 0.5 ml of PDB plus 50 μl chloroform. This 0.5-ml phage stock is the working stock and will contain the desired clone plus several adjacent clones.
10. Titer the working stock (Protocol 6) and plate ≈100 plaques.
11. Repeat the screening process described above (steps 2–9). For secondary screening the phage should be plated at much lower density so that each plaque is clearly separate.
12. Plug "secondary" isolated positives with the small end of a sterile Pasteur pipette and again put into 0.5 ml of PDB plus 50 μl chloroform. This is the stock from which you will isolate DNA in Protocol 8. Check clones by agarose gel electrophoresis, restriction analysis, and autoradiography (Figure 6; see also Chapter 8).

Protocol 8: Miniprep Isolation of λ DNA

(Time: day 1 [steps 1–2], 30 min plus incubation time [6–8 hr]; day 2, ≈4 hr)

Once the λ clone of interest has been isolated, large quantities of the cloned DNA can be grown and purified. Although it is possible to sequence the cloned DNA in the

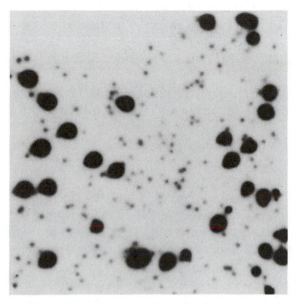

FIGURE 5 Autoradiograph of a filter lift. The dark circles correspond to positive plaques and the small light marks correspond to negative plaques.

λ vector directly, it is usually desirable to subclone the DNA into a plasmid or M13 vector, because of greater ease of sequencing and DNA preparation. For most λ vectors, the cloned DNA must be isolated prior to the subcloning steps. Newly developed λ vectors (such as the λZAP vectors of Stratagene Cloning Systems) contain a plasmid within the λ vector, and allow an in vivo excision of the plasmid using a helper phage, thus bypassing the subcloning steps. For other λ vectors, the following simple protocol can be used to isolate the cloned DNA for subcloning. For other protocols or for large-scale isolation of phage DNA, see Miller (1987).

1. Add 450 μl of host bacterial cells (prepared as described in Protocol 6, steps 1–3) to enough λ stock to contain approximately 50,000–100,000 phage-forming units (PFU) in a sterile culture tube. Incubate at 37°C for 15 min with gentle shaking (≈100 rpm). Then add 7 ml top agar (or top agarose) at 48°C (as in Protocol 6, steps 6–7) and plate on a 150 mm L-broth + $MgSO_4$ + maltose plate (Appendix). Some batches of agar contain contaminants that inhibit restriction enzyme activity; therefore, agarose plates are preferable to agar plates for isolation of phage DNA. Grow 6–8 hr at 37°C. The plaques should be confluent or nearly so.
2. Add 5 ml of PDB to the plate and shake gently at 4°C overnight.
3. Remove the PDB with a Pasteur pipette and transfer it to a glass or polypropylene centrifuge tube. Add 200 μl of chloroform and mix.
4. Spin down the debris at 7500 g for 10 min at 4°C.

5. Collect the supernatant, transfer it to a clean glass or polypropylene centrifuge tube, and add 1 μg/ml of DNase I and RNase A (normally kept as 1 mg/ml stocks).
6. Incubate 30 min at 37°C.
7. Add an equal volume of PEG stock (Appendix) and mix gently.
8. Incubate 1 hr on ice.
9. Pellet the precipitated phage by centrifugation at 12,000 g for 20 min at 4°C.
10. Decant the supernatant and allow the inverted tube to drain thoroughly. (Note: A white precipitate should be clearly visible.)
11. Resuspend the pellet in 0.5 ml of PDB in a 1.5-ml microcentrifuge tube. Add 5 μl of 0.5 M EDTA.
12. Incubate at 65°C for 15 min.
13. Extract twice with an equal volume of PCI as described in Protocol 1 (steps 7–9). A large amount of PEG will collect at the interface during these extractions.
14. Extract twice with an equal volume of CI as described in Protocol 1 (steps 11–13).
15. Add 50 μl of 2 M NaCl and 1 ml of ethanol to precipitate the DNA.
16. Centrifuge at 7500 g for 10 min to pellet the DNA.
17. Decant the ethanol, dry the pellet in a vacuum centrifuge, and resuspend the DNA in 250 μl of 1× TE. Check concentration and purity spectrophotometrically as described in Protocol 1, step 19. Ten microliters of this stock should be ample for a test restriction or a subcloning experiment.

Protocol 9: Subcloning into Plasmids or M13

(Time: steps 2–10, ≈6 hr)

The following protocol assumes that the target DNA sequence is flanked by appropriate restriction sites for the vector of choice. If not, then linkers need to be added to the target sequence (see Helfman et al., 1987, for protocol and additional information).

1. Isolate DNA from the λ clone of interest (Protocol 8) or from PCR amplification (Protocol 16).
2. Digest 1 μg of the DNA with the appropriate restriction enzyme(s) to cut out the sequence of interest.
3. Digest 0.5 μg of plasmid or M13 vector DNA with a restriction enzyme that produces compatible ends (e.g., BamHI and BglII produce compatible ends).
4. Add 1/10 volume of 2 M NaCl and 2.5 volumes of cold absolute ethanol to precipitate the DNA. Incubate for 2 hr at −20°C.
5. Centrifuge for 15 min at 12,000 g to pellet the DNA.
6. Decant the ethanol, and dry the pellet in a vacuum centrifuge.
7. Resuspend the pellets in 10 μl of TE and assay 2–3 μl on a minigel. This will verify that the restriction digest and subsequent ethanol precipitation were successful.
8. Mix the target and vector DNA in a 2:1 molar ratio of ligatable ends. (Use the size of the cloning vector and of the targeted insert to determine the molar ratio.)
9. Bring the volume of the DNA solution to 39 μl with water. Add 5 μl of 10× ligation buffer (Appendix), 5 μl of 5 mM ATP, and 1 μl (4 Weiss units) of T4 DNA ligase.

10. Mix the ligation reaction and incubate overnight at 4°C.
11. Transform the ligation (Protocol 11) and screen for the desired clone.

Protocol 10: Preparation of Frozen Competent Cells for Transformation

(Time: day 1, 10 min; day 2, ≈3–4 hr; day 3, ≈30 min)

For the plasmid clones produced in Protocol 9 to be grown in quantity, they must be introduced into bacterial host strains. This is accomplished at high efficiency by making the host bacteria competent for transformation. Production of competent cells requires careful attention to detail (especially with regard to maintaining low temperature and the density of cells at harvest), and all sterile tubes, glassware, and solutions. There will likely be considerable variation in transformation efficiency from batch to batch of competent cells, so if one preparation produces poor results, try the procedure again. Commercial preparations of competent cells are also available and are usually of reliable quality. For additional information, see Miller (1987).

1. Inoculate 10 ml of L-broth with a loopful of an appropriate strain of E. coli cells from a single colony. Grow overnight at 37°C.
2. Subculture 5 ml of the overnight culture into 500 ml of L-broth in a 2 L flask.
3. Grow to OD_{600} = 0.4–0.5, as measured with a spectrophotometer (usually between 2 and 3 hr).
4. Pour the culture into sterile 250-ml plastic bottles. Centrifuge at 2500 g for 5 min. Decant the supernatant.
5. Resuspend the pellets in 100 ml of cold (0–4°C) 0.1 M $MgCl_2$ (total volume). Transfer the cell suspension to two 50-ml Oak Ridge tubes. (Note: From this point in the protocol, the cells must be kept between 0 and 4°C).
6. Incubate the cells on ice for 5 min.
7. Centrifuge the cells at 2500 g for 5 min at 4°C. Decant the supernatant.
8. Wash the cell pellets with cold 0.1 M $CaCl_2$. Do not vortex. Centrifuge at 2500 g for 5 min at 4°C. Decant the supernatant.
9. Resuspend each pellet in 7 ml of cold (0–4°C) 0.1 M $CaCl_2$.
10. Incubate the cells on ice overnight.
11. Add 3 ml of ice cold 50% v/v glycerol/50 mM $CaCl_2$ to each tube. Mix gently.
12. Aliquot 0.5 ml of cells/tube into prechilled tubes and quick-freeze in liquid nitrogen. Store the frozen cells at −80°C. Cells prepared in this manner retain >90% of their original competency for up to 1 year.

Protocol 11: Transformation of E. coli with Plasmid DNA

(Time: ≈2 hr to step 9)

The following protocol is used to isolate and screen plasmid clones created in Protocol 9. The plasmids are introduced into competent E. coli cells (produced in Protocol 10). Because the plasmid carries a gene for antibiotic resistance (typically ampicillin or tetracycline), the transformed bacteria can be isolated by growing the cells with the appropriate antibiotic. However, cells with both recombinant as well as non-recombinant plasmids will grow under these conditions, so a second screening condition is usually imposed. For some plasmids, this involves a second gene for a different antibiotic resistance that is disrupted by cloning into the target site. Recombinant plasmids are then separated from nonrecombinant plasmids by replicate plating on plates with one and with both antibiotics. Most plasmid vectors, however, use color screening for recombinant plasmids. The most common system involves a β-galactosidase gene that bridges the cloning site. By adding appropriate substrates to the plates (X-Gal and IPTG), bacterial colonies that contain nonrecombinant plasmids will produce blue colonies, whereas colonies with recombinant plasmids (which have nonfunctional β-galactosidase genes) will produce white colonies. The following protocol assumes that a plasmid with blue/white screening is used (for information on alternative screening methods, see Berger and Kimmel, 1987). Ten 100-mm plates will be sufficient for a transformation involving up to 0.1 μg of vector DNA.

1. Thaw frozen competent cells on ice.
2. Aliquot 200 μl of cells into a sterile tube on ice. This should be enough cells to allow efficient transformation with the DNA from a subcloning experiment.
3. Add ligated DNA in up to 50 μl total volume and sufficient 1.0 M $CaCl_2$ to keep the Ca^{2+} concentration at 0.1 M.
4. Mix thoroughly and incubate on ice for 30 min.
5. Heat shock the cells at exactly 42°C for 2–3 min.
6. Allow the cells to cool to room temperature, then add 1 ml of L-broth.
7. Incubate the cells at 37°C for 30 min to allow the expression of drug resistance.
8. Spread the cells on L-broth + 1% agar plates. The plates should contain the appropriate antibiotic for the plasmid (e.g., 100 mg ampicillin/liter), as well as 50 mg IPTG and 40 mg X-Gal per liter of broth.
9. Incubate plates overnight at 37°C. Colonies that contain recombinant plasmids will be white; colonies that contain nonrecombinant colonies will be blue. DNA can be isolated from white colonies for screening by using a scaled-down version of Protocol 13, Part A; after the correct clone is identified (Figure 6), it should be streaked onto a new plate (with the appropriate antibiotic), grown in volume for DNA isolation (Protocol 13), and frozen for permanent storage (Protocol 15).

Protocol 12: Transformation of M13 Bacteriophage DNA

(Time: ≈1 hr to step 9)

This protocol should be followed for transformation of *E. coli* with M13 clones. Blue/white screening for recombinant DNA (as described in Protocol 11) is used for M13 phage. One-tenth of a subcloning reaction involving 1 μg of M13 DNA will yield sufficient recombinant phage for analysis.

1. Thaw frozen competent cells on ice.
2. Aliquot 200 μl of cells into a sterile tube on ice.

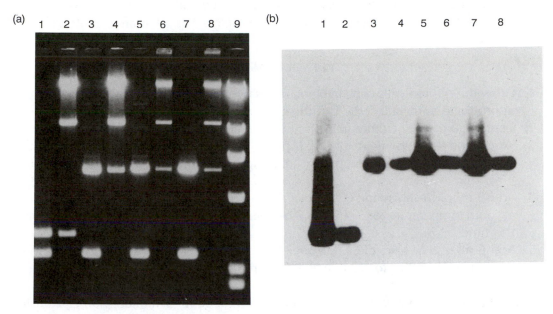

FIGURE 6. (a) Check gel for a series of lambda bacteriophage clones (even lanes) and their plasimid subclones (odd lanes) digested with *Eco*RI. Lane 9 is lambda DNA digested with *Hind* III. The two larger fragments in the even lanes correspond to the two arms of the lambda bacteriophage; the smaller fragment in each of the odd lanes is the linearized plasmid vector. (b) Autoradiograph of Southern blot from check gel shown in Figure 6a, hybridized with an homologous probe to verify clones.

3. Add the ligated DNA and sufficient 1.0 *M* CaCl$_2$ to keep the Ca^{2+} concentration at 0.1 *M*.
4. Mix thoroughly and incubate on ice for 30 min.
5. Heat shock the cells at exactly 42°C for 2–3 min.
6. Allow the cells to cool to room temperature, then aliquot the transformation into a number of sterile culture tubes equal to the number of plates desired.
7. Add 100 μl of a fresh overnight culture of an appropriate strain of *E. coli* to each tube.
8. Add 4 ml of warm (48°C) top agar and immediately spread on a L-broth + 1% agar plate containing 50 mg IPTG and 40 mg X-Gal per liter of medium.
9. Incubate plates overnight at 37°C. See comments under Protocol 11, step 9.

Protocol 13: Isolation of Plasmid DNA

(Time: Part A: day 1, ≈10 min; day 2, ≈2 hr. Part B: day 1, ≈10 min; day 2, ≈3 hr; day 3, ≈6 hr)

The following protocol contains two parts. Clean preparations of plasmid DNA suitable for most purposes (including sequencing) can be obtained by following Part A of the protocol. If further purification is necessary, the CsCl protocol (Part B) can be used, but Part B requires an ultracentrifuge. Either part can be scaled up or down as needed; just be sure to scale all reagents by the same factor. For alternative protocols and modifications, see Miller (1987).

Part A. Basic Method

1. Grow the desired cells containing plasmids overnight in 300 ml of L-broth plus the appropriate antibiotic at 37°C with vigorous shaking ($\approx$300 rpm).
2. Centrifuge the cell culture in 250-ml bottles at 2500 g for 10 min at 4°C.
3. Resuspend the cells in 4 ml of GTE (Appendix) plus 1 mg/ml lysozyme. Transfer the suspension to 50-ml polypropylene centrifuge tubes. Incubate at room temperature for 5 min.
4. Add 8 ml of freshly made 0.2 M NaOH plus 1% SDS. Mix by hand and incubate at room temperature for 5–15 min. The solution should become less viscous during this time.
5. Add 6 ml of 5 M KAc, pH 4.8. Vortex thoroughly, then incubate on ice for 5 min.
6. Centrifuge at 7500 g for 10 min at 4°C. Carefully transfer the supernatant to a new tube.
7. Add an equal volume of PCI and vortex. Centrifuge for 1 min at 7500 g. Transfer the aqueous (top) phase to a new tube.
8. Add an equal volume of ether and vortex. Centrifuge for 10–20 sec. Remove ether (top layer), and save lower layer in tube.
9. Add 2.5 volumes of cold absolute ethanol, mix thoroughly, and incubate 10 min or longer at -20°C.
10. Centrifuge at 8000 g for 5 min at 4°C to pellet the plasmid DNA. Decant the ethanol. Add 1 ml of 70% ethanol and transfer the DNA pellet plus 70% ethanol to a microcentrifuge tube. Spin in microcentrifuge for 1 min, decant ethanol, and dry the DNA in a vacuum centrifuge until the ethanol has just evaporated.
11. Dissolve DNA in 1 ml of 1× TE. Add 10 µg/ml of RNase A and incubate at 37°C for 30 min.
12. Add 0.1 ml 5 M KAc. Repeat steps 7–10.
13. Dissolve DNA in up to 1 ml of 1× TE, and check concentration and purity.

Part B. CsCl Gradient Purification

1. Follow steps 1–10 of part A.
2. Resuspend the pellet in 5 ml of 1.7 g/ml CsCl in 1× TE.
3. Add 250 µl of 4 mg/ml ethidium bromide. Check the density of this solution and adjust it to 1.60 g/ml by adding more CsCl or TE. (Note: ethidium bromide is a mutagen and suspected carcinogen.)
4. Centrifuge for 18 hr at $\approx$150,000 g at 20°C in an ultracentrifuge.
5. Collect the plasmid band (the lower of the two bands) by side puncture with a syringe. Add an equal volume of 1× TE to dilute the CsCl.
6. Extract repeatedly with CsCl-saturated isoamyl alcohol until the aqueous phase is colorless. (The alcohol is the upper phase; keep the lower phase.)

7. Dialyze the DNA solution against 4 liters of 10× TE for ≈4 hr at 4°C (see Chapter 8 for preparation of dialysis tubing).

8. Ethanol-precipitate the DNA, resuspend it in 200 μl of 1× TE, and check concentration and purity.

Protocol 14: Miniprep Isolation of M13 DNA

(Time: day 1, 10 min; day 2, 2–3 hr)

   M13 DNA can be isolated using Protocol 13 (without the antibiotic in step 1 or lysozyme treatment in step 3), or small amounts of M13 DNA can be isolated using the following protocol. This protocol is useful as a first step if large numbers of M13 clones are to be screened; larger quantities of the desired DNA can then be prepared using Protocol 13.

1. Inoculate 2 ml of L-broth with 1 drop of an overnight host bacterial culture and a single white plaque.
2. Incubate 12–16 hr at 37°C with shaking (≈300 rpm).
3. Remove 1.5 ml of the culture and centrifuge for 5 min at 7000 g.
4. Remove 1 ml of the supernatant and place it in a clean microcentrifuge tube. Be careful to avoid the bacterial pellet.
5. Add 150 μl of PEG stock and mix thoroughly.
6. Incubate for 30 min on ice.
7. Centrifuge for 5 min at 7000 g (there should be a clearly visible pellet).
8. Remove the supernatant, recentrifuge for 30 sec (7000 g), and remove any residual supernatant.
9. Resuspend the phage pellet in 100 μl of 10× TE.
10. Extract with 200 μl of PCI as described in Protocol 1 (steps 7–9). (Note: Sacrifice yield to avoid the interface.)
11. Extract with 200 μl of chloroform as described in Protocol 1 (steps 11–13).
12. Extract with 500 μl of ether. (Note: The ether will be the upper phase and the DNA will be in the lower phase.)
13. Add 10 μl of 2 M NaCl and 250 μl of absolute ethanol to precipitate the DNA.
14. Centrifuge at 7000 g for 10 min to pellet the DNA. Rinse the pellet once with 70% ethanol and recentrifuge briefly at 7000 g.
15. Decant the supernatant, dry the pellet in a vacuum centrifuge, and resuspend it in 20 μl of 1× TE.

Protocol 15: Preparing Permanent Frozen Stocks of Plasmid Clones

(Time: day 1, ≈10 min; day 2, ≈10 min)

   Stocks of bacteriophage clones are best stored at 4°C in PDB with 0.4% chloroform added. The chloroform will prevent bacterial growth and preparations are stable for years. However, it may be necessary to amplify the stocks after prolonged storage. Plasmid clones can be stored indefinitely at −80°C using the following protocol.

1. Grow a fresh overnight culture of the desired clone in liquid media plus antibiotics.
2. Combine 0.85 ml of the overnight culture with 0.15 ml of sterile glycerol and mix well by vortexing.
3. Flash-freeze in liquid nitrogen and store at −80°C. Cell stocks prepared in this manner will last for years if they are not allowed to thaw. To access the frozen cells, simply scrape the top of the frozen culture with a sterile loop and spread on a plate with appropriate antibiotics.

Protocol 16: DNA Amplification by the Polymerase Chain Reaction

(Time: ≈5 hr for double-stranded template, ≈10 hr for single-stranded template)

The following protocol can be accomplished by manually cycling the sample among waterbaths at the appropriate temperatures, or one of a number of automatic thermal cycling machines can be used. Protocols for PCR are being rapidly refined; among parameters that can be varied to optimize amplification are the concentrations of DNA template, primers, dNTPs, Mg^{2+}, KCl, and *Taq* polymerase, and well as the length and temperature of the annealing and extension cycles (see Gyllensten and Erlich, 1988; Lawyer et al., 1989; White et al., 1989; Kocher and White, 1989; Innis et al., 1990).

Careful design of primers is critical to the success of PCR amplification. The two primers should be complementary to opposite strands, and should flank the target sequence at a distance of up to 4 kb (larger fragments can be amplified with diminishing success). If the amplified fragment is to be cloned into a sequencing vector (often required for fragments larger than ≈600 bp), it is helpful to incorporate a restriction enzyme recognition site on the 5′ end of the primers. A primer so constructed should have an additional two to four bases 5′ to the restriction site. Mismatches at the 5′end of the primer will not usually impede the amplification process, although an absolute match of the primer to the target DNA is preferable. For protein-coding genes, primers should not end on the third position of a codon. Although primers as short as 17 bp have been used effectively, it is usually desirable to use primers in the range of 25–35 bp. Care should be taken to match the melting temperatures (T_m) of the two primers: $T_m = [4 \times (No.\ G's + C's)] + [2 \times (No.\ A's + T's)]$. Primer mixtures with up to 256-fold degeneracy have been used successfully in PCR amplification, although more than 32-fold primer degeneracy often results in highly heterogeneous amplification products. Degeneracy should be no more than 2-fold at any one site.

It is essential that strong measures be taken to avoid contamination of PCR reactions with foreign DNA, as even a single foreign molecule can be amplified and result in substantial contamination. This is especially important in systematic studies in which homologous genes from many species are to be studied. A separate lab area (preferably a separate room or containment hood) should be used for amplification. DNA-free controls must be run with each amplification, and these negative controls should be assayed along with the amplified DNA samples on an agarose gel (positive controls that include DNA known to amplify with a given set of primers are also advisable). Additional safeguards are described by White et al. (1989).

Several methods have been developed for sequencing DNA from PCR reactions. Part A of this protocol describes the basic PCR procedure; Parts B and C are alternate methods for sequencing the amplified product. Sequencing single-stranded template through assymmetric reamplification (Part B; Gyllensten and Erlich, 1988) is usually limited to short (600 bp or less) amplified fragments; longer fragments can be sequenced directly using Part C (DuBose and Hartl, 1990). Both direct sequencing options (Parts B and C) require homogeneous amplification product. If the PCR product is hetero-geneous, or a clone is desired, then the DNA should be inserted into a sequencing vector for analysis (see Protocols 9–14).

Part A: Initial Amplification

1. Mix
   10.0 μl genomic DNA (working stock = 0.05 μg/μl)
   5.0 μl 10× *Taq* salts (see Appendix)
   5.0 μl 10 m*M* primer 1 (2 pmol/μl)
   5.0 μl 10 m*M* primer 2
   5.0 μl 8 m*M* dNTP mix (8 m*M* of each dNTP, 40 m*M* Tris-HCl, pH 7.9)
   0.25 μl *Taq* polymerase (4 μ/μl)
   sterile, doubled-distilled H_2O to a final volume of 50 μl
   (Mixes lacking DNA and primers can be premixed in bulk and stored for 1–2 months at $-20°C$.)
2. Vortex, spin, and overlay with mineral oil.
3. Denature template for 1–2 min at 94°C. Then cycle for 20–40 iterations:
   Annealing: 2–3 min at 37–42°C (low stringency) or $T_m-5°C$ (high stringency)
   Extension: 1–3 min at 72°C (approximately 1 min per 2000 bp)
   Denaturation: 1 min at 94°C
   At the end of the last cycle, omit the heat denaturation step and allow the extension step to proceed for an additional 7 min. For samples in which amplification of unwanted contaminants could be a problem, use the lower number of cycles and shorter extension cycle times.
4. Extract the completed reactions once with an equal volume of chloroform to remove the mineral oil. Dilute the samples to a final volume of 2 ml with TE.
5. Check a 10–20 μl aliquot of the reaction mixture (and the negative control) by agarose gel electrophoresis to determine the success of the reaction.

Part B: Asymmetric Reamplification

1. Purify amplified DNA samples using low melting point agarose gels (Sambrook et al., 1989) or Centricon 30 cartridges (Amicon Corp., Danvers, MA). Wash the Centricon 30 cartridges by applying 2 ml of TE and centrifuging at 4800 *g* for 10 min at 4°C. Then add the 2 ml of amplified DNA solution to the cartridge and centrifuge as above for 15 min. Discard the solution in the reservoir. Collect the purified DNA sample by

inverting the cartridge and centrifuging at 200 g for 2 min. The final volume of DNA sample should be approximately 100 μl. The yield of DNA should be approximately 7–10 μg.

2. Repeat steps 1–5 of Part A, but with a 1:100 ratio of the two primers (some experimentation in primer ratio may be necessary). After asymmetric reamplification, the low concentration primer can be used to sequence the fragment (see Protocols 17, 18, and 20).

Part C: Isolation of Double-Stranded DNA for Sequencing

1. Concentrate the PCR product to ≈25 μl total volume in a vacuum centrifuge/concentrator.
2. Prepare a Sepharose CL-6B column (Boehringer Mannheim) by mixing thoroughly. Remove the top cap, then the bottom cap, and drain excess buffer from the column. Spin the column 2.5 min at 1100 g .
3. Using a new collection tube, add the sample from step 1 (≈35 μl) to the middle of the column. Spin for 10.5 min at 1100 g to recover the purified DNA.
4. To prepare DNA template for sequencing, use ≈1–2 μg of the purified DNA (≈10 μl) in Protocol 18, Part A. Sequence the product using modified T7 DNA polymerase (Tabor and Richardson, 1987) as described in Protocol 18.

Protocol 17: Preparing a Sequencing Gel

(Time: 1–2 hr)

The details of the following protocol will vary depending on the style of sequencing apparatus used; a simple sequencing apparatus is shown in Figure 7. The gel spacers can vary in thickness from 0.2 to 0.8 mm. If spacers of uniform thickness are used, the bands at the bottom of the gel will be widely spaced, whereas those at the top will be very close together. Much longer sequences can be read from gels that take advantage of field gradients produced with wedge-shaped spacers (Ansorge and Labeit, 1984). With wedge-shaped spacers, bands will be much more evenly spaced along the length of the gel. Wedge-shaped spacers can be obtained commercially, but are expensive and often not uniform. An effective alternative is to combine two layers of spacers at the bottom of a gel, with only a single layer of spacers at the top. Experimentation will be required to find the optimal gradient for a particular sequencing system, but a gradient of 0.2–0.8 mm is usually quite effective.

Reading long sequences (>600 bp) requires long sequencing gels (>80 cm), wedge-shaped spacers, and use of ^{35}S-labeled (rather than ^{32}P-labeled) nucleotides. Pouring and handling very long gels present additional difficulties. An alternative pouring strategy to the one given below is to slide the plates together, pouring the acrylamide gel mixture ahead of the leading edge of the top plate. With practice, gels without any bubbles can be prepared from very long plates with this technique. Another technique for pouring long gels involves injecting the acrylamide through a small hole in the bottom of one of the two glass plates (Slightom et al., 1987). For handling long gels, it is preferable to bind

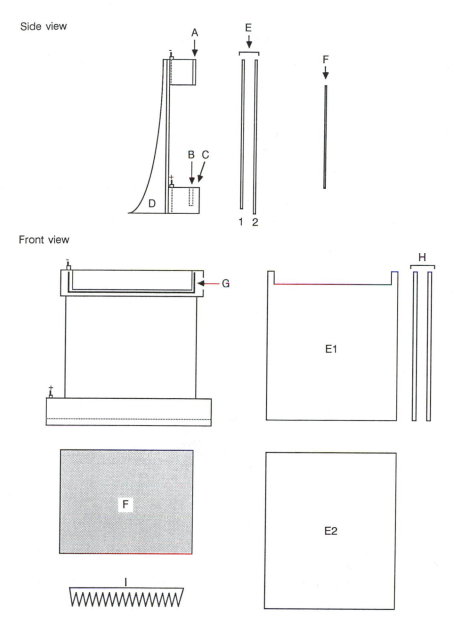

FIGURE 7. A basic sequencing gel unit. The gel is poured between the two glass plates (E1 and E2), which are separated by the teflon spacers (H). Note that the front plate (E2) is slightly longer than the back plate (E1) to allow contact between the gel and buffer in the lower tray. A sharkstooth comb (I) is inserted at the top of the gel (see Protocol 17). The two plates are held together by clamps (heavy duty paper clamps work well) and the gel is inserted into the lower well (C) where it is held in place by an acrylic bar (B). The top of the gel is clamped to the side ears of the upper tank (A); note that the front of the upper tank is open to allow contact between the buffer and the gel. A rubber gasket (G) forms a seal between the upper tank and the earred glass plate (E1). An aluminium plate (F) is clamped to the glass plates to insure even heating. The electrodes are constructed from platinum wire to prevent corrosion. The stand (D) can be modified to permit height adjustment of the upper buffer tray so that gels of many different lengths can be accommodated. Sequencing gels are typically 40–100 cm long and 20–40 cm wide.

the gel to one of the plates (using bind-silane [γ-methacryloxypropyltrimethoxysilane]) rather than to transfer the gel to filter paper for vacuum drying. Gels attached to the glass can be dried with a hot-air blower or in a drying cabinet (see Protocol 20).

1. Prepare the inner surfaces of the gel plates (after cleaning) using 2% dimethyldichlorosilane solution in 1,1,1-trichloroethane (add ≈1 full Pasteur pipette of silane per surface and spread with a lab tissue; polish surface until smooth). Wear gloves and prepare the plates in a fume hood, as the silane solution is highly toxic.

2. Clamp the plates together with spacers between them. Be sure that the spacer covers the complete length of the gel plate (including the "ears"). Do not use a spacer across the bottom of the plates.

3. Tape all sides of the gel plates except the top, making sure that all edges are tightly sealed. Reclamp the sides of the taped plates.

4. Mix the gel solution in a 500-ml flask (for 4% gel):
   60 ml urea mix (Appendix)
   20 ml 20% acrylamide (Appendix)
   20 ml 1× TBE buffer (Appendix)
   400 μl 10% ammonium persulfate
   50 μl TEMED
   (Note: The concentration of acrylamide for DNA gels should be 4–6%. For RNA gels, use 8% acrylamide).

5. Pour the gel solution between the plates using a 25-ml pipette and a regulating pipette bulb. Allow the solution to run down one edge and fill from the bottom. Avoid forming bubbles between the plates.

6. Insert a sharkstooth comb backward between the plates at the top, aligning the holes in the comb with the edge of the back plate (shorter one). Allow gel solution to cover the outer surface of the comb. Clamp into place and allow to rest for 1 hr at an incline. Remove the clamps after the gel sets.

7. Pour diluted electrode buffer on the comb with a Pasteur pipette. Remove the comb and rinse it clean in distilled water. Reinsert the comb with the teeth pointing inward so that the tips of the teeth barely touch the surface of the gel.

8. Cut the tape from the bottom edge of the gel with a razor blade.

9. Clamp the gel onto the gel-running apparatus.

10. Fill the upper and lower reservoirs with 1× TBE buffer.

11. Use a syringe or micropipetter to clear the wells formed by the sharkstooth comb. The wells are the spaces between the teeth.

12. Fill the wells with 4.5 μl of stop buffer. Prerun the gel, setting the current not to exceed 25 mA and the voltage not to exceed 2000 V. Use a micropipetter with microthin tips or capillary tubes drawn to a fine tip to load the gel. (Length of prerun = 15–30 min.)

Protocol 18: DNA Sequencing Reactions

(Time: Part A, ≈1.5 hr; Part B, ≈30 min; Part C, ≈1 hr)

The conditions of DNA sequencing reactions can be varied according to (1) the length of sequence to be determined, (2) whether single- or double-stranded DNA is to be sequenced, (3) the base composition of the primer sequence, (4) the base composition of the target sequence, and (5) the sequencing enzyme to be used. In general, these variations are noted in the following protocol, except that particular conditions for the various sequencing enzymes should follow the manufacturer's recommendations. The common sequencing enzymes are Klenow fragment, modified bacteriophage T7 DNA polymerase (Tabor and Richardson, 1987), and *Taq* polymerase (Brow, 1990). Although good results can be obtained with any of these enzymes, the latter two usually provide superior results, especially in regions of strong secondary structure. If problems arise in sequencing regions of high G-C content, it may also be desirable to substitute dITP or 7-deaza dGTP for dGTP in step 1 of Part B (Barnes et al., 1983; Gough and Murray, 1983; Mizusawa et al., 1986). For double-stranded DNA, denature the DNA (Part A; Haltiner et al., 1985) before starting Part B; single-stranded DNA (e.g., M13) can be used directly in Part C.

Part A. Denature and Neutralize DNA Template (Double-Stranded)

1. Bring 1–3 µg of RNased plasmid DNA (from Protocol 13) or amplified linear DNA (from Protocol 16, Part C) to a volume of 20 µl with deionized, distilled water. Add 2 µl of 2 N NaOH. The exact amount of DNA will vary depending on the size of the template. A 1:1 molar ratio should be maintained between primer and template.
2. Incubate at 65°C for 5 min, then place on ice and allow to cool.
3. Add neutralizing salt mix of
   2 µl 8 M NH$_4$Ac
   3 µl 3 M NaOAc
   20 µl ddH$_2$O
   For amplified DNA from Protocol 16, Part C, add 4 µl 1% acrylamide.
4. Add 150 µl absolute ethanol and mix.
5. Precipitate DNA at −80°C for 45 min or more.
6. Pellet DNA for 20 min in refrigerated microcentrifuge.
7. Wash pellet with 70% ethanol (approximately 150 µl), then spin 10 min in refrigerated microcentrifuge.
8. Wash pellet with absolute ethanol (approximately 150 µl), then spin 10 min in refrigerated microcentrifuge.
9. Dry the pellet in a vacuum centrifuge. Note: This DNA may be stored dry in a freezer before proceeding to next step.

Part B. Preparation of Solutions and Termination Tubes

1. Label four tubes per reaction with G, A, T, and C. To each tube add 2.5 µl of the respective ddNTP mixture: All four mixtures contain 80 µM dGTP, 80 µM dATP, 80 µM dCTP, 80 µM dTTP, and 50 µM NaCl. In addition, each contains 8 µM of the respective ddNTP. For dITP sequencing, substitute 160 µM dITP for 80 µM dGTP in each mixture.

2. Prepare labeling mix depending on sequencing distance from primer: Stock: 7.5 μM dGTP (or 15 μM dITP), 7.5 μM dCTP, 7.5 μM dTTP. Dilute stock 1:10 for sequencing close to primer, 1:5 for sequencing 25–300 bp from primer, and use undiluted for greater than 300 bp from primer.
3. Prepare DNA polymerase according to manufacturer's directions.

Part C. Primer Annealing and Sequencing Reaction

1. Resuspend DNA template in 8 μl of primer (2.5 ng/μl) in a 0.5 ml microcentrifuge tube.
2. Add 2 μl of 5× sequencing buffer (e.g., 200 mM Tris-HCl, pH 7.5, 100 mM $MgCl_2$, 250 mM NaCl; this may vary with DNA polymerase used).
3. If the GC/AT ratio of the primer is approximately 0.5 or more, heat the tube to 65°C for 2 min, then allow the tube to cool down at a rate of approximately 1°C/min to 35°C. If the GC/AT ratio is less than 0.5, hold the tube at 37°C for 15 min. Some experimentation will be required for specific primers.
4. Add 1 μl 0.1 M dithiothreitol, 2 μl labeling mix (1:10, 1:5, or undiluted, as prepared in Part B, step 2), 0.5 μl [α-^{32}P]dATP (or 0.5 μl [α-^{35}S]thio-dATP), and 2 μl DNA polymerase. Mix.
5. Incubate at room temperature for 2 min for close to primer, 5 min for intermediate range, or 10 min for long range. Prewarm termination tubes at 37°C during the last minute of this incubation.
6. Add 3.5 μl of this reaction mixture to the termination tubes prepared in Part B, step 1. Mix and incubate at 37°C for 2 min.
7. Add 7 μl stop buffer, mix, and place on ice.
8. Prior to loading on gel, heat samples at 80°C for 2 min.

Protocol 19: RNA Sequencing Reactions

(Time: ≈3 hr)

A common problem in sequencing rRNA with reverse transcriptase is sequencing through regions of strong secondary structure. One method that may help resolve such sequences involves the addition of terminal deoxynucleotidyltransferase (TdT) following the completion of the reverse transcriptase extension reactions (DeBorde et al., 1986). This procedure is indicated below as an optional step 14.

1. Add 6 μl of solution of the RNA to be sequenced to a microcentrifuge tube.
2. Heat the RNA to >90°C for 5 min in a heating block. Cool in ice water, and then spin for several seconds in a microcentrifuge.
3. Add 1 ml of 20× reverse transcription buffer (see Appendix) and 2 ml of labeled primer (working stock = 0.5 pmol/μl) and 1.5 μl of RNasin (2000 U/ml) in this order, vortex, and then spin for several seconds in a microcentrifuge.
4. Incubate at 42°C for 30 min. Spin for several seconds in a microcentrifuge following incubation. (Note: to save time, the next three steps are performed during this incubation.)

5. For each RNA sample to be sequenced, prepare four microcentrifuge tubes marked G, A, T, and C.

6. To the G tube add 1 μl of 1.5 mM ddGTP, to the A tube add 1 μl of 8 mM ddATP, to the T tube add 1 μl of 5 mM ddTTP, and to the C tube add 1 μl of 2 mM ddCTP.

7. Prepare "Reaction Mixture 1" in the following manner for each RNA sample to be sequenced:

    3 μl dNTP mix (5 mM each dATP, dCTP, dGTP, and dTTP)

    3 μl ddH$_2$O

    3 μl reverse transcriptase

    Vortex, spin in a microcentrifuge for several seconds, and store on ice until needed.

8. Add 2.1 μl of the solution from step 4 to each of the four tubes.

9. Add 2 μl of "Reaction Mixture 1" to each tube (G, A, T, and C), vortex, and spin for several seconds in a microcentrifuge.

10. Incubate at 48°C for 40 min.

11. Prepare "Reaction Mixture 2" (during step 10).

    3.0 μl dNTP mix

    3.0 μl reverse transcriptase

    Vortex, spin in a microcentrifuge for several seconds, and store on ice until needed.

12. Add 1 ml of "Reaction Mix 2" to each tube. Vortex and then spin for several seconds in a microcentrifuge.

13. Incubate at 48°C for 40 min. Spin for several seconds in a microcentrifuge following incubation.

14. (Optional; see comments above) Add 1 μl of a mixture of dATP, dCTP, dTTP, and dGTP (each at 1 mM) and 10 units of terminal deoxynucleotidyltransferase to each tube. Incubate at 37°C for 30 min.

15. Add 4 ml of stop buffer (see Protocol 18, Part C, step 7) to each tube.

16. Heat for 5 min to >90°C. Cool on ice, vortex, and then spin for several seconds in a microcentrifuge. Store on ice until use.

Protocol 20: Running a Sequencing Gel

(Time: 3–6 hr to step 18)

Sequencing is best accomplished at approximately 50°C, so that the DNA or RNA remains denatured and relatively few secondary structures form. The temperature must be constant across the width of the gel, or the lanes will migrate at different rates and "smiling" of the bands will occur (Figure 8). Uniform gel temperature is usually maintained in one of two ways: (1) by placement of an aluminum plate against the glass surface of the sequencing gel for even dispersal of heat, and regulating the current to maintain the desired gel temperature; or (2) placement of a thermostatic plate with circulating temperature-controlled water against the sequencing gel. The second option is preferable, because of greater control of gel temperature, but requires more equipment (the thermostatic plate and a circulating waterbath).

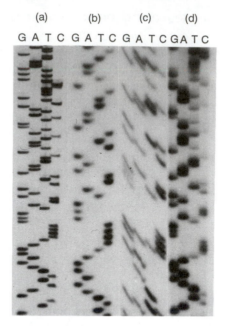

FIGURE 8. Sequencing gel autoradiographs and troubleshooting. (a) Readable control reaction. (b) Same sequence as in (a), but with the lanes overloaded and the gel run at high voltage (resulting in poorly defined bands). (c) Same sequence as in (a), but with smiling effect because of uneven gel temperature (the left lanes are closer to the outside of the gel than are the right lanes). (d) Same sequence as (a), but with RNA contamination. Note the darker background and poorly defined bands.

1. Turn off the prerun (from Protocol 17, step 12).
2. Clear each of the wells by flushing with buffer.
3. Label the reactions to be loaded on the front plate of the gel. Use the order G, A, T, and C for each set of reactions.
4. Load 4.5–5 μl of each reaction (from Protocols 18 or 19) in the wells as marked, using a micropipetter with microthin tips. Store unused portion of the reactions at −20°C.
5. Wipe off lane markings for first load to prevent accidently loading them a second time.
6. If using a thermostatic plate, then set voltage at 2000 V (the constant temperature of the gel will also hold the amperage relatively constant). If an aluminum plate is used to maintain even gel temperature, run the gel at the amperage required to maintain desired gel temperature (measure using a surface thermometer). As the run progresses, the amperage will drop as the voltage rises and the gel heats. Some experimentation will be required for particular gel systems. In general, a voltage of approximately 1500–2000 V and a temperature of approximately 50°C is desirable.
7. Allow the gel to run until the bromophenol blue marker reaches the bottom of the gel. If sequence far from the primer is to be read, a second set of reactions can be

loaded at this point and run until the bromophenol blue marker reaches the bottom of the gel. The distant sequence can then be read from the lanes loaded first, and the closer sequence from the lanes loaded last.

8. Turn off the power and remove electrode cables.

9. Unclamp the gel plate from the gel apparatus. CAUTION: Bottom tank now contains radioactive buffer and must be handled with appropriate care.

10. Dry the gel plates and cut the tape on both sides using a razor blade.

11. Pull the spacers out from between the plates.

12. Use a spatula to pry the plates apart. The gel should adhere to only one of the plates. (Steps 13–15 apply only if the gel is to be removed from the glass for drying.)

13. Cut a sheet of No. 1 filter paper to fit over the gel plate.

14. Place the filter paper over the gel and press it onto the gel. Place the other gel plate over the filter paper and press firmly again.

15. Remove the top plate and pull the filter paper (with gel attached) from the lower plate.

16. If the gel was bound to the glass plate using bind-silane, wash the gel by immersing the entire plate in 10% acetic acid and gently shake for 15 min. If the gel was removed from the glass plates, cover the gel with plastic wrap. Place several sheets of filter paper beneath the gel and put on a gel dryer at 80°C for 1 hr. (If the gel was bound to the glass plates using bind-silane, dry with a hot-air blower or in a drying cabinet.)

17. Place the gel in a film cassette with autoradiography film for 24 hr. (Note: the exact length of exposure will vary).

18. Develop the film for 5 min in developer tank. Rinse in stop bath and place in fixer tank for 5 min. Wash well in running tap water and hang to dry.

INTERPRETATION AND TROUBLESHOOTING
Autoradiograph Interpretation

Although reading autoradiographs of sequence data is relatively straightforward (see Figures 3, 4, and 8a), some practice is required to accurately record the data and to identify and solve problems. When sequencing DNA, it is strongly advisable to sequence both strands, as this provides a check against reading errors. When sequencing RNA, only one strand can be sequenced, so it is necessary to sequence broadly overlapping regions to verify the sequence.

Reading sequences from autoradiographs is greatly simplified by use of one of various gel readers (digitizers coupled directly to a computer). Use of a gel reader reduces human error compared to recording a sequence and then inputting the sequence via a keyboard. Most gel readers and software packages allow previously input sequences to be verified, thus further reducing error. Various automated gel readers have been and continue to be developed, and eventually may replace manual reading of autoradiographs altogether. However, experience in gel reading usually allows higher accuracy of manual sequence interpretation compared to the present automated sequence-reading technology.

The length of readable sequence depends on a number of factors. With use of wedge-shaped spacers (Ansorge and Labeit, 1984), long gels (50–60 cm), ^{35}S rather than ^{32}P, and bacteriophage T7 DNA polymerase (Tabor and Richardson, 1987) or *Taq* polymerase, it is possible to obtain greater than 600 bp from a single sequencing reaction on a single autoradiograph. Wedge-shaped spacers produce a gel that is thicker on the bottom than at the top; thus, the smaller DNA fragments slow down as they approach the bottom, thereby enabling longer electrophoresis runs and improved resolution of the larger fragments. The use of ^{35}S produces sharper bands, so it is possible to deduce sequence from bands that are quite close together. Improved DNA polymerases with higher temperature optima and fidelity (e.g., Tabor and Richardson, 1987) also allow accurate sequencing of long DNA sequences.

One of the most common problems encountered with autoradiographs of sequencing gels is a dark background in the lanes (Figure 8d). This is usually caused by impure template DNA; RNA is the most common contaminant. This can usually be corrected by further purification of the DNA sample, including retreatment with RNase (see Protocol 13 and Figure 8a). If bands are present in the same position in more than one lane, there are several possible problems. If the resolution of the gel is generally poor and there are no bands in the upper portion of the gel, it is likely that there has been a loss of activity of the sequencing enzyme. If ghost bands are apparent in adjacent lanes, then too large a sample may have been loaded into the gel, the loading syringe or pipette tip was not rinsed between loading samples, or there was a poor fit of the sharkstooth comb in the gel.

Another problem sometimes observed is that the bands are too faint on the autoradiograph. This can be caused by old radionucleotides, insufficient exposure time, or salt contamination in the DNA template. If the bands are too dark, this can be corrected simply by exposing the autoradiograph for a shorter interval. Diffuse bands (Figure 8b) are usually caused by poor contact between the film and gel during exposure, but can also be caused by loading too much sample, running gels at too high a voltage, and poor washing and/or fixation of the gel.

"Smiling" is the phenomenon of samples in the outside lanes of a gel running slower than samples in the middle of a gel (Figure 8c). This is caused by uneven gel temperature, and can be corrected by good contact between the glass plates of the gel and an aluminum plate, or, even better, by use of a thermostatic glass plate through which heated water (approximately 50°C) is circulated. If the samples are horizontal with respect to one another but the individual bands are not straight, the problem is likely that urea was present in the sample wells when they were loaded. This can be corrected by thoroughly rinsing the wells with buffer prior to loading.

If bands appear on the autoradiograph only on the areas that correspond to the lower portions of the gel, there are at least three possible causes: (1) the template DNA may not have been added, (2) one or more dNTPs may have been omitted from the reaction, and (3) there may be no recognition sequence in the template DNA for the primer being used. In the first two cases, the reaction must be carried out again. If the primer site is absent in the template DNA, the only remedy is to use a different primer.

Secondary structure of DNA or RNA can cause difficulties in sequencing, especially in GC-rich regions. This problem is especially prevalent in sequencing rRNA. The secondary structure commonly produces a stop in the sequencing reactions, so that bands are present in all four lanes. There are several ways to correct this problem. Higher gel temperatures help prevent formation of secondary structures, so running the gel at a higher temperature will help combat the problem. In sequencing RNA, the addition of a terminal deoxynucleotidyltransferase (TdT) "chase" following the completion of the reverse transcriptase extension reactions is helpful (see Protocol 19 and DeBorde et al., 1986). In sequencing DNA, use of bacteriophage T7 DNA polymerase or *Taq* polymerase rather than Klenow fragment resolves many problems with secondary structure; extreme cases can be resolved by using dITP rather than dGTP in the sequencing reactions (see Protocol 18; also Barnes et al., 1983; Gough and Murray, 1983).

Sequence Comparison and Alignment

Once the sequence has been obtained, it must be related to other sequences to be of use in systematics. Sequences either can be aligned with known orthologs, or similarity searches (often incorrectly termed homology searches) can be performed by matching the sequence to all other sequences in a databank such as GenBank. Davison (1985) reviewed the various algorithms that have been developed for nucleotide sequence alignment, and divided the procedures into four general methods: matrix plots, global alignments, local alignments, and visual inspection. As there is no clear algorithm for the last method, only the first three will be considered here. It should be noted that little headway has been made in procedures for aligning several sequences simultaneously; usually, pairwise alignments must be made, and then combined to produce alignments across several taxa (but see Johnson and Doolittle, 1986).

Of the three major methods used for comparing sequence similarity, matrix comparisons (Figures 9 and 10) are useful for quick determination of major regions of similarity (not necessarily homology; see Chapter 1) and for visual portrayal of these similarities. In the simplest form of this procedure, two sequences are portrayed along the x and y axes of a graph, and every nucleotide in one sequence is compared to every nucleotide in the other sequence. If the nucleotides are the same, then a dot (or some other symbol) is shown in the corresponding row and column of the match; otherwise, the space is left blank. Because there are only four possible nucleotides in DNA sequences, approximately 25% of the comparisons will be matches in random sequences if the four bases are present in equal frequencies. Therefore, usually comparisons are

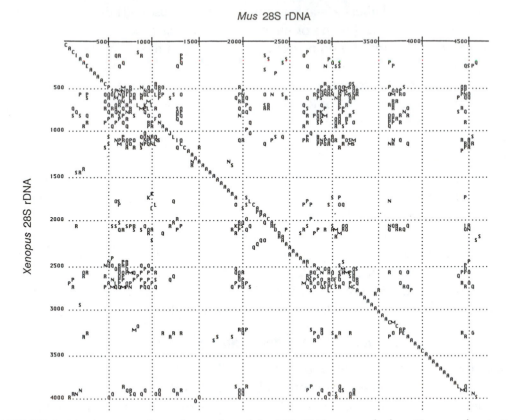

FIGURE 9. Matrix comparison of a portion of the 28S rRNA genes of a frog (*Xenopus laevis*; Ware et al., 1983; vertical axis) and a mouse (*Mus musculus*; Hassouna et al., 1984; horizontal axis). The deflections along the diagonal represent insertion/deletion events (see Hillis and Davis, 1987). The letters represent percent similarity over blocks of 30 bp; A: 100%, B: 98–99%, C: 96–97%, etc. All matches of 65% or higher similarity are shown. Note the regions of similarity between GC-rich regions at positions 500–1,000 and 2,500–3,200.

made between several adjacent nucleotides simultaneously, rather than on a base-by-base basis. Usually, some kind of weighting scheme is employed, so that the percentage of identical bases can be taken into account by use of different symbols (Figures 9 and 10). These kinds of comparisons are often helpful as a first step in comparing sequences, and are especially useful in identifying insertion/deletion events that may have occurred between orthologs.

Global alignments seek to align two entire homologous regions, using a balance between matches and gaps. The introduction of gaps is necessary to account for insertion/deletion events (see Figure 11), but because any two sequences could be aligned perfectly if enough gaps were introduced, gaps must be penalized. Gap penalties can be a combination of the number of gaps and the size of the gaps; in general, number

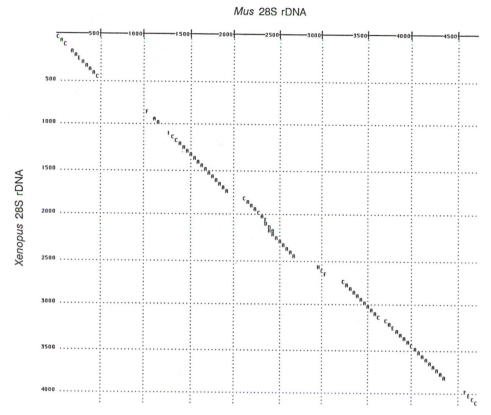

FIGURE 10. Another matrix comparison of the two sequences compared in Figure 9, but filtered so that only matches of 85% or higher are shown. Analyses in Figures 9 and 10 were conducted with the IBI Sequence Analysis Programs, © J. Pustell.

of gaps should be penalized more heavily than the size of the gaps, because there is no a priori reason to think that insertion/deletion events are more likely to involve short sequences. An algorithm for global alignments was developed by Needleman and Wunsch (1970), and is implemented in most DNA analysis software packages. It is also possible to use information from inferred secondary structure to fine-tune alignments (see additional discussion of global alignment in Chapter 11).

Local alignment algorithms find all subsequence matches above a certain defined threshold, both among and within comparison sequences (Figure 11). Search of data banks makes use of these algorithms, the most widely used of which is that of Lipman and Pearson (1985). Multiple local alignment is usually the best way to align orthologous sequences. It has also become commonplace to search a data base for the best match of virtually any sequenced segment of DNA, and to report on the results of the similarity search. Because some sequence will always represent a best match, it is desirable to know if the match is significantly different from a random match. Lipman and Pearson

(a)

	22|00	22|20	22|40

Mus   GTCAGCCAGGACTCTCTACCCGCTCACGGCAAGGCTTCCCTGCCCGCTACCGGAGGCAAC
Rattus   GTCAGCCAGGACTCTCTACCCGCTCACGGCAAGGCTTCCCTGCCCGCTACCGGAGGCAAC
Homo   GTCAGCCAGGACTCTCTACCCGCTCGCGGCAAGGCTTCCCTGCCCGCTACCGGAGGCAAC
Rhineura   GTCAGCCAGGATTCTCTATCCGCTCGCGGCAAGGCTTCCCTGCCCGCTACCGGAGGCAAC
Cacatua   GTCAGCCAGGATTCGCTATCCGCTCGCGGCAAGCCTTCCCTGCCCGCTACCGGAGGCAAC
Xenopus   GTCAGCCAGGATTCTCTACCCGCTCGCGGCAAGCCTTCCCTGCCCGCTACCGGAGGCAGC
Rhyacotriton GTCAGCCAGGATTCTCTATCCGCTCGCGGCAAGCCTTCCCTGCCCGCTACCGGAGGCAAC
Typhlonectes GTCAGCCAGGATTCTCTATCCGCTCGCGGCAAGCCTTCCCTGCCCGCTACCGGAGGCAAC
Latimeria GTCAGCCAGGATTCTCTACCCGCTTGCGGCAAGGCTTCCCTGCCCGCTACCGGAGGCAGC
Cyprinella GTCAGTCCAGGATTCCTACCCGCTGGCGGTCAAGCCTTCCCTCCGGCTACCGGAGGCAGC
                * ** ** **   *   **   ** ** * *   **   *   *

(b)

	22|00	22|20	22|40

Mus   GTCAG-CCAGGACTCTCTACCCGCTCACGG-CAAGGCTTCCCTGCCCGCTACCGGAGGCAAC
Rattus   GTCAG-CCAGGACTCTCTACCCGCTCACGG-CAAGGCTTCCCTGCCCGCTACCGGAGGCAAC
Homo   GTCAG-CCAGGACTCTCTACCCGCTCGCGG-CAAGGCTTCCCTGCCCGCTACCGGAGGCAAC
Rhineura   GTCAG-CCAGGATTCTCTATCCGCTCGCGG-CAAGGCTTCCCTGCCCGCTACCGGAGGCAAC
Cacatua   GTCAG-CCAGGATTCGCTATCCGCTCGCGG-CAAGCCTTCCCTGCCCGCTACCGGAGGCAAC
Xenopus   GTCAG-CCAGGATTCTCTACCCGCTCGCGG-CAAGCCTTCCCTGCCCGCTACCGGAGGCAGC
Rhyacotriton GTCAG-CCAGGATTCTCTATCCGCTCGCGG-CAAGCCTTCCCTGCCCGCTACCGGAGGCAAC
Typhlonectes GTCAG-CCAGGATTCTCTATCCGCTCGCGG-CAAGCCTTCCCTGCCCGCTACCGGAGGCAAC
Latimeria GTCAG-CCAGGATTCTCTACCCGCTTGCGG-CAAGGCTTCCCTGCCCGCTACCGGAGGCAGC
Cyprinella GTCAGTCCAGGATTC-CTACCCGCTGGCGGTCAAGCCTTCCCT-CCGGCTACCGGAGGCAGC
                *   * *   *   **   *   *   *   *

FIGURE 11. Alignment of a portion of the 28S rRNA genes of various species of vertebrates sequenced by Ware et al. (1983), Hadjiolov et al. (1984), Hassouna et al. (1984), Gonzalez et al. (1985), Hillis and Dixon (1989), Larson and Wilson (1989), and Hillis et al. (1990). The numbers refer to the nucleotide positions of the *Mus* 28S gene; sites that are variable among species are marked with an asterisk. Insertions are indicated by dashes. (a) Alignment with no insertions added. Note that this alignment requires 20 variable sites. (b) Alignment with four insertion/deletion events. There are only 10 variable sites in this alignment, including the insertions.

(1985) described the z statistic for this purpose, which is derived from the particular similarity score used in the search procedure. Briefly, the z statistic equals the difference between the similarity score and the mean similarity score from the data base scan, divided by the standard deviation of the similarity scores from the data base scan. They suggested the following guidelines: $z > 3$, possibly significant; $z > 6$, probably significant; and $z > 10$, significant. Other approaches to similarity significance testing have been described by Kanehisa (1984), Lipman et al. (1984), and Smith et al. (1985).

All of the above procedures are usually used to produce alignments prior to phylogenetic analysis. However, the principle used to align sequences is usually the same principle used to infer phylogeny, namely parsimony. Therefore, it has been suggested that alignment of sequences should be part of phylogeny inference, rather than prior to it (Sankoff et al., 1973). However, in practice, it is computationally intractable to carry out such an analysis for any more than a few sequences simultaneously. On the other hand, alternative alignments are often equally good, and phylogenetic congruence offers a means of choosing between otherwise equally parsimonious alignments.

APPENDIX: STOCK SOLUTIONS

20% Acrylamide solution

96.5 g acrylamide
3.35 g bisacrylamide
233.5 g urea (ultrapure)
50.0 ml 10× TBE
distilled H_2O (150 ml start) to 500 ml

Filter before storage; store in brown bottle.

ATE

100 mM sodium acetate, pH 5.2
10 mM Tris-HCl, pH 7.4
1 mM EDTA

Borate buffer

0.2 M Na borate, pH 9
30 mM [ethylenebis(oxyethylenenitrilo)] tetraacetic acid
5 mM dithiothreitol
1% SDS

CI

A solution of chloroform and isoamyl alcohol, in the ratio (v/v) of 24:1.

2× CTAB extraction buffer

10 g CTAB [hexadecyltrimethylammonium bromide]
140 ml 5 M NaCl
25 ml 2 M Tris-HCl, pH 8.0
20 ml 0.5 M EDTA

Denaturing solution

1.5 M NaCl
0.5 M NaOH

DEP-treated water

Add 0.1% diethylpyrocarbonate to water, wait 12 hr, autoclave. (Used to inhibit RNase.)

EDTA (ethylenediaminetetraacetic acid)-disodium salt

Use as 0.5 M stock, pH 8.0 (186.1 g disodium EDTA/liter water; dissolve and pH with NaOH; autoclave).

GTE

0.05 M glucose
0.025 M Tris-HCl, pH 8.0
0.01 M EDTA

KAc

For 100 ml 5 M KAc, pH 4.8:

60 ml 5 M potassium acetate
11.5 ml glacial acetic acid
28.5 ml water

L-Broth

10 g tryptone
10 g NaCl
5 g yeast extract
1 liter H_2O

Adjust pH to 7.2 with NaOH, autoclave. For L-broth plates, add 15 g agar/liter before autoclaving. For L-broth + $MgSO_4$ + maltose plates, add sterile $MgSO_4$ to 10 mM and sterile-filtered maltose to 0.2% after autoclaving. Sterile-filtered antibiotics, IPTG, and X-Gal should also be added after autoclaving, after agar has cooled down to below 50°C. For L-broth top agarose, add 7 g/liter agarose to L-broth before autoclaving.

10× Ligation buffer

500 mM Tris-HCl, pH 8.0
70 mM $MgCl_2$
10 mM dithiothreitol

Neutralizing solution

1.5 M NaCl
0.5 M Tris-HCl, pH 8.0.

PCI

This is a solution of phenol, chloroform, and isoamyl alcohol, in a ratio of 25:24:1. A layer of water will form on the surface; the PCI is the lower layer.

PDB (phage dilution buffer)

0.01 M Tris-HCl, pH 7.4
0.1 M NaCl
0.01 M MgCl$_2$

Autoclave.

PEG (Polyethylene glycol)

Use as a 20% solution.

20× Reverse transcriptase buffer

400 mM Tris-HCl, pH 8.3
150 mM MgCl$_2$
150 mM KCl
40 mM dithiothreitol

SDS (Sodium lauryl sulfate)

Use as a 20% solution. Do not refrigerate or autoclave.

20× SSC

3 M NaCl
0.3 M citrate (trisodium)

Adjust pH to 7.0 (with HCl).

STE

0.1 M NaCl
0.05 M Tris-HCl, pH 7.5
0.001 M EDTA

Autoclave.

Stop buffer

95% formamide
20 mM EDTA
0.05% bromophenol blue
0.05% xylene cyanol FF

10× *Taq* salts

For 1 ml:

125 μl 4M KCl
100 μl 1M Tris-HCl, pH 8.4
25 μl 1M MgCl$_2$
1 mg gelatin
750 μl of DEP-dH$_2$0

10× TBE

0.89 M Tris
0.89 M boric acid
0.01 M disodium EDTA

Filter and adjust pH to 8.3 with NaOH or HCl.

10× TE

0.01 M Tris-HCl, pH 7.5
0.001 M EDTA

Autoclave.

Urea mix

233.5 g urea (ultrapure)
50 ml 10× TBE
distilled H$_2$O to 500 ml

III
ANALYSIS

CHAPTER 10

INTRASPECIFIC DIFFERENTIATION

B. S. Weir

BIOLOGICAL CONTEXT

This chapter considers genetic variation within species. The general goals of population genetic studies are to characterize the extent of genetic variation, within species in this case, and to account for this variation. Variation provides the raw material for future evolutionary change and different levels of variation in different populations may provide evidence for distinct evolutionary events in the past.

Of course the first goal is by far the easiest to meet, and characterizations of variation rest on observations of phenotypic classes in the data set. It will be assumed here that there is a direct relation between phenotype and genotype. The crosses necessary to demonstrate that a band on an electrophoretic gel (for example) does indeed correspond to an allelic form of a single gene have been carried out. It will further be assumed here that the different genetic entities, whether allozymes, restriction site patterns, or nucleotide subsequences, are substantially independent. Although means for accommodating associations between Mendelizing units are available (e.g., Weir and Cockerham, 1989a), they are beyond the scope of this chapter.

The first analyses of genetic data are those that rest simply on the genotypic state. Avise et al. (1987) were able to make statements about the phylogeny of species on the basis of mitochondrial genotypes observed in different populations (see also Chapter 8). At the next level, counts of genotypes lead to simple measures of variation such as the number of alleles per locus or the associated allelic frequencies. The numbers of alleles may be sufficient for the purpose of establishing that there is variation, but it is difficult to use them to make inferences such as comparing levels of variation between different populations. For such purposes, allelic frequencies are better suited and appropriate statistics will be discussed. More complex functions of frequencies, such as gene diversity (e.g., Weir, 1989), begin to address the mechanisms for the maintenance of variation.

The main theme of this chapter is that statistical analyses must be based on biological models. The classic model of an infinite population mating at random for a locus at which there are no disturbing forces of mutation, migration, or selection (i.e., the Hardy–Weinberg formulation) allows statements to be made about statistics such as gene diversity. This quantity is one minus the sum of squared allelic frequencies and is expected to be the same as the frequency of heterozygotes under the classic model. One of the analyses to be treated below compares diversity and heterozygosity—the thought being that a significant difference between the two quantities will indicate the violation of some assumption of the model.

One of the first steps to introduce realism into the classic model is to suppose that the population is finite, although still mating at random. This model allows quantities such as the frequency of heterozygotes to be related to population size and may suggest a means for estimating population size (e.g., Laurie-Ahlberg and Weir, 1979). Unfortunately, the large sampling variance of heterozygosity makes such attempts of limited use in samples that are measured only in the hundreds. Statistics constructed from the joint frequencies of pairs of loci have even more severe problems. **Linkage disequilibrium** refers to the departure of such joint frequencies from the products of single frequencies, and theory relates the expected value of squared linkage disequilibrium to population size and recombination rates between the loci. Although it would be of great advantage to be able to estimate either of these quantities from frequency data, once again the large sampling variances of linkage disequilibria make this unlikely (Hill and Weir, 1988).

Further refinements of the classic model allow for specified mechanisms of selection, migration, or mutation. Each of these leads to predictable changes in allelic frequencies over time. When data are available from several generations, it may be possible to make inferences about the evolutionary events acting within the species. Good discussions have appeared previously in the literature. Prout (1965) detailed the difficulties in estimating the strengths of selection acting at different stages of the life cycle, while Christiansen and Frydenberg (1973) showed the power of having data collected for mother–offspring pairs. Estimating migration or mutation rates in a population generally proceeds under the assumption that the population is at equilibrium (e.g., Chakraborty and Leimar, 1987), so that these forces are causing no further changes over time. Although theory relates functions of allelic frequencies to the rates, care must be taken in the analyses as will be shown later. Estimation of migration rates between local populations, for example, should not use models that assume frequencies in different populations are independent. Easteal (1986) was able to estimate migration rates without an assumption of equilibrium by taking advantage of the known history of different populations of giant toads, *Bufo marinus*. Comparisons among allelic frequencies between introgressed populations and the original populations allowed rates of admixture to be estimated.

This chapter is concerned with the most common type of genetic data available to systematists—survey data from one or several natural populations where neither the population nor the environment has been manipulated. It is becoming increasingly easy to collect data on many loci for many individuals in many populations. Temporal information is not generally available so that direct evidence for selection, for example, is not obtained. What the data do allow, however, is a characterization of the relationships between genes at various levels in a hierarchy. The degrees of relatedness of genes within individuals, between individuals within subpopulations, between subpopulations within populations, and so on, can all be estimated. Comparisons between estimates allow inferences to be made about the forces acting within a species, at least at a gross level. If genes within individuals appear to be related in a population, there may be departures from random mating. If genes within individuals appear related to different extents in different populations, the possibility of different mating systems or population sizes in those populations needs to be investigated. If different loci show different degrees of relatedness for genes within individuals within the same population, the possibility of selective forces acting on those loci needs to be considered. Note that although the degrees of relationship may be expressed in terms of measures of inbreeding, the most general interpretation regards them as correlations (Cockerham, 1973).

Estimation of levels of relatedness of genes in a hierarchical sampling scheme has been considered by many authors for a wide variety of species. Schoen (1982) looked at populations of the annual plant *Gilia achilleifolia* and Guries and Ledig (1982) considered the longer living tree *Pinus rigida*. Foltz and Hoogland (1983) looked at populations of prairie dogs, *Cynomys ludovicianus,* and Chesser (1983) looked at a nested sampling scheme of regions and populations within regions for the same species. These authors were all concerned either with aspects of genetic heterogeneity among populations within a single species or with variation in mating structure.

Genetic and Statistical Sampling

Unless a population is absolutely uniform for the loci being studied, different samples from the population will show different levels of genetic variation. This is simply a consequence of the "statistical sampling" that results in each sample having a different set of individuals. Analyses need to be set up that allow statements to be made about the population, based on the sample at hand, with this sampling variation accommodated.

A basic activity in statistics is estimating the mean, μ, for some variable. If measurements of the variable are denoted by X, and the mean by $\bar{X}$, then μ is estimated by $\bar{X}$. Variation between samples is anticipated by assigning a variance of σ^2/n to this estimate, where σ^2 is the variance of the original variable. As the sample size gets larger, the population is better represented by the sample and different samples become more similar. The variance among samples of the sample mean decreases.

In population genetics there is another level of sampling to be considered. Each generation of a population is formed by the union of gametes chosen from among those produced by the previous generation. This "genetic sampling" process would cause the population to look different if the formation of a new generation was replicated. Population genetic theory depends on the concept of replicate populations that are maintained under the same conditions, but that will differ because of genetic sampling. It is possible to derive variances for statistics of interest that include both types of variation.

One use for such total variances is in predicting future values. From a specified population, it is possible to predict the expected value of a statistic, such as allelic frequency or heterozygosity or linkage disequilibrium, in a future population but not to specify the exact value of the statistic. For a neutral gene in a finite population, for example, it is known that the allelic frequencies have constant *expected* values over time although in any particular population the frequency may have drifted to any value between zero and one. Statements about the statistic in some future sample must therefore take account of the variation between replicate populations, as well as between replicate samples from any one population.

A difficulty arises in that the magnitude of between-population variation cannot be estimated with a sample from a single population. One way around this problem is sometimes afforded by the availability of several unlinked loci in the data set. Although genes at different loci are never completely independent, since they are carried on the same gametes between generations, it may be that they have frequencies that are nearly independent so that they play the role of separate populations. Genes at these loci have each been exposed to the same genetic sampling forces between generations, but can have different pedigrees in the same way as do genes in replicate populations.

Fixed and Random Models

The previous discussion can also be phrased in terms of fixed and random effects, to show how the intended scope of inference affects the sampling properties of genetic statistics. If there is interest only in the one particular population sampled, then the genetic sampling between generations is not of consequence. It is necessary to take account of the statistical sampling for repeated samples only from this one "fixed" population. Future samples would be taken from the population as it is presently constituted. Comparisons between different fixed populations can be phrased in terms of means. It will be shown that procedures of numerical resampling are of use in estimating variances for means from a single population and in comparing different populations.

A different situation arises when the sample is to be used to make inferences about the species as a whole. There is then less interest in the particular population sampled, which can now be regarded as being "random." Future samples may very well be drawn from a different population, so that both statistical and genetic sampling variation need to be considered. The distinction between fixed and random effects arises in statistics. In the analysis of variance context it is easier to detect differences between means in a fixed-effects situation because a smaller variance is used in the denominator for the F test statistic. It is only one specific set of means (fixed effects) that is being compared, and not some population of means (random effects) for which the means at hand are just a sample.

The distinction between fixed and random models in the genetic context has been made previously by Cockerham and Weir (1986). They stress that the random model considers each population to be a replicate sample of the evolutionary process.

As an example of the differences in ranges of inference that follow from the choice of fixed or random models, consider the study of Baker et al. (1982). These authors sampled sparrows, *Zonotrichia leucophrys nuttalli,* from areas known to have different song dialects. They were seeking genetic evidence for lack of mating between different dialects. A fixed-model analysis would have restricted them to making statements about their particular set of four dialects in California. By adopting a random-model analysis, however, they were able to make statements about dialect groups for the species as a whole.

STATISTICAL METHODS

Most population genetic data sets are based on the genotypes of diploid individuals. Exceptions include cases where homozygous lines of *Drosophila* are used so that the data are essentially on gametes rather than on genotypes. Haploid data are also obtained for nonnuclear genes, i.e., mitochondrial and chloroplast genes (Chapters 8 and 9). In human studies, with data collected for family pedigrees, it is often possible to infer the gametic arrangement of the genes studied, and the resulting haplotypic data are analyzed. Even here, though, a proper analysis should take account of the fact that the basic sampling unit is the individual and that it is the genotype that is recorded (or inferred from a phenotype).

One-locus genotypic frequencies will be written as Ps, with subscripts indicating the alleles as in P_{AA} for AA or P_{ij} for $A_i A_j$, ($j \neq i$). Allelic frequencies will be written with lower-case ps. For allele A_i then

$$p_i = P_{ii} + \frac{1}{2}\sum_{j \neq i} P_{ij}$$

The sum over heterozygous classes involves every heterozygote that has allele A_i.

Sample genotypic values will be distinguished from the population values they estimate by tildes, $\tilde{P}$. The population value refers to the particular population in the fixed model, or to all replicate populations in the random case. Expected values will be indicated by the symbol $\mathcal{E}$.

Fixed Populations

Allelic Frequencies   With data collected for genotypes, the first set of descriptors of a population is simply the genotypic frequencies. When the population is sampled in such a way that every member of the population has an equal chance of being sampled, and individuals are sampled independently, the genotypic counts are multinomially distributed. This distribution gives the expected frequencies with which the counts take their possible values in repeated samples from the same population. As a consequence, the count for any particular genotype has a binomial distribution. Writing counts as n_{ij} for genotypes $A_i A_j$, and the sample size as n, the binomial property is expressed as

$$n_{ij} \sim B(n, P_{ij})$$

The average or expected counts, $\mathcal{E}n_{ij}$, over all samples from a population in which genotypic frequencies are P_{ij}, are therefore

$$\mathcal{E}n_{ij} = nP_{ij}$$

and the variances are

$$\text{Var}(n_{ij}) = nP_{ij}(1 - P_{ij})$$

Sample genotypic frequencies are obtained by dividing by sample size, n,

$$\tilde{P}_{ij} = \frac{n_{ij}}{n}$$

so that

$$\mathcal{E}\tilde{P}_{ij} = P_{ij}, \qquad \text{Var}(\tilde{P}_{ij}) = \frac{P_{ij}(1 - P_{ij})}{n}$$

and these variances, or their square roots (the standard deviations), can be presented along with the sample frequencies.

If the locus has several alleles, the number of genotypic classes is large and the data may better be summarized with allelic frequencies. For co-dominant alleles such as those found for allozyme markers (Chapter 4), the allelic numbers can be found directly from the genotypic numbers:

$$n_i = 2n_{ii} + \sum_{j \neq i} n_{ij}$$

and allelic frequencies found by dividing by the number of genes, $2n$:

$$\tilde{p}_i = \tilde{P}_{ii} + \frac{1}{2} \sum_{j \neq i} \tilde{P}_{ij}$$

Taking averages over all samples from the population does give

$$\mathcal{E}\tilde{p}_i = p_i$$

but the variances depend on genotypic frequencies as well as allelic frequencies

$$\text{Var}(\tilde{p}_i) = \frac{1}{2n}(p_i + P_{ii} - 2p_i^2)$$

as noted by Kempthorne (1957). Allelic frequencies do not have binomial distributions.

Only when Hardy–Weinberg equilibrium (HWE) holds, so that

$$P_{ii} = p_i^2, \qquad P_{ij} = 2p_i p_j$$

does the variance of an allelic frequency reduce to the form for the binomial distribution

$$\text{Var}(\tilde{p}_i) = \frac{p_i(1 - p_i)}{2n}$$

In the HWE case, allelic counts are themselves binomially distributed, and the population can be completely characterized by allelic frequencies. Otherwise, allelic frequencies should be presented along with standard deviations calculated according to the genotypic frequencies. Evidently, then, a test for HWE should be one of the first steps in studying intraspecific differentiation.

Note that HWE testing is being suggested here to see if populations can be characterized by allelic instead of genotypic frequencies. Although the finding of HWE frequencies therefore simplifies further analyses, there are also biological ramifications. A demonstration that a population does not have HWE genotypic frequencies means that at least one of the assumptions that lead to such frequencies does not hold for that population. The population may not be large and mating at random, or there may be forces such as selection or migration acting. Noncompliance with HWE does not, by itself, indicate the reason for noncompliance. Unfortunately, compliance with HWE does not mean that all the assumptions are being met. Lewontin and Cockerham (1959) showed that certain patterns of selection can lead to HWE frequencies, and Li (1988) presented a similar argument for non-random mating.

Testing for Hardy–Weinberg Equilibrium   A recent review of the many procedures for testing for Hardy–Weinberg genotypic proportions in a population was given by Hernández and Weir (1989). The simplest approach is to measure departures from HWE with a set of **disequilibrium coefficients**, one for each heterozygous class at the locus in question. For heterozygote $A_i A_j$, define a disequilibrium coefficient D_{ij} by

$$2D_{ij} = 2p_i p_j - P_{ij}$$

which means that homozygote frequencies can be expressed as

$$P_{ii} = p_i^2 + \sum_{j \neq i} D_{ij}$$

The term "disequilibrium" simply means the difference between a joint frequency of two or more alleles and the product of the frequencies of the separate alleles. If there are only two alleles, A and a, the three genotypic frequencies can be written in terms of two allelic frequencies and one disequilibrium coefficient $D_A = D_{Aa}$

$$P_{AA} = p_A^2 + D_A$$
$$P_{Aa} = 2p_A p_a - 2D_A$$
$$P_{aa} = p_a^2 + D_A$$

Sample disequilibria are found by substituting sample allelic and genotypic frequencies in the definition formulas. To test the hypothesis that there is no disequilibrium for heterozygote $A_i A_j$, the estimated disequilibrium coefficient is squared and divided by an estimate of its variance under that hypothesis. The ratio

$$X_{ij}^2 = \frac{2n\tilde{D}_{ij}^2}{\tilde{p}_i \tilde{p}_j [(1 - \tilde{p}_i)(1 - \tilde{p}_j) + \tilde{p}_i \tilde{p}_j] + \Sigma_{k \neq i,j} (\tilde{p}_i^2 \tilde{D}_{jk} + \tilde{p}_j^2 \tilde{D}_{ik})}$$

has a χ^2 distribution with 1 degree of freedom when the hypothesis is true

$$X_{ij}^2 \sim \chi_{(1)}^2$$

which gives a statistic for testing the hypothesis. For a locus with two alleles, A_1 and A_2, this statistic simplifies greatly as there will not be any disequilibria in the denominator

$$X_{12}^2 = \frac{n\tilde{D}_{12}^2}{\tilde{p}_1^2 \tilde{p}_2^2}$$

For a locus with three alleles, A_1, A_2 and A_3,

$$X_{12}^2 = \frac{2n\tilde{D}_{12}^2}{\tilde{p}_1\tilde{p}_2[(1-\tilde{p}_1)(1-\tilde{p}_2)+\tilde{p}_1\tilde{p}_2]+(\tilde{p}_1^2\tilde{D}_{23}+\tilde{p}_2^2\tilde{D}_{13})}$$

$$X_{13}^2 = \frac{2n\tilde{D}_{13}^2}{\tilde{p}_1\tilde{p}_3[(1-\tilde{p}_1)(1-\tilde{p}_3)+\tilde{p}_1\tilde{p}_3]+(\tilde{p}_1^2\tilde{D}_{23}+\tilde{p}_3^2\tilde{D}_{12})}$$

$$X_{23}^2 = \frac{2n\tilde{D}_{23}^2}{\tilde{p}_2\tilde{p}_3[(1-\tilde{p}_2)(1-\tilde{p}_3)+\tilde{p}_2\tilde{p}_3]+(\tilde{p}_2^2\tilde{D}_{13}+\tilde{p}_3^2\tilde{D}_{12})}$$

Each heterozygote can be tested separately for departures from HWE. The procedure is not quite as powerful as that given by likelihood ratio tests, but Hernández and Weir (1989) found it to be satisfactory for preliminary analyses, and it does have the great advantage of simplicity. Sample sizes should be moderately large, certainly at least 20 individuals.

Instead of looking for a detailed picture of the departures from HWE, it is possible to set up an overall χ^2 test. With v alleles at a locus, the quantity

$$X_T^2 = \sum_{i=1}^{v} \frac{n(\tilde{P}_{ii}-\tilde{p}_i^2)^2}{n\tilde{p}_i^2} + \sum_{i=1}^{v-1}\sum_{j=i+1}^{v} \frac{n(\tilde{P}_{ij}-2\tilde{p}_i\tilde{p}_j)^2}{2\tilde{p}_i\tilde{p}_j}$$

has a χ^2 distribution with $v(v-1)/2$ degrees of freedom when there is HWE. It gives a test that all the $v(v-1)/2$ disequilibrium coefficients are zero simultaneously.

For two alleles, A and a, the individual and the overall test statistics both reduce to

$$X_A^2 = \frac{n\tilde{D}_A^2}{\tilde{p}_A^2\tilde{p}_a^2}$$

which is algebraically the same as the goodness-of-fit test for the three genotypic counts against the expected values under HWE.

Another approach to testing for overall disequilibrium can be phrased in terms of the within-population **inbreeding coefficient** f. This coefficient may be defined as the parameter that allows genotypic frequencies to be written as

382 CHAPTER 10 / WEIR

$$P_{ii} = p_i^2 + fp_i(1 - p_i)$$

$$P_{ij} = 2p_ip_j(1 - f)$$

It can be estimated from the method of moments as

$$\hat{f} = \frac{(\tilde{d} - \tilde{h}) + (1/n)\tilde{h}}{\tilde{d} + (1/n)\tilde{h}}$$

where $\tilde{h}$ is the total frequency of heterozygotes in the sample

$$\tilde{h} = 1 - \sum_i \tilde{P}_{ii}$$

and $\tilde{d}$ is the sample gene diversity, sometimes referred to as the frequency of heterozygotes expected under HWE,

$$\tilde{d} = 1 - \sum_i \tilde{p}_i^2$$

The method of moments requires that a statistic be constructed that has expectation equal to the parameter being estimated. The statistic is then regarded as an estimator of that parameter.

The hypothesis that $f = 0$ can be tested by dividing the square of the estimate by an estimate of its variance (the variance referring to repeated samples from the same population), and regarding the ratio X_f^2 as a χ^2 with 1 degree of freedom:

$$X_f^2 = \frac{\tilde{f}^2}{\text{Var}(\tilde{f})}$$

$$\approx \frac{n(\tilde{d} - \tilde{h})^2}{\Sigma_i \tilde{p}_i^2 + (\Sigma_i \tilde{p}_i^2)^2 - 2\Sigma_i \tilde{p}_i^3}$$

Note that use of the single degree of freedom for f is not strictly a test for HWE. Instead, it tests that the sum of heterozygote (or homozygote) frequencies matches the sum expected under HWE. Individual heterozygotes could have frequencies that differ from HWE values, but these differences could cancel out to leave the sum having the HWE values. The disequilibrium coefficient approach appears to have the advantage in that departures from HWE can be identified with specific genotypes. Apart from this identification, there is a statistical advantage to using disequilibrium coefficients as opposed to ratio statistics such as f. Means and variances can be found for sample disequilibria from multinomial theory, whereas there are no exact expressions for the means and variances of ratios. (The variance used for $\tilde{f}$ in setting up the test statistic X_f^2 was an approximation for large sample sizes.) Another advantage of the disequilibrium coefficient approach is that it leads directly into a treatment of linkage disequilibrium (Weir and Cockerham, 1989b).

Differentiation between Populations   Under the fixed-population approach, different populations for the same species are compared simply by comparing mean frequencies. When HWE cannot be assumed, this requires the comparison of genotypic frequencies.

Contingency Tables.   The most straightforward procedure is to use contingency table χ^2 tests. With v alleles at a locus, the genotypic counts in each of r samples are arranged in an $v(v + 1)/2 \times r$ contingency table and a χ^2 statistic with $[v(v + 1)/2 - 1] \times (r - 1)$ degrees of freedom is calculated. In practice, the method has problems when some cells have small expected counts since this can give test statistics that are spuriously large, and it may be necessary to collapse the least frequent classes. This problem increases with the number of alleles, but even for two alleles a sample of size 100 is expected to have only four individuals in the *aa* class when allelic frequencies are $P_A = 0.8$, $p_a = 0.2$. Conventional wisdom says that goodness-of-fit χ^2 should not be performed on classes with expected counts less than five.

When HWE is assumed, it is sufficient to compare the allelic arrays in each of the populations. The contingency table is then only $v \times r$, and problems of small expected counts are less likely.

Numerical Resampling.   An alternative to the contingency table approach is provided by numerical resampling (Efron, 1982). This is a means of making inferences about the population allelic frequencies from the sample frequencies. Variances can be estimated and confidence intervals can be constructed, both referring to repeated sampling from the same populations. Briefly, numerical resampling mimics the drawing of new samples by resampling the one sample at hand from each population. Two methods are commonly used: **jackknifing** and **bootstrapping**. They were described in a genetic context by Dodds (1986).

If some parameter, ϕ, is to be estimated from a sample of size n, the jackknifing procedure makes use of the fact that almost as good an estimate could be obtained from a sample of size $n - 1$. Since there are n ways of dropping one of the observations to form a sample of size $n - 1$, there are n new estimates of ϕ. The variance among these new estimates provides an estimate of the variance of original estimate. Specifically, suppose the original estimate, based on all n observations, is $\hat{\phi}$. The estimate obtained when observation i is dropped is written as $\hat{\phi}_{(i)}$, and the variance of $\hat{\phi}$ is estimated as

$$\text{Var}\,(\hat{\phi}) \triangleq \frac{n-1}{n} \sum_{i=1}^{n} \left(\hat{\phi}_{(i)} - \frac{1}{n} \sum_{i=1}^{n} \hat{\phi}_{(i)} \right)^2$$

$$= \frac{n-1}{n} \sum_{i=1}^{n} (\hat{\phi}_{(i)} - \hat{\phi}_{(\cdot)})^2$$

where

$$\hat{\phi}_{(\cdot)} = \frac{1}{n} \sum_{i=1}^{n} \hat{\phi}_{(i)}$$

Besides giving an estimate of the variance of an estimate, jackknifing also gives a refined estimate that is expected to be less biased than the original. This new estimate is

$$\hat{\phi}^* = n\hat{\phi} - (n-1)\hat{\phi}_{(\cdot)}$$

With Hardy–Weinberg equilibrium, for allelic frequencies p_A the original estimates $\tilde{p}_A$ are already unbiased, and have sample variances that can be estimated by $\tilde{p}_A(1 - \tilde{p}_A)/n$ so that jackknifing is not providing any new information. Indeed, for the HWE situation the allelic frequencies are known to be binomially distributed and the properties of the sample frequencies are therefore known. If there is not HWE, the distribution of allelic frequencies is not known, however, and for more complicated statistics such as f, jackknifing can be invaluable.

The numerical resampling technique of bootstrapping proceeds slightly differently, although it still refers to repeated samples from a fixed population. From the original set of n observations, a new sample of the same size is constructed by random sampling with replacement. In other words, each of the original observations is equally likely to be selected to constitute one of the members of the new sample. The bootstrap sample is therefore likely to have some of the original observations represented many times, and some of them not represented at all. Drawing the sample requires the use of random numbers, as shown in the programs listed in the Appendix. The parameter ϕ is estimated from the new sample, and the process repeated many times, perhaps 1000 or more. In place of the single estimate from the original sample, bootstrapping provides as many new estimates as are desired. Although this collection of new estimates could be used to provide an estimated variance for $\hat{\phi}$, it is more informative to work with the whole distribution of estimates. For example, a 95% confidence interval for ϕ can be constructed as the limits between which the middle 95% of the bootstrap estimates lie.

Bootstrapping within each of the r samples will therefore provide confidence intervals for the population allelic frequencies, without having to invoke the binomial theorem and therefore without having to assume Hardy–Weinberg equilibrium. Two populations can be judged to have different allelic frequencies if the estimated frequencies have nonoverlapping confidence intervals.

It does not appear that either jackknifing or bootstrapping has yet been reported in the literature for the comparison of populations in a fixed-model context. The methods do provide a convenient way of making inferences when there is not the basis for expectations over populations provided by random models.

Random Populations

Under the random model, the populations sampled are considered to represent the species and therefore they have a common evolutionary ancestry. Even though the populations may have been distinct for some time, the analysis is built on the assumption that there is a single ancestral population. The expectations referred to by means and variances now refer to repeated samples from the populations *and* to replicate populations. In the absence of disturbing forces, such as selection, the expected allelic frequencies of all the populations are the same.

Underlying the analysis of differentiation in the random model is the notion that genetic sampling causes different genes in a population to be dependent, or related. Even though individuals, or genes, may be sampled randomly the process of taking expectations must recognize that they are dependent through their shared ancestry. Another essential concept for the analysis is that the relationships between various genes are relative to the least-related genes in the data. It is generally assumed that these least-related genes are independent—the data do not allow measures of relationship to be estimated otherwise.

Interest is generally centered on the extent to which different populations within the species have differentiated over the time since the ancestral population. The process of genetic sampling, e.g., drift, between successive generations will result in intraspecific differentiation, and this differentiation is conveniently quantified with the F-statistics of Wright (1951), or the analogous measures of Cockerham (1969, 1973). These quantities measure the degree of relatedness of various pairs of genes. Cockerham (1969) described the three basic quantities in the situation when diploid individuals are sampled from a series of populations as follows: the overall **inbreeding coefficient** F, Wright's F_{IT}, is the correlation of genes within individuals; the **coancestry** θ, Wright's F_{ST}, is the correlation of genes of different individuals in the same populations; and f, Wright's F_{IS}, is the correlation of genes within individuals within populations. Because the populations are assumed to have been isolated since the ancestral population, genes in one population are independent of those in another.

Haploid Data   If data are available on genes directly, then the analysis is in terms of allelic frequencies and is phrased conveniently in terms of a set of indicator variables. For gene j in a sample from population i, a variable x_{ij} can be defined by

$$x_{ij} = \begin{cases} 1 & \text{if gene is allele } A \\ 0 & \text{if gene is not allele } A \end{cases}$$

The expected value of x_{ij}, over samples and replicate populations, is therefore the allelic frequency p_A common to these populations. As the model on which this analysis is based assumes no forces such as mutation or selection, this frequency is also that in the ancestral population. The sample allelic frequency $\tilde{p}_{Ai}$ in a sample of n_i genes from the ith population can be written as

$$\tilde{p}_{Ai} = \frac{1}{n_i} \sum_{j=1}^{n_i} x_{ij}$$

For haploid data, there is only one F-statistic. It measures the relationship between different genes in the same population, relative to the zero relationship between genes of different populations. The quantity is written here as θ, and was termed F_{ST} by Wright (1951). It allows the expected value of a sample allelic frequency to be written as

$$\mathcal{E}\tilde{p}_{Ai}^2 = p_A^2 + p_A(1 - p_A)\theta + \frac{1}{n_i} p_A(1 - p_A)(1 - \theta)$$

Although the allelic frequency p_A is assumed to remain constant, for finite populations the coancestry will increase over time as inbreeding accrues in each population. In other words, θ measures the extent of differentiation between populations. It is worth stressing that this measure of between-population differentiation is a consequence of the relatedness of genes within populations.

Estimation of θ proceeds by the method of moments, with the various statistics being conveniently organized in an analysis of variance format (Table 1). Data are supposed to be available from r populations, with different numbers of genes n_i from each being allowed. The weighted average frequency of allele A over all the samples is written as

$$\tilde{p}_{A\cdot} = \frac{\sum_{i=1}^{r} n_i \tilde{p}_{Ai}}{\sum_{i=1}^{r} n_i}$$

For the method of moments, the parameters are estimated by equating the observed and the expected mean squares. Writing the observed mean square for between populations as MSP and for genes within populations as MSG:

$$p_A(1 - p_A)(1 - \theta) \triangleq \frac{1}{\Sigma_i(n_i - 1)} \sum_i n_i \tilde{p}_{Ai}(1 - \tilde{p}_{Ai})$$
$$= MSG$$

$$p_A(1 - p_A)(1 - \theta) + n_c p_A(1 - p_A)\theta \triangleq \frac{1}{r - 1} \sum_i n_i(\tilde{p}_{Ai} - \tilde{p}_{A\cdot})^2$$
$$= MSP$$

It may be convenient to identify two terms as the components of variance for genes within populations

$$\sigma_g^2 = p_A(1 - p_A)(1 - \theta)$$

and for populations

$$\sigma_p^2 = p_A(1 - p_A)\theta$$

Note that the variance component between populations is the same as the covariance between frequencies of genes in different individuals within the same population (Cockerham, 1969). The two variance components, within and between populations, reflect all the factors that lead to variation in gene frequencies. The sum of these two components involves only the allelic frequency p_A, and this allows the unknown quantity $p_A(1 - p_A)$ to be eliminated. An estimate of θ can then be found as

$$\hat{\theta} = \frac{MSP - MSG}{MSP + (n_c - 1) MSG}$$

Table 1. Analysis of variance layout for haploid data

Source	df	Sum of squares	Expected mean square[a]
Between populations	$r - 1$	$\Sigma_{i=1}^r \frac{x_{i\cdot}^2}{n_i} - \frac{x_{\cdot\cdot}^2}{n_\cdot}$ $= \Sigma_{i=1}^r n_i(\tilde{p}_{Ai} - \tilde{p}_{A\cdot})^2$	$p_A(1 - p_A)[(1 - \theta) + n_c\theta]$
Within populations	$\Sigma_{i=1}^r (n_i - 1)$ $= n_\cdot - r$	$\Sigma_{i=1}^r \Sigma_{j=1}^{n_i} x_{ij}^2 - \Sigma_i \frac{x_i^2}{n_i}$ $= \Sigma_i n_i \tilde{p}_{Ai}(1 - \tilde{p}_{Ai})$	$p_A(1 - p_A)(1 - \theta)$

[a] $n_c = \frac{1}{r - 1}\left(\sum_{i=1}^r n_i - \frac{\Sigma_i n_i^2}{\Sigma_i n_i} \right)$

This can also be expressed in terms of the variance of allelic frequencies over populations

$$s_A^2 = \frac{1}{(r-1)n} \sum_i n_i (\tilde{p}_{Ai} - \tilde{p}_{A\cdot})^2$$

where

$$\bar{n} = \frac{\sum_{i=1}^r n_i}{r}$$

Using either of these two expressions, the estimator is a ratio of two functions of the data, whose expectations differ by the factor θ. The ratio may be denoted

$$\hat{\theta} = \frac{T_1}{T_2}$$

with

$$T_1 = s_A^2 - \frac{1}{\bar{n}-1}\left[\tilde{p}_{A\cdot}(1-\tilde{p}_{A\cdot}) - \frac{r-1}{r}\, s_A^2\right]$$

$$T_2 = \frac{n_c-1}{\bar{n}-1}\,\tilde{p}_{A\cdot}(1-\tilde{p}_{A\cdot}) + \left[1 + \frac{(r+1)(\bar{n}-n_c)}{\bar{n}-1}\right]\frac{s_A^2}{r}$$

For a large number of large samples, when both $1/\bar{n}$ and $1/r$ can be ignored, this estimate reduces to

$$\hat{\theta} = \frac{s_A^2}{\tilde{p}_{A\cdot}(1-\tilde{p}_{A\cdot})}$$

Although this result often appears in the literature, with the availability of computer programs (see Appendix) there seems to be no need to use it instead of the more general result.

The estimation of θ has been presented in terms of one of the alleles, A, at a locus. If there are only two alleles at the locus, then the same estimate would be obtained if the other allele is used. For more than two alleles, however, a different estimate will result with every allele. Since the parameter θ is the same for every allele under the basic model of no disturbing forces, all these estimates refer to the same quantity and an appropriate average is needed to give the best single estimate. For the uth allele, the estimate could be written as

$$\hat{\theta}_u = \frac{T_{1u}}{T_{2u}}$$

and then an overall estimate with the desirable properties of low bias and small variance is given by combining the information from all v alleles

$$\hat{\theta} = \frac{\Sigma^v_{u=1} T_{1u}}{\Sigma^v_{u=1} T_{2u}}$$

There is an additional extension to cover the case where several loci are scored. Once again, under the basic model, every allele at every locus provides an estimate of the same quantity. Indexing the loci by l and alleles within loci by u, then the individual estimates are

$$\hat{\theta}_{lu} = \frac{T_{1lu}}{T_{2lu}}$$

and the overall estimate from all m loci is

$$\hat{\theta} = \frac{\Sigma^m_{l=1} \Sigma^v_{u=1} T_{1lu}}{\Sigma^m_{l=1} \Sigma^v_{u=1} T_{2lu}}$$

With θ estimated, there is a quantification of the degree of divergence among a set of r populations. For a pair of populations, θ may serve as a measure of genetic distance. With genetic drift but no other disturbing forces, $\ln(1 - \theta)$ will increase linearly with the time since divergence of the two populations from the ancestral population. When all populations are fixed at all loci scored, the estimator is undefined, since the equation becomes the ratio of zeros. Indeed, there is no information on the time of divergence of the populations. There is no information in the present sample on how long the populations have been fixed.

If the indicator variables x_{ij} were normally distributed, then the hypothesis H_0: $\theta = 0$ could be tested with the ratio MSP/MSG in the analysis of variance layout (Table 1). Since the x_{ij}s are either 0 or 1, however, this ratio will not have an F distribution, and will not provide a test statistic. Making inferences about θ beyond simply estimating it may instead be accomplished by numerical resampling. Variances for the estimates can be found from jackknifing, whereas confidence intervals follow from bootstrapping.

Because θ is a parameter appropriate for the random model, numerical resampling cannot be performed by resampling genes at a locus within populations, as was done in the fixed model. The resampling must mimic both the genetic sampling that causes replicate populations to differ, and the sampling of genes for observation from each population. Two possibilities are suggested.

In the first place resampling may be done over loci. For a study in which m loci are scored, jackknifing can be performed by omitting each locus, L, in turn to provide m new estimates. The Lth of these is

$$\hat{\theta}_{(L)} = \frac{\Sigma_{l \neq L} \Sigma_u T_{1lu}}{\Sigma_{l \neq L} \Sigma_u T_{2lu}}$$

and the variance of $\hat{\theta}$ is estimated as

$$\text{Var}(\hat{\theta}) \doteq \frac{m-1}{m} \sum_{L=1}^{m} \left(\hat{\theta}_{(L)} - \frac{1}{m} \sum_{L=1}^{m} \hat{\theta}_{(L)} \right)^2$$

This procedure makes use of the fact that loci are expected to provide nearly independent replicates of the genetic sampling process. Jackknifing therefore provides an estimate of the variance of $\hat{\theta}$. Since the estimates are not being assumed to be normally distributed, the variances do not provide an immediate means for testing hypotheses, or for comparing different estimates. These activities will follow from bootstrapping over loci to construct confidence intervals for θ. Each bootstrap sample consists of a set of m pairs of values $\Sigma_u T_{1lu}$, $\Sigma_u T_{2lu}$ drawn with replacement from the m calculated values, and the combined estimate formed from this new collection of Ts. As before, the middle 95% of these new estimates will provide a 95% confidence interval. The hypothesis that θ has some specified value can be rejected, at the 5% significance level, if the confidence interval does not include that value. Data sets with overlapping confidence intervals provide no evidence that the corresponding θ values differ.

It is also possible to jackknife over populations, and this may be done for each locus separately. In that way the estimates of θ for each locus could be compared. Loci that did not give estimates for θ within two standard deviations of each other may indicate the presence of differential disturbing forces such as selection. Jackknifing over populations requires that there are several populations, just as resampling over loci supposes that several loci have been scored. In practice, at least five populations or loci appear to be necessary.

Finally, it is necessary to comment on the possibility that the estimate of θ may be negative. There are two situations that are likely to give this outcome. It may be that the true value of θ is positive but small. Since the estimate has fairly low bias, the estimate is about as likely to be below as above the true value. In this case, estimates less than the true value will often be negative. The second situation is that the parameter may be negative. In statistical language, this corresponds to a negative intraclass correlation. In genetic language, it means that genes are more related between than within populations. This could result from some forms of migration that violate the assumption of the populations having remained distinct since an ancestral population.

A more complete discussion of biological causes for a negative component of variance between populations was given by Cockerham (1973) for the analysis of diploid data. Any mating system, such as the avoidance of self-mating, that causes genes to be more alike between individuals than within individuals can cause this phenomenon.

Mutation and Migration. If other forces are involved, such as mutation, then the allelic frequencies no longer remain constant over time, and it is not appropriate to separate p_A and θ in the expectation of $\tilde{p}_{Ai}^2$. In this case it is sufficient to work with measures of allelic similarity within and between populations. Migration is allowed between the populations. Using notation from Cockerham and Weir (1987), Q_2 is the probability that two genes within a population have the same allelic state, and Q_3 is the chance that two genes, one in each of two separate populations, are in the same state. The expected mean squares in the analysis of variance layout, after summing over alleles, become

$$\mathcal{E}MSP = (1 - Q_2) + n_c(Q_2 - Q_3)$$
$$\mathcal{E}MSG = (1 - Q_2)$$

These similarity measures, in turn, can be related to descent measures— θ_2 for genes within populations and θ_3 for genes between populations. (Until now populations have been assumed to be independent, so that there was no need for a θ_3, and θ_2 was written as θ). The only estimable quantity is

$$\beta = \frac{\theta_2 - \theta_3}{1 - \theta_3} = \frac{Q_2 - Q_3}{1 - Q_3}$$

which is analogous to $f = F_{IS}$. Cockerham and Weir (1987) discussed how β depends on the population size, number of populations, mutation rate, and migration rate.

Suppose first that there is no migration. Under the infinite-alleles mutation model, every mutation is to a new allelic type. In a finite population of size N, an equilibrium will be established between the loss of variation by drift and the introduction of variation by mutation. At such an equilibrium, if the mutation rate is μ, the value of θ is given by

$$\theta = \frac{1}{1 + 4N\mu}$$

(Kimura and Crow, 1964). Note that different populations are considered here to be independent.

It has often been suggested that migration rates could be estimated from θ values (e.g., Slatkin, 1985). There is an infinite-island model, corresponding to the infinite-alleles mutation model. Each generation, any gene sampled from a population has probability m of having migrated from any one of an infinite number of other populations. When these various "islands" are of finite size N, an equilibrium is again established between loss of variation due to drift within islands, and gain of variation by migration from other islands. The equilibrium value of θ is simply

$$\theta = \frac{1}{1 + 4Nm}$$

Although this suggests a means of estimating Nm, there are complications in practice because the infinite-island model is unrealistic. As soon as a finite number of islands are postulated, there is a nonzero probability that two islands receive migrant genes from the same island. The island populations are not independent and there is need to distinguish between θ_2 and θ_3. The quantity β can be estimated (although it should not be called θ or F_{ST}), and this could be assumed to have expectation $1/(1 + 4Nm)$. This result requires assumptions of small mutation and migration rates, with mutation being less frequent than migration, and large numbers of populations. Cockerham (1984) pointed to similar problems with a system of mutation among a finite number of alleles.

Diploid Data   When observations are made on diploid individuals, the analysis should be performed at the diploid level. The same general approach is followed, but it is now possible to estimate the degree of relationship F between genes within individuals, as well as the degree θ between genes of different individuals. The preceding haploid analysis essentially dropped the distinction between F and θ. Differentiation between independent populations is still measured in terms of θ, reflecting the relatedness of individuals within populations, but the analysis also provides an estimate of the overall inbreeding coefficient F. In Wright's notation, $F = F_{IT}$, $\theta = F_{ST}$. The degree of inbreeding within populations, f, or F_{IS}, can be expressed as

$$f = \frac{F - \theta}{1 - \theta}$$

Only f can be estimated from data from a single population, as detailed in the section on fixed populations.

The expected value of the squared sample allelic frequencies now must reflect the two levels of relatedness of different genes within populations; both are a consequence of prior genetic sampling:

$$\mathcal{E}\tilde{p}_{Ai}^2 = p_A^2 + p_A(1 - p_A)\theta + \frac{1}{2n_i} p_A(1 - p_A)(1 + F - 2\theta)$$

Once again, the estimation procedure follows naturally from an analysis of variance layout. There are now three sources of variation: populations, individuals within populations, and genes within populations. The sums of squares are constructed with gene *and* genotypic frequencies, as shown in Table 2.

The expected mean squares could have been written in terms of variance components for populations

$$\sigma_p^2 = p_A(1 - p_A)\theta$$

for individuals within populations

$$\sigma_i^2 = p_A(1 - p_A)(F - \theta)$$

and for genes within individuals

$$\sigma_g^2 = p_A(1 - p_A)(1 - F)$$

In Table 2, the sample sizes n_i are for the numbers of individuals in each sample—in Table 1 they were the numbers of genes. Also in Table 2, $\tilde{P}_{AAi}$ is the frequency of AA homozygotes in the ith sample. In the whole data set, $\tilde{H}_{A\cdot}$ is the frequency of heterozygous individuals that have allele A. In other words

$$\tilde{H}_{A\cdot} = \frac{\Sigma_i n_i \tilde{H}_{Ai}}{\Sigma_i n_i}$$

$$= \frac{1}{\Sigma_i n_i} \sum_i 2 n_i (\tilde{p}_{Ai} - \tilde{P}_{AAi})$$

Table 2. Analysis of variance layout for diploid data

Source	df	Sum of squares	Expected mean square
Between populations	$r - 1$	$2 \Sigma_{i=1}^r n_i(\tilde{p}_{Ai} - \bar{p}_{A\cdot})^2$ $= 2(r - 1)ns_A^2$	$p_A(1 - p_A)\,[(1 - F)$ $+ 2(F - \theta) +$ $2n_c\theta]$
Individuals in populations	$\Sigma_{i=1}^r (n_i - 1)$ $= n_\cdot - r$	$\Sigma_{i=1}^r n_i(\tilde{p}_{Ai} + P_{AAi} - 2\tilde{p}_{Ai}^2)$ $= 2r\bar{n}\,\tilde{p}_{A\cdot}(1 - \tilde{p}_{A\cdot}) - 2(r - 1)\bar{n}$ $s_A^2 - \frac{1}{2}r\bar{n}\tilde{H}_{A\cdot}$	$p_A(1 - p_A)[(1 - F)$ $+ 2(F - \theta)]$
Genes in individuals	$\Sigma_{i=1}^r n_i$ $= n_\cdot$	$\Sigma_{i=1}^r n_i(\tilde{p}_{Ai} - \tilde{P}_{AAi})$ $= \frac{1}{2}r\bar{n}\tilde{H}_{A\cdot}$	$p_A(1 - p_A)(1 - F)$

If the observed mean squares are written as *MSP*, *MSI*, and *MSG* for populations, individuals, and genes, respectively, then the measures of differentiation can be estimated as

$$\hat{F} = 1 - \frac{2n_c \, MSG}{MSP + (n_c - 1)MSI + n_c \, MSG}$$

$$= 1 - \frac{S_3}{S_2}$$

$$\hat{\theta} = \frac{MSP - MSI}{MSP + (n_c - 1)MSI + n_c \, MSG}$$

$$= \frac{S_1}{S_2}$$

These are probably the most convenient computing formulas, but it is possible to give explicit expressions for the terms S_1, S_2, and S_3:

$$S_1 = s_A^2 - \frac{1}{\bar{n} - 1}\left[\tilde{p}_{A\cdot}(1 - \tilde{p}_{A\cdot}) - \frac{r - 1}{r}\ s_A^2 - \frac{1}{4}\tilde{H}_{A\cdot}\right]$$

$$S_2 = \tilde{p}_{A\cdot}(1 - \tilde{p}_{A\cdot}) - \frac{\bar{n}}{r(\bar{n} - 1)}\left[\frac{r(\bar{n} - n_c)}{\bar{n}}\tilde{p}_{A\cdot}(1 - \tilde{p}_{A\cdot})\right.$$

$$\left. - \frac{1}{\bar{n}}[(\bar{n} - 1) + (r - 1)(\bar{n} - n_c)]s_A^2 - \frac{\bar{n} - n_c}{4n_c^2}\tilde{H}_{A\cdot}\right]$$

$$S_3 = \frac{n_c}{\bar{n}}\tilde{H}_{A\cdot}$$

These estimates have all been presented in terms of one particular allele *A*. In practice, several alleles at several loci will be available and each will provide an estimate of the same parameters under the basic model of neutrality. To combine estimates over all alleles, numerators and denominators are combined separately. For the *u*th allele at the *l*th locus, the estimates may be expressed as

$$\hat{F}_{l_u} = 1 - \frac{S_{3lu}}{S_{2lu}}$$

$$\hat{\theta}_{lu} = \frac{S_{1lu}}{S_{2lu}}$$

and then the combined estimates are

$$\hat{F} = 1 - \frac{\Sigma_l \Sigma_u S_{3lu}}{\Sigma_l \Sigma_u S_{2lu}}$$

$$\hat{\theta} = \frac{\Sigma_l \Sigma_u S_{1lu}}{\Sigma_l \Sigma_u S_{2lu}}$$

Also, as in the haploid case, numerical resampling over loci provides the means for making inferences about the parameters F and θ, whereas resampling over populations allows comparisons between loci.

Note that large numbers of large samples allow approximate expressions to be found for the estimates from each allele:

$$\hat{F} = 1 - \frac{\tilde{H}_{A\cdot}}{\tilde{p}_{A\cdot}(1 - \tilde{p}_{A\cdot})}$$

$$\hat{\theta} = \frac{s_A^2}{\tilde{p}_{A\cdot}(1 - \tilde{p}_{A\cdot})}$$

but the use of computers for data analysis makes these common levels of approximation unnecessary.

For populations mating at random, genes are equally related whether they are within or between individuals. In this case $F = \theta$ or $f = 0$. Estimates of F and θ that differ significantly therefore indicate departures from random mating. Any avoidance of mating of relatives will cause θ to exceed F and f to be negative. In the language of variance components, the component for individuals within populations will then be negative. Recall that this component is actually the difference of two positive quantities—it is not being claimed that there is a variance which is negative. More commonly, f is positive ($F > \theta$), which can be interpreted as evidence for inbreeding, i.e., a bias toward mating among relatives. Different patterns of differences for the two estimates at different loci indicate that there are forces other than nonrandom mating affecting these loci.

The effects of selection on the F-statistics were also detailed by Cockerham (1973). Different selective forces in different populations, tending to increase their differences, will increase the value of θ. Within a population, Lewontin and Cockerham (1959) showed that selection at a locus gives negative f values unless the viability of a heterozygote is less than or equal to the geometric mean of the viabilities of the two homozygotes.

Population Subdivision   It is often the case that individuals are sampled in a nested sampling scheme. J. S. F. Barker et al. (1986) sampled *Drosophila buzzatii* from transects and sites within transects while Ferrari and Taylor (1981) sampled *Drosophila subobscura* from subdivisions, regions within subdivisions, and demes within regions. Each recognizable level in such hierarchies adds another level to the degrees of relatedness of genes. The analysis of intraspecific differentiation can be carried out by looking at the hierarchy of relationships between pairs of genes, from the most closely related pairs (within individuals) to the most distantly related (between largest sampling units). This hierarchy is conveniently recognized and organized in nested analyses of variance layouts. These layouts then provide the means for estimating the various measures of differentiation. Details will be given now for a three-level hierarchy, and the extension to larger numbers of levels is straightforward. As before, genes in different units of the largest sampling unit must be assumed to be independent for these analyses to be valid.

Suppose that n_{ij} individuals are sampled from the jth subpopulation of the ith population, and that there are s_i subpopulations sampled from the ith population. There are still r populations. Pairs of genes may be related by virtue of being within individuals, between individuals within subpopulations, or between subpopulations within populations. The three degrees of relatedness are quantified by F, θ_s, and θ_p for differentiation among individuals, subpopulations, or populations, respectively. Genes in different populations are assumed to be unrelated and all genes sampled have the same expectation, p_A, of being of allelic type A. The analysis of variance layout is summarized in Table 3.

The four mean squares use the allelic frequencies, $\tilde{p}_{Aij}$, and the heterozygote frequencies, $\tilde{H}_{Aij}$, for the sample from subpopulation j of population i.

$$MSP = \frac{2}{r-1} \sum_{i=1}^{r} n_{i\cdot}(\tilde{p}_{Ai\cdot} - \tilde{p}_{A\cdot\cdot})^2$$

$$MSS = \frac{2}{s_\cdot - r} \sum_{i=1}^{r} \sum_{j=1}^{s_i} n_{ij}(\tilde{p}_{Aij} - \tilde{p}_{Ai\cdot})^2$$

$$MSI = \frac{2}{n_{\cdot\cdot} - s_\cdot} \left\{ \sum_{i=1}^{r} \sum_{j=1}^{s_i} n_{ij}\left[\tilde{p}_{Aij}(1 - \tilde{p}_{Aij}) - \tfrac{1}{4}\tilde{H}_{Aij}\right] \right\}$$

$$MSG = \frac{1}{2n_{\cdot\cdot}} \sum_{i=1}^{r} \sum_{j=1}^{s_i} n_{ij}\tilde{H}_{Aij}$$

Table 3.   Analysis of variance layout for three-level hierarchy

Source	df	Mean square	Expected mean square[a]
Between populations	$r - 1$	MSP	$p_A(1 - p_A)[(1 - F) + 2(F - \theta_s)$ $+ 2n_{c1}(\theta_s - \theta_p) + 2n_{c2}\theta_p]$
Subpopulations in populations	$\sum_{i=1}^{r} (s_i - 1)$ $= s_. - r$	MSS	$p_A(1 - p_A)[(1 - F) + 2(F - \theta_s)$ $+ 2n_{c3}(\theta_s - \theta_p)]$
Individuals in subpopulations	$\sum_{i=1}^{r}\sum_{j=1}^{si}(n_{ij} - 1)$ $n = n_{..} - s_.$	MSI	$p_A(1 - p_A)[(1 - F) + 2(F - \theta_s)]$
Genes in individuals	$\sum_{i=1}^{r}\sum_{j=1}^{si} n_{ij}$ $= n_{..}$	MSG	$p_A(1 - p_A)(1 - F)$

$$^a n_{i.} = \sum_{j=1}^{s_i} n_{ij} \qquad n_{..} = \sum_{i=1}^{r} n_{i.} \qquad s_. = \sum_{i=1}^{r} s_i$$

$$n_{c1} = \frac{1}{r-1} \sum_{i=1}^{r} \sum_{j=1}^{s_i} \left[\frac{(n_{..} - n_{i.})n_{ij}^2}{n_{i.} n_{..}} \right]$$

$$n_{c2} = \frac{1}{r-1} \left(n_{..} - \frac{1}{n_{..}} \sum_{i=1}^{r} n_{i.}^2 \right)$$

$$n_{c3} = \frac{1}{s_. - r} \left[n_{..} - \sum_{i=1}^{r} \sum_{j=1}^{s_i} \left(\frac{n_{ij}^2}{n_{i.}} \right) \right]$$

$$\bar{p}_{Ai.} = \frac{1}{n_{i.}} \sum_{j=1}^{s_i} n_{ij} \bar{p}_{Aij}$$

$$\bar{p}_{A..} = \frac{1}{n_{..}} \sum_{i=1}^{r} \sum_{j=1}^{s_i} n_{ij} \bar{p}_{Aij}$$

$$= \frac{1}{n_{..}} \sum_{i=1}^{r} n_{i.} \bar{p}_{Ai.}$$

The method of moments provides estimators for each of the three measures of differentiation:

$$\hat{F} = 1 - \frac{R_4}{R_2}$$

$$\hat{\theta}_s = \frac{R_3}{R_2}$$

$$\hat{\theta}_p = \frac{R_1}{R_2}$$

where

$$R_1 = \frac{MSP - MSI}{2n_{c2}} - \frac{n_{c1}(MSS - MSI)}{2n_{c2}n_{c3}}$$

$$R_2 = \frac{MSP - MSI}{2n_{c2}} + \frac{(n_{c2} - n_{c1})(MSS - MSI)}{2n_{c2}n_{c3}} + \frac{MSG - MSI}{2}$$

$$R_3 = \frac{MSP - MSI}{2n_{c2}} + \frac{(n_{c2} - n_{c1})(MSS - MSI)}{2n_{c2}n_{c3}}$$

$$R_4 = MSG$$

The combination over alleles and loci proceeds as before, as does numerical resampling over loci or populations.

IMPLEMENTATION

Sampling

The theory presented above assumes random sampling of individuals from each population, at least with respect to the genetic units being scored. Such randomness is generally guaranteed by the methods used to capture animals or select plants from a natural stand.

Sampling of populations may not be as clearly random. Because the genetic sampling involved in the transmission of genes between generations is random, any extant population is a random representative of all the replicate populations that may arise under the same set of conditions. The extent to which it can be regarded as representative of the species will depend on the extent to which the environmental conditions for that population are representative of those faced by the species. If a population clearly has evolved under unique conditions then it may be appropriate to restrict conclusions to that one population. A fixed analysis is necessary then, and the F-statistics should not be calculated. In general though, natural populations would seem better regarded as random samples in space and time.

In the absence of knowledge of the sampling distributions of parameter estimates, it is not easy to give expressions for the sample sizes necessary for desired levels of precision. A discussion of sizes required for allelic frequencies is contained in Chapter 2. As always, pilot studies are invaluable (Chapter 2), and variances of estimates obtained by numerical resampling in such pilot studies will indicate the precision to be expected in the full-scale study.

In a series of simulation experiments for populations undergoing drift, Reynolds et al. (1983) and Weir and Cockerham (1984) showed that the method of moments gives estimates of low bias, whereas the method suggested here for combining estimates over alleles and loci has sufficiently low variance that the mean square errors compared favorably with alternative methods of combining.

Analysis

As in all population genetic studies, it is important to record the genotypes at all loci scored on an individual basis. Although summary tables may very well present data on a per-locus basis, data should be kept as multilocus genotypes.

There are two principal ways of performing analyses once the data are in hand. The first is to use the equations and tables presented in this chapter. Tables 1, 2, and 3 can all be constructed with allelic and genotypic frequencies from the populations sampled. For moderate-sized samples this can be done by hand, and the various F-statistics calculated from the equations given. This work becomes tedious when there are several loci, or several alleles per locus, and then the analysis should be performed by computer. A computer approach is necessary when numerical resampling is to be used to estimate variances or confidence intervals for the estimates.

Two programs, written in FORTRAN 77, are listed in the Appendix. No claims for optimal performance are made, but the programs have been written using the notation presented in this chapter or in Weir and Cockerham (1984). They are given for haploid and diploid data, and for the simplest hierarchies in each case. The programs perform jackknifing over populations and over loci, and bootstrapping over loci. The only limitations on the size of the data sets are those imposed by the computer used. Sample data sets are listed, together with output from the programs.

The second method of analysis takes advantage of general-purpose statistical packages. As has been stressed, the analyses of population structure can be phrased in terms of nested analyses of variance of indicator variables. Accordingly, the genotypic data can be transformed to values for the indicator variables, and these values then subjected to a nested analysis of variance. For haploid data, analysis in terms of allele A proceeds by coding every occurrence of that allele by "1" and every occurrence of alternative alleles by "0." These values are subjected to a nested analysis of variance with as many levels as there are sampling levels, and the components of variance for each sampling level are estimated. In the SAS package (SAS Institute, 1985), for example, the NESTED procedure is appropriate. If X is the indicator variable, then for a simple hierarchy with populations and genes within populations, the SAS program would include the statements

```
PROC NESTED
CLASS POPN
VAR X
```

with the ERROR mean square providing the variance component for genes within populations. Variance components need to be estimated for every allele for loci with more than two alleles. The estimate of θ follows from this and the POPN variance component as described in this chapter.

A similar procedure holds for diploid data. For an analysis of allele A, homozygotes are replaced by "11," heterozygotes by "10," and all other genotypes by "00." For a three-level hierarchy, an SAS program would contain

```
PROC NESTED
CLASS POPN INDIV
VAR X
```

and the ERROR mean square now provides the variance component for genes within individuals. Note that these analyses should be limited to estimation of variance components. Any F or t test statistics that are calculated by the statistics package will have no meaning since the indicator variables are not normally distributed.

Appendix I: Program for Haploid Data

The program HAPLOID.FOR calculates the quantity θ for any number of loci in any number of samples. A small sample data set HAPLOID.DAT provides the output HAPLOID.OUT.

Note that loci 1 and 5, being invariant in this data set, do not contribute to the estimate of θ. Jackknifing over populations provides estimated standard deviations for each locus separately, and there do not appear to be significant differences between the estimates for the loci. Jackknifing over loci does modify the overall estimate, and provides an estimate of its variance. Evidently the true value of θ is close to zero. Bootstrapping over loci gives a confidence interval that is not symmetric about the estimated value of θ.

```
001  C
002  C     HAPLOID.FOR
003  C
004  C     ESTIMATION OF THETA FOR HAPLOID DATA
005  C
006        IMPLICIT REAL*8(A-H,O-Z)
007        COMMON MB,X(1000)
008        REAL RAN,X
009        DIMENSION AN(x,y),P(x,y,z)
010        DIMENSION IA(y)
011        DIMENSION ANBAR(y),ANNBAR(y),ANC(y)
012        DIMENSION TERM1(y),TERM2(y)
013        DIMENSION PBAR(y,z),PPBAR(y,z),VARP(y,z)
014        DIMENSION THT(y,z)
015        DIMENSION THTL(y),THTB(1000)
016  C
017  C     NP POPULATIONS, INDEXED BY IP.
018  C       Replace x by at least NP in DIMENSION statements.
019  C     NL LOCI, INDEXED BY IL.
020  C       Replace y by at least NL in DIMENSION statements.
021  C     UP TO NU ALLELES PER LOCUS, INDEXED BY IU
022  C       Repalce z by at least NU in DIMENSION statements.
023  C
024  C     DATA IS READ IN FROM FILE HAPLOID.DAT
025  C     FIRST LINE CONTAINS THREE PARAMETERS NP, NL AND NU
026  C       (IN FORMAT 3I3)
027  C     EACH SUCCEEDING LINE CONTAINS POPULATION IDENTIFIER
028  C       (IN FORMAT I3)
029  C     AND THEN THE NL ALLELES, ONE PER LOCUS
030  C       (IN FORMAT I3 FOR EACH LOCUS)
031  C     ALLELES ARE CALLED IA(IL) FOR LOCUS IL
032  C     USE A ZERO, OR BLANK, FOR LOCUS THAT IS NOT SCORED
033  C
034        OPEN(1,FILE='HAPLOID.DAT',STATUS='OLD')
035        OPEN(3,FILE='HAPLOID.OUT',STATUS='UNKNOWN')
036        READ(1,100)NP,NL,NU
037    100 FORMAT(3I3)
038        ANP=NP
039        ANL=NL
040        NB=1000
041  C
042  C     SET COUNTS TO ZERO
043  C     FOR LOCUS IL IN POPULATION IP:
044  C       AN(IP,IL) IS SAMPLE SIZE
045  C       P(IP,IL,IU) IS NUMBER AF ALLELES OF TYPE IU
046  C
047        DO 10 IP=1,NP
048        DO 10 IL=1,NL
049        AN(IP,IL)=0D0
050        DO 10 IU=1,NU
051        P(IP,IL,IU)=0D0
052     10 CONTINUE
053  C
054     20 READ(1,200,END=22)IP,(IA(IL),IL=1,NL)
055    200 FORMAT(I3,15I3)
056  C
061  C
062  C     INCREASE COUNTS OF INDIVIDUALS AND ALLELES
063  C
064        DO 21 IL=1,NL
065        IF(IA(IL).EQ.0)GO TO 21
066        AN(IP,IL)=AN(IP,IL)+1D0
067        IU=IA(IL)
068        P(IP,IL,IU)=P(IP,IL,IU)+1D0
069     21 CONTINUE
070        GO TO 20
071     22 CONTINUE
072  C
073  C     CHANGE COUNTS TO FREQUENCIES
074  C     FORM AVERAGE SAMPLE SIZES FOR EACH LOCUS
075  C     FORM AVERAGES AND VARIANCES OF ALLELIC FREQUENCIES
076  C
077        DO 30 IP=1,NP
078        DO 30 IL=1,NL
079        DO 30 IU=1,NU
080        P(IP,IL,IU)=P(IP,IL,IU)/AN(IP,IL)
081     30 CONTINUE
082  C
083        DO 35 IL=1,NL
084        ANBAR(IL)=0D0
085        ANNBAR(IL)=0D0
086        DO 31 IP=1,NP
087        ANBAR(IL)=ANBAR(IL)+AN(IP,IL)
088        ANNBAR(IL)=ANNBAR(IL)+AN(IP,IL)**2
089     31 CONTINUE
090        IF(ANP.LE.1D0)GO TO 99
091        ANBAR(IL)=ANBAR(IL)/ANP
092        ANC(IL)=(ANP*ANBAR(IL)-ANNBAR(IL)/(ANP*ANBAR(IL)))/(ANP-1D0)
093  C
094        DO 33 IU=1,NU
095        PBAR(IL,IU)=0D0
096        PPBAR(IL,IU)=0D0
097        DO 32 IP=1,NP
098        PBAR(IL,IU)=PBAR(IL,IU)+AN(IP,IL)*P(IP,IL,IU)
099        PPBAR(IL,IU)=PPBAR(IL,IU)+AN(IP,IL)*P(IP,IL,IU)**2
100     32 CONTINUE
101        PBAR(IL,IU)=PBAR(IL,IU)/(ANP*ANBAR(IL))
102        VARP(IL,IU)=PPBAR(IL,IU)-ANP*ANBAR(IL)*PBAR(IL,IU)
103        VARP(IL,IU)=VARP(IL,IU)/((ANP-1D0)*ANBAR(IL))
104     33 CONTINUE
105     35 CONTINUE
106  C
107  C     FORM VARIANCE COMPONENTS FOR EACH ALLELE AT EACH LOCUS
108  C     ESTIMATE THETA FOR EACH ALLELE
109  C     THEN COMBINE OVER ALLELES
110  C     CHECK FOR FIXED LOCI
111  C     WRITE ESTIMATES
112  C
113        TTERM1=0D0
114        TTERM2=0D0
115        DO 42 IL=1,NL
```

```
        TERM1(IL)=0D0                                                        121
        TERM2(IL)=0D0                                                        122
        DO 40 IU=1,NU                                                        123
        IF(PBAR(IL,IU).LE.1D-4.OR.PBAR(IL,IU).GT.9999D-4)GO TO 40            124
        IF(FTEST.EQ.2)GO TO 40                                               125
        D=ANP*ANBAR(IL)*PBAR(IL,IU)*(1D0-PBAR(IL,IU))                        126
        D=D-(ANP-1D0)*ANBAR(IL)*VARP(IL,IU)                                  127
        D=D/(ANP*(ANBAR(IL)-1D0))                                            128
        A=ANBAR(IL)*VARP(IL,IU)                                              129
        A=(A-D)/ANC(IL)                                                      130
        THT(IL,IU)=A/(A+D)                                                   131
        TERM1(IL)=TERM1(IL)+A                                                132
        TERM2(IL)=TERM2(IL)+A+D                                              133
40      CONTINUE                                                             134
        IF(TERM2(IL).EQ.0D0)GO TO 42                                         135
        WRITE(3,300)                                                         136
300     FORMAT('  ')                                                         137
        WRITE(3,400)IL                                                       138
400     FORMAT('  FOR LOCUS ',I3)                                            139
410     FORMAT('  ALLELE ',3X,'THETA')                                       140
        WRITE(3,410)                                                         141
        THTL(IL)=TERM1(IL)/TERM2(IL)                                         142
c                                                                            143
        DO 41 IU=1,NU                                                        144
        WRITE(3,420)IU,THT(IL,IU)                                            145
420     FORMAT(I5,F10.4)                                                     146
41      CONTINUE                                                             147
        WRITE(3,430)THTL(IL)                                                 148
430     FORMAT('  ALL  ',F9.4)                                              149
        TTERM1=TTERM1+TERM1(IL)                                              150
        TTERM2=TTERM2+TERM2(IL)                                              151
42      CONTINUE                                                             152
        THTT=TTERM1/TTERM2                                                   153
        WRITE(3,300)                                                         154
        WRITE(3,300)                                                         155
        WRITE(3,440)                                                         156
440     FORMAT('  OVER ALL LOCI')                                            157
        WRITE(3,450)THTT                                                     158
450     FORMAT('  THETA',F10.4)                                             159
c                                                                            160
c                                                                            161
c       JACKKNIFE OVER POPULATIONS                                           162
c       LEAVE OUT EACH POPULATION IN TURN                                    163
c                                                                            164
        WRITE(3,300)                                                         165
        WRITE(3,300)                                                         166
        WRITE(3,600)                                                         167
600     FORMAT('  JACKKNIFING OVER POPULATIONS')                             168
        DO 65 IL=1,NL                                                        169
        ANBAR(IL)=ANBAR(IL)*ANP                                              170
        ANP1=ANP-1D0                                                         171
        ANP2=ANP1-1D0                                                        172
        THTJ=0D0                                                             173
        THTJJ=0D0                                                            174
        DO 61 IU=1,NU                                                        175
        PBAR(IL,IU)=PBAR(IL,IU)*ANBAR(IL)                                    176
61      CONTINUE                                                             177
        DO 63 IP=1,NP                                                        178
        ANBARJ=(ANBAR(IL)-AN(IP,IL))/ANP1                                    179
                                                                             180
        ANNBAJ=ANNBAR(IL)-AN(IP,IL)**2                                       181
        ANCJ=(ANP1*ANBARJ-ANBAJ-(ANP1*ANBARJ))/(ANP1-1D0)                    182
c                                                                            183
        TERM1J=0D0                                                           184
        TERM2J=0D0                                                           185
        DO 62 IU=1,NU                                                        186
        PBARJ=PBAR(IL,IU)-AN(IP,IL)*P(IP,IL,IU)                              187
        PPBARJ=PPBAR(IL,IU)-AN(IP,IL)*P(IP,IL,IU)**2                         188
c                                                                            189
        PBARJ=PBARJ/(ANP1*ANBARJ)                                            190
        VARPJ=(PPBARJ-ANP1*ANBARJ*PBARJ**2)/(ANP2*ANBARJ)                    191
        IF(PBARJ.LE.1D-4.OR.PBARJ.GE.9999D-4)GO TO 62                        192
c                                                                            193
        D=ANP1*ANBARJ*PBARJ*(1D0-PBARJ)                                      194
        D=D-(ANP1-1D0)*ANBARJ*VARPJ                                          195
        D=D/(ANP1*(ANBARJ-1D0))                                             196
        A=ANBARJ*VARPJ                                                       197
        A=(A-D)/ANCJ                                                         198
c                                                                            199
        TERM1J=TERM1J+A                                                      200
        TERM2J=TERM2J+A+D                                                    201
62      CONTINUE                                                             202
        IF(TERM2J.EQ.0)GO TO 63                                             203
        THTJ=THTJ+TERM1J/TERM2J                                              204
        THTJJ=THTJJ+(TERM1J/TERM2J)**2                                       205
63      CONTINUE                                                             206
        IF(TERM2J.EQ.0D0)GO TO 65                                           207
        WRITE(3,610)                                                         208
610     FORMAT(2X,' LOCUS',' THETA')                                        209
c                                                                            210
        THTJJ=DSQRT(ANP1*(THTJJ-THTJ**2/ANP)/ANP)                            211
        THTJ=ANP*THTL(IL)-ANP1*THTJ/ANP                                      212
c                                                                            213
        WRITE(3,620)IL,THTJ                                                  214
620     FORMAT(2X,I3,F10.4,' MEAN')                                         215
        WRITE(3,630)THTJJ                                                    216
630     FORMAT(5X,F10.4,' STD. DEV.')                                       217
65      CONTINUE                                                             218
c                                                                            219
c                                                                            220
c       JACKKNIFE OVER LOCI                                                  221
c       LEAVE OUT EACH LOCUS IN TURN                                         222
c                                                                            223
        WRITE(3,300)                                                         224
        WRITE(3,300)                                                         225
        WRITE(3,700)                                                         226
700     FORMAT('  JACKKNIFING OVER LOCI')                                    227
        ANL1=ANL-1D0                                                         228
c                                                                            229
        THTJ=0D0                                                             230
        THTJJ=0D0                                                            231
c                                                                            232
        DO 75 IL=1,NL                                                        233
        TERM1J=TTERM1-TERM1(IL)                                              234
        TERM2J=TTERM2-TERM2(IL)                                              235
c                                                                            236
        IF(TERM2J.EQ.0D0)GO TO 75                                           237
        THTJ=THTJ+TERM1J/TERM2J                                              238
        THTJJ=THTJJ+(TERM1J/TERM2J)**2                                       239
                                                                             240
```

```fortran
C
75    CONTINUE
      WRITE(3,710)
710   FORMAT(10X,'THETA')
      THTJU=DSQRT(ANL1*(THTJJ-THTJ**2/ANL)/ANL)
      THTJ=ANL*THTT-ANL1*THTJ/ANL
C
      WRITE(3,720)THTJ
720   FORMAT(' ALL ',F9.4,' MEAN')
      WRITE(3,730)THTJJ
730   FORMAT(5X,F10.4,' STD. DEV.')
C
99    CONTINUE
C
C     BOOTSTRAP OVER LOCI
C     NB=BOOTSTRAP SAMPLES, INDEXED BY IB
C     IXX IS SEED FOR RANDOM NUMBER GENERATOR
C
      WRITE(3,300)
      WRITE(3,300)
      WRITE(3,800)
800   FORMAT(' BOOTSTRAPPING OVER LOCI')
      WRITE(3,300)
C
      IXX=890613
      IL1=RAN(IXX)
C
      MB=0
      DO 85 IB=1,NB
      TERM1B=0D0
      TERM2B=0D0
      DO 80 IL=1,NL
      IL1=INT(ANL*RAN(IL1))+1
      TERM1B=TERM1B+TERM1(IL1)
      TERM2B=TERM2B+TERM2(IL1)
80    CONTINUE
C
      IF(TERM2B.EQ.0)GO TO 85
      MB=MB+1
      THTB(MB)=TERM1B/TERM2B
C
85    CONTINUE
      MBL=INT(FLOAT(MB)/40D0)
      MBU=INT(3900*FLOAT(MB)/40D0)+1
C
      DO 86 IB=1,MB
      X(IB)=THTB(IB)
86    CONTINUE
      CALL SORT
      THTBL=X(MBL)
      THTBU=X(MBU)
C
      WRITE(3,810)THTBL,THTBU
810   FORMAT(' THETA 95% CI:',2F10.4)
C
      STOP
      END
C
C     SORTING SUBROUTINE
C     ORDERS A SET OF NUMBERS X(I), I=1,...,N FROM
C     SMALLEST TO LARGEST
C
      SUBROUTINE SORT
C
      COMMON MB,X(1000)
      DIMENSION Y(1000)
C
      DO 15 I=1,MB
      Y(I)=X(1)
      INDEX=1
      M=MB-I+1
      DO 10 J=2,M
      IF(X(J).GE.Y(I))GO TO 10
      Y(I)=X(J)
      INDEX=J
10    CONTINUE
      K=1
      DO 15 J=1,M
      IF(J.EQ.INDEX)GO TO 15
      X(K)=X(J)
      K=K+1
15    CONTINUE
C
      DO 20 I=1,MB
      X(I)=Y(I)
20    CONTINUE
C
      RETURN
      END
C
C     UNIFORM PSEUDORANDOM NUMBER GENERATOR
C     WRITTEN BY J.MONAHAN, STATISTICS DEPT, N.C.STATE UNIVERISTY
C     FROM LEWIS ET AL, ACM TOMS V.5 (1979) P132
C     FIRST CALL SETS SEED TO IXX, LATER IXX IGNORED
C
      REAL FUNCTION RAN(IXX)
C
      INTEGER A,P,IX,B15,B16,XHI,XALO,LEFTLO,FHI,K
      DATA A/16807/,B15/32768/,B16/65536/,P/2147483647/
      DATA IX/0/
C
      IF(IX.EQ.0)IX=IXX
      XHI=IX/B16
      XALO=(IX-XHI*B16)*A
      LEFTLO=XALO/B16
      FHI=XHI*A+LEFTLO
      K=FHI/B15
      IX=((XALO-LEFTLO*B16)-P)+(FHI-K*B15)*B16)+K
      IF(IX.LT.0)IX=IX+P
      RAN=FLOAT(IX)*4.656612875E-10
      RETURN
      END
```

```
        HAPLOID.DAT                    HAPLOID.OUT

6  5  4                        FOR LOCUS   2
1  4  3  3  3  4               ALLELE   THETA
1  4  4  3  3  4                   1    0.0000
1  4  4  3  3  4                   2    0.0000
1  4  4     3  4                   3   -0.0185
1  4  4  2  4  4                   4   -0.0185
1  4  4  4  4  4               ALL     -0.0185
1  4  4  3  4  4
1  4  4  3  4  4               FOR LOCUS   3
1  4  4     3  4               ALLELE   THETA
1  4  4  3  3  4                   1    0.0000
2  4  4  2  2  4                   2   -0.0650
2  4  3  4  3  4                   3    0.0780
2  4  4  3  3  4                   4    0.0673
2  4  4  4  4  4               ALL      0.0560
2  4  3  4  4  4
2  4  4  4  2  4               FOR LOCUS   4
2  4  4  3  3  4               ALLELE   THETA
2  4  4  4  4  4                   1   -0.0202
3  4  4  4  3  4                   2    0.0756
3  4  4  4  3  4                   3   -0.0654
3  4  3  4  4  4                   4   -0.0744
3  4  4  4  4  4               ALL     -0.0487
3  4  4  3  1  4
3  4  4  3  3  4
3  4  4  3  1  4               OVER ALL LOCI
3  4  4  4  4  4               THETA   -0.0003
3  4  4  4  3  4
4  4  4  3  4  4
4  4  4  3  3  4               JACKKNIFING OVER POPULATIONS
4  4  4  3  3  4                  LOCUS  THETA
4  4  4  3  4  4                   2  -0.0185 MEAN
4  4  4     3  4                       0.0646 STD. DEV.
4  4  4  4  4  4                  LOCUS  THETA
4  4  4  4  3  4                   3   0.0572 MEAN
4  4  3  3  3  4                       0.0708 STD. DEV.
4  4  4  3  3  4                  LOCUS  THETA
4  4  4  3  1  4                   4  -0.0498 MEAN
5  4  4  4  1  4                       0.0208 STD. DEV.
5  4  4  4  3  4
5  4  4  2  3  4
5  4  4  3  3  4               JACKKNIFING OVER LOCI
5  4  4  3  3  4                         THETA
5  4  4  4  4  4               ALL    -0.0006 MEAN
5  4  4  3  3  4                       0.0517 STD. DEV.
5  4  4  4     4
6  4  4  4  3  4
6  4  4  3  3  4               BOOTSTRAPPING OVER LOCI
6  4  4  2  2  4
6  4  4  3  1  4                  THETA 95% CI:   -0.0487    0.0560
6  4  4  4  4  4
6  4  4  4  4  4
6  4  4  4  1  4
6  4  4  4  3  4
6  4  4  4  3  4
6  4  4  3  3  4
```

Appendix II: Program for Diploid Data

The program DIPLOID.FOR calculates the quantities F, θ, f for any number of loci in any number of individuals from any number of samples. A small sample data set DIPLOID.DAT provides the output DIPLOID.OUT.

As in the haploid case, invariant loci do not contribute to the F-statistic estimates. For locus 4, the coancestry exceeds the overall inbreeding coefficient, so that the within-population inbreeding coefficient is negative. This reflects a high frequency of heterozygotes for that locus. Jackknifing over populations reveals no significant differences among loci. Bootstrapping over loci gives confidence intervals that are not symmetric about the estimates.

```
001  C DIPLOID.FOR
002  C ESTIMATION OF F-STATISTICS FOR ONE-LEVEL HIERARCHY
003  C
004  C NP POPULATIONS, INDEXED BY IP
005  C    Replace x by at least NP in DIMENSION statements
006  C NL LOCI, INDEXED BY IL
007  C    Replace y by at least NL in DIMENSION statements
008  C UP TO NU ALLELES PER LOCUS, INDEXED BY IU
009  C    Replace z by at least NU in DIMENSION statements
010  C
011        IMPLICIT REAL*8(A-H,O-Z)
012        COMMON MB,X(1000)
013        REAL RAN,X
014        DIMENSION AN(x,y),H(x,y,z),P(x,y,z)
015        DIMENSION IA1(y),IA2(y)
016        DIMENSION ANBAR(y),ANNBAR(y),ANC(y)
017        DIMENSION TERM1(y),TERM2(y),TERM3(y),TERM4(y),TERM5(y)
018        DIMENSION PBAR(y,z),PPBAR(y,z),VARP(y,z),HBAR(y,z)
019        DIMENSION CAPF(y,z),THETA(y,z),SMALLF(y)
020        DIMENSION CAPFL(y),THETAL(y),SMALLFL(y)
021        DIMENSION CAPFB(1000),THETB(1000),SMLFB(1000)
022  C
023        OPEN(1,FILE='DIPLOID.DAT',STATUS='OLD')
024        OPEN(3,FILE='DIPLOID.OUT',STATUS='UNKNOWN')
025  C
026  C DATA IS READ IN FROM FILE DIPLOID.DAT
027  C FIRST LINE CONTAINS THREE PARAMETERS: NP, NL AND NU
028  C    (IN FORMAT 3I3)
029  C EACH SUCEEDING LINE CONTAINS POPULATION IDENTIFIER
030  C    (IN FORMAT I3)
031  C AND THEN THE NL GENOTYPES
032  C    (IN FORMAT 2I1, WITH A BLANK BETWEEN EACH LOCUS)
033  C ALLELES ARE CALLED IA1(IL) AND IA2(IL) FOR LOCUS IL
034  C USE A ZERO OR A BLANK FOR A LOCUS THAT IS NOT SCORED
035  C
036  100   READ(1,100)NP,NL,NU
037  100   FORMAT(3I3)
038        ANP=NP
039        ANL=NL
040        NB=1000
041  C
042  C SET COUNTS TO ZERO
043  C FOR LOCUS IL IN POPULATION IP:
044  C    AN(IP,IL) IS SAMPLE SIZE
045  C    H(IP,IL,IU) IS NUMBER OF HETEROZYGOTES FOR ALLELE IU
046  C    P(IP,IL,IU) IS NUMBER OF ALLELES OF TYPE IU
047  C
048        DO 10 IP=1,NP
049        DO 10 IL=1,NL
050        AN(IP,IL)=0D0
051        DO 10 IU=1,NU
052        H(IP,IL,IU)=0D0
053        P(IP,IL,IU)=0D0
054  10    CONTINUE
055
056
057
058
059
060

061  C
062  20    READ(1,200,END=25)IP,(IA1(IL),IA2(IL),IL=1,NL)
063  200   FORMAT(I3,6(1X,2I1))
064  C
065  C INCREASE COUNTS OF INDIVIDUALS, HETEROZYGOTES AND ALLELES
066  C
067        DO 21 IL=1,NL
068        IF(IA1(IL).EQ.0.OR.IA2(IL).EQ.0)GO TO 21
069        AN(IP,IL)=AN(IP,IL)+1D0
070        IU=IA1(IL)
071        IV=IA2(IL)
072        P(IP,IL,IU)=P(IP,IL,IU)+1D0
073        P(IP,IL,IV)=P(IP,IL,IV)+1D0
074        IF(IU.EQ.IV)GO TO 21
075        H(IP,IL,IU)=H(IP,IL,IU)+1D0
076        H(IP,IL,IV)=H(IP,IL,IV)+1D0
077  21    CONTINUE
078        GO TO 20
079  25    CONTINUE
080  C
081  C CHANGE COUNTS TO FREQUENCIES
082  C FORM AVERAGE SAMPLE SIZES FOR EACH LOCUS
083  C FORM AVERAGE AND VARIANCE OF ALLELIC FREQUENCIES
084  C FORM AVERAGE HETEROZYGOSITIES
085  C
086        DO 30 IP=1,NP
087        DO 30 IL=1,NL
088        DO 30 IU=1,NU
089        P(IP,IL,IU)=P(IP,IL,IU)/(2D0*AN(IP,IL))
090        H(IP,IL,IU)=H(IP,IL,IU)/AN(IP,IL)
091  30    CONTINUE
092  C
093        DO 35 IL=1,NL
094        ANBAR(IL)=0D0
095        ANNBAR(IL)=0D0
096        DO 31 IP=1,NP
097        ANBAR(IL)=ANBAR(IL)+AN(IP,IL)
098        ANNBAR(IL)=ANNBAR(IL)+AN(IP,IL)**2
099  31    CONTINUE
100        IF(ANP.LE.1D0)GO TO 99
101        ANBAR(IL)=ANBAR(IL)/ANP
102        ANC(IL)=(ANP*ANBAR(IL)-ANNBAR(IL))/(ANP*ANBAR(IL)))/(ANP-1D0)
103  C
104        DO 33 IU=1,NU
105        PBAR(IL,IU)=0D0
106        PPBAR(IL,IU)=0D0
107        HBAR(IL,IU)=0D0
108        DO 32 IP=1,NP
109        PBAR(IL,IU)=PBAR(IL,IU)+AN(IP,IL)*P(IP,IL,IU)
110        PPBAR(IL,IU)=PPBAR(IL,IU)+AN(IP,IL)*P(IP,IL,IU)**2
111        HBAR(IL,IU)=HBAR(IL,IU)+AN(IP,IL)*H(IP,IL,IU)
112  32    CONTINUE
113        PBAR(IL,IU)=PBAR(IL,IU)/(ANP*ANBAR(IL))
114        PPBAR(IL,IU)=PPBAR(IL,IU)-ANP*ANBAR(IL)*PBAR(IL,IU)**2
115        VARP(IL,IU)=VARP(IL,IU)/((ANP-1D0)*ANBAR(IL))
116        HBAR(IL,IU)=HBAR(IL,IU)/(ANP*ANBAR(IL))
117
118
119
120
```

```
 33   CONTINUE
 35   CONTINUE
c
c     FORM VARIANCE COMPONENTS FOR EACH ALLELE AT EACH LOCUS
c     ESTIMATE F-STATISTICS FOR EACH ALLELE
c     THEN COMBINE OVER ALLELES
c     TEST FOR FIXED LOCI
c     WRITE ESTIMATES
c
      TTERM1=0D0                                                       121
      TTERM2=0D0                                                       122
      TTERM3=0D0                                                       123
      TTERM4=0D0                                                       124
      TTERM5=0D0                                                       125
      DO 42 IL=1,NL                                                    126
      TERM1(IL)=0D0                                                    127
      TERM2(IL)=0D0                                                    128
      TERM3(IL)=0D0                                                    129
      TERM4(IL)=0D0                                                    130
      TERM5(IL)=0D0                                                    131
      DO 40 IU=1,NU                                                    132
      IF(PBAR(IL,IU).LE.1D-4.OR.PBAR(IL,IU).GT.9999D-4)GO TO 40        133
      A=PBAR(IL,IU)*(1D0-PBAR(IL,IU))-(ANP-1D0)*VARP(IL,IU)/ANP        134
      B=A                                                              135
c                                                                      136
      A=A-HBAR(IL,IU)/4D0                                              137
      A=ANBAR(IL)*(VARP(IL,IU)-A/(ANBAR(IL)-1D0))/ANC(IL)              138
      B=B-(2D0*ANBAR(IL)-1D0)*HBAR(IL,IU)/(4D0*ANBAR(IL))              139
      B=ANBAR(IL)*B/(ANBAR(IL)-1D0)                                    140
      C=HBAR(IL,IU)/2D0                                                141
      CAPF(IL,IU)=(A+B)/(A+B+C)                                        142
      THETA(IL,IU)=A/(A+B+C)                                           143
      SMALLF(IL,IU)=B/(B+C)                                            144
      TERM1(IL)=TERM1(IL)+A+B                                          145
      TERM2(IL)=TERM2(IL)+A+B+C                                        146
      TERM3(IL)=TERM3(IL)+A                                            147
      TERM4(IL)=TERM4(IL)+B                                            148
      TERM5(IL)=TERM5(IL)+B+C                                          149
 40   CONTINUE                                                         150
      IF(TERM2(IL).EQ.0D0)GO TO 42                                     151
      IF(TERM5(IL).EQ.0D0)GO TO 42                                     152
      WRITE(3,300)                                                     153
300   FORMAT(' ')                                                      154
      WRITE(3,305)IL                                                   155
305   FORMAT(' FOR LOCUS ',I3)                                         156
      WRITE(3,310)                                                     157
310   FORMAT(' ALLELE',3X,'CAPF',6X,'THETA',4X,'SMALLF')               158
      CAPFL(IL)=TERM1(IL)/TERM2(IL)                                    159
      THETAL(IL)=TERM3(IL)/TERM2(IL)                                   160
      SMALFL(IL)=TERM4(IL)/TERM5(IL)                                   161
c                                                                      162
      DO 41 IU=1,NU                                                    163
      WRITE(3,330)IU,CAPF(IL,IU),THETA(IL,IU),SMALLF(IL,IU)            164
330   FORMAT(I5,3F10.4)                                                165
 41   CONTINUE                                                         166
      WRITE(3,340)CAPFL(IL),THETAL(IL),SMALFL(IL)                      167
340   FORMAT(' ALL  ',F9.4,2F10.4)                                     168
      TTERM1=TTERM1+TERM1(IL)                                          169
      TTERM2=TTERM2+TERM2(IL)                                          170
      TTERM3=TTERM3+TERM3(IL)                                          171
      TTERM4=TTERM4+TERM4(IL)                                          172
      TTERM5=TTERM5+TERM5(IL)                                          173
 42   CONTINUE                                                         174
      TCAPF=TTERM1/TTERM2                                              175
      TTHETA=TTERM3/TTERM2                                             176
      TSMALF=TTERM4/TTERM5                                             177
      WRITE(3,300)                                                     178
      WRITE(3,350)                                                     179
350   FORMAT(' OVER ALL LOCI')                                         180
      WRITE(3,360)                                                     181
360   FORMAT(9X,' CAPF',6X,'THETA',4X,'SMALLF')                        182
      WRITE(3,370)TCAPF,TTHETA,TSMALF                                  183
370   FORMAT(5X,3F10.4)                                                184
c                                                                      185
c     JACKKNIFE OVER POPULATIONS                                       186
c     LEAVE OUT EACH POPULATION IN TURN                                187
c                                                                      188
      WRITE(3,300)                                                     189
      WRITE(3,300)                                                     190
      WRITE(3,400)                                                     191
400   FORMAT(' JACKKNIFING OVER POPULATIONS')                          192
      DO 65 IL=1,NL                                                    193
      ANBAR(IL)=ANBAR(IL)*ANP                                          194
      ANP1=ANP-1D0                                                     195
      ANP2=ANP1-1D0                                                    196
      CAPFJ=0D0                                                        197
      CAPFJJ=0D0                                                       198
      THETJ=0D0                                                        199
      THTLJJ=0D0                                                       200
      SMLFJ=0D0                                                        201
      SMFLJJ=0D0                                                       202
      DO 61 IU=1,NU                                                    203
      PBAR(IL,IU)=PBAR(IL,IU)*ANBAR(IL)                                204
      HBAR(IL,IU)=HBAR(IL,IU)*ANBAR(IL)                                205
 61   CONTINUE                                                         206
      DO 63 IP=1,NP                                                    207
      ANBARJ=(ANBAR(IL)-AN(IP,IL))/ANP1                                208
      ANNBARJ=ANNBAR(IL)-AN(IP,IL)**2                                  209
      ANCJ=(ANP1*ANBARJ-ANNBARJ/(ANP1*ANBARJ))/(ANP1-1D0)             210
c                                                                      211
      TERM1J=0D0                                                       212
      TERM2J=0D0                                                       213
      TERM3J=0D0                                                       214
      TERM4J=0D0                                                       215
      TERM5J=0D0                                                       216
      DO 62 IU=1,NU                                                    217
      PBARJ=PBAR(IL,IU)-AN(IP,IL)*P(IP,IL,IU)                          218
      PPBARJ=PPBAR(IL,IU)-AN(IP,IL)*P(IP,IL,IU)**2                     219
      HBARJ=HBAR(IL,IU)-AN(IP,IL)*H(IP,IL,IU)                          220
c                                                                      221
      PBARJ=PBARJ/(ANP1*ANBARJ)                                        222
      VARPJ=(PPBARJ-ANP1*ANBARJ*PBARJ**2)/(ANP2*ANBARJ)               223
      HBARJ=HBARJ/(ANP1*ANBARJ)                                        224
      IF(PBARJ.LE.1D-4.OR.PBARJ.GE.9999D-4)GO TO 62                    225
c                                                                      226
      A=PBARJ*(1D0-PBARJ)-(ANP1-1D0)*VARPJ/ANP1                        227
```

```
241       B=A
          A=ANBARJ*(VARPJ-(A-HBARJ/4D0)/(ANBARJ-1D0))/ANCJ
242       B=B-(2D0*ANBARJ-1D0)*HBARJ/(4D0*ANBARJ)
243       B=ANBARJ*B/(ANBARJ-1D0)
244       C=HBARJ/2D0
245
246    c
247       TERM1J=TERM1J+A+B
248       TERM2J=TERM2J+A+B+C
249       TERM3J=TERM3J+A
250       TERM4J=TERM4J+B
251       TERM5J=TERM5J+B+C
252   62  CONTINUE
253       IF(TERM2J.EQ.0)GO TO 63
254       IF(TERM5J.EQ.0)GO TO 63
255       CAPFLJ=CAPFLJ+TERM1J/TERM2J
256       CAPFJJ=CAPFJJ+(TERM1J/TERM2J)**2
257       THETLJ=THETLJ+TERM3J/TERM2J
258       THTLJJ=THTLJJ+(TERM3J/TERM2J)**2
259       SMLFLJ=SMLFLJ+TERM4J/TERM5J
260       SMLFJJ=SMLFJJ+(TERM4J/TERM5J)**2
261   63  CONTINUE
262       IF(TERM2J.EQ.0D0)GO TO 65
263       IF(TERM5J.EQ.0D0)GO TO 65
264       WRITE(3,300)
265       WRITE(3,305)IL
266       WRITE(3,410)
267  410  FORMAT(10X,'CAPF',6X,'THETA',4X,'SMALLF')
268    c
269       CAPFLJ=DSQRT(ANP1*(CAPFJJ-CAPFLJ**2/ANP)/ANP)
270       CAPFLJ=ANP*CAPFL(IL)-ANP1*CAPFLJ/ANP
271       THTLJJ=DSQRT(ANP1*(THTLJJ-THETLJ**2/ANP)/ANP)
272       THETLJ=ANP*THETAL(IL)-ANP1*THETLJ/ANP
273       SMFLJJ=DSQRT(ANP1*(SMFLJJ-SMLFLJ**2/ANP)/ANP)
274       SMLFLJ=ANP*SMALFL(IL)-ANP1*SMLFLJ/ANP
275    c
276       WRITE(3,420)CAPFLJ,THETLJ,SMLFLJ
277  420  FORMAT(' TOTAL',F9.4,2F10.4,' MEANS')
278       WRITE(3,430)CAPFJJ,THTLJJ,SMFLJJ
279  430  FORMAT(5X,3F10.4,' STD. DEVS.')
280   65  CONTINUE
281    c
282    c
283    c  JACKKNIFE OVER LOCI
284    c  LEAVE OUT EACH LOCUS IN TURN
285    c
286    c
287       WRITE(3,300)
288       WRITE(3,300)
289       WRITE(3,500)
290  500  FORMAT(' JACKKNIFING OVER LOCI')
291       ANL1=ANL-1D0
292    c
293       CAPFLJ=0D0
294       CAPFJJ=0D0
295       THETLJ=0D0
296       THETJJ=0D0
297       SMLFLJ=0D0
298       SMLFJJ=0D0
299    c
300       DO 75 IL=1,NL
301       TERM1J=TTERM1-TERM1(IL)
302       TERM2J=TTERM2-TERM2(IL)
303       TERM3J=TTERM3-TERM3(IL)
304       TERM4J=TTERM4-TERM4(IL)
305       TERM5J=TTERM5-TERM5(IL)
306    c
307       IF(TERM2J.EQ.0)GO TO 75
308       IF(TERM5J.EQ.0)GO TO 75
309       CAPFLJ=CAPFLJ+(TERM1J/TERM2J)
310       CAPFJJ=CAPFJJ+(TERM1J/TERM2J)**2
311       THETLJ=THETLJ+(TERM3J/TERM2J)
312       THETJJ=THETJJ+(TERM3J/TERM2J)**2
313       SMLFLJ=SMLFLJ+(TERM4J/TERM5J)
314       SMLFJJ=SMLFJJ+(TERM4J/TERM5J)**2
315    c
316   75  CONTINUE
317    c
318    c
319       WRITE(3,300)
320       WRITE(3,510)
321  510  FORMAT(10X,'CAPF',6X,'THETA',4X,'SMALLF')
322       CAPFJJ=DSQRT(ANL1*(CAPFJJ-CAPFLJ**2/ANL)/ANL)
323       CAPFLJ=DSQRT(ANL1*TCAPF-ANL1*CAPFLJ/ANL)
324       THETJJ=DSQRT(ANL1*(THETJJ-THETLJ**2/ANL)/ANL)
325       THETLJ=ANL1*THETA-ANL1*THETLJ/ANL
326       SMLFJJ=DSQRT(ANL1*(SMLFJJ-SMLFLJ**2/ANL)/ANL)
327       SMLFLJ=ANL1*SMALF-ANL1*SMLFLJ/ANL
328    c
329       WRITE(3,520)CAPFLJ,THETLJ,SMLFLJ
330  520  FORMAT(' TOTAL',F9.4,2F10.4,' MEANS')
331       WRITE(3,530)CAPFJJ,THETJJ,SMLFJJ
332  530  FORMAT(5X,3F10.4,' STD. DEVS.')
333    c
334    c
335   99  CONTINUE
336    c
337    c
338    c  BOOTSTRAP OVER LOCI
339    c  NB=BOOTSTRAP SAMPLES, INDEXED BY IB
340    c  IXX IS SEED FOR RANDOM NUMBER GENERATOR
341    c
342       WRITE(3,300)
343       WRITE(3,300)
344       WRITE(3,600)
345  600  FORMAT(' BOOTSTRAPPING OVER LOCI')
346    c
347       IXX=890613
348       IL1=RAN(IXX)
349    c
350       MB=0
351       DO 85 IB=1,NB
352       TERM1B=0D0
353       TERM2B=0D0
354       TERM3B=0D0
355       TERM4B=0D0
356       TERM5B=0D0
357       DO 80 IL=1,NL
358       IL1=INT(ANL*RAN(IL))+1
359       TERM1B=TERM1B+TERM1(IL1)
360
```

```
361         TERM2B=TERM2B+TERM2(IL1)
362         TERM3B=TERM3B+TERM3(IL1)
363         TERM4B=TERM4B+TERM4(IL1)
364         TERM5B=TERM5B+TERM5(IL1)
365   80    CONTINUE
366   C
367         IF(TERM2B.EQ.0)GO TO 85
368         IF(TERM5B.EQ.0)GO TO 85
369         MB=MB+1
370         CAPFB(MB)=TERM1B/TERM2B
371         THETB(MB)=TERM3B/TERM2B
372         SMLFB(MB)=TERM4B/TERM5B
373   C
374   85    CONTINUE
375         MBL=INT(FLOAT(MB)/40D0)
376         MBU=INT(3900*FLOAT(MB)/40D0)+1
377   C
378         DO 86 IB=1,MB
379         X(IB)=CAPFB(IB)
380   86    CONTINUE
381         CALL SORT
382         CAFL=X(MBL)
383         CAFU=X(MBU)
384   C
385         DO 87 IB=1,MB
386         X(IB)=THETB(IB)
387   87    CONTINUE
388         CALL SORT
389         THETL=X(MBL)
390         THETU=X(MBU)
391   C
392         DO 88 IB=1,MB
393         X(IB)=SMLFB(IB)
394   88    CONTINUE
395         CALL SORT
396         SMLFL=X(MBL)
397         SMLFU=X(MBU)
398   C
399         WRITE(3,610)CAFL,CAFU
400   610   FORMAT('   CAP-F 95% CI:',2F10.4)
401         WRITE(3,620)THETL,THETU
402   620   FORMAT('   THETA 95% CI:',2F10.4)
403         WRITE(3,630)SMLFL,SMLFU
404   630   FORMAT('  SMALL-F 95% CI:',2F10.4)
405   C
406         STOP
407         END
408   C
409   C     SORTING SUBROUTINE
410   C     ORDERS A SET OF NUMBERS X(I), I=1,...,N FROM
411   C        SMALLEST TO LARGEST
412   C
413   C
414   C
415         SUBROUTINE SORT
416   C
417         COMMON MB,X(1000)
418         DIMENSION Y(1000)
419   C
420         DO 15 I=1,MB
421         Y(I)=X(1)
422         INDEX=1
423         M=MB-I+1
424         DO 10 J=2,M
425         IF(X(J).GE.Y(I))GO TO 10
426         Y(I)=X(J)
427         INDEX=J
428   10    CONTINUE
429         K=1
430         DO 15 J=1,M
431         IF(J.EQ.INDEX)GO TO 15
432         X(K)=X(J)
433         K=K+1
434   15    CONTINUE
435   C
436         DO 20 I=1,MB
437         X(I)=Y(I)
438   20    CONTINUE
439   C
440         RETURN
441         END
442   C
443   C     UNIFORM PSEUDORANDOM NUMBER GENRATOR
444   C     WRITTEN BY J.MONAHAN, STATISTICS DEPT, N.C.STATE UNIVERSITY
445   C     FROM LEWIS ET AL, ACM TOMS V.5 (1979) P132
446   C     FIRST CALL SETS SEED TO IXX, LATER IXX IGNORED
447   C
448         REAL FUNCTION RAN(IXX)
449   C
450         INTEGER A,P,IX,B15,B16,XHI,XALO,LEFTLO,FHI,K
451         DATA A/16807/,B15/32768/,B16/65536/,P/2147483647/
452         DATA IX/0/
453   C
454         IF(IX.EQ.0)IX=IXX
455         XHI=IX/B16
456         XALO=(IX-XHI*B16)*A
457         LEFTLO=XALO/B16
458         FHI=XHI*A+LEFTLO
459         K=FHI/B15
460         IX=((XALO-LEFTLO*B16)-P)+(FHI-K*B15)*B16)+K
461         IF(IX.LT.0)IX=IX+P
462         RAN=FLOAT(IX)*4.656612875E-10
463         RETURN
464         END
```

```
      DIPLOID.DAT                              DIPLOID.OUT

6   5   4                          FOR LOCUS   2
1  44  43  43  33  44             ALLELE   CAPF     THETA    SMALLF
1  44  44  43  33  44               1     0.0000   0.0000   0.0000
1  44  44  43  43  44               2     0.0000   0.0000   0.0000
1  44  44      33  44               3     0.3025   0.0694   0.2505
1  44  44  24  34  44               4     0.3025   0.0694   0.2505
1  44  44      43  44             ALL     0.3025   0.0694   0.2505
1  44  44  43  43  44
1  44  44      43  44             FOR LOCUS   3
2  44  44  33  32  44             ALLELE   CAPF     THETA    SMALLF
2  44  33  44  43  44               1     0.0000   0.0000    0.0000
2  44  43  44  43  44               2     0.0057   0.0369   -0.0324
2  44  44  33  33  44               3    -0.0165  -0.0102   -0.0062
2  44  43  44  44  44               4    -0.0453   0.0171   -0.0635
2  44  44  44  22  44             ALL    -0.0301   0.0046   -0.0348
2  44  44  43  43  44
2  44  44  44  44  44             FOR LOCUS   4
3  44  44  44  43  44             ALLELE   CAPF     THETA    SMALLF
3  44  44  44  44  44               1    -0.0298   0.0465   -0.0800
3  44  44  43  21  44               2     0.1858   0.0111    0.1767
3  44  44  33  43  44               3     0.0072   0.0478   -0.0426
3  44  44  43  21  44               4     0.0268   0.0018    0.0250
4  44  44  43  44  44             ALL     0.0367   0.0243    0.0128
4  44  44  43  43  44
4  44  44  43  43  44             OVER ALL LOCI
4  44  44  43  44  44                      CAPF     THETA    SMALLF
4  44  44  43  44  44                     0.0420   0.0222   0.0202
4  44  44  44  33  44
4  44  44  44  44  44
5  44  44  44  21  44             JACKKNIFING OVER POPULATIONS
5  44  44  44  33  44
5  44  44  43  43  44             FOR LOCUS   2
5  44  44  43  43  44                      CAPF     THETA    SMALLF
5  44  44  44  44  44             TOTAL   0.4607   0.1173   0.3567 MEANS
5  44  44  44  43  44                     0.3030   0.1004   0.2256 STD. DEVS.
5  44  44  43  43  44
5  44  44  44      44             FOR LOCUS   3
5  44  43  44  43  44                      CAPF     THETA    SMALLF
6  44  44  44  43  44             TOTAL  -0.0318   0.0029  -0.0253 MEANS
6  44  44  43  33  44                     0.2219   0.0461   0.2636 STD. DEVS.
6  44  44  44  32  44
6  44  44  43  41  44             FOR LOCUS   4
6  44  44  44  44  44                      CAPF     THETA    SMALLF
6  44  44  44  42  44             TOTAL   0.0346   0.0232   0.0141 MEANS
6  44  44  44  43  44                     0.0978   0.0583   0.1061 STD. DEVS.


                                  JACKKNIFING OVER LOCI

                                           CAPF     THETA    SMALLF
                                  TOTAL   0.0306   0.0209   0.0096 MEANS
                                          0.0462   0.0105   0.0377 STD. DEVS.


                                  BOOTSTRAPPING OVER LOCI

                                    CAP-F 95% ci:   -0.0301    0.3025
                                    THETA 95% ci:    0.0046    0.0694
                                  SMALL-F 95% ci:   -0.0348    0.2505
```

CHAPTER 11

PHYLOGENY RECONSTRUCTION

David L. Swofford and Gary J. Olsen

INTRODUCTION

Inferring phylogenetic relationships from molecular data requires the selection of an appropriate method from the many techniques that have been described. Unfortunately, phylogenetic analysis is frequently treated as a black box into which data are fed and out of which "The Tree" springs. Our goal in this chapter is to provide more than a cursory description of the available analytical methods; rather, we hope to develop a conceptual framework for understanding the theoretical and practical distinctions among alternative methodologies.

Regrettably, within the limits of a single chapter, we cannot accomplish the above objectives and at the same time provide an exhaustive review of the voluminous literature on phylogenetic reconstruction; however, Felsenstein (1982, 1988) has presented recent and generally balanced reviews of methods for inferring phylogenies. Instead, we will focus on methods that are currently in widespread use or that are likely to be used in the foreseeable future. We will also avoid the temptation to cite every relevant paper, generally referring directly only to those papers that are of fundamental importance in the development of each method we discuss. In keeping with the theme of this book, we will pay special attention to practical considerations, the "nuts-and-bolts" issues that confront any investigator faced with the analysis of a new data set.

As any reader even moderately familiar with the current state of affairs in the field of phylogenetic inference already knows, debates among proponents of rival methodologies are often intense and, sometimes, unnecessarily acrimonious. Consequently, we will offer recommendations where we deem appropriate, but will deliberately avoid taking strong positions on or making controversial assertions about issues where there is room for legitimate disagreement. Instead, we hope to provide sufficient background

411

so that readers will be able to make informed decisions regarding the techniques most appropriate for their own data.

Algorithms versus Optimality Criteria

Inferring a phylogeny is really an estimation procedure; we are making a "best estimate" of an evolutionary history based on incomplete information. In the context of molecular systematics, we generally do not have direct information about the past—we have access only to contemporary species and molecules. Because we can postulate evolutionary scenarios by which any chosen phylogeny could have produced the observed data, we must have some basis for selecting one or more preferred trees from among the set of possible phylogenies. Phylogenetic inference methods seek to accomplish this goal in one of two ways: by defining a specific sequence of steps (algorithm) for constructing the best tree, or by defining a criterion for comparing alternative phylogenies to one another and deciding which is better (or that they are equally good).

The first class of methods combines tree inference and the definition of the preferred tree into a single statement. These methods include all forms of pair-group cluster analysis. The methods tend to be fast. In addition to failing to address the underlying evolutionary assumptions directly, no method is generally provided for comparing or ranking suboptimal phylogenies, a serious limitation given the usual degree of uncertainty regarding the correctness of the optimal tree.

The second class of methods has two logical steps. The first is to define the **optimality criterion (objective function)** for evaluating a given tree, i.e., the value that is assigned and subsequently used for comparing one tree to another. The second is to use specific algorithms to compute the value of the objective function and to find the trees that have the best values according to this criterion (a maximum or minimum value, as appropriate). Thus, the evolutionary assumptions made in the first step are decoupled from the computer science of the second step. The price of this logical clarity is that the methods tend to be much slower than those of the first class, a consequence of having to search for the tree(s) with the best score. For data sets containing more than about 8 to 12 taxa, the search for the best tree is usually not exact, and thus we must add caveats regarding the thoroughness of the search for the optimal tree. These issues are dealt with in greater depth below. Because methods of this class generally assign values to every tree examined, alternative phylogenies can be ranked in order of preference.

Implicit versus Explicit Assumptions

If a phylogenetic inference method could be based on a complete knowledge of the evolutionary process, it would be free of systematic errors (i.e., if enough data were obtained, the method would consistently obtain the true phylogeny). Even in the absence of such complete knowledge, hypothetical models of the evolutionary process could be used to derive (or otherwise justify) tree inference methods that would be free of systematic error, *if the assumed model were correct*. A variety of inference techniques have been formulated on the basis of explicit evolutionary assumptions. These methods are not necessarily invalidated when one or more of their assumptions is violated: although the assumptions are sufficient to ensure the validity of a technique, under special circumstances they might not all be necessary. Note, however, that the "special circumstances" themselves constitute alternative assumptions, so that the assumptions are merely being changed, not relaxed.

Other (probably most) methods of phylogenetic inference are not explicitly based on a set of evolutionary assumptions. However, the lack of stated assumptions does not mean that no assumptions are necessary for the method to be valid; they are simply implicit rather than explicit. Unfortunately, identifying the underlying evolutionary assumptions for some techniques can be very difficult. As with explicit assumptions, there can sometimes be alternatives, either of which will suffice.

A Molecular Definition of "Evolutionary History"

Phylogenetic inferences are premised on the inheritance of ancestral characteristics and on the existence of an evolutionary history defined by changes in these characteristics. The stable inheritance of characteristics is mediated by the genome. Differences due to epigenetic or environmental factors do not provide useful phylogenetic information and must be specifically avoided; all characteristics of interest are genetically mediated. Therefore, the data for phylogenetic inference reflect, more or less directly, genomic information. From this reductionistic perspective, a complete evolutionary history is synonymous with an event-by-event accounting of fixed mutations in every genomic lineage of interest. This view of the problem provides a common framework, albeit a purely conceptual one, for analyzing and comparing types of molecular data and analysis techniques.

An assumption implicit in this view concerns the uniqueness of the genomic lineage. Although the potential confusion due to lateral gene transfer has received much recent attention, transfers within the domain of population genetics (Chapter 10) require explicit consideration when analyzing close populations. Thus, our presentation is appropriate for cases in which interspecies differences are large compared to intraspecies variations.

Trees and Roots

Most of the analytical techniques that we will discuss result in the inference of an **unrooted tree** or unrooted phylogeny, that is, a phylogeny in which the earliest point in time (the location of the common ancestor) is not identified. We generally use "tree" and "phylogeny" interchangeably. Also, biologists often refer to an unrooted tree as a "network"; however, this usage conflicts with the definition applied to that term by mathematicians and should probably be avoided. Whenever we find it necessary to distinguish between rooted and unrooted phylogenies or trees, we will do so explicitly.

The parts of a phylogenetic tree go by a variety of names. The contemporary taxa correspond to **terminal nodes,** also called tips, leaves, or external nodes. The branch points within a tree are called **internal nodes.** Nodes are called vertices or points by some authors. The branches connecting (incident to) pairs of nodes are also called links, segments, or edges. We will also use the terms **peripheral branches** to refer to branches that end at a tip and **interior branches** (or, in the case of a tree with four terminal nodes, **central branch**) to refer to branches that are not incident to a tip.

An unrooted, strictly bifurcating tree (one in which every internal node is incident to exactly three branches) has T terminal nodes (corresponding to the taxa) and $T - 2$ internal nodes. The tree has $2T - 3$ branches, of which $T - 3$ are interior and T are peripheral. The total number of distinct unrooted, strictly bifurcating trees for T *taxa* is

$$B(T) = \prod_{i=3}^{T} (2i - 5) \qquad (1)$$

(Felsenstein, 1978b). Adding a root adds one more internal node and one more interior branch. Since the root can be placed along any of the $2T - 3$ branches, the number of possible rooted trees is increased by a factor of $2T - 3$.

TYPES OF DATA

Essentially all of the experimental data gathered by the techniques in this volume fall into one of two broad categories: discrete characters and similarities or distances. A discrete character provides data about an individual species or sequence. A similarity or distance value is a quantitative comparison of two species or sequences; it describes a pairwise relationship.

Character Data

Discrete character data are those for which a data matrix $\mathbf{X}$ assigns a **character state** x_{ij} to each taxon i for each **character** j. Although systematists sometimes disagree regarding the terminological distinction between "character" and "character state," we prefer to think of characters as independent variables whose possible values are collections of mutually exclusive character states.

The assumption of independence among characters is common to virtually all character-based methods of analysis. If we could not assume independence, we would be forced to take covariances among characters into account, and the computational methods would by necessity become vastly more complicated. Furthermore, the assumption of independence enables us to treat each position separately in certain time-consuming stages of computational algorithms, thereby allowing problems to be subdivided into a number of much simpler subproblems. (For example, numbers of substitutions can be minimized separately position by position and then summed over positions in a parsimony algorithm, or probabilities can be multiplied over positions in a maximum likelihood approach.)

A second assumption required of character data is that the characters be homologous. As articulated in Chapter 1, the concept of homology is complicated by the variety of meanings that have been applied to the term. In general, by homology we mean that a character must be defined in such a way that all of the states observed over taxa for that particular character must have been derived, perhaps with modification, from a corresponding state observed in the common ancestor of those taxa. When we are interested in relationships among species rather than among genes, we further restrict this definition to include only orthologous, as opposed to paralogous, genes (see Chapter 1).

A Classification of Character Types   In general, character data are either qualitative, in which the possible states are two or more discrete values, or quantitative, in which the characters vary continuously and are measured on an interval scale. Qualitative characters may be further subdivided into binary (two possible states) and multistate (three or more possible states). Binary characters typically represent the presence or absence of some item, such as the recognition sequence for a restriction endonuclease at a certain map location (restriction site) or a particular allele at an isozyme locus.

Multistate characters may be **ordered** or **unordered**, depending on whether an ordering relationship is imposed on the possible states (Figure 1). For example, nucleotide sequence data are generally treated as unordered multistate characters, since there is no a priori reason to assume, for instance, that state "C" is intermediate between states "A" and "G." In the context of phylogenetic analysis, we say that any state is allowed to transform directly into any other state. If, on the other hand, we are willing to make assumptions involving the relationships among the states of a character, we can rank the character states into an ordered series (i.e., a **linearly ordered character**) or a branching diagram (**partially ordered character** or **character-state tree**.) Multistate ordered characters are not commonly encountered in molecular data sets, but are sometimes used in the analysis of allozyme data (see below).

FIGURE 1. Ordered and unordered characters. (a) Ordered multistate character (transformation between any two states that are not directly connected implies passage through one or more intermediate states). (b) Unordered multistate character (any state can transform directly into any other state). (c) Ordered multistate characters in which the polarity is indicated (the ordering relation is the same in all three cases but the ancestral state differs).

The concepts of character order and character **polarity** must not be confused. The former defines the allowed character-state transformations, whereas the latter refers to the *direction* of character evolution. Estimation of character polarity generally involves an assessment of the observed character state that represents the ancestral condition (i.e., the state found in the most recent common ancestor of the taxa under study). An excellent discussion of character ordering and polarity (in a nonmolecular context) can be found in Mabee (1989). We will return to the subject of character polarity in the discussion of parsimony methods.

Quantitative characters are less commonly used as character data in molecular systematics, the prominent exception occurring when polymorphic characters such as allelic isozymes or mtDNA haplotypes are coded as frequencies.

Sequence Data   In principle, the use of sequence data as characters for phylogenetic analysis is exceedingly straightforward. Given a set of sequences, the "characters" are represented by corresponding positions (offsets) in the sequences and the "character states" are the nucleotide or amino acid residues observed at those positions. For example, if nucleotide "A" is observed to occur at position 139 in a sequence, "position 139" is the character and "A" is the state assigned to that character. (Note that to simplify the exposition, we will usually confine our descriptions to nucleotide sequences unless the distinction is important.)

Unfortunately, this simplicity is deceiving. In addition to requiring the use of homologous molecules (see Chapter 1), phylogenetic analysis of sequence data requires **positional homology**. That is, the nucleotides observed at a given position in the taxa under study should all trace their ancestry to a single position that occurred in a common ancestor of those taxa. Except for highly conserved sequences, insertion and deletion events must nearly always be postulated to make believable the assumption that nucleotides at corresponding positions in the various sequences are in fact homologs. An **alignment** of the sequences is obtained by inserting **gaps,** which correspond to insertions or deletions, into one or more of the sequences to place positions inferred to be homologous into the same column of the data matrix. Alignment is probably the most difficult and least understood component of a phylogenetic analysis from sequence data. Unfortunately, we can hope to provide only general guidelines here, as a complete description of alignment techniques is well beyond the scope of this chapter.

First, we offer the following advice: When regions of the sequences are so divergent that a reasonable alignment cannot be attained by manual methods using a sequence editor ("by eye"), those regions should probably be eliminated from the analysis. Of course, selectively eliminating regions opens the door to the criticism that the researcher is arbitrarily "throwing away data." We would counter this argument with the rebuttal that the researcher has already discarded data, in a sense, by choosing to sequence one molecule rather than another because the latter did not evolve at an appropriate rate for the problem of interest (Chapters 1 and 12). The manual approach to alignment may seem crude and ineffective given the mathematical sophistication and rigor of existing computer algorithms, but it is important to remember that these methods were generally developed for purposes other than the construction of multiple alignments for phylogenetic analysis. For example, "optimal alignments" are often obtained to demonstrate potential homology between regions of two different molecules. But the consequences of violating the assumption that positional homology has been correctly inferred are far less severe for the purpose of detecting similarity than when reconstructing phylogenies.

In spite of the above discussion, computerized alignment algorithms can be profitably used to simplify the task of aligning sequences. Most of the existing computer programs align pairs of sequences using a "dynamic programming" approach inspired by the method of Needleman and Wunsch (1970; see Kruskal, 1983, for a general introduction). In the Needleman–Wunsch method, positive scores (usually 1) are assigned to "matches" (positions containing the same element in both sequences at a given position), zero scores to "mismatches" (positions containing different elements), and negative scores ("penalties") to gaps (positions at which one of the two sequences is included in a gap). For nucleotide sequences, mismatches correspond to substitution events and gaps to insertions or deletions. Gaps are inserted as necessary to maximize the total similarity score when summed over all positions. An alternative, but equivalent, procedure is that of Sellers (1974) as generalized by Waterman et al. (1976). In Sellers' procedure, both gaps and mismatches are assigned positive weights, and an exact algorithm is used to find the alignment that minimizes the total distance between the sequences. Smith et al. (1981) present a rigorous but readable comparison of the two approaches, including a proof of their equivalence. Another useful review of sequence alignment techniques is that of Waterman (1984).

Whether we are attempting to maximize matches (Needleman–Wunsch) or minimize mismatches (Sellers), gaps must be weighted more heavily than mismatches; otherwise the alignment procedure would freely insert gaps to avoid mismatches. Some computer programs treat a gap of length k as k independent insertion–deletion events rather than as a single insertion or deletion having a weight lower than the sum of the independent events. This weighting criterion tends to produce overly dispersed gaps, since, for example, 10 gaps of length 1 are no worse than a single gap of length 10.

Another gap penalty "function" is to simply treat the gap weight as a constant independent of the length of the gap, thereby penalizing each insertion or deletion identically regardless of the number of bases inserted or deleted. This function also tends to have undesirable properties, yielding alignments that contain very long gaps between regions of only moderate similarity. A more useful family of gap penalty functions is of the form

$$w_k = w_G + kw_L \qquad (2)$$

(Smith et al., 1981), where w_k is the penalty for a gap of length k, w_G is a penalty to be enforced regardless of the length of the gap, and w_L is a penalty on the length of the gap. That is, after the first, more heavily penalized, insertion or deletion, each successive insertion or deletion is assigned equal cost. If $w_G = 0$, Equation (2) reduces to the first situation discussed above, where each gap is effectively treated as an independent event. In fact, any monotonically increasing function of the gap length may be used as a gap function. For example, one could (arbitrarily) decide to assign a cost of 2 to a single-base insertion, 3 to a two-base insertion, and 4 to an insertion of three or more bases. Most currently available programs do not provide this level of flexibility in specifying gap penalties, however.

Unfortunately, we can provide little beyond the above qualitative recommendations; there is relatively little theory to guide the choice of gap and mismatch weights. Smith et al. (1981) provide several suggestions, but it is helpful to gain an appreciation of the differences between alignments obtained by using several alternative scoring systems for the sequences being analyzed. In doing so, a researcher is apt to discover that in regions of obvious positional homology, the alignment is relatively insensitive to the choice of weights. Conversely, instability of the alignment with respect to gap and mismatch weights may indicate regions in which positional homology is ambiguous.

One approach to using pairwise alignment algorithms in the construction of a multiple alignment is illustrated in Figure 2. A sequence is chosen,

```
X  =  CGATCAG
Y  =  CGTCAG
Z  =  CGGATCAG

X  =  CGATCAG
Y  ⇒  CG-TCAG

X  ⇒  CG-ATCAG
Z  =  CGGATCAG
Y  ⇒  CG--TCAG
```

FIGURE 2. Alignment of three sequences X, Y, and Z via sequential pairwise alignment. See text for details.

arbitrarily or otherwise, as the "reference." Then, each sequence is aligned in turn to the reference sequence. Whenever a pairwise alignment forces the addition of a gap to the reference sequence, a corresponding gap is placed in each of the nonreference sequences that has been added to the developing alignment up to that point. The final alignment potentially depends on the choice of a reference sequence and the order of addition of the remaining sequences. A strategy used by some workers is to construct all $T(T - 1)/2$ possible pairwise alignments, saving the total alignment score (maximal similarity or minimum distance) for each pair. One of the two members of the closest pair is chosen as the reference. The remaining sequences are then added in order of increasing distance from the reference. As long as the alignment produced by this procedure is taken only as a first approximation, the order in which new sequences are added to the developing alignment is probably not critical.

Restriction Endonuclease Data   Restriction endonuclease analysis provides character data in one of two forms, both of which lead to a set of binary characters for each taxon. Ideally, the characters are map locations and character states are presences or absences of the recognition sequences for particular endonucleases at those locations (restriction-site data). However, because the construction of restriction maps is time consuming (Chapter 8), some workers simply treat the presence or absence of restriction fragments of a given length as character states (restriction-fragment data).

We do not recommend the use of restriction-fragment data for input to phylogenetic analysis, primarily because these data violate the crucial assumption of independence among characters. If a new site evolves between two preexisting sites, one (longer) fragment disappears and two new (shorter) ones appear. Thus, even though two species may share two of the three restriction sites, they have no fragments in common, a potentially serious source of error. Some authors have recognized this difficulty and argue that it can be overcome by looking at "enough" fragment data so that each occurrence of this kind of error will be swamped by other data. We remain unconvinced by this argument, however, because there is no guarantee that if something is done inappropriately enough times, all will work out in the end. A second and related problem with fragment data is that insertions or deletions are difficult to handle. For example, the insertion of a length of DNA long enough to alter the mobility of the fragment (but not containing a restriction site) requires the worker to assert that a species lacks a fragment found in one or more other species, even though the restriction sites responsible for the fragment were at homologous points on the map (see Chapter 8).

Even when sites are mapped, restriction endonuclease data are problematical for phylogenetic analysis due to the asymmetry in the probabilities of gaining and losing sites. If a particular sequence of 6 base pairs is only one substitution away from equaling the recognition sequence of a particular endonuclease (a "one-off" site), then given that a substitution occurs within the 6-base sequence, only one of the 18 possible substitutions of one base for another will convert the sequence to a restriction site. On the other hand, if the 6-base sequence is already a restriction site, then a substitution at any of the six positions will cause the site to be lost. Thus, losing an existing restriction site is much "easier" than gaining a site at a particular location (for more complete discussions, see Templeton, 1983a,b and DeBry and Slade, 1985). Note that this argument applies only to *particular* 6-base sequences; it does not imply a net loss of restriction sites during evolution. Because of these gain–loss asymmetries, special handling may be required for restriction site data, an issue to which we will return.

Isozyme Data   Allozyme (allelic isozyme) data represent the only type of isozyme data routinely used in phylogenetic analysis (but see Buth, 1984, and Chapter 4 for a discussion of other data types). These data are usually presented as a three-dimensional array that specifies the frequency of each allele at each locus in each population or taxon.* Two controversial issues confront the researcher attempting to estimate phylogenies from allozyme data. The first concerns whether to transform the data to genetic similarities or distances (see below). Probably due more to inertia than anything else, the predominant mode of analysis throughout the 1970s and into the 1980s was to compute a matrix of pairwise similarities or distances between taxa that served as the input to cluster analysis or additive tree methods. The stereotypical way in which these data were treated tended to retard the

---

*It is customary to refer to loci as "putative" or "presumptive" and to use the term "electromorphs" rather than "alleles" because of the indirect nature of the data and the usual absence of crossing experiments to confirm the mode of inheritance. For our purpose here, the simpler terms suffice.

development of approaches that made direct use of the character information.

With the development of character-based methods, however, came a second controversy, this one involving the importance of allele frequency information. Some authors (e.g., Mickevich and Johnson, 1976) argued that the presence or absence of an allele was of more fundamental evolutionary importance than was its frequency (which was subject to modification by drift and/or selection), and that frequency information should therefore be discarded. These authors preferred to recast the data into "presence–absence" form.

Defining the "Character."   The earliest attempts to use allozyme characters directly in a phylogenetic analysis generally treated the "allele" as the character and either its presence–absence (e.g., Mickevich and Johnson, 1976) or frequency (e.g., Buth, 1979b; Simon, 1979) as the character state. This procedure, however, is subject to the same criticism leveled at the use of restriction fragment data: the assumption of independence of characters is violated. Specifically, since the frequencies of the alleles at a locus in a given taxon are constrained to sum to one, if the frequency of one allele increases, the frequency of at least one other allele must decrease. This property leads to problems, for example, when "allele-as-character" data are subjected to maximum parsimony analysis, where ancestors are often inferred to contain no alleles at all (presence–absence coding) or frequencies that do not sum to one (frequency coding) for some loci.

Because of these difficulties, Buth (1984) and others have championed an approach that recognizes the *locus* as the character and the allelic composition at the locus in each taxon (i.e., allele or combination of alleles present) as the character state. For example, if some taxa are fixed either for allele *a* or for allele *b,* while others are polymorphic for both alleles, three states would be recognized: "only *a*," "only *b*," and "*a* plus *b*." The resulting discrete characters ("particulate data") are either left unordered or ordered into some logical progression (see Buth, 1984, for details) for subsequent analysis.

Despite its intuitive appeal, several factors limit the utility of the particulate data, locus-as-character approach. When many different alleles occur in various combinations across taxa, the number of unique combinations may approach or even equal the number of taxa. Such characters will contain little or no information if the character states are left unordered. Ordering the character states helps somewhat, but the ordering criteria often seem subjective and arbitrary.

Buth (1984) describes "qualitative" coding, in which observed combinations of alleles are used regardless of frequencies, and "quantitative" coding, in which estimated allele frequencies are used to assess "whether the states expressed by two taxa are statistically identical." Obviously, qualitative coding is extremely susceptible to sampling error. Consider the example in the above paragraph. Taxa that were in reality polymorphic for alleles a and b would often be incorrectly scored as "fixed" if one allele were rare, unless sample sizes were large. (Swofford and Berlocher, 1987, presented a table showing the probability of failing to detect low-frequency alleles in samples of various sizes; see also Chapter 2). Even if allele frequencies could somehow be determined without error, we suggest that it would be unreasonable to argue that allele frequencies are so irrelevant that the distinction between allele frequency arrays of, say, [0.01, 0.99] and [0.99, 0.01] is unimportant.

Quantitative coding presumably makes use of contingency table analysis to test whether two or more samples could have come from a single homogeneous population. In most cases involving interspecific comparisons, however, we know beforehand or from the analysis of other loci that such is not the case, even if the difference between the allele frequency arrays of two taxa at a particular locus is not deemed significant. Furthermore, the power of these tests to detect heterogeneity is weak unless sample sizes are large; therefore, failure to reject the null hypothesis of homogeneity should not usually be taken as evidence that the taxa are "statistically identical." Because of these considerations, we believe that methods that require recoding of allele frequency arrays into discrete states should be used only when levels of polymorphism are extremely low, with problematical loci being excluded from the data set.

Rogers (1984, 1986) and Swofford and Berlocher (1987) have developed methods of analysis that use the observed allele frequencies directly in character-based analyses rather than requiring their recoding as discrete states. Felsenstein's (1981c) maximum likelihood method for continuous characters evolving under a Brownian motion process can also be applied to gene frequency data (after an appropriate transformation).

Similarity and Distance Data

Unlike character data, in which values are assigned to individual taxa, similarity and distance data specify a relationship between *pairs* of taxa or molecules. Some experimental methods, including immunology and nucleic acid hybridization, directly yield data in the form of pairwise similarities. Consequently, distance-based methods provide the only means of estimating evolutionary trees from these data. Other types of data, including macromolecular sequence, restriction map, and allozyme data, can also be treated using distance methods following an appropriate transformation. In some cases, the use of distances may be preferable, even though alternative character-based methods are available.

The terminology used in discussions of similarity/distance data varies from author to author and is sometimes inconsistent within a single paper.

We attempt to rely consistently on the following definitions. **Similarity** is a general term that corresponds in connotation to everyday usage. Similarity values are most frequently represented on a scale from 1 to 0 (100 to 0%). The term "homology" is often incorrectly used as a synonym for similarity; we agree with those who prefer to restrict the meaning of homology to similarity attributable to common ancestry (see above and Chapter 1). The opposite of similarity is **dissimilarity.** It is usually a value from 0 to 1 (0 to 100%). A dissimilarity is frequently defined to be 1 minus a similarity. **Distance** is more ambiguous; identical objects are separated by zero distance, but there is no consensus regarding the maximum value of distance. We use the meaning closest to the word's connotation in English: the largest distance is infinite. However, distance is sometimes treated as synonymous with dissimilarity in the literature.

Three different formulas are commonly used to convert similarities to distances prior to phylogenetic analysis. If the fractional similarity between two taxa is S, then a corresponding distance (d) can be

$$d = 1 - S \tag{3a}$$
$$d = -\ln S \tag{3b}$$
$$d = 1/S - 1 \tag{3c}$$

where ln is the natural logarithm. For very similar molecules, the values from each of the formulas will be nearly equivalent. However, as the divergence increases, the various formulas begin to give different answers. As discussed above, the value calculated by Equation (3a) is merely a dissimilarity and would be the least preferred option. Ideally the selection of a conversion method should be based on a model of the relevant evolutionary and physical (experimental) processes, rather than being arbitrary.

Distances (or similarities) relating pairs of taxa are typically viewed as organized in a T-by-T matrix. For tree inference, the matrix is usually assumed to be symmetric (the distance from A to B is equal to the distance from B to A) and the distance from any taxon to itself is assumed to be zero. Although both halves (triangles) of the matrix need not be calculated and stored, we refer to its elements as though the full matrix existed. Note that missing data also require a unique representation in the data matrix.

A general issue for the distance-based methods described in this chapter is the completeness of the pairwise distance matrix. Distances inferred from sequences are usually available for all pairs of taxa, although difficulties arise when inferring a composite tree from more than one molecule, as in Schwartz and Dayhoff (1978). When pairwise distances are determined experimentally, there is a much greater chance that for practical reasons not all comparisons will be performed. The consequences of these missing data can be substantial. In particular, let A and B be sister taxa, with C being the next closest relative. If the distance between A and B (d_{AB}) is missing from the data matrix, but the distances to C are available, then either A or B will be joined to C (whether it is A or B will depend on random noise), and then the third taxon will be added to the group. This is a fundamental, and frequently avoidable, error in some analyses based on incomplete sets of pairwise data. In addition, although it might appear that this can affect only taxa that are closely related, the artifacts can affect the deepest branchings, including the inferred location of the root.

Immunological and Nucleic Acid Hybridization Data   The immunological measurements in Chapter 5 are discussed in terms of models of the underlying evolutionary and physical processes. The authors conclude that, within certain limits, the immunological distance (ID) increases linearly with the number of amino acid differences in the proteins being compared. The constant of proportionality depends on the number of independent binding domains and on the fraction of amino acid changes that alters a domain sufficiently to inhibit antibody binding. Thus, there is a significant uncertainty in the exact scaling. If we knew the scaling, we would apply a correction for superimposed amino acid replacements (below). However, this is of little practical importance since the amount of divergence being measured is quite small, so any correction would also be small. Consequently we suggest equating evolutionary distance to the immunological distance, that is, assume that $d = $ ID for each pair of proteins.

Hybridization data (Chapter 7) are more varied. The two most common forms are the ΔT_m and the normalized percentage hybridization (NPH). Differences in melting temperature between homoduplex and heteroduplex DNAs are often used directly as a measure of distance, that is,

$$d = \Delta T_m$$

However, an alternative treatment (see Chapter 7) is to convert the value of ΔT_m to an estimate of fractional nucleotide sequence difference (dissimilarity) using the relationship

$$D = \Delta T_m / 110°C^{-1}$$

and then convert this to an estimate of evolutionary distance with a correction for superimposed substitutions (below). Two practical considerations of this procedure are that the conversion constant from ΔT_m to sequence dissimilarity is somewhat uncertain and that this does not immediately apply to the alternative measurements of melting behavior, such as T_{mode}, T_{50H}, or T_{25H} (each behaves differently experimentally, therefore they cannot all require the same correction). In practice, Chapter 7 provides some guidance in selecting the measurement that is most apt to diverge linearly, depending on the evolutionary distance to be spanned.

Percentage DNA incorporated into heteroduplex (NPH) is usually treated as a measure of similarity, and can be converted to distance by either Equation (3b) or (3c).

The treatment of duplicate measurements and reciprocal experiments is discussed in some of the experimental sections, so only general considerations will be discussed here. Unless an alternative treatment is recommended for a particular type of analysis, we suggest that replicate experimental data be averaged. Similarly, in the absence of a specific experimental rationale for doing otherwise, reciprocal measurements can be included in the average. Note that the standard deviation of duplicate measurements provides a very useful empirical measure of certain random errors. With reciprocal measurements, these random errors are potentially compounded with certain systematic errors and additional types of random errors. Unless the number of replicates for each data point is large, the standard deviation of individual determinations is a fairly noisy indicator of error (in particular, getting a few tightly clustered measurements for one particular comparison should not lead to disproportionate confidence in that value). Instead, the average of the standard deviations could be used to provide an overall estimate of the reproducibility of the data.

Transformation of Sequence Data to Distances   Much of the primary data used for molecular systematics consists of sequences. As just discussed, sequences are fundamentally character data. However, phylogenetic analyses based on sequence data are frequently carried out in two steps: reduction of the sequences to distance values relating all pairs of sequences, followed by a phylogenetic analysis of the pairwise distances. Three reasons for this approach are outlined below.

First, a sequence per se has little or no intuitive meaning. It is usually more useful to indicate, say, that one sequence is 93% identical to that of the homologous molecule from a second species and that none of the other known sequences is more than 78% identical. Without formal phylogenetic analysis, this presentation of the data evokes a concrete and usually correct image of a specific relationship between the two more similar sequences. In general, it is relationships between sequences that are of interest in molecular systematics, not the sequences themselves. (However, comparative data generated in systematic studies contribute to our understanding of the evolution of individual molecules, which, in turn, lays the foundation for more sophisticated uses of these molecules in systematic analyses.)

Second, as pointed out above and elaborated on below, as mutations are fixed in the genome, there is an ever-increasing chance of multiple events occurring at a single sequence position. Because we do not have a complete historical accounting of events, later changes can destroy the record of earlier events. By making assumptions about the nature of evolutionary changes, we are able to estimate the number of unseen events (see below). These corrections are applied to evolutionary intervals, not to the sequences themselves. One type of evolutionary interval is the total separation between a pair of contemporary sequences. Therefore, converting the sequence data to a table of distances between pairs of sequences allows these corrections to be applied to the respective distance values. (Maximum likelihood and Lake's method of invariants provide frameworks in which the data are retained as sequences while incorporating corrections for superimposed changes.)

Third, numerous methods for inferring phylogenetic trees from pairwise distance data exist. Many of these methods were developed for the analysis of other forms of similarity and distance data, but they are equally applicable to distances derived from sequences. In fact, sequence data benefit from their more precisely defined relationship to the genome.

The negative side of reducing sequence data to pairwise distances is that information is lost in the transformation. For instance, Penny (1982) has shown examples in which several different sets of sequences yield the same distance matrix, but given only the distances it is impossible to go back to the original sequences . Unfortunately, little is known about how much the apparent loss of information affects either the accuracy or the precision of subsequent phylogenetic analyses. There is, as yet, little reason to believe that the result is devastating when real sequence data are being analyzed. In fact, many sequence data sets yield identical branching patterns with character-based and distance-based analyses (e.g., Olsen, 1987). Another drawback to distance analysis is that it does not lend itself to the combination of different kinds of data into the same analysis, as is possible for character-based analyses (e.g., Miyamoto, 1985). Finally, only through character-based analysis can a researcher identify particularly informative characters (or regions) in order to limit subsequent studies to those characters that are most useful (e.g., the detection of so-called "signature" events).

Sequence Similarity.   Because this chapter is intended to be a practical guide and not a technical discussion (or argument), the various transformations from sequence data to distance estimates will not be presented in terms of the original rationale or assumptions (for which the original articles should be consulted) but will be recast in a more uniform perspective. Similarly, descriptions will be based on general understanding, not on mathematical or logical rigor. For example, it will be assumed that a sequence with "constant base composition" can accept a single base substitution, even though this substitution would modify the base composition by some small amount.

By far the most common method of summarizing the relationship between two sequences is by their fractional (or percentage) similarity. In its simplest form, the sequence similarity is equal to the number of aligned sequence positions containing identical residues divided by the number of sequence positions compared. However, we must explicitly address several subtleties and potential ambiguities: alternatives to limiting the comparison to identical residues, terminal length variation of molecules, alignment gaps, and treatment of ambiguous residues. The following sections assume that the sequence alignment has already been defined (see the section entitled "Character Data," above).

It is frequently of interest to define the similarity of two molecules in terms of a more relaxed criterion than the fraction of identical residues, thereby changing our definition of sequence similarity to the number of aligned sequence positions containing "synonymous" residues divided by the number of sequence positions compared. For example, "conservative substitutions" are commonly ignored when comparing proteins by pooling the amino acids into six groups: acidic (D, E), aromatic (F, W, Y), basic (H, K, R), cysteine, nonpolar (A, G, I, L, P, V), and polar (M, N, Q, S, T). Residues within a group are considered synonymous; residues in different groups are considered nonsynonymous.

Terminal length variation refers to the observation that corresponding molecules from different species (and even within an individual organism) can start and end at different distances from homologous features within the molecules. Many genetic and physiological factors (e.g., substitution mutations, insertions or deletions, or alteration of a processing enzyme) could be responsible for these variations. Regardless of the underlying cause, the result is that some molecules include residues for which there are no homologs in other molecules. Because only homologous features should be compared, these regions of length variation must be omitted from the analysis. Although terminal length variations could be treated as a discrete character, of questionable reliability, this issue is not germane to the current discussion. When several sequences are analyzed simultaneously, all pairwise comparisons should use the same set of sequence positions, those limited by the shortest molecule at each end.

As discussed above, if the evolution of a gene includes insertions and/or deletions, then "gaps" must be inserted to adjust for the internal length changes when aligning the contemporary sequences. An appropriate treatment of the alignment gaps in subsequent analyses is not self-evident. Although the character state "gap" is sometimes treated as a fifth base or twenty-first amino acid, the processes responsible for base substitution and for insertion and deletion are evolutionarily and mechanistically distinct. Because a proper treatment is not obvious, sequence positions with gaps are sometimes omitted from analyses. However, two sequences that are identical except for gaps clearly are not 100% similar, so the analysis cannot merely ignore them. Another reason for not ignoring them is that alignment gaps are usually positioned to maximize the alignment of identical residues in sequences — thus additional genetic events, insertions and deletions, could systematically raise the apparent similarity. The limit on this latter trend is the requirement that a valid analysis be based on homologous residues: if there are large numbers of alignment gaps, or the overall sequence similarity is low (not greater than two times the random level of similarity), then it is unreasonable to assume that the compared residues are homologous and the entire region of sequence should be omitted from analysis. The opposite side of this statement is that the sequence regions retained in analysis will have few insertions and deletions and the treatment of gaps will therefore not have a large effect on the outcome of the analysis (an assertion that is easily tested by applying alternative treatments to the same data).

The most frequently used methods for calculating sequence similarities from aligned sequences are encompassed by the formulas

$$S = M/L \qquad\qquad (4a)$$
$$L = M + U + w_{G}G \qquad\qquad (4b)$$

where S is the similarity, M is the number of alignment positions with synonymous residues, L is an effective sequence length, U is the number of alignment positions with nonsynonymous residues, w_{G} is the weight given to gaps, and G is the number of alignment positions with a gap in one sequence juxtaposed with a residue in the other sequence. The variations are distinguished by the value of the gap weight. Most analyses are based on values of w_{G} in the range from zero (ignoring gaps) to one (treating them as equivalent to substitutions). A value of one-half is a reasonable compromise.

In anticipation of later discussions of "character weighting," it is worth realizing that the calculation of M, U, and G in Equation (4) can easily be based on weighted sums (Olsen, 1988). That is, columns in the sequence alignment can be assigned prior weights that are intended to reflect their phylogenetic usefulness. The most informative columns would be assigned a weight of 1, and less informative positions would get lower values. In the extreme, regions of dubious positional homology would be given zero weight. Another consideration would be to assign nucleotides that base pair in an RNA secondary structure one-half the weight of unpaired positions, thereby reflecting their lack of independence (Stahl et al., 1984; Wheeler and Honeycutt, 1988). The same framework supports weights that might be based on inferred rate of sequence change, to place emphasis on the more conserved positions.

A frequently overlooked issue in pairwise sequence comparison is the treatment of "ambiguities" (i.e., nucleotide or amino acid residues of uncertain identity) in the sequences being compared. This uncertainty is not a difficult problem when the ambiguities are residues of completely un-determined identity: if aligned with a residue in a second sequence, then the ambiguous residue can be safely ignored; if opposite an alignment gap, then it can be included in the value of G. Partially resolved residues (e.g., a purine) are more difficult to treat rigorously. For example, counting a purine as synonymous with A and G and nonsynonymous with C and U will tend to overestimate the similarity. It is possible to correct for this excess similarity and thereby use this partial information, but we are not aware of any publication of these methods. On the whole, it is easiest to pretend that such residues are completely ambiguous (a small amount of information is discarded, but no systematic error is introduced).

The similarity from Equation (4) is an appropriate value for summariz-ing the relationship between sequences. Tables of pairwise similarities are also used directly by some methods of phylogenetic tree construction, most notably cluster analysis. However, it is an inescapable fact that as genes accumulate mutations, there is an ever-increasing likelihood that some of the changes will be at the same sequence location. Because pairwise com-parisons of sequences are based entirely on the identity or nonidentity of residues at corresponding sequence positions, the first substitution at a site will convert identical residues to nonidentical residues. Subsequent changes at the same sequence position cannot further decrease the similarity, though they can raise the similarity by converting the compared residues to similar identities (parallelism or reversion). The net effect of this "superimposi-tion" of substitutions is that similarity does not decline uniformly with the number of events; instead, it declines rapidly at first and more slowly thereafter.

Accounting for Superimposed Events.   Although the relatively unprocessed similarity values are of interest for communicating data to others and are useful for some tree inference procedures, it is frequently preferable to perform analyses that assume that the data are **additive distances** (defined below). We will focus on distances based on the number of substitutions per sequence position (the **evolutionary distance**), but will cover only a fraction of the methods proposed for making this estimation. Two categories of methods will be considered; the first bases the distance estimate on the similarity value and the second uses the sequence data directly.

Many transformations from sequence similarity to evolutionary dis-tance can be represented by the formulas

$$D = 1 - S \qquad\qquad (5a)$$
$$d = -b \ln(1 - D/b) \qquad\qquad (5b)$$

where D is the sequence dissimilarity (fractional difference), S is the frac-tional sequence similarity from Equation (4), d is the estimated evolutionary distance between a pair of sequences, and b is a value that varies with the particular model and data. Formula (5) reflects the expected decay in similarity between randomly diverging sequences in which (1) all substitutions at a given sequence position are independent; (2) all sequence positions are equally subject to change; (3) the base (or amino acid) composition is not shifting over time (the composition is always at equilibrium); (4) the rate of change to each residue type is proportional to the residue's equilibrium abundance but independent of the identity of the starting residue; and (5) no insertions or deletions have occurred. The fourth assumption sounds more complex than it really is: the idea is that all residues are equally likely to change, but what they change *to* is determined by maintenance of the equilibrium composition. The fifth constitutes an assumption about the evolutionary process; it should not be confused with a decision to ignore gaps by setting $w_G = 0$ in Equation (3). Use of this formula requires three decisions: which residue types (if any) will be treated as synonymous, how positions with alignment gaps will be treated, and what value will be used for b.

Equation (5) is very general. It can be applied to nucleotides, amino acids, or groups of residue types (e.g., the amino acid classes discussed above). The groups need not have equal numbers of residue types or add up to equal fractions of the total composition. What does matter is adherence to the third assumption: rates of substitution into a group should depend only on the equilibrium abundance of the residues in the group; rates of substitution within a group are irrelevant.

To use Equation (5) it is not necessary to assume that all sequence positions vary independently. In particular, the presence of Watson–Crick complementarity between pairs of positions does not violate the assumptions. However, base pairing does reduce the number of independent samples of sequence divergence, and hence increases the random noise in the distance estimates. In addition, unless the pair positions are given reduced weights in the calculation of sequence similarities (above), the paired positions will be given disproportionate emphasis in the calculation, potentially increasing the noise unnecessarily (Wheeler and Honeycutt, 1988).

Insertions and deletions are not part of the evolutionary models used in justifying Equation (5) (see assumption 5), so the treatment of alignment gaps varies. Most frequently, similarities are calculated using Equation (4) without special considerations. An alternative is omitting alignment positions at which one or more sequences contain a gap (i.e., skipping the column from all pairwise similarity calculations). Because no treatment of the data can change the evolutionary reality of insertions and deletions, arguing about the above alternatives is unproductive. Once again we emphasize that regions of the sequence alignment that contain substantial numbers of alignment gaps should be omitted from the analysis: positional homology is too uncertain for reliable estimates to be made from these regions.

We are left with the task of defining an appropriate value for b. This variable actually has a concrete physical meaning. Within the context of the above assumption, b is the dissimilarity of infinitely diverged (completely randomized) sequences. For example, if nucleotide sequences are being compared and all four nucleotides are considered equally likely, then $b = 3/4$ (Jukes and Cantor, 1969). If the four nucleotides are not used in equal proportions, then a more general formula,

$$b = 1 - \sum_{i \in R} f_i^2 \tag{6}$$

can be used, where R is the set of possible residue types (e.g., {A,C,G,T} for DNA sequences) and f_i is the frequency of the ith type of residue in the sequences being compared (Tajima and Nei, 1984). The base composition used can be calculated separately for each pair of sequences, or it can be the average for all of the sequences analyzed; we lean toward the latter. If nucleotide types are combined into groups, then the summation is over each group of nucleotides. For example, when looking only at transversions, one combines A with G, and C with T, yielding

$$b = 1 - [(f_A + f_G)^2 + (f_C + f_T)^2]$$

For use with amino acids, the most common correction for multiple substitutions is Equation (5) with $b = 1$, which makes it identical to Equation (3b). This assignment tends to lead to underestimation of the distance between very diverged sequences since it follows from Equation (6) that, if all amino acids are equally frequent, $b = 0.95$ (i.e., 19/20). Accounting for unequal amino acid frequencies gives a value of b closer to 0.93.

For analyses of nucleotide sequences, the preference for **transition substitutions** (between two purines or between two pyrimidines) over **transversion substitutions** (between a purine and a pyrimidine or vice versa) has received much attention. If one is concerned about this effect with a given set of data, at least two practical solutions are possible. If transition substitutions are substantially more frequent than transversions and the sequences are very diverged, then distinguishing between the two purines, or between the two pyrimidines, is unlikely to be useful. Thus, the nucleotides can be reduced to two groups, as described above. Because this approach discards all transition information (motivated by the assumption that such information has been reduced to noise), it cannot be applied to closely related sequences, which will differ by few transversions and hence give statistically inferior conclusions. Alternatively, Kimura (1980) has provided a method for inferring evolutionary distance in which (1) transitions and transversions can occur at unequal rates, (2) all four nucleotides occur with equal frequency, and (3) the remaining assumptions of Equation (5) are satisfied:

$$d = -1/2 \ln[(1 - 2P - Q)\sqrt{1 - 2Q}] \qquad (7a)$$
$$P = U_P/N \qquad (7b)$$
$$Q = U_Q/N \qquad (7c)$$
$$N = M + U_P + U_Q \qquad (7d)$$

where P is the fraction of sequence positions differing by a transition, U_P is the number of sequence positions differing by a transition, N is the number of sequence positions compared in which both sequences contain a nucleotide, Q is the fraction of sequence positions differing by a transversion, U_Q is the number of sequence positions in which the two sequences differ by a transversion, and M is the number of sequence positions in which the two sequences have identical nucleotides. Gaps are never counted for use in Equation (7). They are not part of the model and are not easily worked into Equation (7); ignoring gaps is equivalent to setting $w_G = 0$ in Equation (4).

Protein-Coding DNA Sequences.   In principle, knowledge of the gene sequence should be more informative than the corresponding protein sequence. In practice, two factors call this assertion into question. First, silent substitutions in protein-coding genes are much more frequent than replacement substitutions, thus the third codon position tends to become randomized quickly and conveys very little information about distant phylogenetic relationships. Second, the base composition of the third codon position appears to vary systematically between some species, thereby indicating that it can be subject to an at least moderately strong selective force that is different in different lineages. The presence of directional selection can lead to profound sequence convergences and consequent errors in inferred relationships. With these considerations in mind, three relatively simple strategies can be used to analyze protein-coding sequences, and a host of moderately to extremely complex alternatives exists.

The simplest method of calculating distances between sequences for protein-coding genes is to apply Equations (4) and (5) directly to the gene sequence without special treatment. This method is reasonable, or even preferred, when the total amount of divergence is very small, in which case the resulting trees are based primarily on silent substitutions in the genes. The main drawback is that a systematic undercorrection for superimposed substitutions will result, since the assumption that all positions are equally subject to change will clearly be violated. If the amount of sequence divergence is truly small, then superimposed changes will be rare and the undercorrection will be negligible.

The second approach is to restrict the analysis to the first two nucleotides of each codon. This strategy is more applicable when a substantial sequence divergence is apparent. The rationale is that the third codon position will be largely randomized and hence phylogenetically uninformative. This approach, by definition, also circumvents the problem of the third codon position changing more rapidly than the first two and thereby limiting the applicability of Equation (4).

The third basic method is to infer the protein sequence from the gene sequence and perform the phylogenetic analysis at the protein level. This approach has two merits: (1) the protein is the most biologically relevant aspect of the gene (taken as a whole); and (2) the sequence can be compared with homologous molecules that were sequenced at the protein level, for which nucleotide sequences are therefore unknown.

The more complex methods are generally based on distinguishing sequence positions with silent changes from those with replacement substitutions. Many implicit and explicit assumptions are involved in differentially interpreting 2-fold, 3-fold, and 4-fold degenerate positions and in determining what to do when a position is 2-fold degenerate in one gene and 4-fold degenerate in the other, etc. The reader can refer to Tajima and Nei (1984) for a collection of such approaches.

Transformation of Allozyme and Restriction Endonuclease Data to Distances   A large number of measures have been proposed for transforming allelic and genotypic frequency data to genetic distances (S. Wright, 1978); we will treat only a few of the more commonly used ones here. Historically, the most frequently used genetic distance has been that of Nei (1972, 1978). Let x_i and y_i be the frequencies of the ith allele at a particular locus in taxa X and Y, respectively. Nei's (1972) standard genetic distance can then be defined as

$$D_N = -\ln(J_{XY}/\sqrt{J_X J_Y}) \tag{8}$$

where J_X, J_Y, and J_{XY} are the arithmetic means across loci of Σx_i^2, Σy_i^2, and $\Sigma x_i y_i$, respectively. Equation (8) gives a biased estimate when sample sizes are small; an unbiased estimate of the standard distance is obtained by replacing Σx_i^2 and Σy_i^2 with $(2n\Sigma x_i^2 - 1)/(2n_X - 1)$ and $(2n_Y\Sigma y_i^2 - 1)/(2n_Y - 1)$, respectively (Nei, 1978). D_N is intended to measure the number of codon substitutions per locus that have occurred after divergence between a pair of populations (taxa). However, this interpretation is valid only if the rate of gene substitution per locus is uniform across both loci and lineages, an assumption that is almost certainly unrealistic (Hillis, 1984) for any systematically informative data set. Hillis (1984) demonstrated that violation of the assumption of rate uniformity leads to a peculiar property of D_N when it is applied in systematic studies involving interspecific comparisons. He showed three hypothetical two-locus cases in which for each case, two taxa had identical allele frequencies at one locus and shared no alleles at the second locus. Yet, due to different levels of polymorphism within the two taxa, D_N varied from 0.41 to 1.10. Hillis (1984) consequently recommended the following modification to D_N to alleviate the problems created by nonuniform rates of change:

$$D_N^* = -\ln\left[\sum_L (\Sigma x_i y_i / \sqrt{\Sigma x_i^2 \Sigma y_i^2})/L\right] \qquad (9)$$

where L is the total number of loci; that is, the distance is computed from the arithmetic mean of the single-locus identities. (Although Hillis did not recommend it, an unbiased version of D_N^* could be obtained by a substitution equivalent to that for Nei's original distance.)

Nei's distances (in either their original form or as modified by Hillis, 1984) are nonmetric in that they frequently violate the triangle inequality. Farris (1981) has heavily criticized it for this reason, arguing that when a distance measure is nonmetric, it is meaningless to fit branch lengths under an **additive tree** model in which branch lengths are interpreted as amounts of evolutionary change. Felsenstein (1984) countered that if branch lengths were interpreted as expected, rather than actual, amounts of change, Farris's objections were moot. Although we do not wish to become entangled in this controversy (see also Farris, 1985, 1986; Felsenstein, 1986), we basically agree with Felsenstein, without going so far as to recommend routine usage of Nei's distance. If Nei's model of evolution is correct (which is obviously open to question), then the nonmetricity of his distance is not in itself a reason to shun it.

Another widely used distance measure is that of Rogers (1972):

$$D_R = (1/L)\sum_L \sqrt{\Sigma(x_i - y_i)^2}/2$$

Rogers's measure has the virtues of simplicity and an easily interpretable geometric basis. Except for a scaling factor, it is simply the Euclidean distance between the allele frequency vectors for each locus of the two taxa being compared. However, Rogers's coefficient shares with Nei's the undesirable property of being too heavily influenced by within-taxon heterozygosity (Wright, 1978; Hillis, 1984); the distance between two taxa that are fixed for alternate alleles exceeds that between two taxa in which one or both are heteroallelic but have no alleles in common.

An alternative Euclidean measure that overcomes this limitation is the arc distance of Cavalli-Sforza and Edwards (1967), which is given by

$$D_{arc} = \sqrt{(1/L)\sum_L (2\theta/\pi)^2}$$

where $\theta = \cos^{-1} \Sigma\sqrt{x_i y_i}$. Thus, if no alleles are shared between a pair of taxa, the distance takes its limiting value of one regardless of the variability within either population. Perhaps more importantly, this distance incorporates an angular transformation of gene frequencies in an attempt to make the variances of the transformed frequencies independent of the ranges in which they fall. This transformation has the effect of standardizing the distance with respect to random drift, so that the rate of increase in genetic distance under drift is nearly independent of the initial gene frequencies. The Cavalli-Sforza and Edwards (1967) arc distance and its relative, the chord distance, thus incorporate some realistic assumptions about the nature of evolutionary change in gene frequencies without the undesirable properties of the Nei (1972, 1978) and Rogers (1972) measures.

The simplest distance of all is the Manhattan distance, attributed to Prevosti by S. Wright (1978), which for a single locus equals

$$D_M = (1/2) \, \Sigma \, |x_i - y_i|$$

An arithmetic mean is used to combine distances across loci. Unlike the Cavalli-Sforza and Edwards (1967) distances, this method gives equal weight to a given frequency difference regardless of where it occurs on the scale from zero to one. It is not sensitive to intrataxon variability, however.

To transform restriction-site data to distances, Nei and Li's (1979) method for estimating the number of nucleotide substitutions that have occurred since divergence of a pair of taxa X and Y from a common ancestor is typically used. An estimate of the proportion of ancestral restriction sites that have remained unchanged until the present is given by

$$\hat{S} = 2n_{XY}/(n_X + n_Y)$$

where n_{XY} is the number of identical sites shared by the two taxa, and n_X and n_Y are the total number of restriction sites in taxa X and Y, respectively. From this quantity we can estimate the mean number of substitutions per nucleotide site using either of the following:

$$d = (\ln \hat{S})/r$$

$$d = -(3/2)\ln[(4\hat{S}^{1/2r} - 1)/3]$$

where r is the length of the endonuclease recognition sequence (usually 4 or 6). The researcher should be cautious in interpreting these estimates, however, as they are based on a rather precise probability model for the evolutionary change of restriction sites. If the assumptions of this model are violated, the distance estimates — although perhaps useful as an index of relative divergence — will no longer have a direct biological interpretation.

Nei and Li (1979) also addressed the problem of estimating nucleotide substitutions from restriction fragment data. However, these estimates are reliable only if the actual number of substitutions has been low (e.g., between conspecific populations). Consequently, we will not describe their procedures for dealing with fragment data; the interested reader can consult their paper directly.

GENERAL APPROACHES TO PHYLOGENY ESTIMATION

Having laid the groundwork with respect to data types and transformations, we now turn to the actual descriptions of the methodologies used in the estimation of evolutionary trees. These methods can be broken into four broad categories: distance-matrix, parsimony, invariants, and maximum likelihood. We reiterate our intention to concentrate on methods that are commonly used at present or that we believe will receive more attention in the future.

Methods Based on Pairwise Distances

Distance data are generally characterized by certain mathematical properties. For our purposes, two categories are of special interest: **additive distances** and **ultrametric distances**.

Additive Distances   Mathematically, additive distances satisfy the four-point condition. Specifically, for any four taxa A, B, C, and D:

$$\max(d_{AB}+d_{CD}, d_{AC} + d_{BD}, d_{AD} + d_{BC})$$
$$= \mathrm{mid}(d_{AB} + d_{CD}, d_{AC} + d_{BD}, d_{AD}+ d_{BC})$$

where d_{ij} is the distance between taxa i and j, max is the maximum value function, and mid is the middle value (median) function. Because mid is not a standard function, it might be noted that this expression is equivalent to

$$2\max(d_{AB} + d_{CD}, d_{AC} + d_{BD}, d_{AD} + d_{BC})$$
$$+ \min(d_{AB} + d_{CD}, d_{AC} + d_{BD}, d_{AD} + d_{BC})$$
$$= d_{AB} +d_{CD} + d_{AC} + d_{BD} + d_{AD} + d_{BC}$$

where min is the minimum function.

Additive distances can be fitted to an unrooted tree such that all pairwise distances are equal to the sum of the branch lengths that connect the respective taxa (Figure 3a). As noted earlier, additive distances are an ideal; given systematic and random errors, few if any experimental data constitute additive distances. On the other hand, the total number of fixed mutations (or any other definition based on discrete events) is an example of an additive distance. This sort of concrete image provides motivation for some of the methods of analysis; that is, if the experimental data could be used to infer the number of underlying fixed mutations, then the data could be-

(a)

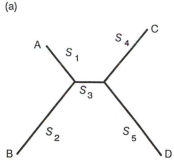

(b)

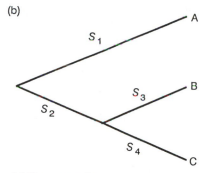

Additive properties:

$$p_{AB} = S_1 + S_2$$

$$p_{AC} = S_1 + S_3 + S_4$$

$$p_{AD} = S_1 + S_3 + S_5$$

$$p_{BC} = S_2 + S_3 + S_4$$

$$p_{BD} = S_2 + S_3 + S_5$$

$$p_{CD} = S_4 + S_5$$

Additive properties:

$$p_{AB} = S_1 + S_2 + S_3$$

$$p_{AC} = S_1 + S_2 + S_4$$

$$p_{BC} = S_3 + S_4$$

Ultrametric properties:

$$S_3 = S_4$$

$$S_1 = S_2 + S_3 = S_2 + S_4$$

FIGURE 3. Additive and ultrametric trees. (a) An additive tree relating four taxa: A, B, C, and D. Also shown are the relationships between the six taxon-to-taxon distances (p_{AB}–p_{CD}) and the five branch lengths (s_1–s_5). Additive distances and trees do not make any assumption about the rooting, hence the relationships are displayed in an unrooted format. All sets of pairwise distances that satisfy the four-point condition (see text) can be represented as a unique additive tree. (b) An ultrametric tree relating three taxa: A, B, and C. In addition to having additive properties (all taxon-to-taxon distances are the total of the branch lengths joining them), every common ancestor is equidistant from all its descendants. Thus, the most recent common ancestor of B and C is s_3 from B and s_4 from C, therefore $s_3 = s_4$. Likewise, the common ancestor of A and B is s_1 from A and $s_2 + s_3$ from B, therefore $s_1 = s_2 + s_3$.

made additive and would precisely fit a unique tree—the problem would be solved.

Many methods have been described that derive the additive tree corresponding to a set of additive distances. We present versions of the neighbor-joining, weighted least-squares, and minimum absolute difference methods below. If the data are perfectly additive, then these methods provide an exact solution. The problem of imperfectly additive data is discussed with the methods themselves.

Ultrametric Distances   Ultrametric distances are the most constrained. Mathematically, ultrametric distances are defined by satisfaction of the three-point condition. The ultrametric inequality requires that for any three taxa A, B, and C,

$$d_{AC} \leq \max(d_{AB}, d_{BC}) \tag{10}$$

An equivalent, but perhaps more intuitive, definition is given by the equation

$$\max(d_{AB}, d_{BC}, d_{AC}) = \text{mid}(d_{AB}, d_{BC}, d_{AC})$$

i.e., the two greatest distances are equal. This equation can be recast to eliminate the mid function:

$$2\max(d_{AB}, d_{BC}, d_{AC}) + \min(d_{AB}, d_{BC}, d_{AC}) = d_{AB} + d_{BC} + d_{AC}$$

Phylogenetically, ultrametric distances will precisely fit a tree so that the distance between any two taxa is equal to the sum of the branches joining them, *and* the tree can be rooted so that all of the taxa are equidistant from the root (Figure 3b). The first half of this description defines an additive tree (and implies that ultrametric distances are additive). The second half of the description corresponds to the concept of a universal molecular clock, so that all lineages are equally diverged. Two potential surprises may emerge, however. First, even with ultrametric data, there is no guarantee that the amount of divergence is *linear* in time. In particular, superimposed sequence changes, which decrease the observed molecular divergence, do not destroy the ultrametric property. Second, obtaining ultrametric data is extremely unlikely; even if the underlying substitution rate is perfectly constant, any finite sample will yield statistical fluctuations in the measured divergences. Consequently, even a universal substitution rate would not give ultrametric data without an infinitely large sample. The closest experimental approximation of an infinite sample is genome hybridization measurements (Chapter 7).

If data are ultrametric by Equation (10), then the use of **cluster analysis** to infer the branching pattern is valid. Colless (1970) provides a precise definition of how much deviation from ultrametricity can be tolerated without introducing systematic errors. However, there is little practical reason to use cluster analysis because related methods (such as the distance Wagner or neighbor-joining methods) are applicable to more general additive distances and require very little additional computation.

Cluster Analysis   Cluster analysis is a family of related techniques for representing similarity or distance data (we will use distances) in the form of an ultrametric tree. If the data themselves are ultrametric, then the representation on the tree will be exact. It should be obvious that if the distance data themselves are not ultrametric, then they *cannot* be fit exactly to such a tree, and therefore errors might be introduced.

The method of cluster analysis is conceptually simple. The raw data are provided as a table of distances between all pairs of taxa. Call d_{ij} the distance between taxa i and j. The tree is constructed by linking the least distant pairs of taxa, followed by successively more distant taxa or groups of taxa. When two taxa are linked, they lose their individual identities and are subsequently referred to as a single cluster. Initially, each taxon constitutes its own cluster. At each stage in the process, as two clusters are merged into one, the number of clusters declines by one. The process is complete when the last two clusters are merged into a single cluster containing all of the original taxa.

The steps of the method are as follows:

1. Given a matrix of pairwise distances, find the clusters (taxa) i and j such that d_{ij} is the minimum value in the table.
2. Define the depth of the branching between i and j (l_{ij}) to be $d_{ij}/2$.
3. If i and j were the last two clusters, the tree is complete. Otherwise, create a new cluster called u.
4. Define the distance from u to each other cluster (k, with $k \neq i$ or j) to be an average of the distances d_{ki} and d_{kj}.
5. Go back to step 1 with one less cluster; clusters i and j have been eliminated, and cluster u has been added.

The variants are primarily in the details of step 4. The most commonly used clustering method is **UPGMA** (unweighted pair group method using arithmetic averages), in which the averaging of the distances in step 4 is based on the total number of taxa in the clusters. That is, if cluster i contains T_i taxa, and cluster j contains T_j taxa, then $d_{ku} = (T_i d_{ki} + T_j d_{kj})/(T_i + T_j)$. If the simple average [$d_{ku} = (d_{ki} + d_{kj})/2$] is used instead, the technique is called **WPGMA** (weighted PGMA). Other variants include using the maximum distance [$d_{ku} = \max(d_{ki}, d_{kj})$, called complete linkage], or the minimum distance [$d_{ku} = \min(d_{ki}, d_{kj})$, called single linkage]. These alternatives all give the same results when the data are ultrametric, but they can differ in their inferences when the data are not ideal.

An example of using UPGMA to infer a tree of five taxa (5S rRNA sequences) is given in Figure 4. The figure presents the upper-right half of the pairwise distance matrix at each stage of the cluster analysis. Starting with the first table, the smallest distance (the 0.1715 substitutions per sequence position separating Bsu and Bst) is indicated in boldface. Thus, the first inferred branching unites these taxa at a depth of $0.1715/2 = 0.0858$. These two taxa are merged into a cluster in the next table, and their distances to all other taxa are averaged. For example, the distance from the Bsu–Bst group to Lvi is $(0.2147 + 0.2991)/2 = 0.2569$. The smallest distance in the second table joins the Bsu–Bst cluster with Mlu at a depth of $0.1096 (= 0.2192/2)$. The distances of the Bsu–Bst–Mlu cluster to the other taxa are then computed by the unweighted method. For example, the distance to Lvi is $(2 \times 0.2569 + 0.3943)/3 = 0.3027$. Notice that this value

	Bsu	Bst	Lvi	Amo	Mlu
Bsu	—	0.1715	0.2147	0.3091	0.2326
Bst		—	0.2991	0.3399	0.2058
Lvi			—	0.2795	0.3943
Amo				—	0.4289
Mlu					—

	Bsu-Bst	Lvi	Amo	Mlu
Bsu-Bst	—	0.2569	0.3245	0.2192
Lvi		—	0.2795	0.3943
Amo			—	0.4289
Mlu				—

	Bsu-Bst-Mlu	Lvi	Amo
Bsu-Bst-Mlu	—	0.3027	0.3593
Lvi		—	0.2795
Amo			—

	Bsu-Bst-Mlu	Lvi-Amo
Bsu-Bst-Mlu	—	0.3310
Lvi-Amo		—

FIGURE 4. Cluster analysis (UPGMA) of 5S rRNA evolutionary distance estimates. Each table represents the pairwise distances (estimated nucleotide substitutions per sequence position) for one round of clustering (only the upper-right half of the symmetrical matrix is shown). The minimum distance value in each table is in boldface. The corresponding pair of taxa (or clusters) is merged into a single cluster in the next table. The boldface distance value is twice the depth of the branch point separating the clusters merged. A diagram of the inferred tree is in Figure 5a. The data are from Olsen (1988). The abbreviations used to identify the taxa are Bsu, *Bacillus subtilis*; Bst, *Bacillus stearothermophilus*; Lvi, *Lactobacillus viridescens*; Amo, *Acholeplasma modicum*; and Mlu, *Micrococcus luteus*. The inferred tree topology is not the same as that inferred with neighbor-joining or least-squares methods (see Figure 5).

is identical to (Bsu:Lvi + Bst:Lvi + Mlu:Lvi)/3, where A:B is the distance from taxon A to taxon B. The fact that each taxon in the original data table contributes equally to the averages is the reason that the method is called unweighted. The smallest distance in the third table unites Lvi and Amo at a depth of 0.1398. The distance between the Bsu–Bst–Mlu and Lvi–Amo clusters is then (3 × 0.3027 + 3 × 0.3593)/6 = 0.3310. Thus, the implied root of the tree joins these two clusters at a depth of 0.1655. The complete tree is shown in Figure 5a.

Note that cluster analysis cannot join two taxa (sometimes called operational taxonomic units or OTUs) unless at least one pairwise distance links them. Thus, missing data within a group can force one or more members out of the group in the inferred tree, a problem discussed in greater detail under "Similarity and Distance Data."

Cluster analysis has been very popular historically for several reasons. First, there are really very few assumptions, merely that the data be ap-

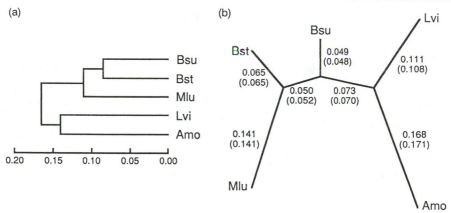

FIGURE 5. Comparison of 5S rRNA phylogenies inferred by different pairwise distance methods (data from Olsen, 1988). (a) Tree obtained by cluster analysis from the analysis in Figure 4. (b) Tree obtained using neighbor-joining and weighted least-squares. The upper branch lengths (expected substitutions per sequence position) are from the analysis in Figure 6, and the parenthetical values are from the analysis in Table 1. Although the tree is unrooted, the *M. luteus* sequence is the outgroup. It can be seen that the neighbor-joining and least-squares procedures produced very similar trees, but the cluster analysis tree is very different. Two of the sequences, those of *L. viridescens* and *A. modicum,* are very much more diverged than are the others, an effect to which cluster analysis is particularly sensitive.

proximately ultrametric. This assumption is of course a very strong one, but it is seductive to believe that a single stringent assumption can be satisfied more easily than a long list of (what might be) less restrictive assumptions. Second, the idea of grouping the taxa that are least different, regardless of any finer points of consideration, has a strong intuitive appeal. The extreme of this view is the phenetic perspective in which it is asserted that nothing but the extent of similarity matters biologically and that consideration of the historical branching order is of purely secondary interest. A third reason is the availability of programs to do cluster analysis and the relative speed of the calculations, thereby enabling large numbers of taxa to be analyzed.

As emphasized above, simple cluster analysis has drawbacks. First, it is just an algorithm (or family of algorithms) with no objective definition of what constitutes an optimal tree when the data are not ideal. In particular, because genes do not diverge uniformly in all organisms or organelles (Chapters 8, 9, and 12), systematic errors are likely to be introduced into cluster analysis reconstructions. Finally, alternative rapid methods are available that will work for all additive trees, not just those that are ultrametric.

Additive Trees Additive tree techniques comprise a relatively broad class of methods that operate under the assumption that the lengths of the branches lying on the path between any pair of taxa can be summed to yield a meaningful quantity (e.g., amount of evolution). We will begin by discussing algorithmic methods. This will be followed by some methods with specific optimality criteria.

A variety of algorithmic methods related to cluster analysis have been proposed that will correctly reconstruct additive trees, whether the data are ultrametric or not. These methods seem to fall into three primary categories. The first transforms any additive distance matrix into an ultrametric matrix and then uses cluster analysis to infer the tree. Included here are the transformed distances method of Li (1981), the present-day ancestor method of Klotz and Blanken (1981), and, in a less obvious sense, the neighbor-joining method of Saitou and Nei (1987). The second category comprises methods that form the clusters consistent with the largest fraction of taxon quartets, using a relaxed definition of additivity for a four-taxon tree. These methods include those of Sattath and Tversky (1977) and Fitch (1981). Methods of the third class, which includes the distance Wagner method (Farris, 1972), build an additive representation of the tree by sequential addition of taxa. The transformed distance approaches all have a computational complexity that is proportional to T^3; therefore, any problem that is tractable with standard cluster analysis can also be solved with these methods. We present a version of the neighbor-joining method below.

Unlike cluster analysis, additive-tree methods yield unrooted trees, which are adequate for some purposes. If a root is to be placed, however, it must be based on an ancillary criterion. Usually, one or more taxa that are assumed to lie outside a monophyletic group of interest are included in the analysis. The location at which these taxa join the tree defines the root with respect to the ingroup. Another method, midpoint rooting, depends on an assumption of rate uniformity that is somewhat weaker than that of a molecular clock: if the two most divergent lineages have evolved at the same rate, then the appropriate root is at the midpoint of the path connecting these taxa.

Neighbor-Joining Method.   Neighbor-joining (Saitou and Nei, 1987) is an algorithm for inferring an additive tree. It is conceptually related to cluster analysis, but it removes the assumption that the data are ultrametric. In practical terms, it does not assume that all lineages have diverged equal amounts. However, it does assume that the data come close to fitting an additive tree, so correction for superimposed substitutions is important for data that might include lineage-to-lineage differences in average rate. In contrast to cluster analysis, neighbor-joining keeps track of nodes on the tree rather than taxa or clusters of taxa. The raw data are provided as a matrix of distances between all pairs of taxa (terminal nodes) on the tree to be inferred. A modified distance matrix is constructed in which the separation between each pair of nodes is adjusted on the basis of their average divergence from all other nodes (conceptually, this adjustment has the effect of normalizing the divergence of each taxon for its average clock rate). The tree is constructed by linking the least distant pair of nodes as defined by this modified matrix. When two nodes are linked, their common ancestral node is added to the tree and the terminal nodes with their respective branches are removed from the tree. This pruning process converts the newly added common ancestor into a terminal node on a tree of reduced size. At each stage in the process, two terminal nodes are replaced by one new node (corresponding to an internal node on the final tree). The process is complete when two nodes remain, separated by a single branch.

The steps of the method (modified from Studier and Keppler, 1988) are as follows:

1. Given a matrix of pairwise distances (d), for each terminal node i calculate its net divergence (r_i) from all other taxa using the formula

$$r_i = \sum_{k=1}^{N} d_{ik} \qquad (11)$$

where N is the number of terminal nodes in the current matrix. Note the assumption that $d_{ii} = 0$, otherwise the summation would need to skip over $k = i$.

2. Create a rate-corrected distance matrix (M) in which the elements are defined by

$$M_{ij} = d_{ij} - (r_i + r_j)/(N - 2) \qquad (12)$$

for all i and with $j > i$ (the matrix is symmetrical, and the case of $i = j$ is not interesting). Only the values i and j for which M_{ij} is minimum need be recorded; saving of the entire matrix is unnecessary.

3. Define a new node u whose three branches join nodes i, j, and the rest of the tree. Define the lengths of the tree branches from u to i and j:

$$s_{iu} = d_{ij}/2 + (r_i - r_j)/[2(N - 2)]$$
$$s_{ju} = d_{ij} - s_{iu}$$

4. Define the distance from u to each other terminal node (for all $k \neq i$ or j)

$$d_{ku} = (d_{ik} + d_{jk} - d_{ij})/2 \qquad (13)$$

5. Remove distances to nodes i and j from the data matrix, and decrease N by 1.

6. If more than two nodes remain, go back to step 1. Otherwise, the tree is fully defined except for the length of the branch joining the two remaining nodes (i and j). Let this remaining branch be

$$s_{ij} = d_{ij}$$

Each step has generated one internal node and has estimated the lengths of two of the branches connected to that node. The tree can be drawn from these data.

An example of using neighbor-joining to infer a tree of five taxa is given in Figure 6. The data are the same as in the cluster analysis example in Figure 4. The pairwise distance estimates are in the upper-right triangle of each matrix (ignoring the last two columns). The distance matrix row totals

	Bsu	Bst	Lvi	Amo	Mlu	R	R/3
Bsu	—	0.1715	0.2147	0.3091	0.2326	0.9279	0.3093
Bst	−0.4766	—	0.2991	0.3399	0.2058	1.0163	0.3388
Lvi	−0.4905	−0.4356	—	**0.2795**	0.3943	1.1876	0.3959
Amo	−0.4527	−0.4514	**−0.5689**	—	0.4289	1.3574	0.4525
Mlu	−0.4972	−0.5535	−0.4221	−0.4441	—	1.2616	0.4205

Lvi to node 1 distance = 0.2795/2 + (0.3959 − 0.4525)/2 = 0.1114
Amo to node 1 distance = 0.2795 − 0.1114 = 0.1681

	Bsu	Bst	Mlu	node 1	R	R/2
Bsu	—	0.1715	0.2326	**0.1222**	0.5263	0.2631
Bst	−0.3701	—	0.2058	0.1798	0.5571	0.2785
Mlu	−0.3856	−0.4278	—	0.2719	0.7103	0.3551
node 1	**−0.4278**	−0.3856	−0.3701	—	0.5739	0.2869

Bsu to node 2 distance = 0.1222/2 + (0.2631 − 0.2869)/2 = 0.0492
node 1 to node 2 distance = 0.1222 − 0.0492 = 0.0730

	Bst	Mlu	node 2	R	R/1
Bst	—	0.2058	**0.1146**	0.3204	0.3204
Mlu	−0.5116	—	0.1912	0.3970	0.3970
node 2	**−0.5116**	−0.5116	—	0.3058	0.3058

Bst to node 3 distance = 0.1146/2 + (0.3204 − 0.3058)/2 = 0.0646
node 2 to node 3 distance = 0.1146 − 0.0646 = 0.0500

	Mlu	node 3
Mlu	—	0.1412
node 3		—

Mlu to node 3 distance = 0.1412

FIGURE 6. Neighbor-joining of 5S rRNA evolutionary distance estimates. The data and abbreviations are as in Figure 4. Each table presents the pairwise distance values input to the round of analysis (upper-right half of the matrix). The rightmost two columns present the row totals for the uncorrected distances [the row being defined based on the full symmetrical matrix; see Equation (11)] and the total divided by the number of terminal nodes minus two. The rate-corrected pairwise distances as defined by Equation (12) are given in the lower-left half of the matrix. The minimum corrected distance value in each table and the corresponding uncorrected pairwise distance are shown in boldfaces. The corresponding pair of taxa (or clusters) is removed from the matrix and replaced by its common ancestral node in the next table and distances based on Equation (13). The inferred tree is diagrammed in Figure 5.

[r from Equation (11)] and $r/(N - 2)$ are given in the last two columns. The rate-corrected distances are in the lower-left triangle of the table. For example, the corrected Bsu to Bst distance is $0.1715 - (0.3093 + 0.3388) = -0.4766$. A general property of these corrected distances is that they are negative; therefore, finding the minimum distance means finding the most negative value. In the first table the minimum value is the -0.5689 relating Amo and Lvi. Both this value and the corresponding uncorrected distance, 0.2795, are in boldface. Thus, Amo and Lvi are joined to one another and to the rest of the taxa through a new node, called "node 1" in this example. The two lines below the table illustrate the calculation of the branch lengths from the two taxa to the node. Amo and Lvi are then removed from the distance table, and the distances from node 1 to the remaining taxa are calculated using Equation (13). For example, the Bsu to node 1 distance is $(0.2147 + 0.3091 - 0.2795)/2 = 0.1222$. The second table, which now relates only four terminal nodes, is treated just as the first table. Looking at the corrected distances, we find two pairs with the lowest value, -0.4278. This is not a coincidence: if Bsu and node 1 are sister nodes, then Bst and Mlu must also be sister groups (if this observation is unclear, try drawing the unrooted tree of four taxa). The remaining arithmetic will yield identical trees regardless of which of these two pairs is joined at this step. In this example, node 2 is added to the tree, joining Bsu, node 1, and the rest of the tree. The branch lengths from Bsu and node 1 to node 2 are calculated below the table. The third table eliminates Bsu and node 1, and adds node 2. In this table, which relates three peripheral nodes, all three rate-corrected distances are identical. As in the previous step, this result is not a coincidence: only one possible unrooted tree can link three taxa. The choice of the pair to be joined is arbitrary; the ultimate outcome will be the same. Adding node 3 to the tree so that it links Bst and node 2 to the rest of the tree (which is only Mlu at this point) gives one more pair of branch lengths and a "tree" containing node 3 and Mlu. Their pairwise distance is used directly as the length of the segment joining them. The tree is completed. The results are shown in Figure 5b.

As the neighbor-joining algorithm seeks to represent the data by an additive tree, it can assign a negative length to a branch. Although negative branch lengths do not appear to be common, the definition of an explicit treatment for this situation would be desirable. However, none of the publications on the method appears to address this issue.

Optimality Criteria in General.   As pointed out in the introductory section of this chapter, there is a conceptual difference between defining an objective and then finding a method to reach that objective versus defining the method (algorithm) *as* the objective. In the first case, the objective is usually related to a (more or less) concrete set of assumptions. In the latter case, the assumptions are rarely evident. For example, when a method is to be used with distances that are not additive, knowing that it will give the correct answer with additive distances is of absolutely no use. The assumption of additivity is not satisfied; therefore, we know nothing about the behavior of the algorithm—it might work very well, but it might function abysmally. The purpose of defining a concrete optimality criterion (objective function) is to describe the behavior of the technique when presented with nonideal data. All of the additive tree techniques discussed will give the same tree when they analyze additive data, but their behavior with other data can be significantly different.

As we pointed out earlier, a complete record of all genetic events would constitute a set of additive distances. Therefore, we will treat the experimentally derived distances as approximations of this ideal. To emphasize the uncertainty in the values, we will call them **distance estimates.** We can now address the situation from the following conceptual perspective: we have uncertain data that we want to fit to a particular mathematical model (an additive tree) and find the optimal value for the adjustable parameters (the branching pattern and the branch lengths).

Fitch–Margoliash and Related Methods.   Given a set of pairwise distance estimates, a large set of techniques is available for fitting these to an additive tree. What is required is a concrete definition of the net disagreement between the tree and the data, i.e., an objective function to be minimized. We will consider criteria of the form

$$E = \sum_{i=1}^{T-1} \sum_{j=i+1}^{T} w_{ij} \left| d_{ij} - p_{ij} \right|^{\alpha} \tag{14}$$

where E defines the error of fitting the distance estimates to the tree, T is the number of taxa, w_{ij} is the weight applied to the separation of taxa i and j, d_{ij} is the pairwise distance estimate, p_{ij} is the length of path connecting i and j in the given tree, the vertical bars represent the absolute value, and α is 1 or 2. A value of α and a weighting scheme must be chosen.

Setting α to 2 represents a weighted least-squares criterion; the weighted square deviation of the tree path lengths from the distance estimates will be minimized. If α is 1, then the weighted absolute differences will be minimized. If the errors in the distance estimates are distributed uniformly across the data, then the least-squares criterion is preferred. If some estimates are apt to be particularly bad, there are two considerations. First, if the identities of the least certain estimates are known, this knowledge can be accommodated in the least-squares method by assigning particularly low weights to these uncertain values. If, however, it is not known a priori which estimates are apt to be erroneous, then using the minimum absolute deviations will reduce the overall perturbation caused by spurious data values. This last condition might pertain to direct experimental determinations of the distance data, a situation in which unrecognized experimental artifacts could substantially flaw some values.

Four reasonably common weighting schemes will be outlined:

$$w_{ij} = 1 \tag{15a}$$
$$w_{ij} = d_{ij}^{-1} \tag{15b}$$
$$w_{ij} = d_{ij}^{-2} \tag{15c}$$
$$w_{ij} = \sigma_{ij}^{-2} \tag{15d}$$

where σ_{ij}^{2} is the expected variance of measurements of d_{ij}. If there is a rational method for estimating σ_{ij}^{2}, then Equation (15d) is preferred. The first three equations amount to implicit assumptions about the uncertainty of the measurements: Equation (15a) (Cavalli-Sforza and Edwards, 1967) assumes that all distance estimates are subject to the same magnitude of error, Equation (15c) (Fitch and Margoliash, 1967) assumes that the estimates are uncertain by the same percentage, and Equation (15b) could be viewed as a compromise that assumes that the uncertainties are proportional to the square roots of the values. Note that missing data are correctly handled by setting the corresponding weight to zero; that is, if d_{ij} is unknown, then set $w_{ij} = 0$.

Providing generalizations about the expected values of σ_{ij}^{2} for experimentally derived data (as opposed to sequences) is difficult. One generally useful technique is to estimate random errors by comparing replicate experiments. Another valuable technique is to use reciprocal comparisons (where appropriate to the experimental protocol; see Chapters 5 and 7) to look into possible biases. These concepts are also discussed in the corresponding experimental chapters.

In the case of sequence comparisons, virtually every distance estimation method includes a model for estimating statistical errors. In general, each model comes down to considering a given sequence to be a random sample of an infinite genome and hence subject to binomial counting error. This process also requires assuming that changes at every site in the molecules are independent. Although this assumption is difficult to evaluate in the case of proteins or mRNAs, it is obviously violated by RNAs with a stable secondary structure (e.g., tRNAs and rRNAs); therefore, statistical errors are likely to be underestimated. However, the underestimation will tend to be the same for all pairs and hence not change the result. More significantly, the paired (i.e., covarying) residues should be given less weight in the calculation of distance estimates since they provide redundant samples of the divergence. All other things being equal, each half of a base pair should be given one-half the weight of an unpaired residue in the distance calculations [see discussion associated with Equation (4)]. If weights are used in calculating the similarities, then the same weights should be applied to counting the sequence positions for use in variance estimates (below). A danger of these statistical estimates of distance uncertainties is that they provide no insights into potential systematic errors and may lead to a false sense of security (systematic errors are discussed in a later section).

The expected variance of a distance estimated with Equation (5b) is

$$\sigma_d^2 = \frac{D(1 - D)}{(1 - D/b)^2 L} \tag{16}$$

(Kimura and Ohta, 1972; Tajima and Nei, 1984), where the symbols and assumptions are as for Equation (5), and an assumption that changes at each position are independent is added. Similarly, Kimura (1980) has shown that the variance of distance estimates based on Equation (7) are expected to be

$$\sigma_d^2 = [c_1^2 P + c_2^2 Q - (c_1 P + c_2 Q)^2]/N \tag{17a}$$

$$c_1 = 1/(1 - 2P - Q) \tag{17b}$$

$$c_2 = [c_1 + 1/(1 - 2Q)]/2 \tag{17c}$$

A very important property of Equations (16) and (17) is that they explicitly state the dependence of uncertainty on the amount of data; the variance is inversely proportional to the sequence length. Consequently, so long as quality (primarily, but not entirely, defined by certainty of positional homology) is not sacrificed, the more data the better. One additional property of these formulas requires comment. If two sequences are identical, the estimated uncertainty will be zero—a questionable conclusion. A practical treatment is to assume that the minimum measurable dissimilarity is one-half of a substitution, yielding $1/(2L^2)$, or $1/(2N^2)$, as a minimum value to be imposed on the estimated variance.

As was noted in the discussion of algorithms and optimality criteria, finding the phylogeny with the lowest value of E, as defined in Equation (14), generally requires two components: optimizing the branch lengths to find the smallest value of E consistent with a given tree topology, and finding the topology with the lowest E of all trees [an interesting alternative proposed by De Soete (1983a,b) combines these into a single step, but this method will not be discussed here]. The following discussion will deal with the first step; the latter issue is common to character-based trees and will be considered in a later section.

To find the branch lengths that minimize E in Equation (14), the equation must be recast explicitly in terms of the branch lengths. For an unrooted tree of T taxa, there are $2T - 3$ independent branches that define the p_{ij} values. To represent mathematically the relationships between the branch lengths, s_k, and the path lengths between pairs of taxa, we need an appropriate representation of the tree topology. Let **A** be a matrix such that the element $A_{(ij)k}$ is equal to 1 if the branch k is part of the path connecting taxon i to taxon j; otherwise $A_{(ij)k}$ is equal to 0. With this definition it follows that

$$p_{ij} = \sum_{k=1}^{2T-3} A_{(ij)k} s_k$$

Substituting into Equation (14) gives

$$E = \sum_{i=1}^{T-1} \sum_{j=i+1}^{T} \left| d_{ij} - \sum_{k=1}^{2T-3} (A_{(ij)k} s_k) \right|^\alpha \qquad (18)$$

Thus, after a weighting function, a value of α, and a tree topology have been chosen, the only undetermined values are the branch lengths. We will not provide a detailed method for minimizing Equation (18) since no issues of biological relevance are involved and the computations must be carried out by computer. Generally, the s_k values that minimize E can be found by linear or quadratic programming, by iterative successive refinement techniques, or—when $\alpha = 2$—by solving a set of simultaneous equations using ordinary linear algebra (e.g., Kidd and Sgaramella-Zonta, 1971; Olsen, 1988). With the first two procedures, the branch lengths can be constrained to be nonnegative (negative-length branches do not correspond to any meaningful biological process and are purely mathematical entities). When using linear algebra, adjustments must be made for negative values after the optimal branch lengths are found but before E is evaluated. A variety of treatments have been proposed for dealing with trees that produce negative-length branches. Allowing branches to have negative values when E is determined does not work because some far from optimal trees can use negative values to produce a low apparent error. Some investigators have chosen to discard any topology that leads to a negative optimal value for any branch. This option sometimes works, but at the risk of throwing out the correct tree in certain realistic situations. The best way to treat negative branches is to set them to zero and then reoptimize the remaining branches, allowing them to compensate for the abolition of negative values. This approach is tedious and would be more easily accomplished by using a different method for optimizing branch lengths in the first place. A compromise approach is to set any negative branches to zero and then calculate E without readjusting the other branches. This method gives exact values of E for trees with no negative branches and overestimates the value of E for trees that contain negative branches. The amount of overestimation is small as long as there were no large negative branch lengths.

Table 1 summarizes the results of a least-squares calculation for a tree of five rRNA sequences. The table presents the pairwise distance estimates with their expected uncertainties, the corresponding path lengths through the inferred tree, and the error contributed by each distance to the overall

Table 1.   Optimal 5S rRNA tree by weighted least-squares criterion[a]

Sequence pair[b]	Estimated distance[c]	Expected distance[d]	Distance difference[e]	Expected uncertainty[f]	Error contribution[g]
Bsu–Bst	0.1717	0.1655	0.0062	0.0522	0.00133
Bsu–Lvi	0.2147	0.2269	−0.0122	0.0600	0.00415
Bsu–Amo	0.3091	0.2895	0.0196	0.0758	0.00667
Bsu–Mlu	0.2326	0.2414	−0.0088	0.0630	0.00194
Bst–Lvi	0.2991	0.2958	0.0033	0.0743	0.00020
Bst–Amo	0.3399	0.3584	−0.0185	0.0809	0.00521
Bst–Mlu	0.2058	0.2058	0.0000	0.0584	0.00000
Lvi–Amo	0.2795	0.2795	0.0000	0.0708	0.00000
Lvi–Mlu	0.3943	0.3716	0.0227	0.0902	0.00633
Amo–Mlu	0.4289	0.4343	−0.0054	0.0906	0.00031

[a]Data are from Olsen (1988). The corresponding tree is illustrated in Figure 5.
[b]Abbreviations are as in Figure 4.
[c]Distance estimate from sequence comparisons, using Equations (4) and (5), with $b = 3/4$.
[d]Sum of appropriate branch lengths along the path joining the taxa in the inferred tree.
[e]Difference of the two previous columns.
[f]Square root of the variance estimate from Equation (16).
[g]The individual terms of the summation in Equation (14), with $\alpha = 2$ and $w_{ij} = \sigma_{ij}^{-2}$.

value of E. As expected for a least-squares methodology, the paths through the best fitting tree will sometimes exceed the corresponding distance estimates (e.g., Bsu to Lvi) and sometimes they will be shorter (e.g., Bsu to Bst). It might be noticed that two distances are fitted exactly. Tree branch lengths assigned by least-squares, minimum length tree, or neighbor-joining will exactly reproduce the distances between sister taxa in a tree (so long as negative numbers are not involved). The inferred tree is shown in Figure 5. The tree topologies inferred by neighbor-joining and least-squares are identical, and the branch lengths are very similar; however, such close agreement is not always the case. This level of agreement contrasts sharply with the very different result given by cluster analysis. In summary, there is little reason to use cluster analysis and the potential pitfalls are great.

The motivation for the least-squares and minimum absolute deviation approaches implicitly assumes that each pairwise distance measurement is independent. Because of the common evolutionary history of the molecules in question, this assumption is not generally true. The primary consequence of violating this assumption is purely statistical; trees will be less well resolved than they would be if the samples were in fact independent. However, a second consequence is that any systematic errors in the distance estimates can also be repeatedly sampled and thus the pairwise methods are potentially more sensitive to undercompensation for homoplasy in the data (see Systematic Errors, below). Felsenstein (1986, 1988) has discussed methods for dealing with interdependencies in pairwise distance data. In practice, none of these methods is used because of their computational complexity and other limitations. Note that neither parsimony nor maximum likelihood suffers from this difficulty.

Distance Wagner and Related Methods. The conceptual perspective of Fitch–Margoliash methods and neighbor-joining is that the estimated pairwise distances are to be fit to an additive tree, with some of the estimates (observations) being greater than the true values and some of them being smaller than the true values. An alternative view is one in which the sequence (or other) differences are not corrected for superimposed changes and thus provide lower bounds for the actual evolutionary distance. In this framework, the length of the path connecting any pair of taxa must equal or exceed the corresponding observed distance. In analogy to character-based parsimony, the desired tree is the one that minimizes the total of all branch lengths in the tree, while using the pairwise distances as lower bounds on the paths. Beyer et al. (1974) and Waterman et al. (1977) have described exact methods for accomplishing the desired minimization on a given tree. Farris's (1972) distance Wagner algorithm is an effective heuristic. Modifications to the distance Wagner procedure have subsequently been proposed by Swofford (1981) and Tateno et al. (1982). As with neighbor-joining, if the experimentally determined distances are additive, then the optimal solution will always be found. However, when the fit is not exact, the behavior is not intuitively obvious.

Parsimony Methods

Of the existing numerical approaches to inferring phylogenies directly from character data, methods based on the principle of **maximum parsimony** have been the most widely used by far. Most biologists are familiar with the usual notion of parsimony in science, which essentially maintains that simpler hypotheses are preferable to more complicated ones and that ad hoc hypotheses should be avoided whenever possible. Methods for estimating trees under the criterion of parsimony equate "simplicity" with the explanation of attributes shared among taxa as due to their inheritance from a common ancestor (e.g., Sober, 1989). When character conflicts occur, however, ad hoc hypotheses cannot be avoided if the observed character distributions are to be explained, and assumptions of **homoplasy** (convergence, parallelism, or reversal) must be invoked.

In general, parsimony methods for inferring phylogenies operate by selecting trees that minimize the total **tree length**: the number of evolutionary "steps" (transformations from one character state to another) required to explain a given set of data. For example, the steps might be base substitutions for nucleotide sequence data or gain and loss events for restriction-site data. Obviously, a tree that minimizes the total number of steps also minimizes the number of "extra" steps (homoplasies) needed to explain the data. Thus, a close connection is established between numerical parsimony methods and the methods traditionally used in nonnumerical "cladistics," which also resolve conflicts by striving to minimize ad hoc hypotheses of homoplasy.

In slightly more mathematical terminology, we can define the general maximum parsimony problem as the following. From the set of all possible trees, find all trees τ such that

$$L(\tau) = \sum_{k=1}^{B} \sum_{j=1}^{N} w_j \cdot \operatorname{diff}(x_{k'j}, x_{k''j}) \tag{19}$$

is minimal, where B is the number of branches, N is the number of characters, k' and k'' are the two nodes incident to each branch k, $x_{k'j}$ and $x_{k''j}$ represent either elements of the input data matrix or optimal character-state assignments made to internal nodes, and $\operatorname{diff}(y,z)$ is a function specifying the cost of a transformation from state y to state z along any branch. The coefficient w_j assigns a weight to each character; it is typically set to 1, but this need not be the case. Note also that $\operatorname{diff}(y,z)$ need not equal $\operatorname{diff}(z,y)$, although for methods that yield unrooted trees, $\operatorname{diff}(y,z) = \operatorname{diff}(z,y)$. As discussed below, the definition of "optimal character-state assignments" may include restrictions on the nature of permissible character-state changes.

Any discussion of parsimony methods must distinguish between the optimality criterion (minimal tree length under a specified set of restrictions on permissible character-state changes) and the actual algorithm used to search for optimal trees. Unfortunately, early descriptions of parsimony methods (e.g., Farris, 1970) tended to be presented in a way that obscured the boundaries between criteria and algorithms. Such descriptions promote bad habits for biologists attempting to understand the methodology: they sometimes become so mired in algorithmic details that they lose track of the underlying biological principles and assumptions (Felsenstein, 1982). Algorithms tend to have short life spans—better ones are constantly being invented. Consequently, students are often taught algorithms that have long been abandoned rather than instilled with an appreciation for what the algorithms are trying to accomplish. For example, Farris's (1970) *algorithm* for estimating minimum-length trees under the Wagner parsimony criterion is not, to our knowledge, used in any modern, widely used parsimony computer program (e.g., Farris's Hennig86, Felsenstein's PHYLIP-MIX, or Swofford's PAUP), but his *criterion* forms the basis for all of them. Yet, the algorithm is often presented as if it were "state of the art" (e.g., Wiley, 1981). Although perhaps not a grievous error, such descriptions often lead students and researchers alike, when asked to describe "Wagner parsimony," to recite the sequence of steps outlined in Farris's paper rather than to describe exactly what quantities are being minimized.

For the reasons just outlined, the conceptual framework in which we will discuss parsimony methods assumes that the (difficult) problem of finding optimal trees is not at issue; we assume, for the moment, that every possible tree can be evaluated, optimizing each one according to our chosen criterion and ranking them according to that criterion. We will take up the matter of searching for optimal trees in a subsequent section.

A common misconception regarding the use of parsimony methods is that they require a priori determination of character polarities (see above). In morphologically based studies, character polarity is often inferred using the method of **outgroup comparison,** and the resulting "polarized" characters form the basis of the analysis. Furthermore, since a "hypothetical ancestor" is implied by the polarity assignments, the output of an analysis of polarized characters is a rooted tree. Although specification of polarities provides a sufficient basis for obtaining rooted (rather than unrooted) trees, it is by no means prerequisite to the use of "cladistic analysis" or parsimony methods. This circumstance is fortunate, since the estimation of character polarity is both more difficult and less meaningful for most kinds of molecular data. All that is required to obtain rooted trees from parsimony analysis is to include in the data set one or more assumed outgroup taxa. The location at which the outgroup joins the unrooted tree implies a root with respect to the ingroup. We emphasize, however, that the assignment of taxa to the outgroup constitutes an assumption that the remaining taxa (the **ingroup** taxa) are **monophyletic** (hopefully justified by evidence extrinsic to the data at hand); if this assumption is wrong, the tree will be rooted incorrectly.

Parsimony analysis actually comprises a group of related methods, united by the goal of minimizing some evolutionarily significant quantity but differing in their underlying evolutionary models. We will now address each of these methods in turn. (The methods are presented in a logical progression rather than in chronological order of their introduction into the literature.) In describing the procedures used to compute the minimum length required by a tree under a particular optimality criterion, we will consider a single character (position) in isolation from the rest. Because of the assumption of independence among characters, we can compute the overall tree length by summing, over all characters, the lengths required by each individual character. For the simplest procedures (Wagner and Fitch parsimony), we provide pencil-and-paper algorithms for computing tree lengths and determining optimal character-state assignments. Again, we are concerned only with calculating the length of a single tree, which is taken as a "given"; this tree may not be a most parsimonious arrangement for our example character (or even over all characters); it is simply a tree that we wish to evaluate.

Fitch and Wagner Parsimony   These are the simplest parsimony methods, imposing no (Fitch) or minimal (Wagner) constraints on permissible character-state changes. The Wagner method, formalized by Kluge and Farris (1969) and Farris (1970), assumes that characters are measured on an interval scale. Thus, it is appropriate for binary, ordered multistate, and continuous characters. Fitch (1971b) generalized the method to allow un-ordered multistate characters (e.g., nucleotide and protein sequences). Wagner parsimony assumes that any transformation from one character state to another also implies a transformation through any intervening states, as defined by the ordering relationship. Fitch parsimony allows any state to transform directly to any other state. Both methods permit free **reversibility;** that is, change of character states in either direction is assumed to be equally probable, and character states may transform from one state to another and back again. A consequence of reversibility is that the tree may be rooted at any point with no change in the tree length.

To determine the minimum length required by a given character j under either the Wagner or Fitch criteria, only a single pass over the tree is required, proceeding from the tips toward the root. Computer scientists call this pass a postorder traversal. Although the computation can be performed in other ways, we recommend rooting the tree at one of the terminal taxa, denoted r, as shown in Figure 7. The algorithm for computing the length of a strictly bifurcating tree under the Wagner parsimony criterion then proceeds as follows (see Swofford and Maddison, 1987, for a more rigorous presentation):

1. To each terminal node i (including the one at the root), assign a **state set** S_i containing the character state assigned to the corresponding taxon in the input data matrix (= x_{ij}). Initialize the tree length to zero.
2. Visit an internal node k for which a state set S_k has not been defined but for which the state sets of k's two immediate descendants has been defined. Let i and j represent k's two immediate descendants. Assign to k a state set S_k according to the following rules:
   a. If the intersection of the state sets assigned to nodes i and j is non-empty ($S_i \cap S_j \neq \emptyset$), let k's state set equal this intersection (i.e., $S_k = S_i \cap S_j$). The intersection can be represented as a closed interval $[a_k, b_k]$.
   b. Otherwise ($S_i \cap S_j = \emptyset$), let k's state set equal the smallest closed interval $[a_k, b_k]$ containing an element from each of the state sets assigned to i and j. Increase the tree length by $b_k - a_k$.
3. If node k is located at the basal fork of the tree (i.e., the immediate descendant of the terminal node placed at the root), the traversal has been completed; proceed to step 4. Otherwise, return to step 2.
4. If the state assigned to the terminal node at the root of the tree (x_r) is not contained in the state set just assigned to the node at the basal fork of the tree (S_k), increase the tree length by the distance from x_r to S_k. (This distance equals $a_k - x_r$ if $x_r < a_k$ or $x_r - a_k$ if $x_r > b_k$.)

An application of the above algorithm is presented in Figure 7. We wish to compute the length of the unrooted tree of Figure 7a. (Although the more usual situation for molecular data would involve binary rather than multistate characters, we treat the multistate case to demonstrate the generality of the algorithm. Binary characters are simply a special case.) We first reroot the tree at node A (although we could have chosen any node), yielding the rooted tree shown in Figure 7b. Also shown in Figure 7b are the state sets assigned to the terminal nodes according to step 1 of the algorithm. Visiting internal node X in the first invocation of step 2, we observe that $S_B \cap S_C = \{0\} \cap \{2\} = \emptyset$, and hence assign the interval [0,2] to S_X, adding $2 - 0 = 2$ to the tree length. Similarly, we let $S_Y = [1,3]$ in the second invocation of step 2, and add $3 - 1 = 2$ to the length, which is now 4. In the third and final invocation of step 2, we observe that the intersection $S_X \cap S_Y = [0,2] \cap [1,3]$ is not empty, and therefore assign the interval [1,2] to S_Z. The situation as we arrive at step 4 is shown in Figure 7c. Since $x_r = 0$ is not an element of $S_Z = [1,2]$, we add an additional $1 - 0 = 1$ to the length. Thus, evolution of this character requires a minimum of five steps on our given tree.

The procedure outlined above is sufficient to obtain the minimal length required by any character on a given tree; however, it does not actually assign optimal character states to the hypothetical ancestors (internal nodes) of the tree to yield a **most parsimonious reconstruction (MPR)**. To obtain such a reconstruction we can make a second pass over the tree, this time proceeding from the root toward the tips (a preorder traversal):

5. Visit an internal node k for which an optimal state assignment x_k has not yet been made but for which such an assignment has been made to k's immediate ancestor, denoted m. (Note that the first time this step is invoked, k corresponds to the node at the basal fork of the tree and $m = r$, the terminal taxon at the root of the tree.)
6. Assign to k the state from the state set computed in the first pass, S_k ($= [a_k, b_k]$), that is closest to x_m. Specifically, if x_m is contained in S_k, we let $x_k = x_m$. Otherwise, we let $x_k = a_k$ if $x_m < a_k$ or $x_k = b_k$ if $x_m > b_k$.
7. If all internal nodes have been visited, stop. Otherwise return to step 1.

Applying steps 5–7 to the example of Figure 7, we first assign state 1 (the closest state in [1,2] to 0) to node Z. We then assign state 1 (the closest state in [0,2] to 1) to node X; likewise we assign state 1 (the closest state in [1,3] to 1) to node Y. The resulting reconstruction is shown in Figure 7d, and confirms the value of 5 as the minimum length for this character.

It is important to remember that this method finds only a single MPR, although others may exist. For instance, the reconstruction in Figure 7e also requires five steps. Swofford and Maddison (1987) described an exact algorithm for obtaining all MPRs for discrete character data under the Wagner parsimony criterion.

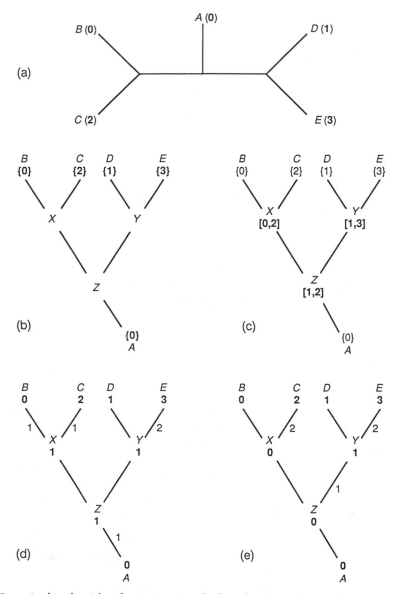

FIGURE 7. Steps in the algorithm for computing the length of an ordered character under Wagner parsimony. (a) The unrooted tree and character states. (b) Tree obtained by rooting at terminal node "A" and initial state sets assigned to terminal nodes. (c) State sets computed for interior node (boldface). (d) Reconstruction obtained according to the algorithm described in the text. (e) An alternative, equally parsimonious reconstruction.

Simple modifications of the above algorithm provide for the treatment of multistate unordered characters (e.g., nucleotide sequence positions) under the Fitch (1971b) parsimony criterion. In the initial pass (computation of state sets and tree lengths), modify steps 2 and 4 as follows:

2a'. If the intersection of the state sets assigned to nodes i and j is nonempty $(S_i \cap S_j \neq \emptyset)$, let k's state set equal this intersection (i.e., $S_k = S_i \cap S_j$).

2b'. Otherwise $(S_i \cap S_j = \emptyset)$, let k's state set equal the union of the state sets assigned to nodes i and j $(S_i \cup S_j)$, and increase the tree length by 1.

4'. If the state assigned to the terminal node at the root of the tree (x_r) is not contained in the state set just assigned to the node at the basal fork of the tree (S_k), increase the tree length by 1.

In order to obtain an MPR, modify step 6 above as follows:

6'. If x_m is contained in the state set assigned to k in the first pass (S_k), assign this state to k as well. Otherwise, arbitrarily assign any state from S_k to k.

An example of the application of the above algorithm is shown in Figure 8. We are interested in computing the length required by a single character on the unrooted tree of Figure 8b. As before, we reroot the tree arbitrarily at node A, yielding the tree shown in Figure 8b. The state sets assigned to the terminal nodes are indicated on the figure. Visiting node X in the first invocation of step 2', we see that $\{A\} \cap \{C\} = \emptyset$, and hence assign the union $\{A,C\}$ as the state set S_X, and set the tree length for this character to 1. Moving to node Y, we assign $\{A,C\} \cap \{A\} = \{A\}$ to S_Y. Finally, since $\{A\} \cap \{G\} = \emptyset$, we assign the state set $\{A,G\}$ to node Z, again adding 1 to the tree length. Thus, at the beginning of step 4', the state sets are as indicated in Figure 8c. Since $x_r = C$ is not an element of $S_Z = \{A,G\}$, we add an additional step to the length, so that a total of three steps (nucleotide substitutions) is required on this tree.

If we wish to obtain one of the MPRs, we observe that the state C taken by the terminal taxon at the root of the tree is not contained in the set $\{A,G\}$ assigned to the node at the first fork, and we may arbitrarily choose to assign state A to this node. We then assign state A to node Y as well (since the state set was singleton no decision need be made). Finally, since state A is contained in node X's state set $\{A,C\}$, we assign it to the node, yielding the reconstruction shown in Figure 8d.

As was the case for the ordered character example, more than one MPR exists. For example, if we had chosen to assign state G rather than state A to node Z, we would have obtained the reconstruction shown in Figure 8e. Still another MPR exists, however, in which state C is assigned to all three internal nodes. That C was a possible state for node Z was not readily

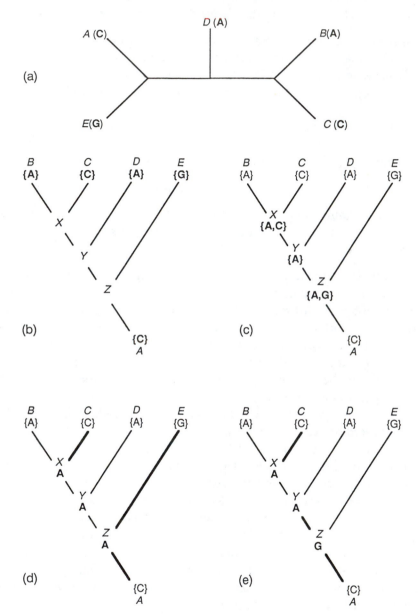

FIGURE 8. Steps in the algorithm for computing the length of an unordered character under Fitch parsimony. (a) The unrooted tree and character states. (b) Tree obtained by rooting at terminal node "A" and initial state sets assigned to terminal nodes. (c) States sets computed for interior nodes (boldface). (d) Reconstruction obtained according to the algorithm described in the text. (e) An alternative, equally parsimonious reconstruction.

apparent from the state set {A,G} originally assigned to that node. In fact, a second pass over the tree is necessary to obtain all of the possible state assignments to each interior node. Fitch (1971b) described one such method and gave an algorithm for enumerating all of the possible MPRs.

Although all the algorithms described above are restricted to strictly bifurcating trees, they easily can be modified to handle multifurcations (polytomies). Maddison (1989) reviews algorithms for obtaining MPRs on polytomous trees under a variety of evolutionary models, including the introduction of some novel approaches.

Dollo Parsimony   The Wagner and Fitch parsimony criteria are appropriate under the assumption that probabilities of character change are symmetrical (i.e., the probability of a transformation from state 0 to state 1 in some small unit of evolutionary time is equivalent to that of a change from state 1 to state 0). As discussed above, this assumption is probably unreasonable for restriction-site characters, since it is much "easier" to lose an existing restriction site than to gain a site at any particular location.

Because of this asymmetry, DeBry and Slade (1985) and others have suggested that the Dollo parsimony model (Farris, 1977) is more appropriate for restriction-site data. The Dollo parsimony criterion can be applied to binary or linearly ordered multistate characters for which we can reasonably hypothesize an ancestral condition (polarity). As for Wagner and Fitch parsimony, the preferred tree is the one requiring the fewest steps, but the character state reconstruction (and hence the tree length assigned) must be consistent with the constraint that every derived character state be *uniquely* derived. If a "hypothetical ancestor" (an artificial taxon to which the assumed ancestral states for each character have been assigned) is included in the analysis, this definition corresponds to the traditional Dollo model (Farris, 1977): each character state is allowed to originate but once on the tree and any required homoplasy takes the form of reversals to a more ancestral condition (i.e., parallel or convergent gains of the derived condition are not allowed). In the context of restriction-site data, each site may be gained once, with as many parallel losses of the site being assumed as necessary to explain the data. For example, for the tree and character states shown in Figure 9 and with state 0 (site absent) assumed to be ancestral, the reconstruction of Figure 9a, requiring only two steps, is not acceptable under the Dollo model because two gains are indicated. Consequently, three steps would be required under the Dollo criterion (Figure 9b): a single gain followed by two losses.

Use of the Dollo parsimony criterion does not require inclusion of a hypothetical ancestor; it can be applied to unrooted trees as well. Stated another way, although the Dollo criterion requires specification of character polarity in a universal sense, it does not require us to know the state occurring in the most recent ancestor of the ingroup taxa. Specifically, the

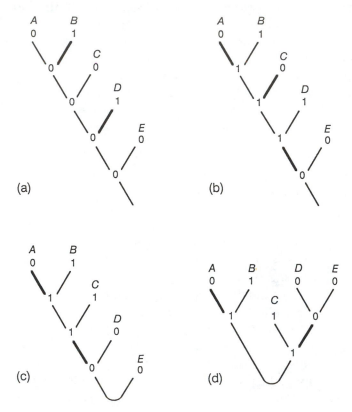

FIGURE 9. Character-state reconstruction demonstrating the Dollo parsimony criterion. Branches on which character-state changes occur are indicated in boldface. (a) Most parsimonious reconstruction if multiple originations of state 1 are allowed. (b) Most parsimonious reconstruction under Dollo parsimony, in which only single origination of state 1 is permitted. (c,d) Reconstruction obtained under unrooted Dollo model. Either rooting of the tree implies a minimum of two character-state changes and only a single origination of state 1.

unrooted Dollo model forces us to assign character states to the interior nodes of the tree such that if a path is traced from any terminal taxon to any other, a backward change (from a more derived state to a more ancestral state) is never followed by a forward change (from a more ancestral state to a more derived state) . Under this definition, the position of the root affects neither the assignment of character states to interior nodes nor the length of the tree. For example, both of the trees shown in Figure 9c and 9d, which differ only in the placement of the root, require two steps under the unrooted Dollo model (assuming that state 1 is the derived state); neither tree requires more than a single origination of state 1. [Note that in the tree of Figure 9d, the derived state (1) is assumed to be ancestral with respect to the group ABCD but derived relative to some more inclusive group.]

The unrooted Dollo approach is particularly convenient for restriction-site characters since it does not require the construction of a hypothetical ancestor, only the inclusion of one or more outgroup taxa. If a site is present in some of the ingroup taxa and in one or more of the outgroup taxa as well, then the most recent common ancestor of the ingroup is assumed to have had the site. The analysis will then seek to minimize the number of losses of the site over the full tree (ingroup and outgroup). If, on the other hand, the site is found only in some of the ingroup taxa but not in the outgroup, then the site is assumed to be ancestrally absent with respect to the ingroup, and a single gain will be postulated at an optimal location within the ingroup. Remember that the specification of "site absent" as the ancestral condition does not imply that the site was absent in the most recent common ancestor of the ingroup taxa, only that the site was absent in some, perhaps quite distant, ancestor.

The drawback to use of Dollo parsimony for restriction-site characters is demonstrated in Figure 10. If, despite its unlikelihood, a particular restriction site does originate independently in two lineages (Figure 10a), then the actual number of evolutionary changes can be drastically overestimated (Figure 10b) due to the strict enforcement of the requirement for unique originations. This pathological behavior may occur more often than the reader might suspect. Suppose one particular position within the restriction site were less constrained than the others, and further suppose that transition substitutions at this position were much more likely to occur than transversions. Then it is easy to imagine that the nucleotide at this position would, on an evolutionary time scale, toggle between the two purines (or pyrimidines). The site would then "blink" on and off, depending on which base was present at any particular point on a lineage. If we permitted only a single origination of the site, the number of losses we would be forced to postulate could become large.

One way to avoid the problem noted above is to adopt a "relaxed" Dollo criterion. For example, we might prefer one gain and two losses to two independent gains, but we might prefer two independent gains to one gain and 10 losses. The "generalized parsimony" method, discussed below, provides a mechanism for implementing a relaxed Dollo model.

Camin–Sokal Parsimony The method of Camin and Sokal (1965) was actually the first discrete-character parsimony approach to be described. It makes the strongest assumption of any of the methods discussed so far, namely that evolution is irreversible.* We mention it here only for the sake of completeness, since it is highly unlikely that the assumption of irreversibility could be justified for any type of molecular data.

---

*Some readers, familiar with "Dollo's Law of Irreversibility," may be confused at this point. The Dollo parsimony model does not assume complete irreversibility, only that a derived character state cannot be lost and then regained. The Camin-Sokal model does not permit a derived character state to return to the ancestral condition.

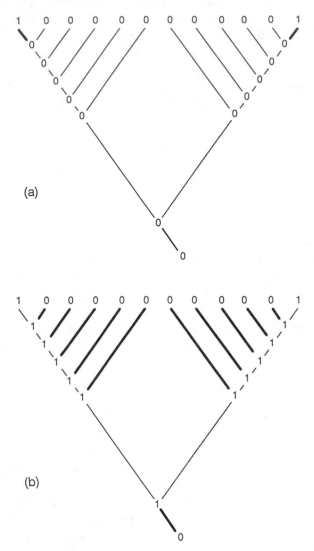

FIGURE 10. Demonstration of problems affecting Dollo parsimony if multiple originations of the derived state actually occur. (a) "True" tree has two steps due to independent derivations of state 1. (b) Reconstruction obtained under Dollo parsimony requires 11 steps (one derivation of state 1 and 10 reversals to the ancestral state 0).

Transversion Parsimony   Several authors (e.g., Brown et al., 1982) have observed that transition substitutions occur more frequently than transversions. For some molecules, it might even be argued that transitions occur so frequently that they quickly degenerate into noise and should therefore be ignored altogether. A simple method for ignoring transitions is to recode the four nucleotides as either R (purine; A or G) or Y (pyrimidine; C or T).

Standard Wagner parsimony may then be applied to the resulting binary-coded matrix.

A disadvantage to the complete rejection of information on transitions is that although transitions may become saturated over long evolutionary distances, they may nonetheless be highly informative with respect to relationships among closely related taxa. One way around the dilemma is to assign greater weight to transversions than transitions, without going so far as to give transitions zero weight, as does transversion parsimony. Generalized parsimony can also be used for this purpose, as outlined below.

Note that some authors (e.g., Lake, 1987) use the term "transversion parsimony" in a different sense than that described here.

Generalized Parsimony All of the above parsimony variants can be subsumed into a generalized method that assigns a "cost" for the transformation of each character state to the other possible states (Sankoff, 1975; but see Sankoff and Cedergren, 1983, for a somewhat more consumer-oriented treatment; note that "generalized parsimony" is our term, not theirs). The costs can be represented as an m-by-m matrix S, where S_{ij} represents the increase in tree length (weight) associated with a transformation from state i to state j, and m is the total number of possible states. Three such weighting matrices, corresponding to the Wagner, Fitch, and Dollo parsimony criteria, are shown in Figure 11a–c. An exact, dynamic programming algorithm can be used to determine the minimum length required on a given tree for any particular choice of costs and to obtain one or all of the MPRs that yield this length (Sankoff and Cedergren, 1983); because of the complexity of this algorithm, we will not attempt to describe it here.

Unfortunately, the generalized parsimony approach is much more computationally expensive than the algorithms described above for certain special cases. Its advantage lies in its generality. For instance, S is not required to be symmetric. Relaxation of this requirement provides a means of implementing a relaxed Dollo criterion: by making the cost of a forward transformation greater than that of a backward transformation, we can prefer single-gain, multiple-loss scenarios until the number of losses becomes great enough that we are willing to allow independent gains. For example, the stepmatrix shown in Figure 11d would prefer one gain and *two* losses over two gains but would prefer two gains over one gain and *four* losses.

Generalized parsimony can also be used to attach greater importance to transversions than to transitions by assigning costs such that changes between two purines or between two pyrimidines receive lower weight than changes from a purine to a pyrimidine or vice versa (e.g., Figure 11e).

Perhaps the most troublesome aspect of generalized parsimony is determining how to choose the costs for different kinds of transformations.

	(a)		(b)		(c)
	a b c d		a b c d		a b c d
a	- 1 2 3	a	- 1 1 1	a	- M 2M 3M
b	1 - 1 2	b	1 - 1 1	b	1 - M 2M
c	2 1 - 1	c	1 1 - 1	c	2 1 - M
d	3 2 1 -	d	1 1 1 -	d	3 2 1 -

	(d)		(e)
	0 1		A C G T
0	- 3	A	- 5 1 5
1	1 -	C	5 - 5 1
		G	1 5 - 5
		T	5 1 5 -

FIGURE 11. Weighting matrices for generalized parsimony. (a) Weighting matrix equivalent to Wagner parsimony (ordered characters). (b) Weighting matrix equivalent to Fitch parsimony (unordered characters). (c) Weighting matrix equivalent to Dollo parsimony. M is an arbitrarily large number, guaranteeing that only one transformation to each derived state will be permitted. (d) Weighting matrix that assigns greater cost to gains ($0 \rightarrow 1$ changes) than to losses ($1 \rightarrow 0$ changes). (e) Weighting matrix that assigns greater weight to transversions than to transitions.

One approach is to assign weights consistent with the researcher's assumptions about the relative frequency of different kinds of events. As a matter of general principle, we disagree with those who argue that a priori weighting of different kinds of changes introduces an unacceptable level of subjectivity into the analysis; an assumption of equal weights is itself a strong assumption. If, for example, we examined an alignment and observed that of 200 variable positions (columns), 80 contained only A and G, 80 contained only C and T, and only 40 contained a mixture of purines and pyrimidines, the conclusion that transitions occur much more frequently than transversions would not be controversial. In this case, a transversion:transition weighting of 1:1 would certainly represent a stronger assumption than a 2:1 weighting. Even if we have no idea how much more frequently transitions occur than transversions, a transversion:transition weight such as a 1.1:1 weighting may be desirable. Suppose that under equal weighting one tree required five homoplastic transversions and three homoplastic transitions, whereas another tree required one homoplastic transversion and seven homoplastic transitions. Whether the "optimal" transversion:transition weighting is 2:1, 3:1, or 20:1, the tree requiring only one "extra" transversion would be preferable and would be chosen as superior under the 1.1:1 weighting. Similar arguments can also be advanced for the use of gain:loss weights other than 1:1 for restriction sites.

An alternative to assuming a particular set of costs based on extrinsic criteria is to estimate the appropriate weights from the data themselves. Williams and Fitch (1989) discuss methods for choosing initial weights and for refining them by iterative improvement. Unfortunately, these methods may be sensitive to the starting point, a frequent drawback to successive approximation methods. Iterative approximation of optimal weights remains an area of active research, and further developments may be expected in the near future.

The methods developed by Sankoff and his colleagues were also designed to construct optimal alignments on a given tree by incorporating insertion–deletion weights (with insertions of gaps as appropriate) in addition to the substitution weights. This strategy is very appealing in that it effectively merges the problems of alignment and tree selection into a single problem; insertions and deletions are treated as events localized to particular branches on the tree to maximize the overall parsimony. The alternative method, construction of a multiple alignment prior to the phylogenetic analysis, is vastly inferior since the topology of the tree cannot not be ignored when deciding where to place gaps.

Unfortunately, rigorous application of Sankoff's method is computationally impractical for more than three sequences and one interior node, an obvious limitation for most systematic applications. However, Sankoff et al. (1976) describe an iterative procedure that rigorously aligns within local regions of a tree (three sequences adjacent to a single interior node), sacrificing the guarantee of global optimality but providing some degree of tractability. Even this procedure is too slow for all but small problems (both in number of taxa and sequence length). Nanney et al. (1989) have described and programmed a more approximate, but much faster, procedure that operates by assuming that lengths of insertions and deletions are sufficiently small to allow alignment within a local "window" rather than obtaining a global alignment for any triplet of sequences. Hein (1989a,b) has very recently developed a sophisticated program that addresses both the problems of alignment on a tree and of searching for optimal trees, using effective heuristics to speed the computations.

Parsimony on Protein Sequences   Because this book does not specifically deal with amino acid sequencing, our discussion of parsimony methods for treating them will be brief. Three general procedures have been used. The first, and simplest, is to minimize the number of amino acid replacements using Fitch parsimony as described above (i.e., each position in the aligned sequences is a multistate unordered character of which the possible states are the 20 possible amino acid residues). This approach, apparently used first by Eck and Dayhoff (1966), ignores the genetic code, failing to consider the minimal number of nucleotide required for the replacement of one amino acid by another (i.e., some replacements require a single nucleotide substitution and others two or even three).

Goodman, Moore, and their colleagues have developed a more sophisticated approach (reviewed by Goodman, 1981) that seeks trees requiring the fewest number of nucleotide substitutions at the mRNA level. They have developed an algorithm that generalizes the Fitch parsimony approach to codons, taking into account the degeneracy of the genetic code and guaranteeing to obtain the minimum number of nucleotide substitutions required by any given tree. (A highly readable presentation of the algorithm, including a worked example, was given by Moore, 1976; see also Goodman et al., 1979.) A more recent modification to their algorithm by Czelusniak permits the mixture of amino acid and nucleotide sequences (when available) in the same analysis (Goodman, 1981). Despite its elegance, the Moore–Goodman–Czelusniak algorithm may be "overkill" in the sense that it pays too much attention to silent substitutions (e.g., substitutions at third positions in codons that do not change the corresponding amino acid). If silent substitutions occur so frequently that the information from third positions quickly reaches saturation, then these positions would contribute mainly noise (or worse, systematic errors; see below) and should therefore be ignored. Weighting methods (see below) could presumably be used to minimize the contribution of third positions without ignoring them entirely. To our knowledge, however, these have not been used.

A third approach that is intermediate between the first two has been implemented by Joseph Felsenstein in his PROTPARS program from the PHYLIP package, but has yet to be formally described in the literature. Unlike the Eck–Dayhoff approach, it does consider the genetic code, but it also deviates from the Moore–Goodman–Czelusniak method by ignoring silent substitutions. Although ignoring silent substitutions sounds like extra work, the required "bookkeeping" is in fact simplified considerably because the program does not need to consider all of the potential mRNA codons responsible for a particular amino acid residue or all of the potential synonymous codon assignments to the interior nodes. For example, PROTPARS would assign one step to a change from lysine to arginine (e.g., AAA →AGA) but two steps to a change from lysine to proline [e.g., AAA → CAA (glutamine) → CCA]. Changes such as phenylalanine to glutamine require three nucleotide substitutions [e.g., AAA → GAA (leucine) → GAT (leucine) → GTT] but are counted as only two steps, since the middle substitution is silent.

One could, of course, take Felsenstein's argument a step further. Because of the biochemical properties of the various amino acids, there is often little selection against changes between amino acids having similar properties (e.g., between aspartic and glutamic acids). If changes between similar residues occur very frequently, perhaps we should ignore them as well (or at least give them less weight). Although no one seems to have suggested it, the generalized parsimony method described above could perhaps be used to implement this strategy, with the weights perhaps being derived from the PAM250 matrix (Dayhoff, 1978).

Parsimony on Allozyme Data The problems with treating allele frequencies or presence–absences as characters in a phylogenetic analysis were discussed above (see "Character Data"). To clarify these issues within the context of parsimony analysis, consider the example in Figure 12. If the alleles indicated in Figure 12a are scored as present or absent and then treated as independent characters, the most parsimonious reconstruction

(a)

Taxon	Allele Frequencies					Presence-Absences				
	a	b	c	d	e	a	b	c	d	e
A	0.8	0	0	0	0.2	1	0	0	0	1
B	1.0	0	0	0	0	1	0	0	0	0
C	0	0.5	0	0.5	0	0	1	0	1	0
D	0	0	1.0	0	0	0	0	1	0	0
E	0	0	1.0	0	0	0	0	1	0	0

(b)

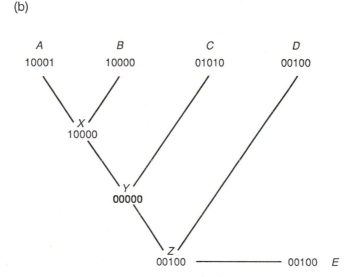

FIGURE 12. Demonstration of one problem with "independent alleles" coding. (a) Allele frequencies and data matrix resulting from presence–absence coding. (b) The most parsimonious reconstruction for the tree indicated assigns no alleles to ancestor *Y*.

under Wagner parsimony (Figure 12b) assigns no alleles to ancestor Y, an outcome that most biologists would find unacceptable. A similar example could have been constructed using allele frequencies rather than presence–absences in which the most parsimonious reconstruction assigned ancestral frequencies that summed to a value less than one.

Rogers (1984, 1986) and Swofford and Berlocher (1987) have developed methods for minimizing the total amount of frequency change on a given tree, subject to the constraint that the array of allele frequencies (for a particular locus) assigned to each interior node of the tree must exist in "allele-frequency space" (a hyperplane in which the sum of the frequencies for all alleles is one). These methods differ only in the choice of distances used for measuring branch lengths. The original method of Rogers (1984) was derived for his earlier (1972) distance measure (see above); he later extended it to a variety of other (mostly Euclidean) distance measures. His procedure uses the optimization technique of "hyperboloid approximation," which requires that the distance measure be representable as a differentiable function. Swofford and Berlocher (1987) argued for the superiority of the Manhattan metric and were forced to solve the problem via linear programming.

Methods that use allele frequencies rather than presence or absence are often criticized on the grounds that the allele frequencies are too easily modified by random drift and/or selection, and therefore do not provide reliable information for phylogenetic analysis (e.g., Mickevich and Johnson, 1976). In some cases, allele frequencies are known to vary temporally over the span of a few years, and this observation has also been used to question their relevance to phylogeny (Brian Crother, personal communication). We would argue, however, that even if the information contained in allele frequencies is somewhat unreliable, the frequencies at least provide a way to weight the presence or absence of particular alleles. For example, if an allele were detected sporadically in the taxa being analyzed but never at frequencies higher than 0.04, then we would be hesitant to attach much importance to the shared presence of that allele in some of the taxa; it could easily be present in other taxa at similar frequencies, but missed due to sampling error. On the other hand, an allele that is either fixed or nearly fixed whenever it occurs is probably more indicative of relationship. It should be emphasized that adopting a cutoff frequency (typically 0.05) does not solve the problem unless a researcher is willing to assert that an allele known to occur in a sample at an estimated frequency of, say, 0.04 is "not present."

Although both the Rogers and the Swofford–Berlocher methods are conceptually simple, the computer algorithms used to implement them are quite complex; the interested reader should refer to the original papers for details.

Rooting Revisited   Most of the methods discussed above do not specify the location of the root. If, as is generally the case, a rooted tree is desired, the root must be located using extrinsic information. As mentioned above, the most commonly used method is to include one or more taxa that are assumed to lie cladistically outside a presumed monophyletic group. We recommend including more than one outgroup taxon as a means of testing the assumption of ingroup monophyly. If there is a single branch on the unrooted tree that partitions the ingroup taxa from the outgroup taxa (e.g., Figure 13a), then the tree is consistent with the assumption of ingroup monophyly. If, on the other hand, there is no such branch (Figure 13b), then we have rejected the monophyletic-ingroup hypothesis (at least in a nonstatistical sense). Of course, this test is one-sided; the existence of a branch that partitions the assumed ingroup versus outgroup taxa still does not guarantee that the root does not lie somewhere within the "ingroup." But at least the attempt to reject the hypothesis of ingroup monophyly failed, and one can feel somewhat more confident about the assumption for that reason.

Lake's Method of Invariants

Rationale   Chance dictates that when a phylogeny contains two particularly diverged sequences, some of the changes introduced will be the same,

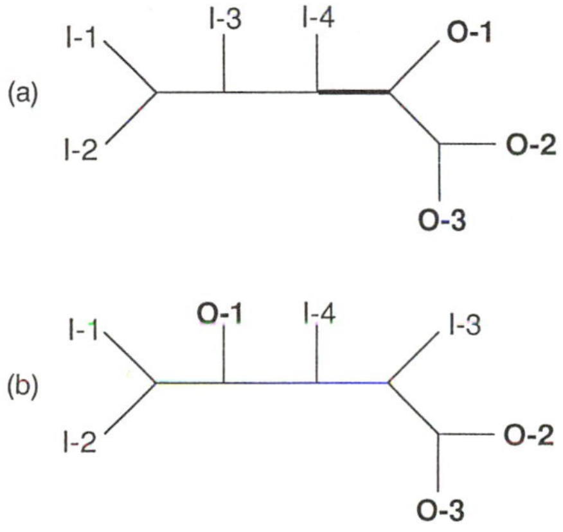

FIGURE 13. Use of multiple outgroup taxa to infer the location of the root of a tree. (a) The branch indicated in boldface partitions the ingroup taxa from the outgroup taxa, yielding an unambiguous root for the ingroup portion of the tree. (b) No single branch partitions the ingroup taxa from the outgroup taxa. The data do not support the assumption of ingroup monophyly.

even though the evolution along the lineages was independent. In the parlance of cladistic analysis, these are homoplasies. Homoplasies systematically tend to bring the diverged sequences together as sister groups (specific relatives) in an unrooted phylogenetic tree. The effect is particularly evident in parsimony analysis (Felsenstein, 1978a) and in distance-based analyses with inadequate compensation for superimposed substitutions (perfect compensation would eliminate the problem; Felsenstein, 1986). This error is a systematic one; if it is of sufficient magnitude, then no amount of sequence data will yield the correct answer.

Ideally, we would like to distinguish informative changes from homoplasies. In parsimony and maximum likelihood analyses, the addition of new sequences whose branch points subdivide the longest lineages (i.e., representation of taxa that are specifically related to the most divergent taxa already in the tree) will tend to accomplish this goal. The effect is illustrated in Figure 14 where adding sequences A′ and B′ to the tree would reduce the effects of homoplasies along the branches leading to A and B. Of course, the practical utility of this approach requires that appropriate taxa exist, that their identities are known, and that the corresponding sequence data exist or can be generated. A second method of reducing the effects of homoplasy is to confine the analysis to the most conserved sequences (both on the basis of the overall conservation of the molecule and by selecting the most conserved portions of the molecule). In distance-based analyses, estimates of the superimposed substitutions (which include the homoplasies) can also be included.

Lake (1987) has suggested an alternative method, dubbed "evolutionary parsimony," for analyzing the branching pattern linking four nucleotide sequences. The analysis can be derived from the following assumptions: (1)

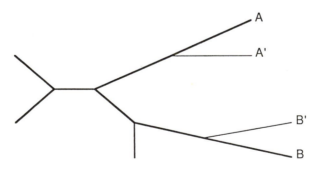

FIGURE 14. Adding new taxa to a parsimony or maximum likelihood tree to reduce the effects of homoplasy. Given the unrooted tree shown in heavy lines, the long lineages leading to A and B would have the greatest tendency to artifactually group due to parallel or convergent changes in sequence. Adding taxa A′ and B′ would reduce this effect by subdividing the long lines.

substitutions at a given sequence position are independent; (2) a balance exists among specific classes of transversions (a sufficient condition for this balance is that transversions are equally likely to yield each of the two possible substitution products, so that C is equally likely to change to A or G, etc.); (3) a balance exists among specific classes of transitions (a sufficient condition for this balance is that transition substitutions in opposite directions have equivalent frequencies so that changes from A to G and from G to A are equally likely, etc.); and (4) insertions or deletions can be safely ignored.

If the assumptions were satisfied, then parallel transversions in the two branches of a tree would give equal amounts of similar (type 1 in Figure 15) and dissimilar (type 2 in Figure 15) nucleotides. Thus, the net effect of peripheral branch transversions could be cancelled if the type 2 events were subtracted from the type 1 events. A complete accounting of possible transversions and transitions yields the scoring system in Table 2.

The Method   Lake's method can be described by the following sequence of steps:

1. Choose a quartet of aligned sequences; call them A, B, C, and D.
2. Find the alignment positions in which two sequences have purines and two have pyrimidines.
3. Consider the three possible groupings of sequences (see Figure 16): AB/CD, (A with B, C with D), AC/BD and AD/BC. Call these branching patterns X, Y, and Z, respectively.
4. Using the sequence positions at which sequences A and B are *both* purines or *both* pyrimidines (and sequences C and D are both of the opposite class of base), use the rules in Table 2 to count the number of positions that support and the number that counter branching order X. Call these totals X^+ and X^-, respectively. Similarly, find the support (Y^+) and countersupport (Y^-) for branching order Y, using the sequence positions at which sequences A and C have the same class of base, and B and D have the opposite class. Finally, find the support (Z^+) and countersupport (Z^-) for branching pattern Z. If the counting has been done correctly, the total of X^+, X^-, Y^+, Y^-, Z^+, and Z^- will be equal to the total number of positions with two purines and two pyrimidines, as found in the second step.
5. The net supports for branching patterns X, Y, and Z are

$$X = X^+ - X^- \tag{20a}$$
$$Y = Y^+ - Y^- \tag{20b}$$
$$Z = Z^+ - Z^- \tag{20c}$$

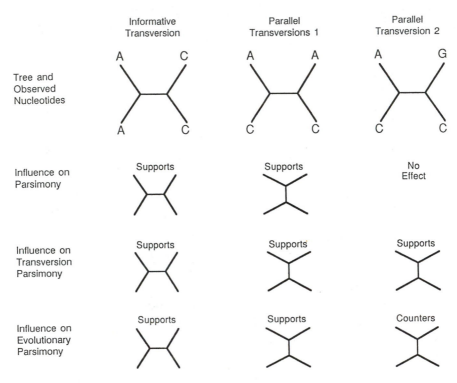

FIGURE 15. Nucleotide substitution patterns and their effects on different methods of phylogenetic tree inference. The first pattern, informative transversion, represents the effect of a single nucleotide substitution that is in the internal (central) branch of the tree. It is an example of the informative characters on which parsimony depends. Transversion parsimony and Lake's method of invariants rely entirely on transversions for informative events. The second pattern portrays a possible outcome of two peripheral branch transversions. Because the results are indistinguishable from the first pattern (two As and two Cs), all methods will mistake this as support for an incorrect phylogeny. The third pattern illustrates the possibility that independent transversions in two peripheral branches will yield different nucleotides. The pattern is uninformative to traditional parsimony (two substitutions would be required regardless of the assumed branching order). Transversion parsimony will consider this pattern to be support for the incorrect tree since the outcome looks like a central branch transversion (in an incorrect tree) combined with a peripheral branch transition (which is ignored). Lake's method treats this third pattern as an estimator of multiple substitutions in peripheral branches and subtracts it from the support for the incorrect tree.

Table 2. Interpretation of nucleotide patterns observed among four aligned sequences using evolutionary parsimony[a]

Nucleotide pattern observed at alignment position	Influence of position on support[b]
Two identical purines and two identical pyrimidines	Support
Two dissimilar purines and two identical pyrimidines	Countersupport
Two identical purines and two dissimilar pyrimidines	Countersupport
Two dissimilar purines and two dissimilar pyrimidines	Support
Any other combination of nucleotides	No effect

[a]The process is performed for every position (i.e., column) in the alignment, keeping running totals of the support and countersupport for each of the three possible branching patterns (Lake, 1987; Olsen, 1987).
[b]For each sequence position, the support or countersupport applies only to the topology that would group the two purines together, and the two pyrimidines together.

The support for two of the branching patterns should be near zero, while the remaining branching pattern may or may not be supported by a significantly nonzero score.

6. Lake (1987) suggested that statistical significance be evaluated by a one degree of freedom χ^2 test:

$$\chi_X^2 = X^2/(X^+ + X^-)$$
$$\chi_Y^2 = Y^2/(Y^+ + Y^-)$$
$$\chi_Z^2 = Z^2/(Z^+ + Z^-)$$

Therefore the outcome of interest is two values of χ^2 that do not differ significantly from zero and one value that does. Holmquist et al. (1988a) correctly pointed out that the χ^2 approximation is inadequate when counts are low and recommend the use of the exact binomial test instead.

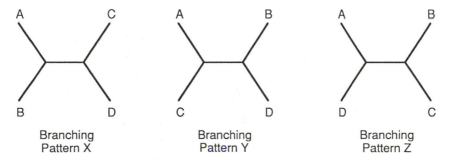

FIGURE 16. The three unrooted branching patterns with four sequences.

Several additional issues are relevant to the use of Lake's method.

Negative Values   The net support of a tree can be negative and yet significant (e.g., X is negative and χ^2_X is significantly large). Lake (1987) suggested that this result could be interpreted as positive evidence for the corresponding branching pattern, if no other pattern has significant support. However, significantly negative values should be viewed with extreme caution, because such outcomes are most likely to be the result of selective pressure or some other nonrandom process.

Moving Beyond Four Taxa   Lake's method was defined for the analysis of four sequences. It has been applied to larger groups of sequences in two distinct fashions. The first is to infer a relationship between groups of sequences (Lake, 1987). That is, prior information suggests that the sequences of interest can be divided into four disjoint groups (relationships within these groups need not be known) and the object is to discover the branching pattern that relates these four groups. Lake (1987) suggested a complex method for evaluating the significance of the support for each of the three trees relating four groups of sequences. The essence of the problem is to evaluate the extent to which the scores obtained (by using each of the sequences in turn) are independent, to estimate the expected variance of the composite score (that score resulting from a summation over all combinations of sequences) for the null hypothesis (no specific relationship). Superficially, this would seem to be an ideal application of the bootstrap or jackknife, using resampling of the sequence positions to empirically assess the significance of any nonzero scores. This approach does not appear to have been tried.

The second application to more taxa is to infer a fully resolved branching of many species (Lake, 1988); however, it has yet to be described in sufficient detail to be reproduced.

Transitions and Transversions   The phylogenetic information provided by Lake's method is based entirely on transversion substitutions, so positions with two purines and two pyrimidines are required. If there are no transversions, there will be no signal. On the other hand, transition substitutions decrease the signal. In particular, peripheral branch transitions convert informative (supportive) positions into countersupport, suggesting that the method might be particularly sensitive to the ratio of transitions to transversions. If transitions are indeed substantially more frequent than transversions, then we will have difficulty accumulating significant numbers of transversions to infer the branching pattern without having the signal randomized by transitions (see Li et al., 1987). As noted above, generalized parsimony (intracharacter weighting), transversion parsimony, and transversion-based distance methods provide alternative methods of coping with a high transition/transversion ratio.

Interestingly, transversion parsimony (as defined in this chapter, which differs from Lake's use of the term) applied to four sequences seeks the tree, X, Y, or Z, with the largest value of $X^+ + X^-$, $Y^+ + Y^-$, and $Z^+ + Z^-$. By examining Equation (20), it can be seen that transversion parsimony uses the same data but *adds* the terms that look like a peripheral branch transition (and a central branch transversion) rather than subtracting them as does Lake's method.

Maximum Likelihood Phylogenies for Nucleotide Sequences

Several areas of biological research, notably genetic mapping and clinical testing, routinely use maximum likelihood methods for testing hypotheses. However, the perceived and actual complexities of obtaining maximum likelihood solutions to problems that involve numerous alternative hypotheses have inhibited the more general use of these techniques. The following discussion attempts to outline the elements of a maximum likelihood formulation of phylogenetic inference.

Objective   Phylogenetic analysis seeks to infer the history (or set of histories) that is most consistent with a set of observed data. In the present case, the data are observed nucleotide sequences and the unknowns are the branching order and the tree branch lengths. To apply a maximum likelihood approach, a concrete model of the evolutionary process that converts one sequence into another must be specified. This model may be fully defined; alternatively, it may contain many parameters that are to be estimated from the data. A maximum likelihood approach to phylogenetic inference evaluates the net likelihood that the given evolutionary model will yield the observed sequences; the inferred phylogenies are those with the highest likelihood.

Models of Sequence Evolution   The starting point for using maximum likelihood to infer a sequence phylogeny is defining a model of sequence evolution. Although this can be made arbitrarily complex, the programs in current use are all based on simple models that could be called the Jukes–Cantor model, the Kimura two-parameter model, and the generalized two-parameter model. The Jukes–Cantor model makes the assumptions that all four nucleotides are equally frequent, all substitution types are equally likely, and the remaining assumptions associated with Equation (5). Similarly, the Kimura two-parameter model makes the assumptions associated with Equation (7), in particular that all four nucleotides are equally frequent and that there are independent rates for transition substitutions and for transversion substitutions. The generalization of the Kimura model permits each nucleotide to be present at a different frequency, so long as the substitution rates are balanced so as to maintain the equilibrium abundance of the nucleotides. This model, which encompasses both other models, is used in Felsenstein's DNAML program. None of the models treats insertions or deletions; therefore, columns in a sequence alignment that include any alignment gaps are usually omitted from the analysis.

One feature of those models that permit differing transition and transversion rates is worth noting. The Kimura (1980) formula for estimating evolutionary distance between sequences (Equation 7) independently estimates (optimizes) the ratio of transversion to transition rates for each pair of taxa. This is not necessarily a consistent approach, since various pairs of sequences share portions of their evolutionary history and consequently had the same transition/transversion ratio for that portion of their history. As it is being applied in the present discussion, this ratio is defined for the evolutionary tree as a whole, and therefore will necessarily be self-consistent.

The mathematical expression of a model of evolution is a table of rates (substitutions per position per unit evolutionary distance) at which each nucleotide is replaced by each alternative nucleotide. These rates are conveniently summarized in a 4-by-4 matrix, $\mathbf{R}$, in which the element R_{ij} is the rate at which a residue of state i changes to state j. For example, the Jukes–Cantor change matrix is

$$
\begin{bmatrix}
-3\alpha & \alpha & \alpha & \alpha \\
\alpha & -3\alpha & \alpha & \alpha \\
\alpha & \alpha & -3\alpha & \alpha \\
\alpha & \alpha & \alpha & -3\alpha
\end{bmatrix}
$$

where the residues are in the order A, C, G, T. Thus, each residue type is equally likely to change to any other type and the diagonal elements, which represent the loss of the initial residue type, exactly balance the conversion to other types. If $3\alpha = 1$, then the average rate of change at the position is one substitution per unit distance. The equivalent matrix for the Kimura (1980) two-parameter model is

$$
\begin{bmatrix}
-\alpha-2\beta & \beta & \alpha & \beta \\
\beta & -\alpha-2\beta & \beta & \alpha \\
\alpha & \beta & -\alpha-2\beta & \beta \\
\beta & \alpha & \beta & -\alpha-2\beta
\end{bmatrix}
$$

where α is the rate of transitions and β is the rate of transversions. Setting $\alpha + 2\beta = 1$ will normalize the rate of change at the sequence position to one substitution per unit distance. The parameter that expresses the underlying ratio of transition to transversion rates is α/β. The generalized two-parameter model is a straightforward extension (from Kishino and Hasegawa 1989):

$$\begin{bmatrix} -[k/(f_A+f_G)+3]\beta f_A & \beta f_A & [k/(f_A+f_G)+1]\beta f_A & \beta f_A \\ \beta f_C & -[k/(f_C+f_T)+3]\beta f_C & \beta f_C & [k/(f_C+f_T)+1]\beta f_C \\ [k/(f_A+f_G)+1]\beta f_G & \beta f_G & -[k/(f_A+f_G)+3]\beta f_G & \beta f_G \\ \beta f_T & [k/(f_C+f_T)+1]\beta f_T & \beta f_T & -[k/(f_C+f_T)+3]\beta f_T \end{bmatrix}$$

where k defines the relationship between transitions and transversions and f_i is the frequency of base i. When $k = 0$, the rate for a transversion is equal to the rate for a transition.

In order to convert rates to a model of sequence evolution over a finite interval, it is necessary to view each matrix as defining a set of four simultaneous, linear differential equations. In matrix form the equations are

$$\frac{\partial}{\partial d}\begin{bmatrix} A \\ C \\ G \\ T \end{bmatrix} = \begin{bmatrix} R_{AA} & R_{CA} & R_{GA} & R_{TA} \\ R_{AC} & R_{CC} & R_{GC} & R_{TC} \\ R_{AG} & R_{CG} & R_{GG} & R_{TG} \\ R_{AT} & R_{CT} & R_{GT} & R_{TT} \end{bmatrix}\begin{bmatrix} A \\ C \\ G \\ T \end{bmatrix}$$

where A, C, G, and T are the probabilities that the nucleotide at the given site is A, G, C, or T, respectively. Illustrating this with the Kimura two-parameter rate matrix yields

$$\frac{\partial A}{\partial d} = (-\alpha-2\beta)A + \beta C + \alpha G + \beta T$$

$$\frac{\partial C}{\partial d} = \beta A + (-\alpha-2\beta)C + \beta G + \alpha T$$

$$\frac{\partial G}{\partial d} = \alpha A + \beta C + (-\alpha-2\beta)G + \beta T$$

$$\frac{\partial T}{\partial d} = \beta A + \alpha C + \beta G + (-\alpha-2\beta)T$$

For simple cases, these equations can be analytically integrated with respect to evolutionary distance to give a matrix, $\mathbf{M}(d)$, in which element $M_{ij}(d)$ is the probability that a nucleotide of initial identity i has identity j after evolving through a distance d. This integration accounts for all possible series of substitutions linking the initial and final residues, providing an intrinsic correction for multiple substitutions. For the Jukes–Cantor rate matrix the integrated substitution matrix is

$$\mathbf{M}(d) = \begin{bmatrix} a & b & b & b \\ b & a & b & b \\ b & b & a & b \\ b & b & b & a \end{bmatrix} \tag{21}$$

$$\text{where } a = \frac{1+3e^{-4\alpha d}}{4} \quad \text{and } b = \frac{1-e^{-4\alpha d}}{4}$$

Similarly, the change matrix for the Kimura (1980) two-parameter model is

$$\mathbf{M}(d) = \begin{bmatrix} a & b & c & b \\ b & a & b & c \\ c & b & a & b \\ b & c & b & a \end{bmatrix}$$

$$\text{where } a = \frac{1+2e^{-2(\alpha+\beta)d}+e^{-4\beta d}}{4} \quad b = \frac{1-e^{-4\beta d}}{4} \quad \text{and } c = \frac{1-2e^{-2(\alpha+\beta)d}+e^{-4\beta d}}{4}$$

The generalized two-parameter rate model can also be integrated analytically (not shown). With still more general rate models it may be necessary to integrate the differential equations numerically.

In summary, for a given model of sequence evolution, one constructs formulas that describe the probabilities that an initial nucleotide will yield a given final nucleotide at the end of an evolutionary interval d.

Evolutionary Relationship of Two Sequences   The next level of complexity is to consider the relationship between two sequences under a given model of sequence change. With only two sequences the tree is trivial: it is a single branch of length s that joins the sequences. The object is to find the length of the branch. This task requires finding the likelihood that at the end of an evolutionary distance d the first sequence will be converted exactly into the second sequence; s is then the value of d at which the likelihood is maximized. If x_{ij} is the identity of nucleotide j in sequence i, then

$$L(x_{1j}, x_{2j}; d) = f_{x1j} M_{x1j, x2j}(d)$$

where $L(a,b; d)$ is the likelihood of observing nucleotide a in the first sequence and b in the second sequence, under the condition that the sequences are separated by distance d. The equation says that this likelihood is equal to the prior probability of finding the initial nucleotide at the sequence position times the probability that it is transformed into the second nucleotide in the interval of length d. If all the sequence positions are independent, then the overall likelihood of finding sequence x_1 and sequence x_2 separated by distance d is the products of the likelihoods at each position. That is,

$$L(x_{1j}, x_{2j}; d) = \prod_{j=1}^{N} f_{x1j} M_{x1j, x2j}(d)$$

where N is the number of sequence positions. In practice it is usual to deal with the logarithm of the likelihood, thereby converting the product into a sum:

$$\ln L(x_{1j}, x_{2j}; d) = \sum_{j=1}^{N} \ln[f_{x1j} M_{x1j, x2j}(d)]$$

$$= \sum_{j=1}^{N} \{\ln(f_{x1j}) + \ln[M_{x1j, x2j}(d)]\} \qquad (22)$$

$$= \sum_{j=1}^{N} \ln[M_{x1j, x2j}(d)] + \sum_{j=1}^{N} \ln(f_{x1j})$$

One of the merits of this representation is that the final term is independent of d, and thus will go to zero when taking the partial derivative with respect to d in the process of finding the maximum of the likelihood. In general, Equation (22) can be used to obtain maximum likelihood estimates of pairwise distances. If one adds the assumption that all sequence positions change at the same rate and uses Equation (21) as the change matrix, then the maximum likelihood distance will be identical to that in Equation (5) with $b = 3/4$. This may seem like a lot of work, but Equation (22) is much more flexible. For example, it is possible to allow the substitution rate α to be defined separately for each sequence position.

Maximum Likelihood Tree for Multiple Sequences To move on to trees of sequences, it is necessary to consider the likelihoods of each residue occurring at each node in the tree as a function of tree branching order and branch lengths. As with other methods that define the optimal tree in terms of an optimality criterion (e.g., least-squares and parsimony), we will assume that the tree is given, and that the present task is to determine how good the tree is. The method for evaluating the likelihood of a given tree proceeds from a hypothetical root node at any convenient location in the tree, and combines the likelihoods of each of its daughter trees (i.e., descendant lineages). For all the models of evolution that have been discussed, the choice of root location will not change the likelihood of the tree. If A is an ancestor that gave rise to sequences B and C, then the likelihood of residue i at sequence position j in A is

$$L(x_{Aj} = i) = \left[\sum_{k} M_{ik}(d_{AB}) L(x_{Bj} = k)\right]\left[\sum_{l} M_{il}(d_{AC}) L(x_{Cj} = l)\right] \qquad (23)$$

where d_{AB} is the evolutionary distance from sequence A to sequence B. (Note that the prior probability of i, f_i, has been omitted for the time being, so this formula is not quite correct. The omission will be rectified later.) In words, the likelihood that A has residue i is the product of the likelihoods that the i could have given rise to the outcomes in B and C. The first term on the right-hand side is the probability of i changing to k in the interval d_{AB}, $M_{ik}(d_{AB})$, times the likelihood that sequence B has residue k at the corresponding position, summed over all identities of k. If B is a known sequence, then the likelihood that position j has identity k is 1 if k is equal to the observed nucleotide in the sequence; otherwise the likelihood is zero. On the other hand, if B is an ancestor, then the likelihoods of it having nucleotide k are derived recursively, by inserting another copy of the right hand side of Equation (23) into the equation. The second term in Equation (23) is analogous to the first, but refers to the lineage leading to C. The likelihood of the entire evolutionary tree at sequence position j requires multiplying the likelihood of each possible nucleotide at the root node, $L(x_{Aj} = i)$, by its prior probability, f_i, and summing over all ancestral nucleotides, i. Usually the root node will be made coincident with one of the other nodes in the tree, eliminating one branch and one summation. The product of the position-specific likelihoods is the overall likelihood of the tree. Again, this is usually expressed as a sum of position-specific log-likelihoods.

There is a very close relationship between finding the likelihood of the tree and finding the number of changes (the cost) of a tree under the general parsimony criterion. Pointing out some of the corresponding features might clarify this point. The cost of a given change under parsimony is analogous to the likelihood of the given change from the substitution matrix, **M**. In parsimony, the cost of placing a given residue at an internal node is the sum of the costs of deriving both of the daughter trees from that residue, whereas the likelihood of an ancestral nucleotide is the product of the likelihoods of the residue giving rise to the daughter trees. In parsimony, the total cost of the tree is the sum of the costs at each position, whereas the net log-likelihood of a tree is the sum of the log-likelihoods of the evolution at each sequence position. Essential differences between the general parsimony approach and the maximum likelihood approach are that the cost of a change in general parsimony is not a function of branch length, whereas it is in maximum likelihood; and that maximum parsimony looks only at the single, lowest cost solution, whereas maximum likelihood looks at the total likelihood for all solutions (ancestral nucleotides) consistent with the tree and branch lengths. Felsenstein has used the relationship between likelihood and parsimony to gain several insights into the parsimony procedure, including the discovery of the potential convergence of long branches (Felsenstein, 1978a) and the inference of a character-weighting rationale (Felsenstein, 1981b).

Figure 17 illustrates a tree of five sequences. The corresponding likelihood for a position, j, is

$$L(j) = \sum_m f_m \left[\sum_k M_{m,k}(d_{GH}) M_{m,xAj}(d_{AF}) M_{k,xBj}(d_{BF}) \right]$$

$$\times \left[\sum_l M_{m,l}(d_{GH}) M_{l,xDj}(d_{DH}) M_{l,xEj}(d_{EH}) \right] M_{m,xCj}(d_{CG})$$

where A, B, C, D, and E are the sequences, F, G, and H are the labels of the internal nodes, and the hypothetical root has been placed at node G. The overall likelihood would be the product over positions. The four factors of the outer summation are (1) the prior probability of a nucleotide with identity m; (2) the likelihood of nucleotide m at node G giving rise to nucleotide k at node F, and k giving rise to x_{Aj} at node A, and k giving rise to x_{Bj} at node B; (3) the likelihood of m giving rise to l at node H, and l giving rise to x_{Dj} at node D, and l giving rise to x_{Ej} at node E; and (4) the likelihood of m giving rise to x_{Cj} at node C. This basic pattern can be expanded to trees of indefinite complexity, but the computational difficulty of finding the optimal branch lengths increases rapidly.

The methods for finding the optimal branch lengths are beyond the scope of this chapter (see Felsenstein, 1981a). However, it has been demonstrated that for the Jukes–Cantor model of evolution, there do not appear to be problems with local minima while finding the optimal branch lengths (Fukami and Tateno, 1989). This is perhaps a small consolation given that the overall computational complexity is very high. In practice, the largest maximum likelihood trees published involve a small number of taxa relative to parsimony or least-squares distance trees.

After optimizing the branch lengths for a tree topology of interest, one then obtains the likelihood value for the topology. Although it is seductive to assume that the ratios of the likelihoods of two trees correspond directly to how many times better one tree is than the other, this is not the case.

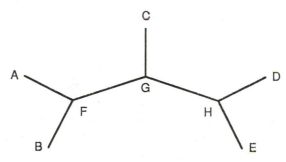

FIGURE 17. An evolutionary tree of five sequences. The known sequences are at the terminal nodes and are labeled A, B, C, D, and E. The nodes F, G, and H represent ancestral sequences.

Several methods for evaluating the significance of the likelihood values have been proposed. Felsenstein (1988) suggested that a one-degree of freedom χ^2 test on twice the difference in log-likelihoods is a conservative test. Kishino and Hasegawa (1989) provide an interesting test that might be less sensitive to details of the evolutionary model. They get the same confidence interval as with a bootstrap analysis, but with a miniscule fraction of the computation.

SEARCHING FOR OPTIMAL TREES

As emphasized above, methods that have explicit optimality criteria (e.g., maximum parsimony, least-squares additive trees, and maximum like-lihood) separate the problem of evaluating a particular tree under the selected criterion from that of finding the optimal tree(s). Most of our presentation to this point has dealt with the former problem; in this section, we address the latter. For data sets of small to moderate size (8 to 20 taxa, depending on the criterion), exact methods that guarantee the discovery of all optimal trees may be used. For larger data sets, exact solutions require a prohibitive amount of computing time; consequently, approximate methods that do not guarantee optimality must be used.

Exact Algorithms

Exhaustive Search   The conceptually simplest approach to the search for optimal trees is simply to evaluate every possible tree. Assuming that exact methods exist for evaluating a particular tree, we need only a method for enumerating all possible (strictly bifurcating) trees to find a globally optimal solution. A simple algorithm, outlined in Figure 18, can be used to perform this enumeration. Initially, we connect the first three taxa in the data set to form the only possible unrooted tree for these taxa (Figure 18, row 1). In the next step, we add the fourth taxon to each of the three branches of the three-taxon tree, thereby generating all three possible unrooted trees for the first four taxa (Figure 18, row 2). We continue in a similar fashion: adding the ith taxon to each branch of every tree (containing $i - 1$ taxa) generated during a previous step. Thus, for example, trees 3A–3O of Figure 18 are all 15 possible trees for the first five taxa, obtained by adding the fifth taxon to each of the five possible branches for the three trees obtained at the four-taxon stage. This makes clear the rationale for expression (1) for counting the number of possible unrooted bifurcating trees for T taxa: for each of the possible trees for $i - 1$ taxa, there are $2(i - 1) - 3 = 2i - 5$ branches to which the ith taxon can be connected. Note that the order of addition is immaterial; we could just as easily have chosen taxa at random for the next addition at each step.

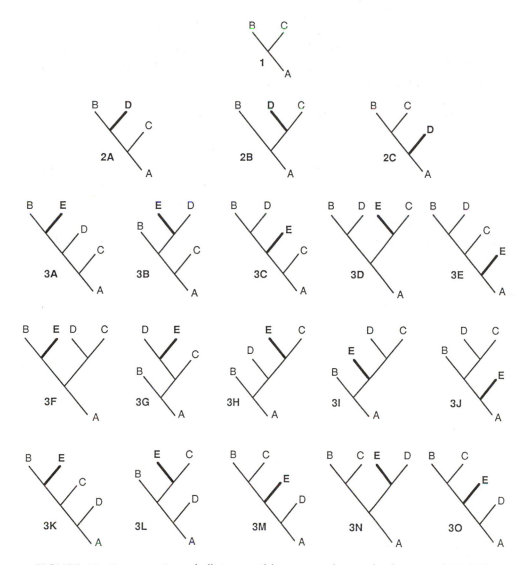

FIGURE 18. Enumeration of all 15 possible unrooted trees for five taxa (see text).

Evaluation of expression (1) for several possible values of T quickly reveals why exhaustive search procedures are useful only for small numbers of taxa. There are 945 possible trees for only 7 taxa, over 2×10^6 trees for 10 taxa, and over 2×10^{20} possible trees for 20 taxa (Felsenstein, 1978b)! Thus, exhaustive enumeration of all possible trees is currently feasible only for 11 or fewer taxa (34,459,425 trees).

Branch-and-Bound Methods   Fortunately, an exact algorithm for identifying all optimal trees that does not require exhaustive enumeration is
available for any criterion whose value is known to be nondecreasing as
additional taxa are connected to a tree. The branch-and-bound method,
frequently used to solve problems in combinatorial optimization, apparently was first applied to evolutionary trees by Hendy and Penny (1982).
The branch-and-bound method closely resembles the exhaustive search
algorithm described above. In this procedure, we traverse a "search tree" in
a "depth-first" sequence, as illustrated in Figure 19. The root of the search
tree (A) contains the only possible tree for the first three taxa. We first
construct one of the three possible trees obtained by connecting taxon 4 to
tree A, yielding tree B1. Then, to this tree, we connect taxon 5, yielding tree
C1.1. (If there were more than five terminal taxa, we would continue to join
additional taxa in this manner until a tree containing all T taxa had been
completed.) Now, we backtrack one node on the search tree (i.e., back to

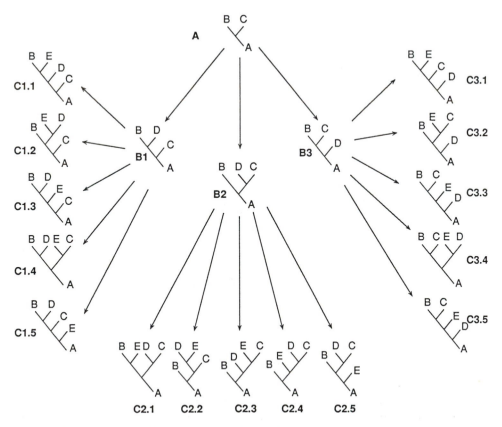

FIGURE 19. Search tree for branch-and-bound algorithm (see text).

tree B1) and generate the second tree resulting from the addition of taxon 5 to tree B1 (= tree C1.2). When all five of the trees derivable from tree B1 (C1.1–C1.5) have been constructed, we backtrack all the way to tree A of the search tree and take the second path away from this node, leading to tree B2. As before, all five trees derivable from tree B2 (C2.1–C2.5) are constructed in turn. Then we backtrack once again to tree A and proceed down the third path toward trees C3.1–C3.5. Eventually, we will have constructed all of the possible trees culminating with tree C3.5. If the score of each tree containing all five taxa were evaluated at the time of its construction, we will have performed an exhaustive search equivalent to that described in the above section.

Let L represent an upper bound on the optimal value of the chosen optimality criterion. (We assume that we want to minimize this criterion, such as minimizing the tree length under a parsimony criterion or minimizing the sum of squared deviations in an additive-tree distance method.) For the present, we can obtain L, for example, by evaluating a random tree; if we know that a tree of score L exists, then the score of the optimal tree(s) cannot exceed this value. As we are moving along a path of the search tree toward its tips (containing all T taxa), if we encounter a tree whose score exceeds L, then there is no need to proceed further along this path; connecting additional taxa cannot possibly decrease the score. Thus, we can dispense with the evaluation of all (phylogenetic) trees that descend from this node in the search tree and immediately backtrack and proceed down a different path. By cutting off portions of the search tree in this manner, we can greatly reduce the number of trees that must actually be evaluated.

If we reach the end of a path on the search tree and obtain a tree whose score is equal to the upper bound L, then this tree is a candidate for optimality. If instead this score is less than L, then this is the best tree found so far, and we have improved the upper bound on the score of the optimal tree(s). This improvement is important, as it may enable other search paths to be terminated more quickly. When the entire search tree has been traversed, all optimal trees will have been identified.

The branch-and-bound method is extremely effective for parsimony criteria, permitting exact solutions for 20 or more taxa, depending on the efficiency of the implementation, the speed of the available computer, and the "messiness" of the data. The utility of branch-and-bound methods for additive-tree fitting is unknown; we are unaware of any attempts to implement them for distance methods. A potential difficulty seems likely: if most of the increase in the score of the phylogenetic trees being evaluated occurs near the tips of the search tree (i.e., not until most of the taxa have been connected), relatively little computation will be avoided in comparison to an exhaustive search.

The above presentation of the branch-and-bound method, while correct, is an oversimplification of the algorithms actually used in state-of-the-art computer programs. Refinements in the algorithms that greatly speed the computations are usually implemented. These refinements, designed to promote earlier cut-offs in the traversal of the search tree, include (1) using **heuristic methods** (see below) to obtain a near-optimal tree whose score is used as the initial upper bound; (2) designing the search tree so that divergent taxa are added early, thereby increasing the length of the initial trees in the search path; and (3) using pairwise incompatibility to improve the *lower* bound on the length that will ultimately be required by trees descending from a tree at a given node of the search tree. These methods are discussed in more detail in Hendy and Penny (1982) and Swofford (1990).

An obvious question may have occurred to the reader at this point. Since the branch-and-bound method requires evaluation of all trees as its worst possible case, why would we ever want to perform an exhaustive search? In fact, if we were interested only in the optimal trees, the branch-and-bound algorithm would indeed be the preferred means of finding them. However, exhaustive searches permit the researcher to examine the frequency distribution of tree lengths. It is often useful to know, for example, whether there are few or many near-optimal trees, or where some tree of prior interest lies in the distribution of tree lengths.

Heuristic Approaches

When a data set is too large to permit the use of exact methods, optimal trees must be sought via heuristic approaches that sacrifice the guarantee of optimality in favor of reduced computing time. The task of searching for an optimal tree by approximate methods is somewhat analogous to the plight of a myopic pilot who loses his glasses when forced to parachute from his airplane into a mountainous region. He suspects that there is a manned outpost at the top of the highest peak in the area, but he must somehow grope his way there to have any hope of rescue. Obviously, simply walking uphill from the point of landing will not necessarily lead him to his goal, since he may not have started on a slope of the highest peak. Suppose that he reaches a summit and finds no outpost. Two possibilities remain: (1) he is, in fact, at the top of the highest peak, but was wrong about the existence of the outpost, or (2) he has climbed the wrong hill. Although rather absurd, the analogy is quite appropriate.

Heuristic tree searches generally operate by hill-climbing methods. An initial tree is used to start the process; we then seek to improve the tree by rearranging it in a way that improves its score under our chosen optimality criterion (e.g., minimum length). When we can find no way to further improve the tree, we stop. Like the downed pilot, however, we generally have no way of knowing whether we ended up at the top of the highest hill.

That is, we do not know whether we have arrived at a global or merely a local optimum.

The details of heuristic search procedures vary considerably from one implementation to the next. In addition, better methods are often invented. Consequently, we prefer to leave the specifics to the documentation of the computer program used to perform the search, and will concentrate on more general concepts.

Stepwise Addition.   The most commonly used method for obtaining a starting point for further rearrangement is by stepwise addition of taxa to a growing tree. First, three taxa are chosen for the initial tree. Next, one of the unplaced taxa is selected for next addition. Each of the three trees that would result from joining the unplaced taxon to the tree along one of its (three) branches is evaluated, and the one whose score is optimal is saved for the next round. In this next round, yet another unplaced taxon is connected to the tree, this time to one of the five possible branches on the tree saved from the previous round. The process terminates when all taxa have been joined to the tree.

Of course, the above description is oversimplified in that several decisions are required, none of which has a straightforward answer. Which three taxa should be used initially? How do we decide which unplaced taxon to connect to the tree next? One approach is to simply add the taxa in the same order in which they are presented in the data matrix, starting with the first three and sequentially adding the rest. This strategy, for example, is the one used in Felsenstein's PHYLIP package. Another approach, optionally available in Swofford's PAUP, is to check all triplets of taxa and start with the one that yields the shortest tree. At each successive step, all remaining unplaced taxa are considered for connection to every branch of the tree, and the taxon–branch combination that requires the smallest increase in tree length is chosen. Still another approach, suggested by Farris (1970), is to prespecify an addition sequence based on each taxon's distance to a reference taxon (called a "hypothetical ancestor" by Farris, but it could just as well be any taxon in the data matrix). Unfortunately, there seems to be no strategy that works best for all data sets; the best approach is to try as many alternatives as possible, each of which may potentially provide a different starting point for branch swapping (below).

The biggest drawback of stepwise addition algorithms is that they are too "greedy." Like the nearsighted pilot who is unable to scan the horizon and instead must simply proceed up the nearest hill, these methods strive for optimality given the current situation rather than attempting to look more broadly into the future. Thus, one placement of a taxon may be best given the taxa currently on the tree, but that placement may become suboptimal upon the addition of subsequent taxa. Once a decision has been made to connect a taxon to a certain point, however, we must usually accept the consequences of that decision for the remainder of the stepwise addition process, perhaps ending up in a local optimum as a result.

Branch Swapping.   Because of the excessive greediness and susceptibility to local-optima problems, stepwise-addition algorithms generally do not find optimal trees unless the data are very clean. However, it may be possible to improve the initial estimate by performing sets of predefined rearrangements, a technique commonly referred to as "branch swapping." In general, any one of these rearrangements amounts to a "stab in the dark," but the hope is that if a better tree exists, one of the rearrangements will find it. Examples of three kinds of rearrangements used in current branch-swapping algorithms are shown in Figures 20 through 22.

Of course, the globally optimal tree(s) may be several rearrangements away from the starting tree. If a rearrangement is successful in finding a better tree, a round of rearrangements is initiated on this new tree. So long as each round of rearrangements is successful in finding an improved tree (according to their score under the optimality criterion), then we will eventually arrive at the global optimum. However, if the intermediate trees would require us to pass through trees that are inferior to the one(s) already obtained, we will once again find ourselves trapped in a local optimum unless an option is provided for branch-swapping on suboptimal trees (e.g.,

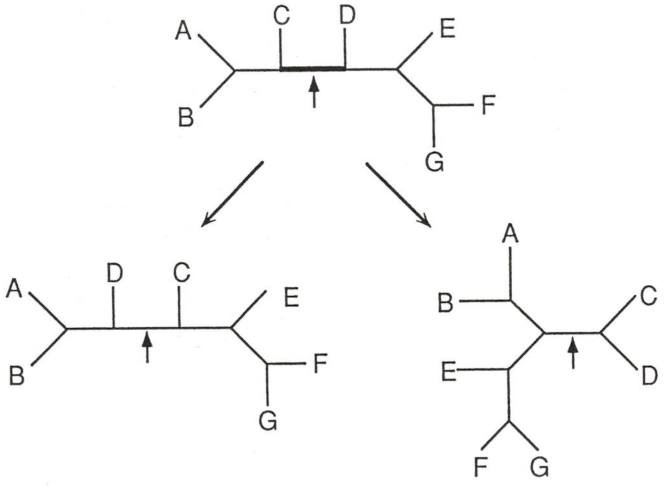

FIGURE 20. Branch swapping by "nearest-neighbor interchanges" (NNIs). Each interior branch of the tree defines a local region of four subtrees connected by the interior branch. Interchanging a subtree on one side of the branch with one from the other constitutes an NNI. Two such rearrangements are possible for each interior branch.

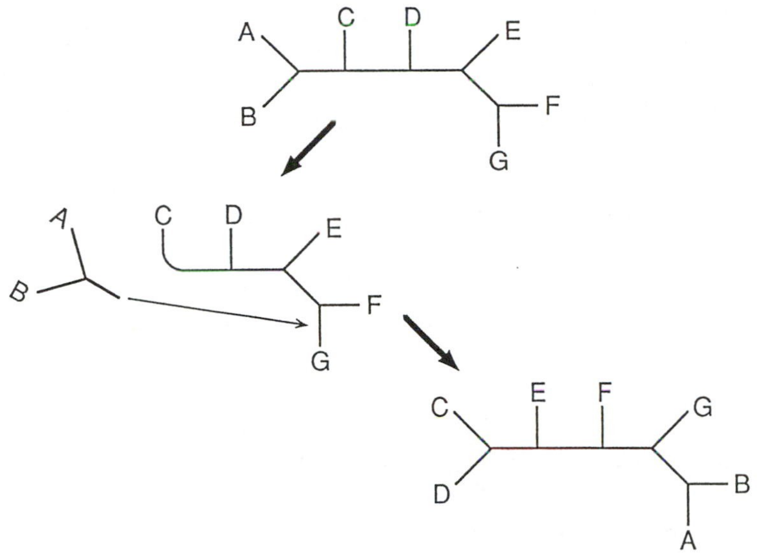

FIGURE 21. Branch swapping by "subtree pruning and regrafting." A subtree is pruned from the tree (e.g., the subtree containing terminal nodes A and B as indicated). The subtree is then regrafted to a different location on the tree. All possible subtree removals and reattachment points are evaluated.

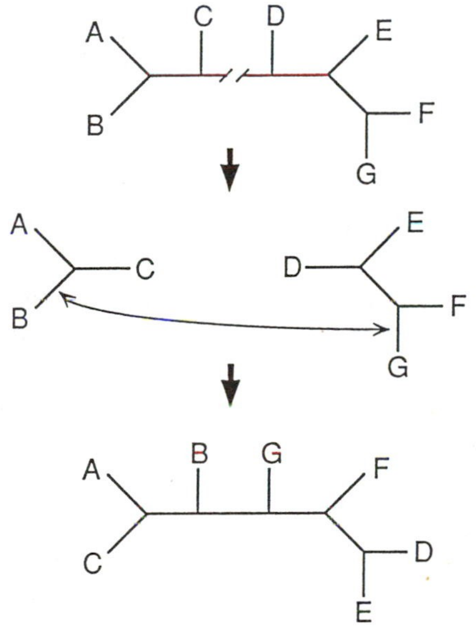

FIGURE 22. Branch swapping by "tree bisection and reconnection." The tree is bisected along a branch, yielding two disjoint subtrees. The subtrees are then reconnected by joining a pair of branches, one from each subtree. All possible bisections and pairwise reconnections are evaluated.

the "KEEP" option in PAUP; Swofford, 1990). A related problem concerns "plateaus" on the optimality surface. It may be the case, for example, that an optimal tree lies several rearrangements away from the current tree, and that these rearrangements all correspond to trees having equal scores under the optimality criterion. If the intermediate trees are discarded because they are "not better," then the optimal tree will not be found. A few programs do not retain equally good trees because they have no protection against cycling (alternation between two trees, each of which can be rearranged to yield the other); these programs will not be effective if plateaus are encountered since they are unable to traverse the plateau.

Testing for Convergence.   Because of the limitations of heuristic approaches discussed above, some way of evaluating the success of the chosen method in obtaining a globally optimal solution is needed. The obvious strategy in this regard is to begin from different starting points and ask whether the same result is always obtained. For example, a set of random sequences for addition of taxa can be used to generate initial trees for input to branch swapping. Since, at least for reasonably noisy data, the starting trees will vary depending on the addition sequence, convergence to a common optimal tree (or set of trees) is encouraging. (A more extreme approach—using random trees rather than random addition sequences—could be adopted; however, the starting trees are, on average, so far from the optimal trees that this strategy seems to be less effective.) Even if rearrangements of different starting trees do not converge to the same end point, the use of several starting trees is a good idea; if multiple peaks on the optimality surface exist, we will be more likely to find them.

RELIABILITY OF INFERRED TREES
Random Errors

Even if evolution works exactly in the way for which a particular analytical method is appropriate, with finite data an incorrect tree may be inferred due to chance events. For example, convergent substitutions might all occur in a way that leads us to conclude that two taxa are closely related when in fact they are not. This misconception can happen even when the presumed model is correct. By analogy, the observation of 20 consecutive "heads" in a coin tossing experiment might lead us to conclude that the coin is two-headed, but of course this outcome has a finite probability of occurring (approximately 10^{-6}) even if the coin is fair. In inferential statistics, we generally choose a certain probability (typically 0.05) below which an outcome is improbable enough to warrant rejection of a null hypothesis. Unfortunately, except for explicitly statistical methods based on a precise probability model of evolution such as maximum likelihood, the statistical testing of hypotheses concerning phylogenetic relationships is not easily accomplished.

The best way to avoid random errors is to obtain an infinite amount of data; this practice will guarantee the correct result so long as the method is consistent (see below). Since this option is unavailable to most researchers, other methods must be used to estimate the reliability of the results of a phylogenetic analysis. Penny and Hendy (1986) and Felsenstein (1988) have presented detailed and readable treatments on the estimation of reliability; consequently, we will provide only a brief overview of two techniques that are available to any method: the bootstrap and the jackknife.

Bootstrap and Jackknife Methods   The bootstrap and the jackknife (Efron, 1982; Efron and Gong, 1983) can be used to estimate the variability associated with a statistic for which the underlying sampling distribution is either unknown or difficult to derive analytically. These methods are called "resampling techniques" because they operate by estimating the form of the sampling distribution by repeatedly resampling data from the original data set; under certain reasonable assumptions (Efron, 1982) the distribution of the statistic of interest can be approximated from the distribution of the sample estimate over replications of the resampling process. The jackknife was first used in a phylogenetic context by Mueller and Ayala (1982). Felsenstein (1985) discussed the potential application of the bootstrap to the estimation of confidence intervals for phylogenies. Note that the presentation below is somewhat nonrigorous; interested readers should consult the above-cited papers for details.

The bootstrap and the jackknife differ in the way in which resampling is performed. In the bootstrap, data points are sampled randomly, with replacement, from the original data set until a new data set containing the original number of observations is obtained. Thus, some data points will not be included at all in a given bootstrap replication, others will be included once, and still others twice or more. For each replication, the statistic of interest is computed. If the sample data are in fact representative of the underlying population, the confidence interval associated with that statistic can be constructed by choosing the confidence limits so as to enclose the desired percentage (e.g., 95%) of the bootstrap replicate estimates. The jackknife (see Miller, 1974), on the other hand, resamples the original data set by dropping k data points at a time and recomputing the estimate from the remaining $n - k$ observations. Typically, k is set to 1, so that each of the n data points are dropped, in turn, and a "pseudoestimate" is computed from the remaining $n - 1$ points. The change in the estimate resulting from dropping one observation is, on average, $1/n$ as large as the change in the estimate that would result if we took a new sample of n observations from the larger population. Thus, we can estimate the variance of the estimate by extrapolating from the pseudoestimates to the population at large; we omit the computational details.

As applied by Felsenstein (1985), the "data points" are characters (columns of the data matrix) and the "statistic of interest" is a binary variable representing the presence or absence of a prespecified monophyletic group on the tree(s) resulting from each replication. Thus, characters are weighted according to the number of times they appear in each replicate sample; if a particular group occurs in 95% or more of the trees resulting from these replicates, one can conclude that the group is significantly supported at the 95% level.

Felsenstein (1985, 1988) attaches several caveats to the use of the bootstrap in this manner, of which we will discuss only three. First, for the confidence limits to be valid, a single group for which monophyly is to be tested must be specified in advance. Otherwise, we run into a "multiple-tests" problem similar to the one arising in a posteriori comparison of means following an analysis of variance: inflation of the type I error rate above the nominal level. (The problem may be circumvented to some degree if the researcher interprets the frequency in which a group appears in replicate trees as an "index of support" rather than as a statistical statement, but this interpretation is far from satisfactory.) If we are interested in testing the monophyly of more than one group or if we are unable to prespecify the group(s) of interest, we can adjust the significance level to allow for the fact that we are conducting more than one test (e.g., by dividing the significance level by the number of tests implied, analogous to a Bonferroni interval). However, if the groups cannot be prespecified, the number of potential groups is so large that an almost hopelessly low α level would be required to maintain an overall type I error rate of, say, 0.05, so that the resulting confidence interval is distressingly inclusive.

A second caveat is that the bootstrap can assume only that the data at hand are representative of the underlying distribution and thereby estimate the variation that would be obtained by sampling additional data from that distribution. If the data are not representive or if the reconstruction method makes an inconsistent estimate of the phylogeny (see below), then the resulting confidence intervals are not meaningful. Third, the confidence intervals obtained via resampling methods are only approximate unless the original sample size (number of characters) is large. Unfortunately, "large" usually conveys a very different meaning to a statistician than to a biologist; further work, perhaps computer simulation studies, will be needed to assess whether the number of characters typically available to a molecular systematist is "large enough."

Bootstrapping and jackknifing can easily be used either with methods that operate on characters directly or with methods in which character data are first transformed to distances. In character-based methods, weighting vectors corresponding to the number of times each character is sampled can be constructed and input to the analysis. For distance methods, the resampling is conducted prior to calculation of the distance matrix; each replication is then performed using a different input matrix corresponding to the replicate sample.

Sanderson (1989) has pointed out that even when it is impossible to place high confidence in the monophyly of particular groups, the bootstrap can be used to test other interesting hypotheses of relationship. For example, one can ask whether a given group—even when not monophyletic— is confidently contained within a larger monophyletic group. Any reader interested in pursuing bootstrap analyses should consult Sanderson's insightful paper.

Systematic Errors

When the evolutionary process does not satisfy the assumptions crucial to a particular method, a bias may be introduced into the evaluation of alternative phylogenies, favoring some branching patterns and decreasing the support for others. If the bias becomes sufficiently great, it may overcome the legitimate support for the correct tree and lead the researcher to an incorrect conclusion. Because the effect is systematic, the addition of more data will tend to solidify the incorrect conclusion (and the method is said to be "inconsistent" or "positively misleading" under those conditions). For an error to occur, the magnitude of the bias must exceed the valid support for the correct branching order. Furthermore, the bias must be in the direction of an erroneous tree, as it is possible for systematic bias to increase apparent support for the historically correct tree. Thus the presence of a bias does not necessarily lead to wrong answers, but it does cast doubt on the validity of the inference process.

Conditions Leading to Systematic Errors Fortunately, the situations likely to lead to systematic errors under most of the methods described above are relatively well understood.

Parsimony. If the number of actual sequence changes per sequence position in a macromolecule is always small (zero or one), then parsimony will correctly reconstruct the phylogeny given enough data (Felsenstein, 1978a). As the number of changes increases, the proportion of those changes that are homoplastic (parallel, convergent, or reversed) increases. If the tree is relatively "dense" (i.e., branch lengths are short enough so that the expected number of changes on any one branch is small), these homoplastic changes will be detected as such. However, multiple changes cannot be detected on long, unbranched lineages, thereby creating a potential for artifactual association of taxa at the ends of these long branches (Felsenstein, 1978a). Because this potential was first recognized in evolutionary scenarios with at least one particularly diverged lineage (so that a greater than average rate of change would be evident in any rooted representation of the tree), it has sometimes been called the **unequal rate effect** (e.g., Lake, 1987). Unfortunately, this association is misleading: unequal rates are neither necessary nor sufficient for the error; it is long, unbranched lineages that are required.

Cluster Analysis.   If the assumption of ultrametricity is satisfied and no distance values between sister taxa are missing from the data matrix, cluster analysis will be free of systematic errors. But if two lineages are not equally distant from a third, more diverged lineage (i.e., if the pairwise distances are not ultrametric), systematic errors will be introduced. As pointed out above, satisfaction of the three-point condition establishes that the distances are ultrametric.

Additive Tree Techniques.   The additive tree techniques presented are free of systematic error if the distance data are additive (satisfy the four-point condition) and no distance values between sister taxa are missing from the data matrix. This internal consistency of the technique places the burden of accuracy on the estimation and transformation of the distance data as opposed to the actual tree inference procedure. Specifically, the model used to correct for unobserved changes ("multiple hits") must reflect some degree of evolutionary reality. To the extent that it does not, additive-tree methods are susceptible to systematic errors.

Recognizing Systematic Errors   There is no foolproof method for identifying artifacts in phylogenetic trees that are due to systematic errors in the data or tree inference method—if such a prescription existed, we would routinely produce "error-free" trees. There are, however, a few techniques that can help in recognizing whether systematic errors are likely to have occurred.

Sensitivity to Specific Taxa in Tree.   If the data and tree inference technique were ideal, analyzing any two subsets of taxa would yield congruent trees (i.e., the trees would be identical after pruning taxa absent from one or both trees). In practice this is not the case. (Otherwise, finding optimal trees would be almost trivial, since constructing a tree by sequential addition of taxa would always lead directly to the globally optimal tree, regardless of the order of addition.) In practice, both systematic and random errors can distort the tree so that the inferred branching order is a function of the taxa included. Because the total error contains both systematic and random components, variation with the sampling of taxa does not necessarily indicate systematic errors, but it is suggestive. In particular, the amount of variation in branching order can be compared to that resulting from boot-strap resampling of sequence positions. In addition, most sources of systematic error are expected to increase with branch length; therefore, if the changes are specific to the most diverged taxa, then there is again reason to suspect systematic errors.

Lanyon (1985) has described a jackknife method that evaluates taxon stability by computing T trees, each time leaving out one of the taxa. By computing a strict consensus of these trees using a method that allows different subsets of taxa to be contained on each of the rival trees, the worker can determine which relationships are consistent. Felsenstein (1988) suggests that this method may not have the properties of a statistically valid jackknifing procedure, but it nonetheless provides a useful index of which groups are most stable.

Contribution of Individual Taxa to the Optimality Criterion.   If the placement of a particular taxon is problematic (due to systematic errors), removal of that taxon from the tree will frequently make a disproportionate change in a measure of tree quality, such as the least-squares criterion in a distance tree or the estimated homoplasy (e.g., **consistency index**) of a tree derived by parsimony.

Inferences Based on Different Molecules.   Phylogenetic relationships inferred from two or more different molecules should, in theory, be congruent. If they are not, we should ask why not. It is important to avoid confusing differences between the "optimal trees" with the conclusion that the results are incongruent: the former might simply reflect uncertainty in one or both trees, whereas the latter asserts the existence of a significant conflict. One method for deciding between these two possibilities is to fit each data set to the tree(s) derived from the other data set(s). Most modern programs allow the input of user-defined trees for evaluation under a particular optimality criterion. For example, suppose tree 1 is optimal for data set A and tree 2 is optimal for data set B. If tree 2 is nearly as good as tree 1 for data set A, and if tree 1 is nearly as good as tree Y for data set B, then there is no real conflict, just inadequate information. This result can sometimes occur even though the two trees differ substantially in their topologies!

If a conflict cannot be resolved on the basis of known uncertainties, then one of the following possible explanations should be considered: the use of nonorthologous genes (e.g., a tree with mouse and rabbit α-globin and rat β-globin; paralogy), lateral gene transfer (xenology), or the presence of systematic errors in one or both trees.

Nonparametric Approaches.   An additional source of guidance in evaluating a tree inferred from distance data (or for which pairwise distance estimates can be generated from the character data) is nonparametric tests. In practice, the usefulness of these tests is dependent on the details of the tree inferred, and in many circumstances the tests may not be able to distinguish alternatives. An illustration of a case in which they might be useful is provided by the trees in Figure 23. A comparison of the paths from A and B to D yields the expectation that $d_{AD}>d_{BD}$ for all three trees. Let us assume that this trend is significantly supported by the data (for example, the trend is verified by bootstrap samplings of sequence positions). If we now consider the relationships of C to A and B, we expect that $d_{AC}>d_{BC}$ in trees 1 and 3 (an expectation that could also be true of a minor variant of tree 2), whereas $d_{BC}>d_{AC}$ is consistent only with tree 2. Again, we can examine the data directly to see if one of these inequalities is significantly supported. In particular, if we observe that $d_{BC}>d_{AC}$, then we must conclude that trees 1 and 3 are incorrect, leaving tree 2 by elimination. Yet, if tree 2 were historically correct, systematic errors could have biased the tree inference procedure to group the long branches leading to C and D, leading to the incorrect choice of tree 1. The reason that it is possible to infer tree 2 from the data and yet to find certain distances significantly inconsistent with that tree lies in the particular ratios of branches and in the fact that the latter test does not need to examine the most underestimated distance (that separating C and D). In contrast, the tree inference procedures discussed would include the distance from C to D (directly or indirectly) and potentially be misled by this value.

Avoiding Systematic Errors   Several strategies are available to the researcher to minimize the effect of systematic errors on a phylogenetic analysis.

Removing Long Branches.   A practical consideration in the inference of trees from pairwise distance data is that most methods of estimating pairwise distances tend to underestimate long distances. As noted in the dis-

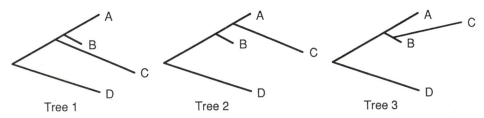

FIGURE 23. Three alternative trees relating four taxa that can be distinguished by a nonparametric test on the distance data. See text for discussion.

cussion of the Fitch and Margoliash technique, pairwise distance methods include all measurements in the calculations as though they were independent. Therefore, having many long distances in a tree will tend to compound errors. To work around this problem, the use of outgroup sequences should be kept to a minimum when using a pairwise distance method. However, the substitution of different outgroup taxa, one or two at a time, can still be used to evaluate the reliability in the position of the root.

Ironically, the effect of multiple outgroups in parsimony is almost exactly the opposite. The use of multiple outgroup taxa will tend to divide the longest branch in the tree, thereby decreasing its tendency to attract other long branches. To be most effective, however, additional outgroups should be chosen so as to divide long branches reasonably evenly; adding an extremely close relative of a very distant outgroup will gain little. Of course, the benefits of adding additional taxa are not limited to the outgroup. Long branches (sparse regions) within the ingroup can also contribute to systematic errors, and multiple substitutions are more easily detected in dense regions. A somewhat paradoxical phenomenon results. With large numbers of taxa, correctly inferring every aspect of the true topology is extremely difficult, but if we were interested in the relationships of, say, only four taxa, we would be much better off to compute a tree for 20 taxa (interspersed among the four of interest) and prune 16 of them from the tree than to compute the tree for only the four taxa. As an aside, we note that for this reason, comparisons of methods that are based on simulated or actual four-taxon data sets are often misleading because the behavior of a method may be quite different for a four-taxon data set than for a larger one.

Eliminating Unreliable Data. Another practical consideration concerns the fact that a branch is long because a large number of substitutions have occurred in the sequences being compared. Limiting an analysis to those sequence regions in which positional homology is most certain tends to exclude the most variable positions in sequences, thereby shortening branches and decreasing the sensitivity of the analysis to multiple substitutions. This concept can be pushed further: if variable regions can be identified in a set of sequences, then they might be eliminated from the analysis, even if their positional homology is not in doubt. This phenomenon provides one motivation for character weighting.

Weighting Characters. Obviously, all characters are not equally informative with respect to the evolutionary history of the taxa under study. Some characters are both informative and reliable; they are telling us the truth about their past. Other characters may be reliable but uninformative: while they are not actively misleading us, they are not telling us anything very useful either. The reason that phylogenetic analysis is so difficult lies in the third category of characters: those that are misinformative. If we could somehow deduce which characters were in fact the unreliable ones, the task of reconstructing evolutionary trees would be greatly simplified, since we could minimize their influence in the analysis by giving them less weight.

Identification of unreliable characters is also an effective way to avoid systematic errors. By assigning lower weight to the characters that either violate the assumptions of a method or are known to predispose the method to inconsistency, we can minimize the likelihood that systematic errors will occur. For instance, parsimony methods will be consistent when probabilities of character change are low, and consequently work best if the events being minimized (homoplastic changes) are in fact the rare events. If the rapidly evolving characters are recognized as such and given little weight in the analysis, the problem of attraction of long branches due to chance convergences will be eliminated. Unfortunately, however, beyond the use of alignment difficulty as a criterion for macromolecular sequences (see above), objective assessments of character reliability are not easily accomplished.

One extreme form of weighting is the elimination of characters, as discussed above. By assigning one set of characters the maximum weight (unity) and another set of characters the minimum weight (zero), we essentially assert that there are two classes of characters, one comprising characters that, at least on an a priori basis, are all equally reliable, the other containing characters that are worthless. If we believed that characters actually behaved this way, we would use a method of analysis known as **character compatibility** (Felsenstein, 1981b), which searches for the largest "clique"—a set of mutually "compatible" characters that can all evolve without homoplasy on the same evolutionary tree (e.g., Le Quesne, 1982; Estabrook, 1983). Compatibility methods are no longer in widespread use, probably because of their implicit adherence to an unrealistic model that asserts that once a character has been excluded from the largest clique, it no longer conveys any useful information whatsoever.

A promising approach that uses compatibility as an objective weighting criterion (rather than to infer phylogeny directly) has been developed by Penny and Hendy (1985, 1986). Sharkey (1989), apparently unaware of the work of Penny and Hendy, has described a related approach, but limited to binary characters. The strategy of these workers is to count the observed number of incompatibilities (O_j) between each character and each other character. (For methods to test the pairwise compatibility of unordered multistate characters, see Estabrook and Landrum, 1975; Fitch, 1975, 1977; Sneath et al., 1975.) To convert this number to a weight, Penny and Hendy recommend computing the number of incompatibilities expected by chance (E_j) if the distribution of states for each character was independent of that for other characters (i.e., free of any nonindependence imposed by their evolution on a common phylogeny). Penny and Hendy (1985) tested several weighting functions, but seem to have settled on the simple relationship

$$w_j = \max[\,1 - (O_j/E_j), 0\,]$$

Thus, a character that is compatible with all other characters gets the maximum weight (unity) whereas a character that is incompatible with as many characters as would be expected by chance alone gets zero weight. More importantly, characters that fall between these two extremes get intermediate weight. (Note that if the observed number of incompatibilities actually exceeded the expected number, a negative weight would be assigned unless the weights are constrained to be nonnegative.)

Another approach to character weighting is to estimate optimal weights by successive approximation (Farris, 1969). An initial set of weights (perhaps uniform weights) is used to obtain an initial estimate of the tree. From some measure of the fit of the characters to this tree, a new set of weights is derived, which are then used to estimate a second tree. The iterative rederivation of weights and recomputation of trees continues until the solution converges (i.e., the tree derived from a new set of weights is identical to the tree that was used to derive those weights). Farris (1969) used reweighting functions based on the consistency index (Kluge and Farris, 1969), defined as r_j/l_j where r_j is the "range" of character j (defined as the minimum number of steps that the character would require on any possible tree) and l_j is the length required by the character on the tree at hand. Thus, characters that change the minimum possible number of times have perfect consistency (1.0), whereas characters that change more often have lower consistencies (approaching zero in the limit).

The danger inherent in any successive approximations (a posteriori) approach to character weighting is circularity (Neff, 1986). It is easy to see that a character inconsistent with the initial tree and downweighted as a result will have less influence in the second iteration than it did in the first. But there are some trees on which the character would have been perfectly consistent, and would therefore have been given maximum weight. Farris (1969) tested the effectiveness of his successive approximations method by adding random noise to a data set containing otherwise compatible characters and testing whether the noise characters were in fact the ones assigned little weight in successive iterations; they were. We would suggest that one should not become overconfident on seeing this kind of result, however, as characters in real data sets do not fall cleanly into "completely reliable" versus "random noise" categories. We illustrate our point with an extreme example. Consider a data set containing two subsets of characters, each of which contains characters that are pairwise compatible within the subset, but incompatible with those of the other subset. If there is even one more character contained in the first subset than in the second, then the weights for the characters in the second subset will receive lower weight in the second iteration. Thus, a researcher could easily be misled into believing that there was substantially greater support for the tree consistent with the first subset than for the tree consistent with the second subset, when clearly this conclusion would be unjustified. For this reason, we are inclined to favor a priori weighting methods such as that of Penny and Hendy (1985, 1986) over a posteriori and/or iterative methods such as that of Farris (1969). There is need for more work on this topic, but with the evaluation of a weighting method's success being based on a more defensible criterion than the method's ability to reduce the number of trees that are considered optimal (e.g., Carpenter, 1988).

SOFTWARE PACKAGES AVAILABLE

We conclude with a brief overview of the major computer software packages for phylogenetic analysis. We have chosen to limit this summary to those programs that are readily available, actively supported, and in widespread use. This decision is not intended to discourage readers from pursuing other sources of useful software; it simply reflects the fact that the number of potentially useful programs is so great that it would not be practical to attempt a listing of them here. (Several of the papers cited in this chapter report the availability of programs that implement methods described therein.)

Due to the rapidity with which software evolves, it would be risky to go into great detail in the descriptions of these programs, as this information might quickly become obsolete or erroneous. Instead, we will merely list some of the major features of each program. Note that for the same reason we do not provide information on cost or availability; interested readers should consult the original authors for this information.

The four most widely distributed programs useful for phylogenetic analysis of molecular data are described below.

1. PHYLIP: Phylogeny Inference Package (Joseph Felsenstein, Department of Genetics SK-50, University of Washington, Seattle, WA 98195, U.S.A.). PHYLIP is a collection of about 30 independent programs implementing maximum likelihood, parsimony, compatibility, distance, and invariants methods, plus some utility programs. Some of the programs provide bootstrap methods for estimating confidence limits. PHYLIP is generally distributed as Pascal source code and is easily implemented on most computer systems. (It is also available from other sources in precompiled form for IBM-PC and Macintosh microcomputers; contact Felsenstein for details.)

2. PAUP: Phylogenetic Analysis Using Parsimony (David L. Swofford,

Illinois Natural History Survey, 607 E. Peabody Dr., Champaign, IL 61820, U.S.A.). PAUP performs parsimony analysis under a variety of models, including the Wagner, Fitch, Dollo, Camin–Sokal, and generalized methods discussed herein. Lake's invariants and their associated statistics are calculated, and bootstrapping routines are available. Also available from the same author are BIOSYS-1 (Swofford and Selander, 1981), which includes cluster analysis and distance Wagner routines for gene frequency data, and FREQPARS, which implements the method of Swofford and Berlocher (1987). PAUP is distributed in precompiled form for IBM-PC and Macintosh microcomputers, and as C source code for workstations, minicomputers and mainframes.

3. Hennig86 (James S. Farris, 41 Admiral Street, Port Jefferson Station, New York, NY 11776, U.S.A.). Hennig86 is a small, fast, and effective program for parsimony analysis under the Wagner (= "additive") and Fitch (= "nonadditive") models. Successive approximations character weighting is available. The program also has rudimentary interactive tree editing and character diagnostic capabilities. The program is distributed as executable program file for IBM-compatible computers only.

4. MacClade (written by Wayne P. Maddison and David R. Maddison; distributed as of summer 1991 by Sinauer Associates, Sunderland, MA 01375). MacClade is useful in the analysis of character evolution and the testing of phylogenetic hypotheses under the same parsimony models described for PAUP plus some additional ones. Although the program is not intended as a sophisticated "tree-finding" program, it provides powerful graphically oriented diagnostic information including several features designed specifically for molecular data (charts of frequencies of base changes on one or more trees, a summary of the changes at each codon position, etc.). MacClade is available only for Macintosh computers. PAUP and MacClade use a common file format, so that data files can be freely interchanged between the two programs.

CHAPTER 12

AN OVERVIEW OF APPLICATIONS OF MOLECULAR SYSTEMATICS

David M. Hillis and Craig Moritz

INTRODUCTION

From the preceding chapters, it should be clear that the diversity of molecular techniques available to systematists is considerable, and that problems can be addressed with these techniques ranging from intrademic relationships to the phylogeny of life. The rapid development and power of these techniques has produced a euphoria in evolutionary biology; because so many new problems can be addressed, it is a commonly held misconception that all evolutionary problems are solvable with molecular data. This clearly is not the case. Worse, inappropriate techniques often are applied (at a considerable waste of time and expense) to particular problems that could be effectively addressed with alternative techniques. In other cases, the technique chosen may not be the most cost-effective choice. Therefore, we provide some guidelines in this chapter to aid in matching techniques to problems.

One common application (and an area of considerable controversy) of molecular systematics is the prediction of time from molecular divergence data (the **molecular clock hypothesis**). It is clear that molecular divergence is roughly correlated with divergence of time; however, there is considerable debate about constancy of rates of divergence and how much error is associated with predictions of divergence times from measures of molecular similarity (Chapter 1). There have been many applications of the molecular clock hypothesis with few serious attempts to determine confidence limits of molecular clocks. Part of the problem is that the combination of potential sources of error at various stages of analysis (e.g., errors of measurement in collecting the data, errors associated with transforming the data into a divergence measure, errors in calibrating the "clock" because of inaccurate paleontological estimates) make confidence estimations highly suspect; any realistic estimation of confidence limits must take all these factors into account (Carlson et al., 1978). In this chapter, we address some of the concerns of this field and suggest that much greater rigor and caution are needed in estimating divergence times from molecular data than are commonly excercised.

Finally, we address some current trends and the possible future of molecular systematics. This is a dangerous thing to do in a rapidly changing field; however, we see much unexplored potential in molecular systematics. We hope that the identification of current limitations and pitfalls throughout this volume will facilitate the further development of molecular systematics as a major tool in evolutionary biology.

CHOOSING A TECHNIQUE FOR A PARTICULAR PROBLEM

Advances in technology often promote various kinds of data chauvinism. When isozyme electrophoresis and microcomplement fixation began to be applied widely to systematic problems in the 1960s and early 1970s, the new biochemical data were immediately promoted by some as "better" than traditional morphological data. The development of DNA hybridization techniques and restriction fragment analysis was accompanied by new assertions of superiority, and individuals who worked with isozyme electrophoresis were chastised (in review of grants and publications, for instance) for being "old-fashioned." Most recently, some proponents of sequencing have suggested that all other techniques are superfluous and outdated. We disagree strongly with all of these assertions; certain techniques are better than others for answering particular problems, but no technique is best under all circumstances. Morphological data are clearly superior to molecular data under certain conditions (e.g., for studies of long-extinct species), just as the reverse is true under other conditions (Hillis, 1987; see Chapter 1). Only by combining data from various morphological and molecular techniques is it possible to obtain a comprehensive view of evolution.

Given the above caveats, we will now address the issue of choosing a molecular technique to address a particular problem. Table 1 lists many of the common applications of molecular techniques in systematics. We roughly classify each technique into one of four categories for each of the problems listed: the technique is either (1) inappropriate for the problem, (2) appropriate under limited conditions, (3) appropriate but not usually cost effective, or (4) appropriate under most conditions. By inappropriate, we mean that considerable time, money, and effort can be wasted by attempting to answer the given problem using a particular technique, with little likely fruition. Assuming that there are no technical barriers, the most common reason for such a failure is that there is too little or too much

Table 1. Applications of various molecular techniques to problems in systematics[a]

Problem	Isozymes	Immunology	Cytogenetics	DNA hybridization	Restriction analysis	DNA/RNA sequencing
Mating systems	+	–	M	–	M	$
Clonal detection	+	–	M	–	+	$
Heterozygosity	+	–	–	–	+	M
Paternity testing	M	–	–	–	+	+
Relatedness	M	–	–	–	+	$
Geographic variation	+	–	M	–	+	+
Hybridization	+	–	+	–	+	M
Species boundaries	+	M	M	M	+	$
Phylogeny (0–5 mya)	+	+	+	+	+	+
Phylogeny (5–50 mya)	M	M	M	M	M	+
Phylogeny (50–500 mya)	M	M	M	M	M	+
Phylogeny (500–3500 mya)	–	–	–	–	–	+

[a]Key: –, Inappropriate use of technique; M, marginally appropriate or appropriate under limited circumstances; $, appropriate use of technique, but unlikely to be cost effective; +, appropriate and effective method.

variation to address the question of interest. A technique is listed in the second category for a particular problem if success has been obtained under some conditions (when levels of variability are appropriate), but alternative techniques are more likely to yield more robust results for the same or less effort. The third category (appropriate but not cost effective) indicates that the given technique may be used to address the problem, but that other techniques will probably be as effective for much less effort and/or money. In other words, except under unusual circumstances, a technique is recommended for a particular problem only if it falls in the fourth category. One final caveat—the times of divergence given in Table 1 are very rough. Because rates of molecular divergence can be very different among lineages and among molecules (see below), the times should be used only as a first approximation.

A quick scan of Table 1 shows that two techniques, namely isozyme electrophoresis and restriction site analysis, are applicable to most studies of intraspecific variation (see Chapters 4 and 8). DNA sequencing is also applicable at this level, but most studies of intraspecific variation and population genetics require examination of large numbers of individuals over large numbers of loci (see Chapter 2). Although it has become easier to obtain sequences from many individuals for certain loci (particularly the mitochondrial genome) by amplifying the DNA (see Chapter 9), it is still inordinately expensive and time consuming to obtain sequence information from numerous nuclear loci for sufficient specimens. Therefore, for most studies at the intraspecific level, isozyme electrophoresis and restriction site analysis (including DNA fingerprinting) remain the usual tools of choice, although DNA sequencing may be needed to resolve particularly difficult problems.

For studies of mating systems, population structure, and heterozygosity estimates, isozyme electrophoresis remains the best technique available. These studies usually require information from many individuals at many loci, and are suited perfectly to the kind of data provided by isozyme electrophoresis. Cytogenetic analysis, particularly of meiotic configurations, can reveal significant changes in the genetic system (e.g., clonal inheritance, polyploidy, interchange heterozygosity) that are important in their own right and also affect interpretation of other types of genetic markers. Studies of individual relatedness require analysis of variation at large numbers of loci as well, and, under certain conditions, isozymes may provide this information. DNA fingerprinting (see Chapter 8) is perhaps the most powerful approach for inferring individual relatedness, but should be restricted to inferences about close relatives (Lynch, 1988). Theoretically DNA sequencing can be used with high precision, but only if many loci are examined from each individual, a feat not yet attempted with regularity.

Geographic variation within species, detection of clonal diversity within unisexual species, the origin of unisexual species, hybridization, and discovery of cryptic species are all effectively studied with both isozyme electrophoresis and restriction site analysis. Analyses of cpDNA and mtDNA, which are maternally inherited, can be combined with studies of nuclear loci (e.g., allozymes) to provide information on both the degree and biases in direction of hybridization. The two kinds of data can also be combined (often with cytogenetic data as well) to determine not only the species involved in initial hybridization events that gave rise to unisexual species, but also the sexes of each species involved in the hybridization event(s).

Detection of morphologically cryptic species is often accidental; with any molecular technique, one should be open to the possibility that previous perceptions of species boundaries may have been wrong. In many cases, systematists choose to investigate a suspicious "polymorphic" taxon; allozyme electrophoresis is most commonly used in these cases. However, many other cryptic species have been discovered accidentally. In addition to examples from studies of isozymes, cryptic species have also been discovered by cytogenetic techniques (e.g., Moritz, 1983) and by immunological techniques (e.g., Scanlan et al., 1980). In the former case, a morphologically variable nominal species of gecko, *Heteronotia binoei,* was shown to consist of several cryptic bisexual species and numerous parthenogenetic lineages of hybrid origin. These conclusions were subsequently supported by analysis of isozymes (Moritz et al., 1989a,b). In the immunological example, Scanlan et al. (1980) showed that some individuals of the nominal frog species *Gastrotheca riobambae* appeared to be more closely related to other species than to other individuals of *G. riobambae.* This led Duellman and Hillis (1987) to examine this group with allozyme electrophoresis, which extended the findings from immunology to suggest that six species in two different species groups had been confused under the name *G. riobambae.* After the species boundaries became clearer, diagnostic morphological traits were found for each of the species. As in this case, information on species boundaries from molecular data is often invaluable for separating intraspecific morphological polymorphisms from diagnostic characters.

Perhaps the most common application of molecular techniques in systematics is for estimation of phylogeny. All of the techniques discussed in this book have been successfully applied to questions of phylogeny, although the appropriate techniques will vary from study to study. For a technique to be useful for reconstructing phylogeny, enough variation must exist among species examined for application of phylogenetic reasoning, but there must be little enough variation that the characters under study are not saturated by change. To a first-order approximation, useful ranges of divergence can be predicted for each major technique (Table 1). However, these ranges are very rough; some groups show much less variation for certain characters, and application of a given technique may be extended further back into time for such groups (see also the section below on predictions of time from molecular data). For example, mtDNA and many commonly examined isozyme loci can be used to study higher taxonomic levels of birds than is possible within most other groups of vertebrates (Kessler and Avise, 1985b). In groups that have never been studied, some experimentation may be required to find a suitable technique for a particular phylogenetic problem. For most studies, however, Table 1 will provide a guide to selection of an appropriate technique, at least for a pilot study.

Closely related species (diverged within the past five million or so years) are best studied by examining relatively fast-evolving isozyme loci (see Chapter 4) or, in animals, the mitochondrial genome (see Chapters 8 and 9). The other techniques have, on occasion, proven useful in this range, but in the majority of cases are not sensitive enough to detect sufficient changes over such a short time scale. Even with these rapidly evolving sequences it remains impractical to resolve very recent (e.g., post-Pleistocene) divergences, because it is difficult to sort uniquely derived character states from random fixation of ancestral polymorphisms (Arnold, 1981; Neigel and Avise, 1986). The most common time frame of divergence studied by systematists (roughly 5–50 million years) is within the range of study of virtually all of the techniques discussed in this book. Further back into time (50–500 million years: Table 1) most of the techniques are relatively ineffective, except for sequencing relatively conserved genes (Chapter 9) and perhaps comparing changes in organization of organelle genomes (Chapter 8). Beyond 500 million years divergence, only sequencing the most conserved genes has been effective for phylogeny reconstruction. In this range, adequate resolution of closely spaced divergence points becomes highly unlikely using any technique.

If several techniques are appropriate for addressing a particular problem, cost and the availability of technology often are paramount. Costs for laboratory setup and operating expenses vary considerably, but in general isozyme electrophoresis and cytogenetics are the least expensive techniques per specimen examined, whereas immunology, DNA hybridization, and restriction analysis are several times as expensive, and nucleic acid sequencing is the most costly approach. However, this does not mean that a given problem will always be answered with less money by the less expensive techniques, because considerable money (and time) can be wasted by trying to apply an inappropriate technique to a particular question. All heritable information is potentially accessible to DNA sequencing, whereas only subsets of this information are accessible to the other techniques. Often, the choice of technique will depend on the resolution required to address the question of interest.

For many problems, it will be useful to use more than one approach. For example, simultaneous examination of chromosomes, allozymes, and mtDNA to investigate population structure, clonal diversity, or hybridization phenomena can provide qualitatively superior information than would be obtained from the use of any one approach. For phylogenetic studies, it may often be valuable to study sequences that evolve at different rates to resolve different parts of the phylogeny. An investigator may choose to compare allozymes to identify groups and to obtain some phylogenetic information within and between groups, rapidly evolving sequences (e.g., animal mtDNA) to resolve relationships within groups, and slowly evolving sequences (e.g., rDNA) to further resolve among group relationships or to root the tree by comparison to outgroup taxa.

PREDICTIONS OF TIME FROM MOLECULAR DATA

A common, although controversial, application of molecular systematic data concerns estimating times of divergence. In the early 1960s, Zuckerkandl and Pauling (1962) suggested that genes and their protein products might evolve at rates constant enough that measures of molecular divergence could be used to calibrate a "molecular clock." Recent advocates of this hypothesis view molecular differentiation not as a metronome, but as a process with regularity of the same order of magnitude as radioactive decay (Wilson et al., 1977, 1987). This has promoted the use of molecular divergence measures to provide a time frame for phylogenies, particularly where there are insufficient data from fossils. This approach is prone to a relatively large margin of error (see below), but in some cases these estimates of time provide important biogeographic and evolutionary insights (e.g., Sarich and Wilson, 1967; Wilson et al., 1974, 1975; Beverley and Wilson, 1985; Bowen and Avise, 1989).

Although it is obvious that molecular divergence is correlated with time, there is considerable disagreement about the constancy of divergence rate and the use of estimates of molecular differentiation to predict times of divergence. Proponents of a molecular clock argue that a sloppy clock is better than no clock at all, whereas detractors argue that estimates of time based on molecular divergence are compounded by too many sources of error and are highly misleading. Perhaps the largest source of error is in paleontological calibration of the molecular clock; even with an outstanding fossil record for a group, it is exceedingly difficult to pinpoint the age of the last common ancestor of a group of living species (Carlson et al., 1978). In addition, rates of divergence are known to differ between many groups of organisms, and accurate paleontological data are rarely available for the group of interest. In some cases, times have been estimated from another type of molecular data to calibrate a clock. This approach is suspect, as it assumes that the initial calibration (based on another group) is valid for the group in question, and because the errors associated with calibration of the two clocks are compounded. There is also the residual error associated with the regression of time on molecular divergence. Finally, there is the measurement error associated with the particular measure of molecular divergence (see also Carlson et al., 1978).

In general, molecular clocks are calibrated by dividing the average estimate of the age of the last common ancestor by the average measure of molecular divergence. The only error that is generally taken into account is that associated with the estimate of molecular divergence. However, this error is usually insignificant in comparison to other sources of error. The regression model that has been used to establish calibrations can also be used to place confidence limits on a new estimate of time; this model is usually a simple weighted linear regression of time on molecular divergence, with the constraint that the intercept of the regression line is the origin. The common calibration technique of dividing the average time of divergence by the average molecular divergence will produce the correct slope of this regression under the assumption that the residual error of the regression is proportional to molecular divergence (Snedecor and Cochran, 1980). In other words, this method is acceptable if there is proportionally greater deviation about the regression line at high levels of molecular divergence than at low levels of molecular divergence. This seems to be a reasonable assumption, and plots of time versus molecular divergence generally follow this pattern (Figures 1–3).

Calculating times that correspond to the 95% confidence limits on the estimates of molecular divergence to produce confidence limits on the predicted value of time is inadequate, because this source of error is trivial compared with the residual error of the regression. It is more rigorous to assume that the error associated with the molecular measure is trivial and calculate confidence limits for predictions of time based on the regression. Figure 1 shows two 95% confidence limits for a regression of time on molecular divergence. The data are based on percentage divergence (corrected for expected multiple substitutions, C%) of silent substitutions in coding regions of several genes compared among various pairs of mammals (from Britten, 1986). A weighted linear regression of time (Y) on divergence (X) gives $Y = 1.39X$ (as represented by line A in Figure 1). The standard deviation of the residuals under this model of regression is given by

$$s_{Y \cdot X} = \{[\Sigma(Y^2/X) - (\Sigma Y)^2/\Sigma X]/(n - 1)\}^{0.5}$$

and the standard deviation of the slope ($b = 1.39$) is given by

$$s_b = s_{Y \cdot X}/\sqrt{\Sigma X}$$

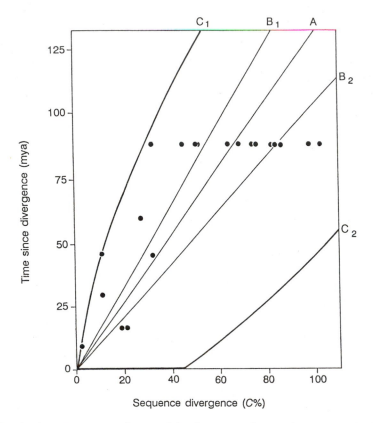

FIGURE 1. Regression (A; see text for model) of estimated time since separation on divergence of silent substitutions (corrected for expected multiple substitutions) in coding regions of DNA in mammals (data from Britten, 1986). B_1 and B_2 are the bounds of the 95% confidence limit of the regression line. C_1 and C_2 are the bounds of the 95% confidence limits for predicted values of time based on new measurements of sequence divergence, except that regions of negative time are collapsed to zero.

(Snedecor and Cochran, 1980). For the data of Britten (1986) shown in Figure 1, $s_b = 0.137$; since Student's t at $\alpha = 0.05$ (with $N - 1 = 19$ degrees of freedom) is 2.093, the 95% confidence interval of b is 1.39 ± 0.287. This interval is bounded by B_1 and B_2 in Figure 1. Although this provides confidence limits for the slope of the regression, it does not provide confidence limits for a new predicted value of time given a known sequence divergence. The standard deviation of a new predicted value of time ($\hat{Y}$) is given by

$$s_{\hat{Y}} = s_{Y \cdot X} \sqrt{X^2 / \Sigma X^2 + X}$$

(Snedecor and Cochran, 1980).* The 95% confidence limits of new estimates of time from the data presented in Figure 1 are represented by lines C_1 and C_2. These limits are quite large; for instance, at C% = 50, the 95% confidence interval is 69.5 ± 65.34 million years.

The above approach can be used to calculate confidence limits for all estimates of time based on molecular clock calibrations. Typically, one will find that the estimates have extremely broad confidence limits. However, if one is interested in applying molecular clock models to questions of time since divergence, then the error associated with the estimates of time cannot be ignored.

The data that have been used to calibrate two other molecular clocks are plotted in Figures 2 and 3. In Figure 2, data on mtDNA sequence divergence are plotted against time since divergence information derived from the fossil record of primates (from Brown et al., 1979). This calibration is widely used as a standard mtDNA clock (Wilson et al., 1985), although Moritz et al. (1987) stressed likely errors associated with paleontological calibrations and from variation among lineages. Although the confidence interval for this calibration is smaller than that in Figure 1, it is still large enough to be quite limiting for most applications of a molecular clock. In addition, it remains to be demonstrated that a calibration based on divergence in primates is generally applicable (Powell et al., 1986; Vawter and Brown, 1986; cf. DeSalle and Templeton, 1988; Shields and Wilson, 1987).

The calibration of albumin divergence based on immunological comparisons among birds (Prager et al., 1974; Figure 3) shows that the confidence limits of new predicted values of time may be so large as to not exclude any reasonable possibility. Note, however, that the confidence limits for the slope of this calibration for birds (B_1 and B_2 in Figure 3) do not include the rate reported for mammals (D in Figure 3), as Prager et al. (1974) correctly concluded. This highlights the necessity of calibrating molecular clocks within the group of interest.

The values of time since divergence used in Figures 1–3 are by no means universally accepted; indeed, the extreme difficulty with which such data may be garnered from the fossil record is a major obstacle to calibrating molecular clocks. We have used the original data on which these calibrations were based to provide confidence limits for estimates derived from the calibrations. New calibrations based on new values of time are possible, but these calibrations should be accompanied by newly calculated confidence limits.

We attempted to provide confidence limits for an allozyme (Nei's genetic distance) clock, but were unable to locate any applicable data. As noted by

---

*It is important in calculating these confidence limits to recall that the regression model assumes the residual error of the regression is proportional to molecular divergence and that the regression line runs through the origin. Confidence limit calculations that assume the residual error is the same for all values of molecular divergence (e.g., Carlson et al., 1978) seriously underestimate the actual confidence limits.

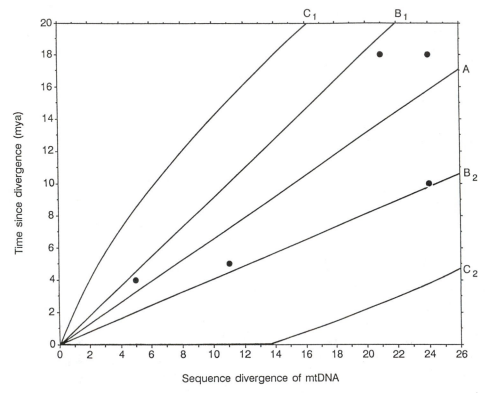

FIGURE 2. Regression (A) of estimated time since separation on sequence divergence of mitochondrial DNA in primates (data from Brown et al., 1979). Key to confidence limits same as for Figure 1.

Avise and Aquadro (1982), "the major obstacle to critical tests of the electrophoretic protein clock is the almost total lack of reliable independent information about times of speciation." This lack of data has resulted in an enormous range of estimated rates for divergence of Nei's D. For instance, the time for accumulation of a Nei's D of 1.0 in various groups of vertebrates has been given as anywhere from 0.7 to 18 million years (Figure 4; Avise and Aquadro, 1982). With such a range of estimated rates, "it is hard to imagine a genetic distance estimate that would not be 'compatible' with almost any fossil or geologic data" (Avise and Aquadro, 1982). Combined with the fact that no confidence limits can be assigned to estimates derived from any of the possible rates because of the lack of calibration data, estimates of time based on Nei's D are probably no better than arbitrary guesses.

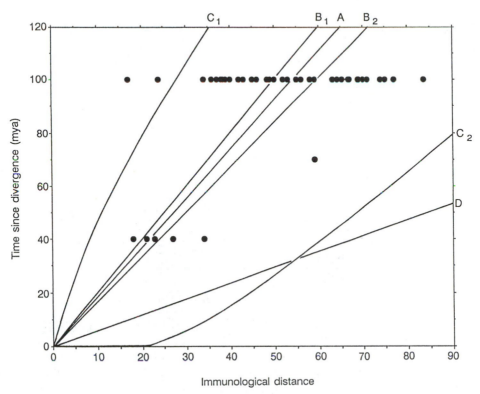

FIGURE 3. Regression of estimated time since separation on immunological distance. The data points are the same that were used by Prager et al. (1974) to calculate the rate of albumin evolution in birds (A); hence, some of the points are averages for comparisons of several species. Key to confidence limits is the same as for Figure 1. D is the reported relationship between time since divergence and albumin immunological distance for mammals (Sarich and Wilson, 1967). Confidence limits of D cannot be calculated because of an insufficient number of data points.

In summary, the following guidelines should be considered when estimating time from values of molecular divergence: (1) For any estimate of time, reference should be made to an explicit calibration of the clock for the particular type of molecular data analyzed. This calibration should be based on independently derived estimates of time since divergence (e.g., the fossil record, but not other molecular data). Many calibrations are based on single points or on no data at all; such calibrations are obviously insufficient. (2) The confidence limits of the calibration must be considered and these limits should be calculated for the new estimate of time. (3) Calibrations should be used only within the group of organisms for which the calibration applies. Extreme caution should be exercised before applying a calibration

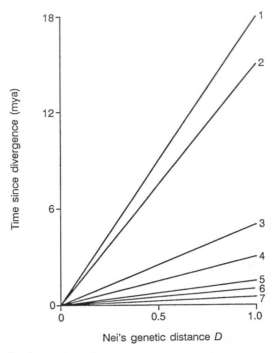

FIGURE 4. Proposed relationships between time since divergence and Nei's D in vertebrates (summarized in Avise and Aquadro, 1982). No confidence limits can be calculated for any of these regression lines because of a paucity of data. Rates 1, 3, and 6 have been suggested for various teleost fishes, rate 2 for plethodontid salamanders, rates 1 and 3 for squamate reptiles, rate 3 for birds, and rates 4, 5, and 7 for various mammals.

derived from one group of organisms to another group of organisms, as rates of molecular divergence often differ markedly among groups.

The above discussion suggests that caution is needed in predicting absolute times of divergence from molecular data. However, this in no way impedes the calculation of *relative* times of divergence, because many methods of phylogeny inference are relatively insensitive to differences in rates of divergence (Chapter 11). As an example, although we may have limited confidence in the absolute time that the orangutan lineage diverged from the common ancestor of humans, chimps, and gorillas, molecular data leave little doubt that this event occurred before the latter three lineages diverged (Slightom et al., 1987). Fortunately, the vast majority of applications of molecular systematics do not depend on calibrations (or even the existence) of molecular clocks. Differences in rates of divergence among lineages detract only from methods of analysis that require clock-like behavior of molecules, and alternative methods of analysis exist for all applications of molecular systematics except for the absolute estimation of time.

THE FUTURE OF MOLECULAR SYSTEMATICS

Molecular systematics has undergone a number of remarkable changes during the past two decades. These changes include not only technological developments and refinements (e.g., discovery and isolation of Type II restriction enzymes, development of DNA sequencing, discovery of a heat-stable DNA polymerase and its use in DNA amplification) but also major advances in issues of analysis. We expect these advances to continue, and we hope that discussions of the current limitations of data collection and analysis presented throughout this book will stimulate consideration of these issues. Improvements can come from technological developments per se, as well as from increased sophistication in the use of current methods. The interplay between our increased understanding of the evolutionary dynamics of molecules and their use as markers in systematic studies is fundamental to developing more efficient approaches to sampling, assaying variation, and interpreting results.

The 1990s will be an exciting decade for molecular systematics. We will almost certainly see entire prokaryote genomes sequenced and compared, which will provide us with opportunities to examine questions of whole genome evolution. The push to sequence the human genome will almost certainly spin off more comparative sequencing projects within eukaryotes. As more data are accumulated on the processes of genome evolution, this information can be incorporated into better and more reliable algorithms for estimating phylogenies. Molecular biology may begin to provide more information on the molecular basis of development, so that a true synthesis of molecular and morphological data can occur.

All levels of systematics are enjoying a renaissance, as the importance of understanding historical relationships in interpreting patterns throughout biology is beginning to be widely appreciated. In addition, the recent emphasis and concern for biodiversity and conservation have also placed more national and international attention on systematics (Wilson, 1985, 1986). This emphasis and attention can have either a positive or a negative effect on systematics. The effect will be negative if, in the rush to obtain systematic information, scientific rigor is abandoned. The effect will be positive if, in the need for accurate information, a premium is placed on rigorous data collection and analysis. We hope that this book will help to stimulate the latter course of action.

Acknowledgments

Chapter 2 / **Peter R. Baverstock and Craig Moritz**

We are grateful to M. Adams, R. Andrews, H. Dessauer, S. Donnellan, M. Mahoney, and R. W. Murphy for their constructive comments on earlier drafts of the manuscript.

Chapter 3 / **Herbert C. Dessauer, Charles J. Cole, and Mark S. Hafner**

We are especially indebted to Sheldon I. Guttman, Mia Molvray, and Elizabeth A. Zimmer for assistance with the botanical section, as our knowledge in this area is limited. Carol R. Townsend developed the folded aluminum foil packets and cardboard sleeves for packaging tissues from small organisms (Figure 1c,d). Robert M. Zink and K. Elaine Hoagland read and gave valuable advice on the manuscript. This chapter is an outgrowth of the Workshop on Frozen Tissue Collections and Management supported by the National Science Foundation (Dessauer and Hafner, 1984).

Chapter 4 / **Robert W. Murphy, Jack W. Sites, Jr., Donald G. Buth, and Christopher H. Haufler**

The refinement of electrophoresis methods was supported by grants to: RWM from the Natural Sciences and Engineering Research Council (NSERC A3148), the Department of Zoology, University of Toronto, and the National Institutes of Health (Minority Biomedical Research Support program NIH RR08156-10, D. J. Morafka [P.I.] and RWM); JWS, Jr. from the National Science Foundation (U.S.A.) (BSR 85-09092), the National Geographic Society (2803-84 and 3088-85), and the College of Biology and Agriculture and the M. L. Beam Life Science Museum, BYU. P. T. Chippindale, K. A. Coates, L. A. Lowcock, R. D. MacCulloch, and J. L. Sites provided assistance in the laboratory in the preparation of this manuscript.

We thank the following persons for supplying information and advice on staining recipes: Donald E. Campton, Paul T. Chippindale, L. W. Frick, Ronald G. Garthwaite, Carla Hass, Gennady P. Manchenko, Ronald H. Matson, Donald C. Morizot, and James B. Shaklee. We are indebted to Donald E. Campton and Ronald H. Matson, who provided us with their own lists of buffer formulas which we have consulted liberally. We also thank Herb Dessauer, Stephen D. Ferris, James B. Shaklee, and Gregory S. Whitt for many helpful comments over the years. Ross MacCulloch, Lisa Gilhooley, Marty Rouse, Cathy Rutland, and Cynthia Horkey greatly assisted with the preparation of the manuscript. Paul Chippindale, Maurice

Ringuette, Gregory S. Whitt and Ronald H. Matson provided valuable editorial comments. Brian Thompson prepared the equipment diagrams.

Chapter 5 / Linda R. Maxson and R. D. Maxson

We thank C. Hass and B. Hedges for suggestions and V. M. Sarich for a very thorough and helpful review of an initial version of our manuscript. Support from the Systematic Biology Program of NSF is gratefully acknowledged.

Chapter 8 / Thomas E. Dowling, Craig Moritz, and Jeffrey D. Palmer

TED and CM are eternally grateful to W. M. Brown for his inspiration and guidance. We thank S. Degnan, S. Lavery, R. Slade, and L. Joseph for comments on the manuscript. Supported by the National Science Foundation (USA), the National Institutes of Health, and the Australian Research Council.

Chapter 9 / David M. Hillis, Allan Larson, Scott K. Davis, and Elizabeth A. Zimmer

We thank L. Ammerman, J. Cracraft, M. Dixon, N. Gieg-Sinclair, M. Goodman, M. M. Miyamoto, C. Moritz, R. de Sá, J. Slightom, and D. Tagle for comments and suggestions on the manuscript. L. Ammerman and M. Dixon assisted in figure preparation. Research support to the authors from the National Science Foundation is gratefully acknowledged.

Chapter 10 / Bruce S. Weir

This is paper number 12105 of the Journal Series of the North Carolina Agricultural Research Service, Raleigh, NC 27695-7601. This investigation was supported in part by NIH Grant GM 11546.

Glossary

AAR   Amino acid replacement.

Acetone powders   Preparations obtained by grinding tissues in ice-cold acetone and allowing the acetone to evaporate from the resultant solids.

Additive distances   A set of distances between pairs of sequences or taxa that will precisely fit a unique, additive phylogenetic tree. Defined mathematically by satisfying the four-point condition (see Chapter 11).

Additive tree   A phylogenetic tree in which the distance between any two points is the sum of the lengths of the branches along the path connecting the two points.

Affinity   The binding strength of an antibody.

Alignment   The juxtaposition of amino acids or nucleotides in homologous molecules that are assumed to be positional homologs. A column in an alignment is assumed to contain residues that are all derived from a single common ancestral residue.

Allele   A particular form of a gene at a particular locus.

Allopatric   Occurring in geographically separate areas (see sympatric, parapatric).

Allozyme   An allele of an enzyme.

Alu-repeat   The most abundant interspersed repeated DNA family occurring in primates.

Anion   A negatively charged molecule.

Anode   The positive electrode in an electrolytic cell (such as an electrophoresis chamber) toward which anions migrate.

ANS   Anilinonaphthalene sulfonate.

Antibody   A large protein made in response to a foreign antigen (generally a protein).

Antigen   Any molecule that elicits an antibody response.

Antigenic site   A region of five to ten amino acids on an antigen to which antibodies can be elicited.

Apomorphy   A derived character state.

ATE   Abbreviation for an acetate-Tris-EDTA buffer (see Appendix, Chapter 9).

Autapomorphy   A derived character state unique to a particular taxon.

Autoradiograph   An image produced on X-ray film by placing a gel or filter that contains radio-actively labeled molecules next to the film in a light-proof cassette.

BCIP   5-Bromo-4-chloro-3-indolyl phosphate.

BGD   Bromcresol green dye.

BN buffer   Bicarbonate-nonidet buffer (see Chapter 6, Appendix).

Bootstrapping   A statistical method based on repeated random sampling with replacement from an original sample to provide a collection of new estimates of some parameter, from which confidence limits can be calculated (see Chapters 10 and 11).

bp   Base pair.

BrdU   Bromodeoxyuridine.

BSA   Bovine serum albumin.

Buffy coat   A thin layer of white blood cells that lies above the erythrocytes after vertebrate blood has been centrifuged.

Cathode   The negative electrode in an electrolytic cell (such as an electrophoresis chamber) toward which cations migrate.

Cation   A positively charged molecule.

C-bands   Dark bands on chromosomes produced by strong alkaline treatment at high temperature, followed by incubation in sodium citrate solution, followed by Giemsa staining. C-bands generally correspond to regions of constitutive heterochromatin.

Central branch   The interior branch connecting the two internal nodes of a four taxa, unrooted phylogenetic tree.

Chaotropic agent   In DNA–DNA hybridization, an agent that reduces thermal stability of base pairing of DNA.

Character   A variable feature that in any given taxon or sequence takes one out of a set of two or more different states (e.g., eye color or the amino acid at position 12).

Character compatibility   A method of phylogenetic analysis that seeks the largest clique of characters that can be fitted to a common tree so that each character state arises only once (see Chapter 11).

Character polarity   The inferred direction of change of a character in the phylogenetic tree; usually by reference to the character state in an outgroup.

Character state The specific value taken by a character in a specific taxon or sequence (e.g., green eyes or glycine at position 12). See character.

Character-state tree A description of the transitions among the states of a multistate character, especially when the transitions do not define a linear series of states.

Chromatid The eukaryotic chromosome prior to replication, or one of the two longitudinal subunits of a chromosome after replication.

Chromomere A region on a chromosome of densely packed chromatid fibers that produces a dark band (as on a polytene chromosome).

CI A mixture of chloroform and isoamyl alcohol, used in DNA extraction protocols (see Chapter 9, Appendix). Also an abbreviation for consistency index.

CIC See cold-induced constriction.

Cladogram Used in two ways by different authors: either a tree produced using the principle of parsimony, or a tree that depicts inferred historical relationships among entities. Generally, the branch lengths in a cladogram are arbitrary; only the branching order is significant. See phylogram.

Cluster analysis A rapid method of hierarchically grouping taxa or sequences on the basis of similarity (or minimum distance).

Coancestry coefficient The correlation of genes of different individuals in the same population; a measure of the relatedness of individuals within populations (symbolized by θ or F_{ST}; see Chapter 10).

Cold-induced constriction A chromosome-specific constriction induced in certain species with large chromosomes by prolonged treatment of the organism at 0.5–2.5°C in the presence of colchicine.

Complement A group of serum proteins that coats antigen-antibody complexes.

Concerted evolution The maintenance of homogeneity among members of a family of DNA repeats within a species.

Conformational isozymes Multiple forms of a single gene product that differ in secondary or tertiary structure.

Consistency index A measure of the amount of homoplasy exhibited by a set of characters or a tree, defined as the sum of the individual character ranges divided by the total length of the tree. If there is no homoplasy, these quantities will be equal, so that the consistency index achieves its maximum value of one.

Constitutive heterochromatin Regions on chromosomes consisting mostly of highly repeated, noncoding sequences.

C_0t Initial concentration of single-stranded DNA in a DNA reassociation experiment (in moles/liter) multiplied by time of incubation in seconds.

C_0t plot A plot of percentage of single-stranded DNA versus log of C_0t.

cpDNA Chloroplast DNA.

Criterion In DNA–DNA hybridization, the stringency of reassociation of single-stranded DNA measured by the difference between the T_m of perfect duplexes in the incubation buffer and the temperature of incubation.

Cryptic allele An undetected (by a particular technique) variant at a gene locus.

CTAB Cetyltrimethylammonium bromide.

C-value A measure of haploid DNA content per cell.

DAB Diaminobenzidine tetrahydrochloride.

DAPI Diamidino-2-phenylindole.

Dendrogram Any branching tree-like diagram.

DEP Diethyl pyrocarbonate.

Disequilibrium coefficient A term that describes the difference between a joint frequency of two or more alleles and the product of the frequencies of the separate alleles.

Dissimilarity A generic measure of the difference between two objects, usually measured on a scale of 0 to 1.

Distance A measure of the difference between two objects, usually measured on a scale of 0 to infinity.

Distance estimates A phrase used to emphasize the potentially imperfect reflection of evolutionary history in distance values inferred from experimental or sequence data.

DMSO Dimethyl sulfoxide.

DNA polymerase An enzyme that catalyzes synthesis of DNA under direction of a single-stranded DNA template.

Driver Unlabeled DNA used in DNA–DNA hybridization experiments (see Tracer).

EB Ethidium bromide.

EDTA Ethylenediaminetetraacetic acid.

Electrodecantation The settling of proteins of high molecular weight toward the bottom of a horizontal gel during electrophoresis.

Electroendosmosis Movement of ionized buffer

solution through a gel caused by gel charge groups.

Electromorph An electrophoretically indistinguishable class of isozymes. Electromorphs represent alleles if all differences between variants result in changes in electrophoretic migration rate.

Electrophoresis The separation of molecules in an electric field.

Epigenetic All processes relating to the expression and interaction of genes.

Evolutionary distance An idealized measure of the evolutionary separation of sequences or taxa, such as the total number of fixed mutational events. They are defined so that the values are additive and hence will precisely fit an additive evolutionary tree.

FCS Fetal calf serum.

F_{IS} See inbreeding coefficient.

F_{IT} See inbreeding coefficient.

FITC Fluorescein isothiocyanate.

Fixation index See inbreeding coefficient.

F_{ST} See coancestry coefficient.

F-statistics A set of coefficients that describe how genetic variation is partitioned within and among populations and individuals (see coancestry coeffient, inbreeding coeffient, and Chapter 10).

Gaps Editing symbols that are inserted into sequences in the process of alignment in order to compensate for presumptive insertion and deletion events.

GARGG Goat anti-rabbit γ-globulin.

G-bands Dark bands on chromosomes produced by Giemsa staining. G-bands occur primarily in AT-rich regions.

Gene conversion A genetic process by which one sequence replaces another at an orthologous or paralogous locus. May result from mismatch repair in heteroduplexes.

Genomic library A mixture of cloned DNA fragments (usually in viral or cosmid vectors) that together represent virtually all of an organism's DNA. Partial or subgenomic libraries contain only restriction fragments of a certain size range.

GTE Glucose-Tris-EDTA buffer (see Chapter 9, Appendix).

HAP Hydroxyapatite. HAP is used in columns to separate single-stranded DNA from double-stranded DNA (see Chapter 7).

Hardy–Weinberg equilibrium An equilibrium of genotypes achieved in populations of infinite size (in which there is no immigration, emigration, selection, or mutation) after at least one generation of panmictic mating. With two alleles A and B of frequency p and q, respectively, the Hardy–Weinberg equilibrium frequencies of the genotypes AA, AB, and BB are p^2, $2pq$, and q^2, respectively.

Heterochromatin Chromosomal segments or whole chromosomes that generally exhibit a condensed state throughout interphase and late replication. See constitutive heterochromatin.

Heteroduplex A hybrid DNA–DNA molecule formed from tracer and driver from different individuals or (usually) species.

Heterologous In reference to a homologous molecular probe from a species other than that being examined.

Heterologous reaction In immunological studies, a reaction involving antigen from one species and antibodies that were produced to the homologous antigen of another species.

Heteroplasmy The containment by one cell or individual of more than one type of a particular organellar DNA (e.g., mtDNA or cpDNA).

Heteropolymer A multimeric protein formed from products of multiple alleles.

Heuristic method Any analysis procedure that does not guarantee finding the optimal solution to a problem (usually used to obtain a large increase in speed over exact methods).

Homoduplex Reassociated DNA in a DNA–DNA hybridization reaction in which the tracer and driver are from the same individual (or sometimes, species).

Homology Common ancestry of two or more genes or gene products (or portions thereof).

Homomeric isozyme An enzyme composed of multiple identical polypeptide chains.

Homoplasy A collection of phenomena that leads to similarities in character states for reasons other than inheritance from a common ancestor. These include convergence, parallelism, and reversal.

Homologous reaction In immunological studies, a reaction in which antibodies are reacted with the same antigen to which the antibodies were produced.

HWE Hardy–Weinberg equilibrium.

Hybrizymes Alleles found in hybrid zones that are rare or absent in populations of the parental (nonhybrid) species.

Immunodiffusion An immunological assay that

monitors reaction of antibody and antigen in agar gels.

Immunoelectrophoresis An immunological assay in which antigens are first separated based on charge in an electric field and then exposed to antibodies to one or more antigens. Immuno-electrophoresis is used to identify antigens that two species have in common and test the purity of an antiserum.

Immunogenic A substance that is capable of eliciting the production of antibodies.

Immunological distance (ID) A measure that estimates the number of amino acid replacements (AAR) between two homologous proteins. For microcomplement fixation with albumin, one ID unit is approximately equal to one AAR.

Inbreeding coefficient The correlation of genes within individuals (symbolized by F or F_{IT}; this is the overall inbreeding coefficient), or the correlation of genes within individuals within populations (symbolized by f or F_{IS}; this is the within-population inbreeding coefficient; see Chapter 10). F_{IS} is also known as the fixation index. Both F_{IS} and F_{IT} are measures of deviation from expected Hardy-Weinberg proportions; positive values indicate a deficiency of heterozygotes while negative values indicate an excess of heterozygotes.

Ingroup An assumed monophyletic group, usually comprising the taxa of primary interest.

In situ hybridization The annealing of a mobile, labeled nucleic acid probe to a stationary nucleic acid target (usually whole chromosomes) to form base-paired duplexes.

Interior branches Branches in a phylogenetic tree that do not connect to a tip of the tree.

Internal nodes The branch points in a phylogenetic tree. If the tree is rooted, the root node is also an internal node.

Isoelectric point The pH at which the positive and negative charges of a protein are equal.

Isoloci Two or more loci of a multilocus enzyme system that produce products of the same electrophoretic mobility.

Isology Sequence similarity of aligned nucleic acids or polypeptides; the similarity may be due to homology or convergence.

Isoschizomer Restriction endonuclease with the same recognition sequence as another restriction endonuclease.

Isozyme An isomer of an enzyme.

Jackknifing A statistical method of numerical re-sampling based on n samples of size $n-1$ used to calculate the variance of an estimate from an original sample of size n (see Chapter 10).

KAc Potassium acetate.

kb Kilobase, or 1000 base pairs of DNA.

Lambda (λ) bacteriophage A virus of the bacterium *E. coli* that is widely used as a cloning vector in molecular biology. It is a double-stranded DNA virus approximately 50 kb in length, with single-stranded complementary ends that allow the virus to circularize after entering its bacterial host. The DNA is packaged into a protein coat in the mature virus.

Lampbrush chromosome A bivalent at diplotene stage in a female meiotic cell; found in the oocytes of most animals.

L-broth Luria broth (see Chapter 9, Appendix).

Linearly ordered character A multistate character in which the allowed transitions between states form a linear chain.

Linkage disequilibrium Departure from the predicted frequencies of multiple locus gamete types assuming alleles are randomly associated.

Lyophilization Drying from the frozen state.

Lysogenic cycle A cycle of phage growth in which the phage become a stable prophage component of the bacterial genome.

Lytic cycle A cycle of phage growth in which the phage are replicated many times, resulting in eventual destruction of the host bacterial cell and release of the progeny phage.

M13 A filamentous bacteriophage of the bacterium *E. coli* that is widely used for cloning and sequencing. The genome of M13 is circular and approximately 6500 bp in length. M13 occurs in both a double-stranded replicative form (used for cloning small fragments) and a single-stranded form (used for Sanger dideoxy sequencing); see Chapter 9.

Maximum parsimony A method of phylogenetic tree inference based on the principle of minimizing the amount of evolutionary change needed to explain the data.

MC′F Microcomplement fixation.

MEM Eagle's minimum essential medium.

Methylation The chemical process of adding a methyl group to a molecule.

Microcomplement fixation A quantitative immunological assay that estimates amino acid replacements between homologous proteins by measuring the fixation of complement in an antigen–antibody reaction.

Mitogen A substance that stimulates mitosis.

Molecular clock hypothesis The hypothesis that molecules evolve in direct proportion to time, so that differences between orthologous molecules in two different species can be used to estimate the time elapsed since the two species last shared a common ancestor.

Molecular systematics The detection, description, and explanation of molecular biological diversity, both within and among species.

Monoclonal antibody A single antibody produced in quantity by cultured lines of hybridoma cells.

Monomeric protein A protein that contains a single polypeptide chain.

Monophyletic A group that contains all of the descendants of the most recent common ancestor of the constituent species.

Most parsimonious reconstruction Any assignment of ancestral states to characters on a tree so that the change of each character is minimized (subject to any constraints being enforced).

MPR See most parsimonious reconstruction.

mRNA Messenger RNA.

mtDNA Mitochondrial DNA.

MTT 3-(4,5-Dimethylthiazol-2-yl)-2,5-diphenyltetrazolium bromide.

Multimeric protein A protein that contains multiple polypeptide chains.

NaAc Sodium acetate.

NAD β-Nicotinamide adenine dinucleotide.

NADH β-Nicotinamide adenine dinucleotide, reduced form.

NADP β-Nicotinamide adenine dinucleotide phosphate.

NBT Nitro blue tetrazolium.

nDNA Nuclear DNA.

NOR Nucleolar organizer region.

NPH Normalized percentage of hybridization (see Chapter 7). Defined as the extent of hybridization in a heteroduplex comparison divided by that for the homoduplex control, expressed as a percentage.

Nuclear genome The portion of the genome contained in the nucleus of eukaryotes, i.e., the chromosomes.

Nucleolar organizer region A region on a chromosome that contains the ribosomal RNA genes and associated spacers.

Null allele An allele that produces either no protein product or a nonfunctional protein product.

Objective function A function that defines the relative quality of any given phylogenetic tree in terms of a set of data.

OD Optical density, as measured in a spectrophotometer. Used to estimate concentration and purity of DNA solutions (see Chapters 7–9).

Oligonucleotide A short chain of nucleotides, often produced in the laboratory.

Optimality criterion Same as objective function.

Ordered character A multistate character for which the changes between states are constrained; not all states can be reached directly from any other.

Organellar genome The DNA contained in cytoplasmic organelles (i.e., mtDNA and cpDNA).

Orthology Homology that arises via speciation.

OTU Operational taxonomic unit. Synonymous with terminal taxon in this book.

Ouchterlony test A specialized immunodiffusion test.

Outgroup One or more taxa assumed to be phylogenetically outside the ingroup.

Outgroup comparison A method that can be used for assigning the direction of change to ordered characters by examining the state in lineages outside of the group of primary interest.

Paracentric inversion An inversion of a region of a chromosome that does not include the centromere.

Paralogy Homology that arises via gene duplication.

Parapatric Adjacent but nonoverlapping distributions (see allopatric, sympatric).

Partially ordered character A multistate character that is ordered, but for which the permitted state transitions do not form a linear series.

PB Phosphate buffer, used in DNA–DNA hybridization experiments (see Chapter 7, Appendix).

PBS Phosphate-buffered saline (see Chapter 6, Appendix).

PCI A mixture of phenol, chloroform, and isoamyl alcohol, used in DNA extraction protocols (see Chapter 9, Appendix).

PCR Polymerase chain reaction.

PDB Phage dilution buffer (see Chapter 9, Appendix).

PEG Polyethylene glycol.

Pericentric inversion An inversion of a region of a chromosome that includes the centromere.

Peripheral branches The branches on a phylogenetic tree that connect to a terminal taxon or sequence.

PERT   Phenol emulsion reassociation technique (see Chapter 7, Protocol 10).

PHA   Phytohemagglutinin, a mitogen.

Phenogram   A branching diagram that links entities by estimates of overall similarity. Usually constructed using UPGMA cluster analysis.

Phylogram   A tree that depicts inferred historical relationships among entities. Differs from cladogram in that the branches are drawn proportional to the amount of inferred character change.

PI   Propidium iodide.

Plaque   A clear spot on a bacterial lawn (in a petri plate) that results from lysis of the resident bacteria by bacteriophage.

Plasmid   A self-replicating extrachromosomal circular DNA.

Plesiomorphy   An ancestral character state.

PMS   Phenazine methosulfate.

Polymerase chain reaction   A series of thermal cycles of denaturation, annealing of primers, and primer extension catalyzed by a thermostable DNA polymerase, in which a target DNA fragment is amplified exponentially.

Polytene chromosome   A somatic chromosome that has undergone many rounds of endoreplication such that each chromosomal element consists of hundreds to thousands of unseparated chromatids.

Positional homology   The property of residues in a set of sequences that are derived from the same ancestral residue, with or without intervening substitutions.

Posttranslation modification   Any process that modifies a polypeptide after its translation from RNA.

PPS   A solution of phenoxyethanol–phosphate–sucrose, used to preserve proteins.

PWM   Pokeweed mitogen.

Q-bands   Fluorescent (under UV light) bands on chromosomes produced by quinacrine staining. Q-bands are brightest in AT-rich regions.

Radioimmunoassay (RIA)   A immunological assay that recognizes major antigenic sites and can be used to study minute quantities of proteins (including proteins recovered from recent fossils).

R-bands   Bands on chromosomes that exhibit the reverse pattern of Q- or G-bands.

rDNA   Ribosomal DNA, which contains the genes for ribosomal RNA and the associated spacer regions.

RE   Restriction enzyme.

Reciprocity   The degree to which reciprocal measures of divergence (e.g., A to B versus B to A) agree.

Restriction endonuclease   An enzyme that cleaves double-stranded DNA. Type I restriction endonucleases are not sequence-specific; type II restriction endonucleases cleave DNA at particular recognition sequences (typically 4–6 bp palindromes).

Restriction fragment length polymorphism (RFLP)   A polymorphism in an individual, population, or species defined by restriction fragments of a distinctive length. Usually caused by gain or loss of a restriction site, but may result from an insertion or deletion of a fragment of DNA between two conserved restriction sites.

Retroposition   Reverse transcription of RNA to DNA with subsequent integration of the DNA at new genomic sites.

Reverse transcriptase   An enzyme that transcribes RNA into DNA; used in direct RNA sequencing.

Reversibility   The ability of a character state to change back to its original state.

RFLP   See restriction fragment length polymorphism.

RIA   Radioimmunoassay.

Robertsonian translocation   Fission or fusion of chromosomes at their centromeres.

rRNA   Ribosomal RNA, the nucleic acid component of ribosomes, which functions in translation of proteins from mRNA.

S1 nuclease   An enzyme that digests single-stranded DNA.

Satellite DNA   Highly repeated DNA sequences that band apart from most nuclear DNA in CsCl ultracentrifugation.

scnDNA   Single copy nuclear DNA.

SCP   Saline citrate–phosphate (see Chapter 6, Appendix).

SDS   Sodium dodecyl sulfate (= sodium lauryl sulfate).

Secondary isozyme   A conformational isozyme.

SEDTA   Saline-EDTA (see Chapter 7, Appendix).

Sequential electrophoresis   The use of a series of different electrophoretic conditions to uncover hidden heterogeneity in isozyme electrophoresis.

Similarity   A generic measure of the resemblance between two objects, usually on a scale from 1 to 0.

Southern blot A membrane onto which DNA has been transferred directly from an electrophoretic gel.

Specificity The degree to which antibodies react with multiple antigenic sites. Initially antibodies are monospecific, but with longer periods of immunization they become more cross reactive (react with more antigenic sites).

SSC Saline sodium citrate (see Chapter 6, Appendix).

SSRBC Sensitized sheep red blood cells used for assays in microcomplement fixation (Chapter 5).

State set A mathematical set of character states used during a parsimony analysis to keep track of the states that are consistent with the minimum amount of change.

STE A buffer used in many DNA procedures; contains sodium chloride, Tris, and EDTA (see Chapter 9, Appendix).

STES A buffer with sodium chloride, Tris, EDTA, and sucrose used in DNA isolation (Chapter 8).

Stringency In DNA–DNA or DNA–RNA hybridization, the conditions of the hybridization (such as temperature and concentration of chemical additives) that determine the degree of similarity that will result in formation of hybrid molecules.

Subbands Nonallelic bands in isozyme electrophoresis that represent the electrophoretic location of conformational isozymes.

Sympatric Occurring in the same place (see allopatric, parapatric).

Symplesiomorphy A shared ancestral character state.

Synapomorphy A shared derived character state.

TAE A Tris–acetic acid–EDTA buffer used for electrophoresis of DNA (Chapter 8, Appendix).

Taq polymerase A thermostable DNA polymerase from *Thermus aquaticus,* a thermophilic bacterium. Used for amplification via the polymerase chain reaction.

TCA Trichloroacetic acid.

TE A Tris-EDTA buffer used to dilute DNA (see Chapter 9, Appendix).

TEACL Tetraethylammonium chloride, a chaotropic agent used to eliminate the effect of base composition on hybrid melting temperature in DNA–DNA hybridization (see Chapter 7).

Terminal nodes Tips of a phylogenetic tree at which contemporary sequences or taxa are placed.

Titer The concentration of a substance as determined by the amount of a known reagent required to bring about a given effect in a test solution. In microcomplement fixation, the concentration of antibody that gives 75% complement fixation under standard conditions.

T_{50H} The interpolated temperature along a heteroduplex DNA melting curve at which 50% of the DNA is double stranded in comparison to the homoduplex control. T_{50H} differs from $_m$ when all DNA in a DNA–DNA hybridization reaction does not form duplexes. The difference in T_{50H} between homoduplex and heteroduplex curves is ΔT_{50H}.

T_m The interpolated temperature along a DNA melting curve at which 50% of the duplex DNA formed in a DNA–DNA hybridization reaction is double stranded. The difference in T_m between homoduplex and heteroduplex curves is ΔT_m.

T_{mode} The interpolated temperature of the peak of a differential plot of a DNA melting curve. The difference in T_{mode} between homoduplex and heteroduplex curves is ΔT_{mode}.

TPBS Tris–phosphate buffered saline.

Tracer Radioactively labeled, usually fractionated single-copy DNA used in DNA–DNA hybridization experiments (see Driver).

Transition substitution A nucleotide substitution that changes one purine into the other purine, or that changes one pyrimidine into the other pyrimidine.

Transposable element A genomic element that can move from site to site in the genome of an organism, either through direct DNA copying (at least in prokaryotes) or reverse transcription from an RNA intermediate (probably the usual mechanism in eukaryotes).

Transposon A segment of DNA flanked by transposable elements that is capable of moving its location in the genome.

Transversion substitution A nucleotide substitution that changes a purine to a pyrimidine, or vice versa.

Tree length The total amount of change in a maximum parsimony tree, or the sum of the branch lengths in a minimum distance tree.

Ultrametric distances Pairwise distance values that precisely fit a rooted tree with a constant molecular clock. Defined mathematically by satisfying the three-point condition (see Chapter 11).

Unequal crossing over Physical cross over between imperfectly aligned repeats of a multigene family, which results in one smaller and one larger DNA molecule.

Unequal rate effect A misnomer for the tendency of homoplasy in long branches of a phylogenetic tree to artifactually group these branches together during tree inference.

Unordered character A character for which any state can change directly to any other state.

Unrooted tree A phylogenetic tree in which the location of the most recent common ancestor of the taxa is unknown (or not indicated).

UPGMA Unweighted pair group method using arithmetic average. A cluster analysis technique.

Variable number tandem repeat loci (VNTR loci) Loci containing variable numbers of short tandemly repeated sequences, resulting in a high frequency of length variation that can be revealed with an appropriate restriction endonuclease (see Chapter 8).

WPGMA Weighted pair group method using arithmetic average. A cluster analysis technique.

Xenology Homology that arises via lateral gene transfer between unrelated species (e.g., by retroviruses).

Zymogram The pattern on an allozyme electrophoresis gel visualized by histochemical staining.

Literature Cited

Aboitiz, F. 1987. Letter to the editor. Cell 51:515–516.

Adams, M., P. R. Baverstock, C. J. S. Watts and T. Reardon. 1987. Electrophoretic resolution of species boundaries in Australian Microchiroptera. II. The *Pipistrellus* group (Chiroptera: Vespertilionidae). Aust. J. Biol. Sci. 40:163–170.

Aebersold, P. B., G. A. Winans, D. J. Teel, G. B. Milner and F. M. Utter. 1987. Manual for starch gel electrophoresis: A method for the detection of genetic variation. NOAA Tech. Report NMFS No. 61.

Aldrich, J., B. Cherney, E. Merlin and J. D. Palmer. 1986a. Sequence of the *rbcL* gene for the large subunit of ribulose bisphosphate carboxylase-oxygenase from petunia. Nucl. Acids Res. 14:9534.

Aldrich, J., B. Cherney, E. Merlin and J. D. Palmer. 1986b. Sequence of the *rbcL* gene for the large subunit of ribulose bisphosphate carboxylase-oxygenase from alfalfa. Nucl. Acids Res. 14:9535.

Allendorf, F. W. 1977. Electromorphs or alleles. Genetics 87:821–822.

Allendorf, F. W. and S. R. Phelps. 1981. Use of allelic frequencies to describe population structure. Can. J. Fish. Aquat. Sci. 38:1507–1514.

Allendorf, F. W. and G. H. Thorgaard. 1984. Tetraploidy and the evolution of salmonid fishes, pp. 1–53. *In* B. J. Turner (ed.), *Evolutionary Genetics of Fishes*. Plenum, New York.

Allendorf, F. W. and F. M. Utter. 1973. Gene duplication within the family Salmonidae: Disomic inheritance of two loci reported to be tetrasomic in rainbow trout. Genetics 74:647–654.

Allendorf, F. W., K. L. Knudsen and R. F. Leary. 1983. Adaptive significance of differences in the tissue-specific expression of a phosphoglucomutase gene in rainbow trout. Proc. Natl. Acad. Sci. U.S.A. 800:1397–1400.

Allendorf, F. W., G. Stahl and N. Ryman. 1984. Silencing of duplicate genes: A null polymorphism for lactate dehydrogenase in rainbow trout. Mol. Biol. Evol. 1:238–248.

Anderson, D. M. and W. R. Folk. 1976. Iodination of DNA. Studies of the reaction and iodination of papovavirus DNA. Biochemistry 15:1022–1030.

Anderson, J. O., J. Nath and E. J. Harner. 1978. Effect of freeze–preservation on some pollen enzymes. Cryobiology 15:469–477.

Anderson, S., A. T. Bankier, B. G. Barrell, M. H. L. DeBruijn, A. R. Coulson, J. Drouin, I. C. Eperon, D. P. Nierlich, B. A. Roe, F. Sanger, P. H. Schreier, A. J. H. Smith, R. Staden and I. G. Young. 1981. Sequence and organization of the human mitochondrial genome. Nature (London) 290:457–465.

Andronico F., S. De Luccini, F. Graziani, I. Nardi, R. Batistoni and G. Barsacchi-Pilone. 1985. Molecular organization of ribosomal RNA genes clustered at variable chromosomal sites in *Triturus vulgaris meridionalis* (Amphibia, Urodela). J. Mol. Biol. 186:219–229.

Angerer, R. C., E. H. Davidson and R. J. Britten. 1976. Single copy DNA and structural gene sequence relationships among four sea urchin species. Chromosoma 56:213–226.

Ansorge, W. and S. Labeit. 1984. Field gradients improve resolution on DNA sequencing gels. J. Biochem. Biophys. Methods 10:237–243.

Appels, R. and J. Dvořák. 1982. The wheat ribosomal DNA spacer region: Its structure and variation in populations and among species. Theoret. Appl. Gen. 63:337–348.

Appels, R. and R. L. Honeycutt. 1987. rDNA: Evolution over a billion years, pp. 81–135. *In* S. K. Dutta (ed.), *DNA Systematics*. CRC Press, Boca Raton, FL.

Aquadro, C. F. and J. C. Avise. 1982a. An assessment of "hidden" heterogeneity within electromorphs at three loci in deer mice. Genetics 102:269–284.

Aquadro, C. F. and J. C. Avise. 1982b. Evolutionary genetics of birds. VI. A reexamination of protein divergence using varied electrophoretic conditions. Evolution 36:1003–1019.

Aquadro, C. F. and B. D. Greenberg. 1983. Human mitochondrial DNA variation and evolution: Analysis of nucleotide sequences from seven individuals. Genetics 103:287–312.

Aquadro, C. F., S. F. Deese, M. M. Bland, C. H. Langley and C. C. Laurie-Ahlberg. 1986. Molecular population genetics of the alcohol dehydrogenase gene region of *Drosophila melanogaster*. Genetics 114:1165–1190.

Archie, J. W., C. Simon and A. Martin. 1989. Small sample size does decrease the stability of dendrograms calculated from allozyme–frequency data. Evolution 43:678–683.

Arctander, P. 1988. Comparative studies of avian DNA by restriction fragment length polymorphism analysis: Convenient procedures based on blood samples from live birds. J. Ornithol. 129:205–216.

Arnason, U. and B. Widegren. 1984. Different rates of divergence in highly repetitive DNA of cetaceans. Hereditas 101:171–177.

Arnheim, N. 1983. Concerted evolution of multigene families, pp. 38–61. In M. Nei and R. K. Koehn (eds.), Evolution of Genes and Proteins. Sinauer, Sunderland, MA.

Arnheim, N., E. M. Prager and A. C. Wilson. 1969. Immunological prediction of sequence differences among proteins. Chemical comparisons of chicken, quail, and pheasant lysozymes. J. Biol. Chem. 244:2085–2094.

Arnheim, N., D. Treco, B. Taylor and E. M. Eicher. 1982. Distribution of ribosomal DNA length variants among mouse chromosomes. Proc. Natl. Acad. Sci. U.S.A. 79:4677–4680.

Arnold, E. N. 1981. Estimating phylogenies at low taxonomic levels. Z. Zool. Syst. Evol.-Forsch. 19:1–35.

Arnold, M. L., N. Conteras and D. D. Shaw. 1988. Biased gene conversion and assymetrical introgression between species. Chromosoma 96:368–371.

Arnold, M. L., D. D. Shaw and N. Contreras. 1987a. Ribosomal RNA-encoding DNA introgression across a narrow hybrid zone between two subspecies of grasshopper. Proc. Natl. Acad. Sci. U.S.A. 84:3946–3950.

Arnold, M. L., P. Wilkinson, D. D. Shaw, A. D. Marchant and N. Contreras. 1987b. Highly repeated DNA and allozyme variation between sibling species: Evidence for introgression. Genome 29:272–279.

Arrand J. E. 1985. Preparation of nucleic acid probes, pp. 17–45. In B. D. Hames and S. J. Higgins (eds.), Nucleic Acid Hybridisation: A Practical Approach. IRL Press, Oxford.

Arulsekar, S., D. E. Parfitt and G. H. McGranahan. 1985. Isozyme gene markers in Juglans species. J. Hered. 76:103–106.

Asher, J. H. 1970. Parthenogenesis and genetic variability. II. One locus model for various diploid populations. Genetics 66:369–391.

Attardi, G. 1985. Animal mitochondrial DNA: An extreme example of genetic economy. Int. Rev. Cytol. 93:93–145.

Ausubel, F. M. (ed.) 1989. Current Protocols in Molecular Biology. Wiley, New York.

Avise, J. C. 1974. Systematic value of electrophoretic data. Syst. Zool. 23:465–481.

Avise, J. C. 1976. Genetic differentiation during speciation, pp. 106–122. In F. J. Ayala (ed.), Molecular Evolution. Sinauer, Sunderland, MA.

Avise, J. C. 1986. Mitochondrial DNA and the evolutionary genetics of higher animals. Phil. Trans. Roy. Soc. London Ser. B 312:325–342.

Avise, J. C. and C. F. Aquadro. 1982. A comparative summary of genetic distances in the vertebrates. Evol. Biol. 15:151–158.

Avise, J. C. and G. B. Kitto. 1973. Phosphoglucose isomerase gene duplication in the bony fishes: An evolutionary history. Biochem. Genet. 8:113–132.

Avise, J. C. and R. A. Lansman. 1983. Polymorphism of mitochondrial DNA in populations of higher animals, pp. 165–190. In M. Nei and R. K. Koehn (eds.), Evolution of Genes and Proteins. Sinauer, Sunderland, MA.

Avise, J. C., J. J. Smith and F. J. Ayala. 1975. Adaptive differentiation with little genic change between two native California minnows. Evolution 29:411–426.

Avise, J. C., C. Giblin-Davidson, J. Laerm, J. C. Patton and R. A. Lansman. 1979a. Mitochondrial DNA clones and matriarchal phylogeny within and among geographic populations of the pocket gopher Geomys pinetis. Proc. Natl. Acad. Sci. U.S.A. 76:6694–6698.

Avise, J. C., R. A. Lansman and R. O. Shade. 1979b. The use of restriction endonucleases to measure mitochondrial DNA sequence relatedness in natural populations. I. Population structure and evolution in the genus Peromyscus. Genetics 92:279–295.

Avise, J. C., J. C. Patton and C. F. Aquadro. 1980. Evolutionary genetics of birds. I. Relationships among North American thrushes and allies. Auk 97: 135–147.

Avise, J. C., J. E. Neigel and J. Arnold. 1984. Demographic influences of mitochondrial DNA lineage survivorship in animal populations. J. Mol. Evol. 20:99–105.

Avise, J. C., J. Arnold, R. M. Ball, E. Bermingham, T. Lamb, J. E. Neigel, C. A. Reeb and N. C. Saunders. 1987. Intraspecific phylogeography: The mitochondrial bridge between population genetics and systematics. Annu. Rev. Ecol. Syst. 18:489–522.

Avise, J. C., R. M. Ball and J. Arnold. 1988. Current versus historical population sizes in vertebrate species with high gene flow: A comparison based on mitochondrial DNA lineages and inbreeding theory for neutral mutations. Mol. Biol. Evol. 5:331–344.

Avise, J. C., B. W. Bowen and T. Lamb. 1989. DNA

fingerprints from hypervariable mitochondrial genotypes. Mol. Biol. Evol. 6:258–269.

Ayala, F. J. 1982. Genetic variation in natural populations: Problem of electrophoretically cryptic alleles. Proc. Natl. Acad. Sci. U.S.A. 79:550–554.

Ayala, F. J., J. R. Powell, M. L. Tracey, C. A. Mourao and S. Perez-Salas. 1972. Enzyme variability in the *Drosophila willistoni* group. IV. Genic variation in natural populations of *Drosophila willistoni*. Genetics 70:113-139.

Baba, M. L., M. Goodman, H. Dene and G. W. Moore. 1975. Origins of the Ceboidea viewed from an immunological perspective. J. Human Evol. 4:89–102.

Baker, A. J. and A. Moeed. 1987. Rapid genetic differentiation and founder effect in colonizing populations of Common Mynas (*Acridotheres tristis*). Evolution 41:523–538.

Baker, M. C., D. B. Thompson, G. L. Sherman, M. A. Cunningham and D. F. Tomback. 1982. Allozyme frequencies in a linear series of song dialect populations. Evolution 36:1020–1029.

Baker, R. J., S. K. Davis, R. D. Bradley, M. J. Hamilton and R. A. Van Den Bussche. 1989. Ribosomal-DNA, mitochondrial-DNA, chromosomal, and allozymic studies on a contact zone in the pocket gopher, *Geomys*. Evolution 43:63–75.

Ball, R. M., Jr., S. Freeman, F. C. James, E. Bermingham and J. C. Avise. 1988. Phylogeographic population structure of red-winged blackbirds assessed by mitochondrial DNA. Proc. Natl. Acad. Sci. U.S.A. 85:1558–1562.

Banks, J. A. and C. W. Birky Jr. 1985. Chloroplast DNA diversity is low in a wild plant, *Lupinus texensis*. Proc. Natl. Acad. Sci. U.S.A. 82:6950–6954.

Barker, J. S. F., P. D. East and B. S. Weir. 1986. Temporal and microgeographic variation in allozyme frequencies in a natural population of *Drosophilia buzzartii*. Genetics 112:577–611.

Barker, P. E., J. R. Testa, N. Z. Parsa and R. Snyder. 1986. High molecular weight DNA from fixed cytogenetic preparations. Am. J. Human Genet. 39:661–668.

Barnes, P. T. and C. C. Laurie-Ahlberg. 1986. Genetic variability of flight metabolism in *Drosophila melanogaster*. III. Effects of GPDH allozymes and environmental temperature on power output. Genetics 112:267–294.

Barnes, W. M. 1987. Sequencing DNA with dideoxyribonucleotides as chain terminators: Hints and strategies for big projects. Methods Enzymol. 152: 538–556.

Barnes, W. M., M. Bevan and P. H. Son. 1983. Kilosequencing: Creation of an ordered nest of asymmetric deletions across a large target sequence carried on phage M13. Methods Enzymol. 101: 98–122.

Barrie, P. A., A. J. Jeffries and A. F. Scott. 1981. Evolution of the β-globin gene cluster in man and the primates. J. Mol. Biol. 149:319–336.

Barrowclough, G. R., N. K. Johnson and R. M. Zink. 1985. On the nature of genic variation in birds, pp. 135–154. *In* R. F. Johnston (ed.), *Current Ornithology*, Vol. 2. Plenum, New York.

Barton, N. H., R. B. Halliday and G. M. Hewitt. 1983. Rare electrophoretic variants in a hybrid zone. Heredity 50:139–146.

Bautz, E. K. and F. A. Bautz. 1964. The influence of noncomplementary bases on the stability of ordered polynucleotides. Proc. Natl. Acad. Sci. U.S.A. 52:1476–1481.

Baverstock, P. R. and M. Adams. 1987. Comparative rates of molecular, chromosomal and morphological evolution in some Australian vertebrates, pp. 175–188. *In* K. S. W. Campbell and M. F. Day (eds.), *Rates of Evolution*, Allen & Unwin, London.

Baverstock, P. R., C. H. S. Watts and S. R. Cole. 1977. Electrophoretic comparisons between the allopatric populations of five Australian pseudomyine rodents (Muridae). Aust. J. Biol. Sci. 30:471–485.

Baverstock, P. R., S. R. Cole, B. J. Richardson and C. H. Watts. 1979. Electrophoresis and cladistics. Syst. Zool. 28:214–219.

Baverstock, P. R., M. Adams and I. Beveridge. 1985. Biochemical differentiation in bile duct cestodes and their marsupial hosts. Mol. Biol. Evol. 2:321–337.

Baverstock, P. R., M. Adams and C. H. S. Watts. 1986. Biochemical differentiation among karyotypic forms of Australian *Rattus*. Genetica 71:11–22.

Benjamin, D. C., J. A. Berzofsky, I. J. East, F. R. N. Gurd, C. Hannum, S. J. Leach, E. Margoliash, J. G. Michael, A. Miller, E. M. Prager, M. Reichlin, E. E. Sercarz, S. J. Smith-Gill, P. E. Todd and A. C. Wilson. 1984. The antigenic structure of proteins: A reappraisal. Annu. Rev. Immunol. 2:67–101.

Benn, P. A. and M. A. Perle. 1986. Chromosome staining and banding techniques, pp. 57–84. *In* D. E. Rooney and B. H. Czepulkowski (eds.), *Human Cytogenetics*. IRL Press, Oxford.

Bennett, M. D. 1972. Nuclear DNA content and minimum mitotic time in herbaceous plants. Proc. Roy. Soc. London Ser. B 181:109–135.

Benveniste, R. E. 1985. The contribution of retroviruses to the study of mammalian evolution, pp. 359–417. *In* R. J. MacIntyre (ed.), *Molecular Evolutionary Genetics,* Plenum, New York.

Benveniste, R. E. and G. J. Todaro. 1976. Evolution of type C viral genes: Evidence for an Asian origin of man. Nature (London) 261:101–108.

Berg, W. J. and D. G. Buth. 1984. Glucose dehydrogenase in teleosts: Tissue distribution and proposed function. Comp. Biochem. Physiol. 77B:285–288.

Berger, S. L. and A. R. Kimmel (eds.) 1987. Guide to molecular cloning techniques. Methods Enzymol. 152:1–812.

Berlocher, S. J. and G. L. Bush. 1982. An electrophoretic analysis of *Rhagoletis* (Diptera: Tephritidae) phylogeny. Syst. Zool. 31:136–155.

Bermingham, E. and J. C. Avise. 1986. Molecular zoogeography of freshwaterfishes in the southeastern United States. Genetics 113:939–965.

Beutler, E. 1969. Electrophoresis of phosphoglycerate kinase. Biochem. Genet. 3:189–195.

Beverley, S. M. and A. C. Wilson. 1982. Molecular evolution in *Drosophila* and higher diptera. I. Micro-complement fixation studies of a larval hemolymph protein. J. Mol. Evol. 18:251–264.

Beverley, S. M. and A. C. Wilson. 1985. Ancient origin for Hawaiian Drosophilinae inferred from protein comparisons. Proc. Natl. Acad. Sci. U.S.A. 82:4753–4757.

Beyer, W. A., M. L. Stein, T. F. Smith and S. M. Ulam. 1974. A molecular–sequence metric and evolutionary trees. Math. Biosci. 19:9–25.

Birky, C. W., Jr. 1983. The partitioning of cytoplasmic organelles at cell division. Int. Rev. Cytol. 15:49–89.

Birky, C. W., Jr., T. Maruyama and P. Fuerst. 1983. An approach to population and evolutionary genetic theory for genes in mitochondria and chloroplasts, and some results. Genetics 103:513–527.

Birley, A. J. and J. H. Croft. 1986. Mitochondrial DNAs and phylogenetic relationships, pp. 107–137. *In* S. K. Dutta (ed.), *DNA Systematics.* CRC Press, Boca Raton, FL.

Bishop, J. G. and J. A. Hunt. 1988. DNA divergence in and around the alcohol dehydrogenase locus in five closely related species of Hawaiian *Drosophila.* Mol. Biol. Evol. 5:415–432.

Bledsoe, A. H. 1987. DNA evolutionary rates in nine-primaried passerine birds. Mol. Biol. Evol. 4:559–571.

Bodmer, M. and M. Ashburner. 1984. Conservation and change in the DNA sequences coding for alcohol dehydrogenase in sibling species of *Drosophila.* Nature (London) 309:421–430.

Boerwinkle, E., W. Xiong, E. Fourest and L. Chan. 1989. Rapid typing of tandemly repeated hypervariable loci by the polymerase chain reaction: Application to the apolipoprotein B 3′ hypervariable region. Proc. Natl. Acad. Sci. U.S.A. 86:212–216.

Bogart, J. P., L. A. Lowcock, C. W. Zeyl and B. K. Mable. 1987. Genome constitution and reproductive biology of hybrid salamanders, genus *Ambystoma,* on Kelleys Island in Lake Erie. Can. J. Zool. 65:2188–2201.

Bonnell, M. T. and R. K. Selander. 1974. Elephant seals: Genetic variation and near extinction. Science 184:908–909.

Bonner, T. I., D. J. Brenner, B. R. Neufeld and R. J. Britten. 1973. Reduction in rate of DNA reassociation by sequence divergence. J. Mol. Biol. 81:123–135.

Bonner, T., R. Heinemann and G. J. Todaro. 1980. Evolution of DNA sequences has been retarded in Malagasy primates. Nature 286:420–423.

Bowen, B. W. and J. C. Avise. 1989. An odyssey of the green sea turtle: Ascension Island revisited. Proc. Natl. Acad. Sci. U.S.A. 86:573–576.

Boyden, A. 1942. Systematic serology: A critical appraisal. Physiol. Zool. 15:109–145.

Boyden, A. (ed.) 1948–1978. Serol. Mus. Bull. Vols. 1–51.

Boyden, A. 1964. Perspectives in systematic serology, pp. 75–99. *In* C. A. Leone (ed.), *Taxonomic Biochemistry and Serology.* Ronald Press, New York.

Boyden, M. G. 1967. It's about time. Ser. Mus. Bull. 37:7–10.

Boyer, S. H. 1961. Alkaline phosphatase in human sera and placentae. Science 134:1002–1004.

Boyer, S. H., D. C. Fainer and E. J. Watson-Williams. 1963. Lactate dehydrogenase variant from human blood: Evidence for molecular subunits. Science 141:642–643.

Brazaitis, P. and M. Watanabe. 1982. The doppler, a new tool for reptile and amphibian hematological studies. J. Herpetol. 16:1–6.

Brewer, G. J. 1970. *An Introduction to Isozyme Techniques.* Academic Press, New York.

Britten, R. J. 1986. Rates of DNA sequence evolution differ between taxonomic groups. Science 231:1393–1398.

Britten, R. J. 1990. Comment on DNA hybridization issues raised at Lake Arrowhead. J. Mol. Evol. (in press).

Britten, R. J. and E. H. Davidson. 1969. Gene regulation for higher cells: A theory. Science 165:349–357.

Britten, R. J. and E. H. Davidson. 1985. Hybridisation strategy, pp. 3–15. *In* B. D. Hames and S. J. Higgins (eds.), *Nucleic Acid Hybridisation: A Practical Approach*. IRL Press, Oxford.

Britten, R. J. and D. E. Kohne. 1967. Nucleotide sequence repetition in DNA. Carnegie Inst. of Wash. Yearbook 65:78–106.

Britten, R. J. and D. E. Kohne. 1968. Repeated sequences in DNA. Science 161:529–540.

Britten, R. J., D. E. Graham and B. R. Neufeld. 1974. Analysis of repeating DNA sequences by reassociation. Methods Enzymol. 29:363–418.

Britten, R. J., A. Cetta and E. H. Davidson. 1978. The single-copy sequence polymorphism of the sea urchin *Strongylocentrotus purpuratus*. Cell 15:1175-1186.

Brooks, D. R. 1981. Hennig's parasitological method: A proposed solution. Syst. Zool. 30:229–249.

Brooks, D. R. 1990. Parsimony analysis in historical biogeography and coevolution: Methodological and theoretical update. Syst. Zool. 39 (in press).

Brow, M.A.D. 1990. Sequencing with *Taq* DNA polymerase, pp. 189–196. *In* M.A. Innis, D.H. Gelfand, J.J. Sninsky and T.J. White (eds.), *PCR Protocols: A Guide to Methods and Applications*. Academic Press, San Diego.

Brown, A. D. H. 1975. Sample sizes needed to detect linkage disequilibrium between two or three loci. Theoret. Pop. Biol. 8:184–201.

Brown, K. L. 1985. Demographic and genetic characteristics of dispersal in the mosquitofish, *Gambusia affinis* (Pisces: Poeciliidae). Copeia 1985:597-612.

Brown, W. M. 1980. Polymorphism in mitochondrial DNA of humans as revealed by restriction endonuclease analysis. Proc. Natl. Acad. Sci. U.S.A. 77:3605–3609.

Brown, W. M. 1983. Evolution of animal mitochondrial DNA, pp. 62-88. *In* M. Nei and R. K. Koehn (eds.), *Evolution of Genes and Proteins*. Sinauer, Sunderland, MA.

Brown, W. M. 1985. The mitochondrial genome of animals, pp. 95–130. *In* R. MacIntyre (ed.), *Molecular Evolutionary Genetics*. Plenum, New York.

Brown, W. M. and J. Wright. 1979. Mitochondrial DNA analyses and the origin and relative age of parthenogenetic lizards (genus *Cnemidophorus*). Science 203:1247–1249.

Brown, W. M., M. George, Jr. and A. C. Wilson. 1979. Rapid evolution of animal mitochondrial DNA. Proc. Natl. Acad. Sci. U.S.A. 76:1967–1971.

Brown, W. M., E. M. Prager, A. Wang and A. C. Wilson. 1982. Mitochondrial DNA sequences of primates: Tempo and mode of evolution. J. Mol. Evol. 18:225–239.

Bruns, T. D. and J. D. Palmer. 1989. Evolution of mushroom mitochondrial DNA: *Suillus* and related genera. J. Mol. Evol. 28:348–362.

Buneman, P. 1971. The recovery of trees from measures of dissimilarity, pp. 387–395. *In* F. R. Hodson, D. G. Kendall and P. Tautu (eds.), *Mathematics in the Archaeological and Historical Sciences*, Edinburgh Univ. Press, Edinburgh.

Burke, T. 1989. DNA fingerprinting and other methods for the study of mating success. Trends Ecol. Evol. 4:139–144.

Burke, T. and M. W. Bruford. 1987. DNA fingerprinting in birds. Nature (London) 327:149–152.

Burke, T., N. B. Davies, M. W. Bruford and B. J. Hatchwell. 1989. Parental care and mating behaviour of polyandrous dunnocks *Prunella modularis* related to paternity by DNA fingerprinting. Nature 338:249–251.

Burkhart, B. D., E. Montgomery, C. H. Langley and R. A. Voelker. 1984. Characterization of allozyme null and low activity alleles from two natural populations of *Drosophila melanogaster*. Genetics 107:295–306.

Busack, S. D., B. G. Jericho, L. R. Maxson and T. Uzzell. 1988. Evolutionary relationships of salamanders in the genus *Triturus:* The view from immunology. Herpetologica 44:307–316.

Buth, D. G. 1979a. Creatine kinase variability in *Moxostoma macrolepidotum* (Cypriniformes: Catostomidae). Copeia 1979:152–154.

Buth, D. G. 1979b. Genetic relationships among the torrent suckers, genus *Thoburnia*. Biochem. Syst. Ecol. 3:311–316.

Buth, D. G. 1980. Staining procedures for D-2-hydroxyacid dehydrogenase as applied to studies of lower vertebrates. Isozyme Bull. 13:115.

Buth, D. G. 1982a. Glucosephosphate-isomerase expression in a tetraploid fish, *Moxostoma lachneri* (Cypriniformes, Catostomidae): Evidence for "re-tetraploidization"? Genetica 57:171–175.

Buth, D. G. 1982b. Locus assignments for general muscle proteins of darters (Etheostomatini). Copeia 1982:217–219.

Buth, D. G. 1983. Duplicate isozyme loci in fishes: Origins, distribution, phyletic consequences, and locus nomenclature, pp. 381–400. *In* M. C. Rattazzi, J. G. Scandalios and G. S. Whitt (eds.), *Isozymes: Current Topics in Biological and Medical Research,* Vol. 10. Liss, New York.

Buth, D. G. 1984. The application of electrophoretic data in systematic studies. Annu. Rev. Ecol. Syst. 15:501–522.

Buth, D. G. and R. W. Murphy. 1980. Use of nicotinamide adenine dinucleotide (NAD)-dependent glucose-6-phosphate dehydrogenase in enzyme staining procedures. Stain Technol. 55:173–176.

Buth, D. G., B. M. Burr and J. R. Schenck. 1980. Electrophoretic evidence for relationships and differentiation among members of the percid subgenus *Microperca.* Biochem. Syst. Ecol. 8:297–304.

Buth, D. G., R. W. Murphy, M. M. Miyamoto and C. S. Lieb. 1985. Creatine kinases of amphibians and reptiles: Evolutionary and systematic aspects of gene expression. Copeia 1985:279–284.

Caccone, A. and J. R. Powell. 1987. Molecular evolutionary divergence among North American cave crickets. II. DNA–DNA hybridization. Evolution 41:1215–1238.

Caccone, A., G. D. Amato and J. R. Powell. 1987. Intraspecific DNA divergence in *Drosophila:* A study on parthenogenetic *D. mercatorum.* Mol. Biol. Evol. 4:343–350.

Caccone, A., G. D. Amato and J. R. Powell. 1988a. Rates and patterns of scnDNA and mtDNA divergence within the *Drosophila melanogaster* subgroup. Genetics 118:671–683.

Caccone, A., R. DeSalle and J. R. Powell. 1988b. Calibration of the change in thermal stability of DNA duplexes and degree of base pair mismatch. J. Mol. Evol. 27:212–216.

Cadle, J. E. 1988. Phylogenetic relationships among advanced snakes: A molecular perspective. Univ. Calif. Pub. Zool. 119:1–77.

Callan, H. G. 1966. Chromosomes and nucleoli of the axolotl, *Ambystoma mexicanum.* J. Cell Sci. 1:85–108.

Callan, H. G. 1986. *Lampbrush Chromosomes.* Springer-Verlag, Berlin.

Callan, H. G., J. G. Gall and C. A. Berg. 1987. The *lampbrush chromosomes of Xenopus laevis:* Preparation, identification, and distribution of 5S DNA sequences. Chromosoma 95:236-250.

Camin, J. H. and R. R. Sokal. 1965. A method for deducing branching sequences in phylogeny. Evolution 19:311–326.

Cann, R. L. and A. C. Wilson. 1983. Length mutations in human mitochondrial DNA. Genetics 104:699–711.

Cann, R. L., W. M. Brown and A. C. Wilson. 1984. Polymorphic sites and the mechanism of evolution in human mitochondrial DNA. Genetics 106:479–499.

Cantatore, P., M. N. Gadaleta, M. Roberti, C. Saccone and A. C. Wilson. 1987. Duplication and remoulding of tRNA genes during the evolutionary rearrangement of mitochondrial genomes. Nature (London) 329:853–854.

Carlson, S. S., A. C. Wilson and R. D. Maxson. 1978. Do albumin clocks run on time? Science 200:1183–1185.

Carpenter, J. M. 1988. Choosing among multiple equally parsimonious cladograms. Cladistics 4:291–296.

Carr, S. M., A. J. Brothers and A. C. Wilson. 1987. Evolutionary inferences from restriction maps of mitochondrial DNA from nine taxa of *Xenopus* frogs. Evolution 41:176–190.

Case, S. M. and M. H. Wake. 1977. Immunological comparisons of Caecilian albumins (Amphibia: Gymnophiona). Herpetologica 33:94–98.

Case, S. M. and E. E. Williams. 1984. Study of a contact zone in the *Anolis distichus* complex in the central Dominican Republic. Herpetologica 40:118–137.

Casillas, E., J. Sundquist and W. E. Ames. 1982. Optimization of assay conditions for, and selected tissue distribution of alanine aminotransferase and aspartate aminotransferase of English sole, *Parophrys vetulus* Girard. J. Fish Biol. 21:197–204.

Castora, F. J., N. Arnheim and M. V. Simpson. 1980. Mitochondrial DNA polymorphism: Evidence that variants detected by restriction enzymes differ in nucleotide sequence rather than in methylation. Proc. Natl. Acad. Sci. U.S.A. 77:6415–6419.

Catzeflis, F. M., F. H. Sheldon, J. E. Ahlquist and C. G. Sibley. 1987. DNA-DNA hybridization evidence of the rapid rate of rodent DNA evolution. Mol. Biol. Evol. 4:242–253.

Cavalier-Smith, T. (ed.) 1985a. *The Evolution of Genome Size.* Wiley, New York.

Cavalier-Smith, T. 1985b. Eukaryotic gene numbers, non-coding DNA, and genome size, pp. 69–103. *In* T. Cavalier-Smith (ed.), *The Evolution of Genome Size.* Wiley, New York.

Cavalli-Sforza, L. L. and A. W. F. Edwards. 1967a. Phylogenetic analysis: Models and estimation procedures. Evolution 21:550–570.

Cavalli-Sforza, L. L. and A. W. F. Edwards. 1967b. Phylogenetic analysis: Models and estimation procedures. Am. J. Hum. Genet. 19:233–257.

Cedergren, R., M. W. Gray, Y. Abel and D. Sankoff. 1988. The evolutionary relationships among known life forms. J. Mol. Evol. 28:98–112.

Cei, J. M. 1972. Archaeobatrachia versus Neobatrachia: A first serological approach. Serolog. Mus. Bull. 48:1–4.

Cei, J. M. and L. P. Castro. 1973. Taxonomic and serological researches on the *Phymaturus patagonicus* complex. J. Herpetol. 7:237–247.

Chakraborty, R. and O. Leimar. 1987. Genetic variation within a subdivided population, pp. 90–120. *In* N. Ryman and F. Utter (eds.), *Population Genetics and Fisheries Management,* Univ. Washington Press, Seattle.

Chakraborty, R. and M. Nei. 1977. Bottleneck effects on average heterozygosity and genetic distance with the stepwise mutation model. Evolution 31:347–356.

Chambers, G. K., W. G. Laver, S. Campbell and J. B. Gibson. 1981. Structural analysis of an electrophoretically cryptic alcohol dehydrogenase variant from an Australian population of *Drosophila melanogaster.* Proc. Natl. Acad. Sci. U.S.A. 78:3103–3107.

Champion, A. B., E. M. Prager, D. Wachter and A. C. Wilson. 1974. Microcomplement fixation, pp. 397–416. *In* C. A. Wright (ed.), *Biochemical and Immunological Taxonomy of Animals.* Academic Press, London.

Champion, A. B., E. L. Barrett, N. J. Palleroni, K. L. Soderberg, R. Kunisawa, R. Contopoulou, A. C. Wilson and M. Duodoroff. 1980. Evolution in *Pseudomonas fluorescens.* J. Gen. Micro. 120:485–511.

Chan, H.-C., W. T. Ruyechan and J. G. Wetmur. 1976. In vitro iodination of low complexity nucleic acids without chain scission. Biochemistry 15:5487–5490.

Chapman, R. W. and D. A. Powers. 1984. A method for rapid isolation of mtDNA from fishes. Maryland Sea Grant Tech. Rep. MD-SG-TS-84-05, 11 pp.

Cheliak, W. M. and J. A. Pitel. 1984. Techniques for starch gel electrophoresis of enzymes from forest trees. Information Report PI-X-42. Petawawa National Forestry Institute, Canadian Forestry Service.

Chen, B.-Y., S.-H. Mao and Y.-H. Ling. 1980. Evolutionary relationships of turtles suggested by immunological cross-reactivity of albumins. Comp. Biochem. Physiol. 66B:421–425.

Chepko-Sade, B. D. and Z. T. Halpin (eds.) 1987. *Mammalian Dispersal Patterns. The Effects of Social Structure on Population Genetics.* Univ. Chicago Press, Chicago.

Cherry, L. M., S. M. Case, J. G. Kunkel, J. S. Wyles and A. C. Wilson. 1982. Body shape metrics and organismal evolution. Evolution 36:914–933.

Chesser, R. K. 1983. Genetic variability within and among populations of the black-tailed prairie dog. Evolution 37:320–331.

Chilson, O. P., L. A. Costello and N. O. Kaplan. 1965. Effects of freezing on enzymes. Fed. Proc. 24 (supplement 15):555–565.

Chippindale, P. 1989. A high-pH discontinuous buffer system for resolution of isozymes in starch-gel electrophoresis. Stain. Technol. 64:61–64.

Chrambach, A. and D. Rodbard. 1971. Polyacrylamide gel electrophoresis. Science 172:440–451.

Christiansen, F. B. and O. Frydenberg. 1973. Selection component analysis of natural polymorphisms using population samples including mother-offspring combinations. Theoret. Pop. Biol. 4:425–445.

Church, G. M. and W. Gilbert. 1984. Genomic sequencing. Proc. Natl. Acad. Sci. U.S.A. 81:1991–1995.

Church, G. M. and S. Kieffer-Higgins. 1988. Multiplex DNA sequencing. Science 240:185–188.

Clayton, J. W. and D. N. Tretiak. 1972. Amine-citrate buffers for pH control in starch gel electrophoresis. J. Fish. Res. Board Canada 29:1169–1172.

Cochrane, B. J. and R. C. Richmond. 1979. Studies of esterase 6 in *Drosophila melanogaster.* 1. The genetics of posttranslational modification. Biochem. Genet. 17:167-183.

Cockerham, C. C. 1969. Variance of gene frequencies. Evolution 23:72–84.

Cockerham, C. C. 1973. Analyses of gene frequencies. Genetics 74:679–700.

Cockerham, C. C. 1984. Drift and mutation with a finite number of allelic states. Proc. Natl. Acad. Sci U.S.A. 81:530–534.

Cockerham, C. C. and B. S. Weir. 1986. Estimation of inbreeding parameters in stratified populations. Ann. Hum. Genet. 50:271–281.

Cockerham, C. C. and B. S. Weir. 1987. Correlations of descent measures: Drift with migration and mutation. Proc. Natl. Acad. Sci. U.S.A. 84:8512–8514.

Cocks, G. T. and A. C. Wilson. 1972. Enzyme evolution in the Enterobacteriaceae. J. Bacteriol. 110:793–802.

Coen, E., T. Strachan and G. Dover. 1982. Dynamics of concerted evolution in regions of ribosomal DNA and histone gene families in the *melano-*

gaster group of *Drosophila*. J. Mol. Biol. 158:17–35.

Colless, D. H. 1970. The phenogram as an estimate of phylogeny. Syst. Zool. 19:352–362.

Collier, G. E. and R. J. MacIntyre. 1977. Microcomplement fixation studies on the evolution of α-glycerophosphate dehydrogenase within the genus *Drosophila*. Proc. Natl. Acad. Sci. U.S.A. 74:684–688.

Comings, D. E. 1978. Mechanisms of chromosome banding and implications for chromosome structure. Annu. Rev. Genet. 12:25–46.

Comings, D. E., E. Avelino, T. A. Okado and H. E. Wyandt. 1973. The mechanism of C- and G-banding of chromosomes. Exp. Cell Res. 77:469–493.

Commorford, S. L. 1971. Iodination of nucleic acids in vitro. Biochemistry 10:1993–2000.

Conger, A. D. and L. M. Fairchild. 1953. A quick-freeze method for making smear slides permanent. Stain Tech. 28:289–293.

Conkle, M. T., P. D. Hodgskiss, L. B. Nunnally and S. C. Hunter. 1982. Starch gel electrophoresis of conifer seeds: A laboratory manual. Gen. Tech. Report PSW-64. Pacific Southwest Forest and Range Exp. Stn., Forest Serv., U.S. Dept. Agric., Berkeley, California.

Coradin, L. and D. E. Giannasi. 1980. The effects of chemical preservatives on plant collections to be used in chemotaxonomic surveys. Taxon 29:33–40.

Coyne, J. 1982. Gel electrophoresis and cryptic protein variation, pp. 1–32. *In* M. Rattazzi, J. Scandalios and G. Whitt (eds.), *Isozymes: Current Topics in Biological and Medical Research,* Vol. 6. Liss, New York.

Cracraft, J. 1987. DNA hybridization and avian phylogenetics. Evol. Biol. 21:47–96.

Crawford, D. J. 1983. Phylogenetic and systematic inferences from electrophoretic studies, pp. 257–287. *In* S.D. Tanksley and T. J. Orton (eds.), *Isozymes in Plant Genetics and Breeding, Part A.* Elsevier, Amsterdam.

Crawford, D. J. 1989. *Plant Molecular Systematics: Macromolecular Approaches.* Wiley, New York.

Crawford, T. J. 1984. What is a population?, pp. 135–174. *In* B. Shorrocks (ed.), *Evolutionary Ecology. The 23rd Symposium of the British Ecological Society, Leeds 1982.* Blackwell, Oxford.

Cremisi, F., R. Vignali, R. Batistoni and G. Barsacchi. 1988. Heterochromatic DNA in *Triturus* (Amphibia, Urodela) II. A centromeric satellite DNA. Chromosoma 97:204–211.

Cronin, J. E. and V. M. Sarich. 1975. Molecular systematics of the New World monkeys. J. Human Evol. 4:357–375.

Cross, T. F., R. D. Ward and A. Abreu-Grobois. 1979. Duplicate loci and allelic variation for mitochondrial malic enzyme in the Atlantic salmon, *Salmo salar* L. Comp. Biochem. Physiol. 62B:403–406.

Crouau-Roy, B. 1986. Genetic divergence between populations of two closely related troglobitic beetle species (*Speonomus:* Bathysciinae, Coleoptera). Genetica 68:97–103.

Crouau-Roy, B. 1988. Genetic structure of cave-dwelling beetles populations: Significant deficiencies of heterozygotes. Heredity 60:321–327.

Crouse, J. and D. Amorese. 1986. Stability of restriction endonucleases during extended digestion. Focus (BRL) 8:1–2.

CSKRN. 1973. Committee for a standardized karyotype of the Norway rat, *Rattus norvegicus.* Cytogenet Cell Genet. 12:199–205.

Dallas, J. F. 1988. Detection of DNA "fingerprints" of cultivated rice by hybridization with a human minisatellite DNA probe. Proc. Natl. Acad. Sci. U.S.A. 85:6831–6835.

Daly, J. C. 1981. Effects of social organization and environmental diversity on determining the genetic structure of a population of the wild rabbit, *Oryctolagus cuniculus.* Evolution 35:689–706.

Dando, P. R., K. B. Storey, P. W. Hochachka and J. M. Storey. 1981. Multiple dehydrogenases in marine molluscs: Electrophoretic analysis of alanopine dehydrogenase, strombine dehydrogenase, octopine dehydrogenase, and lactate dehydrogenase. Marine Biol. Lett. 2:249–257.

Danna, K. J. 1980. Determination of fragment order through partial digests and multiple enzyme digests. Methods Enzymol. 65:449–467.

Danzmann, R. G. and J. P. Bogart. 1982a. Evidence for a polymorphism in gametic segregation using a malate dehydrogenase locus in the tetraploid treefrog *Hyla versicolor.* Genetics 100:287-306.

Danzmann, R. G. and J. P. Bogart. 1982b. Gene dosage effects on MDH isozyme expression in diploid, triploid, and tetraploid treefrogs of the genus *Hyla.* J. Hered. 73:277-280.

Darnell, R., H. Lodish and D. Baltimore. 1986. *Molecular Cell Biology.* Scientific American Books, New York.

Darwin, C. 1859. *On the Origin of Species by Means of Natural Selection.* Murray, London.

Davies, D. H., R. Lawson, S. J. Burch and J. E. Hanson. 1987. Evolutionary relationships of a "primitive" shark (*Heterodontus*) assessed by micro-complement fixation of serum transferrin. J. Mol. Evol. 25:74–80.

Davis, L. G., M. D. Dibner and J. F. Battey. 1986. *Basic Methods in Molecular Biology*. Elsevier, New York.

Davis, M. B. 1973. Labeling of DNA with ^{125}I. Carnegie Institute of Washington Year Book 72:217–221.

Davison, D. 1985. Sequence similarity ('homology') searching for molecular biologists. Bull. Math. Biol. 47:437–474.

Dawley, R. M. and J. P. Bogart (eds.) 1989. *Evolution and Ecology of Unisexual Vertebrates*. Bull. New York State Museum, Albany.

Dawley, R. M., J. H. Graham and R. J. Schultz. 1985. Triploid progeny of pumpkinseed × green sunfish hybrids. J. Hered. 76:251–257.

Dawson, D. M., H. M. Eppenberger and N. O. Kaplan. 1967. The comparative enzymology of creatine kinases. II. Physical and chemical properties. J. Biol. Chem. 25:210–217.

Dayhoff, M. O. 1978. *Atlas of Protein Sequence and Structure*, Vol. 5, Suppl. 3. National Biomedical Research Foundation, Silver Springs, MD.

Debeau, L., L. A. Chandler, J. R. Gralow, P. W. Nichols and P.A. Jones. 1986. Southern blot analysis of DNA extracted from formalin-fixed pathology specimens. Cancer Res. 46:2964–2969.

DeBorde, D. C., C. W. Naeve, M. L. Herlocher and H. F. Maassab. 1986. Resolution of a common RNA sequencing ambiguity by terminal deoxynucleotidyl transferase. Anal. Biochem. 157:275–282.

DeBry, R. W. and N. A. Slade. 1985. Cladistic analysis of restriction endonuclease cleavage maps within a maximum-likelihood framework. Syst. Zool. 34:21–34.

Deininger, P. L. and G. R. Daniels. 1986. The recent evolution of mammalian repetitive DNA elements. Trends Genet. 2:76–80.

DeLorenzo, R. J. and F. H. Ruddle. 1969. Genetic control of two electrophoretic variants of glucosephosphate isomerase in the mouse. Biochem. Genet. 3:151–162.

Dene, H., M. Goodman and W. S. Prychodko. 1978. An immunological examination of the systematics of the Tupaioidea. J. Mammal. 59:697–706.

Densmore, L. D. 1983. Biochemical and immunological systematics of the order Crocodilia. Evol. Biol. 16:397–465.

Densmore, L. D., J. W. Wright and W. M. Brown.

1985. Length variation and heteroplasmy are frequent in mitochondrial DNA from parthenogentic and bisexual lizards (genus *Cnemidophorus*). Genetics 110:698–707.

Derr, J. N., J. W. Bickham, I. F. Greenbaum, A. G. J. Rhodin and R. A. Mittermeier. 1987. Biochemical systematics and evolution in the South American turtle genus *Platemys* (Pleurodira: Chelidae). Copeia 1987:370–375.

DeSalle, R. and A. R. Templeton. 1988. Founder effects accelerate the rate of mitochondrial DNA evolution in Hawaiian *Drosophila*. Evolution 42:1076–1084.

DeSalle, R., L. V. Giddings and A. R. Templeton. 1986. Mitochondrial DNA variability in natural populations of Hawaiian *Drosophila*. I. Methods and levels of variability in *D. silvestris* and *D. heteroneura* populations. Heredity 56:75–85.

DeSalle, R., T. Freedman, E. M. Prager and A. C. Wilson. 1987a. Tempo and mode of sequence evolution in mitochondrial DNA of Hawaiian *Drosophila*. J. Mol. Evol. 26:157-164.

DeSalle, R., A. R. Templeton, I. Mori, S. Pletscher and J. S. Johnson. 1987b. Temporal and spatial heterogeneity of mtDNA polymorphisms in natural populations of *Drosophila mercatorum*. Genetics 116:215233.

De Soete, G. 1983a. A least squares algorithm for fitting additive trees to proximity data. Psychometrica 48:621–626.

De Soete, G. 1983b. On the construction of "optimal" phylogenetic trees. Z. Naturforsch 38:156–158.

Dessauer, H. C. and C. J. Cole. 1984. Influence of gene dosage on electrophoretic phenotypes of proteins from lizards of the genus *Cnemidophorus*. Comp. Biochem. Physiol. 77B:181–189.

Dessauer, H. C. and C. J. Cole. 1986. Clonal inheritance in parthenogenetic whiptail lizards: Biochemical evidence. J. Hered. 77:8–12.

Dessauer, H. C. and M. S. Hafner (eds.) 1984. *Collections of Frozen Tissues: Value, Management, Field and Laboratory Procedures, and Directory of Existing Collections*. Association of Systematics Collections, Univ. Kansas Press, Lawrence.

Dessauer, H. C. and R. A. Menzies. 1984. Stability of macromolecules during longterm storage, pp. 17–20. *In* H. C. Dessauer and M. S. Hafner (eds.), *Collections of Frozen Tissues: Value, Management, Field and Laboratory Procedures, and Directory of Existing Collections*. Association of Systematics Collections, Univ. Kansas Press, Lawrence.

Dessauer, H. C., M. J. Braun and S. Neville. 1983. A simple hand centrifuge for field use. Isozyme Bull. 16:91.

Dessauer, H. C., R. A. Menzies and D. E. Fairbrothers. 1984. Procedures for collecting and preserving tissues for molecular studies, pp. 21–24. *In* H. C. Dessauer and M. S. Hafner (eds.), *Collections of Frozen Tissues: Value, Management, Field and Laboratory Procedures, and Directory of Existing Collections.* Association of Systematics Collections, Univ. Kansas Press, Lawrence.

Dessauer, H. C., J. E. Cadle and R. Lawson. 1987. Patterns of snake evolution suggested by their proteins. Fieldiana Zool. N.S. 34:1–34.

Dessauer, H. C., M. S. Hafner, R. M. Zink and C. J. Cole. 1988. A national program to develop, maintain, and utilize frozen tissue collections for scientific research. Assoc. Syst. Collections Newslett. 16:3,9–10.

Diaz, M. O., G. Barsacchi-Pilone, K. A. Mahon and J. G. Gall. 1981. Transcripts from both strands of a satellite DNA occur on lampbrush chromosome loops of the newt *Notophthalmus.* Cell 24:649–659.

DiLella, A. G. and S. L. C. Woo. 1987. Cloning large segments of genomic DNA using cosmid vectors. Methods Enzymol. 152:199–212.

DiMichele, L. and D. A. Powers. 1982a. LDH-B genotype-specific hatching times of *Fundulus heteroclitus* embryos. Nature (London) 296:563–564.

DiMichele, L. and D. A. Powers. 1982b. Physiological basis for swimming endurance differences between LDH-B genotypes of *Fundulus heteroclitus.* Science 216:1014–1016.

Dimmick, W. W. 1987. Phylogenetic relationships of Notropis hubbsi, N. welaka and *N. emilae* (Cypriniformes: Cyprinidae). Copeia 1987:316–325.

Dodds, K. G. 1986. Resampling methods in genetics and the effect of family structure in genetic data. Institute of Statistics Mimeo. Series 1684T, North Carolina State Univ., Raleigh.

Donnellan, S. C. and K. P. Aplin. 1989. Resolution of cryptic species in the New Guinean lizard, *Sphenomorphus jobiensis* (Scincidae) by electrophoresis. Copeia 1989:81–88.

Doolittle, W. F. 1985. Middle repetitive DNAs, pp. 443–487. *In* T. Cavalier-Smith (ed.), *The Evolution of Genome Size.* Wiley, New York.

Dover, G. 1987. Letter to the editor. Cell 51:515.

Dover, G. A. and D. Tautz. 1986. Conservation and divergence in multigene families: Alternatives to selection and drift. Phil. Trans. Roy. Soc. London Ser. B 312:275–289.

Dover, G. A., S. Brown, E. Coen, J. Dallas, T. Strachan and M. Trick. 1982. The dynamics of genome evolution and species differentiation, pp. 343–372. *In* G. A. Dover and R. B. Flavell (eds.), *Genome Evolution.* Academic Press, New York.

Dowling, T. E. and W. M. Brown. 1989. Allozymes, mitochondrial DNA, and levels of phylogenetic resolution among four species of minnows (*Notropis:* Cyprinidae). Syst. Zool. 38:126–143.

Dowling, H. C., R. Highton, G. C. Maha and L. R. Maxson. 1983. Biochemical evaluation of colubrid snake phylogeny. J. Zool. (London) 201:309–329.

Dowling, T. E., G. R. Smith and W. M. Brown. 1989. Reproductive isolation and introgression between *Notropis cornutus* and *Notropis chrysocephalus* (family Cyprinidae): Comparison of morphology, allozymes, and mitochondrial DNA. Evolution 43:620–634.

Doyle, J. J. and E. E. Dickson. 1987. Preservation of plant samples for DNA restriction endonuclease analysis. Taxon 36:715–722.

Dubin, D. T., C. C. HsuChen and L. E. Tillotson. 1986. Mosquito mitochondrial transfer RNAs for valine, glycine and glutamate: RNA and gene sequences and vicinal genome organization. Curr. Genet. 10:701-707.

DuBose, R. F. and D. L. Hartl. 1990. Rapid purification of PCR products for DNA sequencing using Sepharose CL-6B spin columns. Biotechniques (in press).

Duellman, W. E. and D. M. Hillis. 1987. Marsupial frogs (Anura: Hylidae: *Gastrotheca*) of the Ecuadorian Andes: Resolution of taxonomic problems and phylogenetic relationships. Herpetologica 43:135–167.

Duellman, W. E., L. R. Maxson and C. A. Jesiolowski. 1988. Evolution of marsupial frogs (Hylidae: Hemiphractinae): Immunological evidence. Copeia 1988:527–543.

Dutrillaux, B. 1975. Discontinued treatment with BudR and staining with acridine orange: Observation of R- or Q- or intermediary banding. Chromosoma 52:261–273.

Dyer, A. F. 1979. *Investigating Chromosomes.* Wiley, New York.

Dykhuizen, D. E., C. Mudd, A. Honeycutt and D. L. Hartl. 1985. Polymorphic posttranslational modification of alkaline phosphatase in *Escherichia coli.* Evolution 39:1–7.

Eanes, W. F. and R. K. Koehn. 1978. An analysis of genetic structure in the monarch butterfly, *Danaus plexippus* L. Evolution 32:784797.

Easteal, S. 1985. The ecological genetics of introduced populations of the giant toad *Bufo marinus*. II. Effective population size. Genetics 110:107–122.

Easteal, S. 1986. The ecological genetics of introduced populations of the giant Toad, *Bufo marinus*. IV. Gene flow estimated from admixture in Australian populations. Heredity 56:145–156.

Echelle, A. A., T. E. Dowling, C. Moritz and W. M. Brown. 1989. Mitochondrial DNA diversity and the origin of the *Menidia clarkhubbsi* complex of unisexual fishes (Atherinidae). Evolution 43:984–993.

Echelle, A. F., A. A. Echelle and D. R. Edds. 1989. Conservation genetics of a spring-dwelling desert fish, the Pecos gambusia (*Gambusia nobilis*, Poeciliidae). Conserv. Biol. 3:159–169.

Eck, R. V. and M. O. Dayhoff (eds.). 1966. *Atlas of Protein Sequence and Structure 1966*. National Biomedical Research Foundation, Silver Springs, MD.

Eckert, R. 1987. New vectors for rapid sequencing of DNA fragments by chemical degradation. Gene 51:242–252.

Efron, B. 1982. The jackknife, the bootstrap, and other resampling plans. CBMS-NSF Regional Conference Series in Applied Mathematics, Monograph 38. Society of Industrial and Applied Mathematics, Philadelphia.

Efron, B. and G. Gong. 1983. A leisurely look at the bootstrap, the jackknife, and cross-validation. Am. Stat. 37:36–48.

Elwood, H. J., G. J. Olsen and M. L. Sogin. 1985. The small-subunit ribosomal RNA gene sequences from the hypotrichous ciliates *Oxytricha nova* and *Stylonychia pustulata*. Mol. Biol. Evol. 2:399–410.

Endler, J. A. 1979. Gene flow and life history patterns. Genetics 93:263–284.

Engel, W., J. Schmidtke, W. Vogel and V. Wolf. 1973. Genetic polymorphism of lactate dehydrogenase isoenzymes in the carp (*Cyprinus carpio*) apparently due to "null alleles." Genetica 8:281–289.

Epplen, J. T. 1988. On simple repeated GAC/TA sequences in animal genomes: A critical reappraisal. J. Hered. 79:409–417.

Estabrook, G. F. 1983. The causes of character incompatibility, pp. 279–295. *In* J. Felsenstein (ed.), *Numerical Taxonomy*. NATO ASI Series, Vol. G1, Springer-Verlag, Berlin.

Estabrook, G. F. and L. Landrum. 1975. A simple test for the possible simultaneous evolutionary divergence of two amino acid positions. J. Math. Biol. 4:195–200.

Evarts, S. and C. J. Williams. 1987. Multiple paternity in a wild population of mallards. Auk 104:597–602.

Fairbrothers, D. E. and M. A. Johnson. 1964. Comparative serological studies within the families Cornaceae (Dogwood) and Nyssaceae (Sour Gum), pp. 305–318. *In* C. A. Leone (ed.), *Taxonomic Biochemistry and Serology*. Ronald Press, New York.

Faith, D. P. 1985. Distance methods and the approximation of most-parsimonious trees. Syst. Zool. 34:312–325.

Farris, J. S. 1969. A successive approximations approach to character weighting. Syst. Zool. 18:374–385.

Farris, J. S. 1970. Methods for computing Wagner trees. Syst. Zool. 19:83–92.

Farris, J. S. 1972. Estimating phylogenetic trees from distance matrices. Am. Natur. 106:645–668.

Farris, J. S. 1977. Phylogenetic analysis under Dollo's Law. Syst. Zool. 26:77–88.

Farris, J. S. 1981. Distance data in phylogenetic analysis, pp. 3–23. *In* V. A. Funk and D. R. Brooks (eds.), *Advances in Cladistics: Proceedings of the First Meeting of the Willi Hennig Society*. New York Botanical Garden, Bronx.

Farris, J. S. 1983. The logical basis of phylogenetic systematics, pp. 7–36. *In* N. I. Platnick and V. A. Funk (eds.), *Advances in Cladistics*. Columbia Univ. Press, New York.

Farris, J. S. 1985. Distance data revisited. Cladistics 1:67–85.

Farris, J. S. 1986. Distances and cladistics. Cladistics 2:144–157.

Feinberg, A. P. and B. Vogelstein. 1983. A technique for radiolabelling DNA restriction endonuclease fragments to high specific activity. Anal. Biochem. 132:6–13.

Felsenstein, J. 1978a. Cases in which parsimony and compatibility methods will be positively misleading. Syst. Zool. 27:401–410.

Felsenstein, J. 1978b. The number of evolutionary trees. Syst. Zool. 27:27–33.

Felsenstein, J. 1981a. Evolutionary trees from DNA sequences: A maximum likelihood approach. J. Mol. Evol. 17:368–376.

Felsenstein, J. 1981b. Evolutionary trees from gene frequencies and quantitative characters: Finding maximum likelihood estimates. Evolution 35:1229–1242.

Felsenstein, J. 1981c. A likelihood approach to character weighting and what it tells us about parsimony and compatibility. Biol. J. Linn. Soc. 16:183–196.

Felsenstein, J. 1982. Numerical methods for inferring evolutionary trees. Q. Rev. Biol. 57:379–404.

Felsenstein, J. 1984. Distance methods for inferring phylogenies: A justification. Evolution 38:16–24.

Felsenstein, J. 1985. Confidence limits on phylogenies: An approach using the bootstrap. Evolution 39:783–791.

Felsenstein, J. 1986. Distance methods: A reply to Farris. Cladistics 2:130–143.

Felsenstein, J. 1987. Estimation of hominoid phylogeny from a DNA hybridization data set. J. Mol. Evol. 26:123–131.

Felsenstein, J. 1988. Phylogenies from molecular sequences: Inference and reliability. Annu. Rev. Genet. 22:521–565.

Fernholm, B., K. Bremer and H. Jornvall (eds.) 1989.The Hierarchy of Life. Elsevier, Amsterdam.

Ferrari, J. A. and C. E. Taylor. 1981. Hierarchical patterns of chromosome variation in Drosophilia subobscura. Evolution 35:391–394.

Ferris, S. D. and G. S. Whitt. 1977a. Duplicate gene expression in diploid and tetraploid loaches (Cypriniformes, Cobitidae). Biochem. Genet. 15:1097–1112.

Ferris, S. D. and G. S. Whitt. 1977b. Loss of duplicate gene expression after polyploidization. Nature (London) 265:258–260.

Ferris, S. D. and G. S. Whitt. 1978a. Phylogeny of tetraploid catostomid fishes based on the loss of duplicate gene expression. Syst. Zool. 27:189–203.

Ferris, S. D. and G. S. Whitt. 1978b. Genetic and molecular analysis of non-random dimer assembly of the creatine kinase isozymes of fishes. Biochem. Genet. 16:811–829.

Ferrucci, L., E. Romano and G. F. De Stefano. 1987. The AluI-induced bands in great apes and man: Implication for heterochromatin characterization and satellite DNA distribution. Cytogenet. Cell Genet. 44:53–57.

Field, K. G., G. J. Olsen, D. J. Lane, S. J. Giovannoni, M. T. Ghiselin, E. C. Raff, N. R. Pace and R. A. Raff. 1988. Molecular phylogeny of the animal kingdom. Science 239:748–753.

Figueroa, F., M. Kasahara, H. Tichy, E. Neufeld, U. Ritte and J. Klein. 1987. Polymorphism of unique noncoding DNA sequences in wild and laboratory mice. Genetics 117:101–108.

Fildes, R. A. and H. Harris. 1966. Genetically determined variation of adenylate kinase in man. Nature (London) 209:261–263.

Fink, S. C. and R. W. Brosemer. 1973. Immunochemical studies with glycerol 3-phosphate dehydrogenase in bees and wasps. Arch. Biochem. Biophys. 158:30–35.

Fisher, S. E. and G. S. Whitt. 1978. Evolution of isozyme loci and their differential tissue expression. Creatine kinase as a model system. J. Mol. Evol. 12:25–55.

Fisher, S. E. and G. S. Whitt. 1979. Evolution of the creatine kinase isozyme system in the primitive vertebrates. Occ. Pap. California Acad. Sci. 134:142–159.

Fisher, S. E., J. B. Shaklee, S. D. Ferris and G. S. Whitt. 1980. Evolution of five multilocus isozyme systems in the chordates. Genetica 52/53:73–85.

Fitch, W. M. 1966. An improved method of testing for evolutionary homology. J. Mol. Biol. 16:9–16.

Fitch, W. M. 1970. Distinguishing homologous from analogous proteins. Syst. Zool. 19:99–113.

Fitch, W. M. 1971a. The non-identity of invariant positions in the cytochrome c of different species. Biochem. Genet. 5:231–241.

Fitch, W. M. 1971b. Toward defining the course of evolution: Minimal change for a specific tree topology. Syst. Zool. 20:406–416.

Fitch, W. M. 1975. Toward finding the tree of maximum parsimony, pp. 189–230. In G. F. Estabrook (ed.), Proceedings of the Eighth International Conference on Numerical Taxonomy. Freeman, San Francisco.

Fitch, W. M. 1976a. The molecular evolution of cytochrome c in eukaryotes. J. Mol. Evol. 8:13–40.

Fitch, W. M. 1976b. Molecular evolutionary clocks, pp. 160–178. In F. J. Ayala (ed.), Molecular Evolution. Sinauer, Sunderland, MA.

Fitch, W. M. 1977. On the problem of discovering the most parsimonious tree. Am. Natur. 111:223–257.

Fitch, W. M. 1981. A non–sequential method for constructing trees and hierarchical classifications. J. Mol. Evol. 18:30–37.

Fitch, W. M. 1986. A hidden bias in the estimate of total nucleotide substitutions from pairwise differences, pp. 315–328. In S. Karlin and E. Nevo (eds.), Evolutionary Processes and Theory. Academic Press, Orlando, FL.

Fitch, W. M. and E. Margoliash. 1967. Construction of phylogenetic trees. Science 155:279–284.

Flavell, R. B. 1986. Repetitive DNA and chromosome

evolution in plants. Phil. Trans. Roy. Soc. London Ser. B 312:227–242.

Flavell, R. B., M. O'Dell, P. Sharp, E. Nevo and A. Beiles. 1986. Variation in the intergenic spacer of ribosomal DNA of wild wheat, *Triticum dicoccoides*, in Israel. Mol. Biol. Evol. 3:547–558.

Fleischer, R. C. 1983. A comparison of theoretical and electrophoretic assessments of genetic structure in populations of the house sparrow (*Passer domesticus*). Evolution 37:1001–1009.

Flint, J., A. V. S. Hill, D. K. Bowden, S. J. Oppenheimer, P. R. Sill, S. W. Serjeantson, J. Bana-Koiri, K. Bhatia, M. P. Alpers, A. J. Boyce, D. J. Weatherall and J. B. Clegg. 1986. High frequencies of α-thalassaemia are the result of natural selection by malaria. Nature (London) 321:744–750.

Foltz, D. W. 1986. Null alleles as a possible cause of heterozygote deficiencies in the oyster *Crassostrea virginica* and other bivalves. Evolution 40:869–870.

Foltz, D. W. and J. L. Hoogland. 1983. Genetic evidence of outbreeding in the black-tailed prairie dog (*Cynomys ludovicianus*). Evolution 37:273–281.

Fonatsch, C., G. Gradl, J. Ragoussis and A. Ziegler. 1987. Assignment of the TCP1 locus to the long arm of human chromosome 6 by *in situ* hybridization. Cytogenet. Cell Genet. 45:109–112.

Fox, G. M. and C. W. Schmid. 1980. Related single copy sequences in the human genome. Biochim.Biophys. Acta 609:349–363.

Fox, G. E., E. Stackebrandt, R. B. Hespell, J. Gibson, J. Maniloff, T. A. Dyer, R. S. Wolfe, W. E. Balch, R. S. Tanner, L. J. Magrum, L. B. Zablen, R. Blakemore, R. Gupta, L. Bonen, B. J. Lewis, D. A. Stahl, K. R. Luehrsen, K. N. Chen and C. R. Woese. 1980a. The phylogeny of prokaryotes. Science 209:457–463.

Fox, G. M., J. Umeda, R. K.-Y. Lee and C. W. Schmid. 1980b. A phase diagram of the binding of mismatched duplex DNAs to hydroxyapatite. Biochim. Biophys. Acta 609:364–371.

Frair, W. 1964. Turtle family relationships as determined by serological tests, pp. 535–544. *In* C. A. Leone (ed.), *Taxonomic Biochemistry and Serology*. Ronald Press, New York.

Freifelder, D. 1982. *Physical Biochemistry: Applications to Biochemistry and Molecular Biology,* 2nd ed. Freeman, New York.

Frelin, C. and F. Vuilleumier. 1979. Biochemical methods and reasoning in systematics. Z. Zool. Syst. Evol.-Forsch. 17:1–10.

Freshney, R. I. 1987. *Culture of Animal Cells.* Liss, New York.

Frick, L. W. 1981. A biochemical, phylogenetic and immunological investigation of the cytosolic di- and tripeptidases of fishes. Ph.D. dissertation, Univ. of Hawaii.

Frick, L. W. 1983. An electrophoretic investigation of the cytosolic di- and tripeptidases of fish: Molecular weights, substrate specificities, and tissue and phylogenetic distributions. Biochem. Genet. 21:309–322.

Frischauf, A.-M. 1987. Construction and characterization of a genomic library in lambda. Methods Enzymol. 152:190–199.

Frommer, M., C. Paul and P. C. Vincent. 1988. Localization of satellite DNA sequences on human metaphase chromosomes using bromodeoxyuridine-labelled probes. Chromosoma 97:11–18.

Frost, D. R. and D. M. Hillis. 1990. Species in concept and practice: Herpetological applications. Herpetologica 46:87–104.

Frykman, I. and B. O. Bengtsson. 1984. Genetic differentiation in *Sorex*. III. Electrophoretic analysis of a hybrid zone between two karyotypic races in *Sorex araneus*. Hereditas 70:259–270.

Fukami, K. and Y. Tateno. 1989. On the maximum likelihood method for estimating molecular trees: Uniqueness of the likelihood point. J. Mol. Evol. 28:460–464.

Funk, V. A. 1985. Phylogenetic patterns and hybridization. Ann. Missouri Bot. Gard. 72:681–715.

Futuyma, D. J. 1986. *Evolutionary Biology.* 2nd ed. Sinauer, Sunderland, MA.

Galau, G. A., M. E. Chamberlin, B. R. Hough, R. J. Britten and E. H. Davidson. 1976. Evolution of repetitive and nonrepetitive DNA in two species ofXenopus, pp. 200–224. *In* F. J. Ayala (ed.), *Molecular Evolution*. Sinauer, Sunderland, MA.

Gall, J. G. and M. L. Pardue. 1969. Formation and detection of RNA-DNA hybrid molecules in cytological preparations. Proc. Natl. Acad. Sci. U.S.A. 63:378–383.

Gargouri, A. 1989. A rapid and simple method for extracting yeast mitochondrial DNA. Curr. Genet. 15:235–237.

Gastony, G. J. 1986. Electrophoretic evidence for the origin of fern species by unreduced spores. Am. J. Bot. 73:1563–1569.

Gastony, G. J. 1988. The *Pellaea glabella* complex: Electrophoretic evidence for the derivations of the agamosporous taxa and a revised taxonomy. Am. Fern J. 78:44–67.

Gauthier, J., A. G. Kluge and T. Rowe. 1988. Amniote phylogeny and the importance of fossils. Cladistics 4:105–205.

Gellisen, G., J. Y. Bradfield, B. N. White and G. R. Wyatt. 1983. Mitochondrial DNA sequences in the nuclear genome of a locust. Nature (London) 301:631-634.

Georges, M., A.-S. Lequarre, M. Castelli, R. Hanset and G. Vassart. 1988. DNA fingerprinting in domestic animals using four different minisatellite probes. Cytogenet. Cell Genet. 47:127–131.

Gerbi, S. A. 1985. Evolution of ribosomal DNA, pp. 419–517. In R. J. MacIntyre (ed.), Molecular Evolutionary Genetics. Plenum, New York.

Ghiselin, M. T. 1988. The origin of molluscs in light of molecular evidence. Oxford Surv. Evol. Biol. 5:66–95.

Gillespie, J. H. 1984. The molecular clock may be an episodic clock. Proc. Natl. Acad. Sci. U.S.A. 81:8009–8013.

Gillespie, J. H. 1986a. Natural selection and the molecular clock. Mol. Biol. Evol. 3:138–155.

Gillespie, J. H. 1986b. Variability of evolutionary rates of DNA. Genetics 113:1077–1091.

Gillespie, J. H. 1986c. Rates of molecular evolution. Ann. Rev. Ecol. Syst. 17:637–665.

Gillespie, J. H. 1987. Molecular evolution and the neutral allele theory. Oxford Surv. Evol. Biol. 4:10–37.

Gillespie, J. H. and K. Kojima. 1968. The degree of polymorphism in enzymes involved in energy production compared to that in nonspecific enzymes in two Drosophila ananassae populations. Genetics 61:582-585.

Goelz, S. E., S. R. Hamilton and B. Vogelstein. 1985. Purification of DNA from formaldehyde fixed and paraffin embedded human tissue. Biochem. Biophy. Res. Commun. 30:118–126.

Golding, G. B. 1983. Estimates of DNA and protein sequence divergence: An examination of some assumptions. Mol. Biol. Evol. 1:125–142.

Golding, G. B. and C. Strobeck. 1983. Increased number of alleles found in hybrid populations due to intragenic recombination. Evolution 37:17–29.

Golenberg, E. M. 1987. Estimation of gene flow and genetic neighborhood size by indirect methods in a selfing annual, Triticum dicoccoides. Evolution 41:1326–1334.

Gollmann, G., P. Roth and W. Hodl. 1988. Hybridi-

zation between fire-bellied toads Bombina bombina and Bombina variegata in the Karst regions of Slovakia and Hungary: Morphological and allozyme evidence. J. Evol. Biol. 1:3–14.

Gonzales, I. L., J. L. Gorski, T. J. Campden, D. J. Dorney, J. M. Erickson, J. E. Sylvester and R. D. Schmickel. 1985. Variation among human 28S ribosomal RNA genes. Proc. Natl. Acad. Sci. U.S.A. 82:7666–7670.

Good, D. A. 1989. Hybridization and cryptic species in Dicamptodon (Caudata: Dicamptodontidae). Evolution 43:728–744.

Good, D. A., G. Z. Wurst and D. B. Wake. 1987. Patterns of geographic variation in allozymes of the Olympic salamander, Rhyacotriton olympicus (Caudata: Dicamptodontidae). Fieldiana Zool. 1374:1–15.

Goodman, M. 1961. The role of immunochemical differences in the phyletic development of human behavior. Hum. Biol. 33:131–162.

Goodman, M. 1963. Serological analysis of the systematics of recent hominoids. Hum. Biol. 35:377–424.

Goodman, M. 1981. Decoding the pattern of protein evolution. Prog. Biophys. Mol. Biol. 37:105–164.

Goodman, M. 1985. Rates of molecular evolution: The hominoid slowdown. BioEssays 3:9–14.

Goodman. M. and G. W. Moore. 1971. Immunodiffusion systematics of the primates. I. The Catarrhini. Syst. Zool. 20:19–62.

Goodman, M., J. Czelusniak, G. W. Moore, A. E. Romero-Herrera and G. Matsuda. 1979. Fitting the gene lineage into the species lineage, a parsimony strategy illustrated by cladograms constructed from globin sequences. Syst. Zool. 28:132–163.

Goodman, M., M. M. Miyamoto and J. Czelusniak. 1987. Pattern and process in vertebrate phylogeny revealed by coevolution of molecules and morphologies, pp. 141–176. In C. Patterson (ed.), Molecules and Morphology in Evolution: Conflict or Compromise? Cambridge Univ. Press, Cambridge.

Gorman, G. C. 1971. Evolutionary genetics of island lizard populations. Yearbook Am. Philo. Soc. 1971:318–319.

Gorman, G. and J. Renzi, Jr. 1979. Genetic distance and heterozygosity estimates in electrophoretic studies: Effects of sample size. Copeia 1979:242–249.

Gorman, G. C. and S. Y. Yang. 1975. A low level of back-crossing between the hybridizing Anolis lizards of Trinidad. Herpetologica 31:196–198.

Gorman, G. C., A. C. Wilson and M. Nakanishi. 1971. A biochemical approach towards the study of reptilian phylogeny: Evolution of serum albumin and lactic dehydrogenase. Syst. Zool. 20:167–185.

Gorman, G. C., D. G. Buth and J. S. Wyles. 1980. *Anolis* lizards of the eastern Caribbean: A case study in evolution. III. A cladistic analysis of albumin immunological data, and the definition of species groups. Syst. Zool. 29:143–158.

Gorzula, S., C. L. Arocha-Pinango and C. Salazar. 1976. A method of obtaining blood by caudal vein from large reptiles. Copeia 1976:838–839.

Gottlieb, L. D. 1982a. Conservation and duplication of isozymes in plants. Science 216:373–380.

Gottlieb, L. D. 1982b. Isozyme number and phylogeny, pp. 209-221. *In* U. Jensen and D. E. Fairbrothers (eds.), *Proteins and Nucleic Acids in Plant Systematics.* Springer-Verlag, Berlin.

Gottlieb, L. D. and N. F. Weeden. 1979. Gene duplication and phylogeny in *Clarkia.* Evolution 33:1024–1039.

Gough, J. A. and N. E. Murray. 1983. Sequence diversity among related genes for recognition of specific targets in DNA molecules. J. Mol. Biol. 166:1–19.

Graham, D. E. 1978. The isolation of high molecular weight DNA from whole organisms or large tissue masses. Anal. Biochem. 85:609–613.

Graham, J. H. and J. D. Felley. 1985. Genomic coadaptation and developmental stability within introgressed populations of *Enneacanthus gloriosus* and *E. obesus* (Pisces, Centrarchidae). Evolution 39:104–114.

Gray, G. S. and W. M. Fitch. 1983. Evolution of antibiotic resistance genes: The DNA sequence of a kanamycin resistance gene from *Staphylococcus aureua.* Mol. Biol. Evol. 1:57–66.

Green, D. M., J. P. Bogart and E. H. Anthony. 1980. An interactive, microcomputer-based karyotype analysis system for phylogenetic cytotaxonomy. Comput. Biol. Med. 10:219–227.

Greenbaum, I. F. 1981. Genetic interactions between hybridizing cytotypes of the tent-making bat (*Uroderma bilobatum*). Evolution 35:305–320.

Greenberg, B. D., J. E. Newbold and A. Sugino. 1983. Intraspecific nucleotide sequence variability surrounding the origin of replication in human mitochondrial DNA. Gene 21:33–49.

Groot, G. S. P. and A. M. Kroon. 1979. Mitochondrial DNA from various organisms does not contain internally methylated cytosine in –CCGG– sequences. Biochim. Biophys. Acta 564:355–357.

Gruenbaum, H., T. Naveh-Many, H. Cedar and A. Razin. 1981. Sequence specificity of methylation in higher plant DNA. Nature (London) 292:860–862.

Grula, J. W., T. J. Hall, T. D. Giugni, G. J. Graham, E. H. Davidson and R. J. Britten. 1982. Sea urchin DNA sequence variation and reduced interspecies differences of the less variable DNA sequences. Evolution 36:665–676.

Guillemette, J. G. and P. N. Lewis. 1983. Detection of subnanogram quantities of DNA and RNA on native and denaturing polyacrylamide and agarose gels by silver staining. Electrophoresis 4:92–94.

Guries, R. P. and F. T. Ledig. 1982. Genetic diversity and population structure in pitch pine (*Pinus rigida* Mill). Evolution 36:387–402.

Gutierrez, R. J., R. M. Zink and S. Y. Yang. 1983. Genetic variation, systematic and biogeographic relationships of some Galliform birds. Auk 100:33–40.

Gyllensten, V. and H. Erlich. 1988. Generation of single-stranded DNA by the polymerase chain reaction and its applications to direct sequencing of the HLA DQa locus. Proc. Natl. Acad. Sci. U.S.A. 85:7652–7656.

Hack, M. S. and H. J. Lawce. 1980. *The Association of Cytogenetic Technologists Cytogenetics Laboratory Manual.* Univ. California, San Francisco.

Hadjiolov, A. A., O. I. Georgiev, V. V. Nosikov and L. P. Yavachev. 1984. Primary and secondary structure of rat 28S ribosomal RNA. Nucl. Acids Res. 12:3677–3693.

Haeckel, E. 1866. *Generelly Morphologiy der Organismen-Allgemeiny Grundzugy der organischen Formen-Wissenschaft, Mechanisch begrundet durch die von Charles Darwin reformirte Descendenz-Theorie.* Georg Riemer, Berlin.

Hafner, M. S. and S. A. Nadler. 1988. Phylogenetic trees support the coevolution of parasites and their hosts. Nature (London) 332:258–259.

Halkka, L., V. Soderlund, U. Skaren and J. Keikkila. 1987. Chromosomal polymorphism and racial evolution of *Sorex araneus* L. in Finland. Hereditas 106:257–275.

Hall, T. C., Y. Ma, B. V. Buchbinder, J. W. Pyne, S. M. Sun and F. A. Bliss. 1978. Messenger RNA for G1 protein of French bean seed: Cell-free translation and product characterization. Proc. Natl. Acad. Sci. U.S.A. 75:3196–3200.

Hall, T. J., J. W. Grula, E. H. Davidson and R. J. Britten. 1980. Evolution of sea urchin non-repetitive DNA. J. Mol. Evol. 16:95–110.

Hall, W. P. and R. K. Selander. 1973. Hybridization

of karyotypically differentiated populations in the *Sceloporus grammicus* complex (Iguanidae). Evolution 27:226–242.

Haltiner, M., T. Kempe and R. Tijian. 1985. A novel strategy for constructing clustered point mutations. Nucl. Acids Res. 13:1015–1026.

Hamby, R. K. and E. A. Zimmer. 1988. Ribosomal RNA sequences for inferring phylogeny within the grass family (Poaceae). Plant Syst. Evol. 160:29–37.

Hamby, R. K., L. Sims, L. Issel and E. Zimmer. 1988. Direct ribosomal RNA sequencing: Optimization of extraction and sequencing methods for work with higher plants. Plant Mol. Biol. Rep. 6:175–192.

Hames, B. D. and S. J. Higgins (eds.) 1986. *Nucleic Acid Hybridisation: A Practical Approach.* IRL Press, Oxford.

Hames, B. D. and D. Rickwood (eds.) 1981. *Gel Electrophoresis of Proteins: A Practical Approach.* IRL Press, Oxford.

Hamkalo, B. A. and N. J. Hutchison. 1984. In situ hybridization at the electron microscope level, pp. 97–115. *In* R. S. Sparkes and F. F. de la Cruz (eds.), *Research Perspectives in Cytogenetics.* Univ. Park Press, Baltimore.

Hamlyn, P. H., G. G. Brownlee, C.-C. Cheng, M. J. Gait and C. Milstein. 1978. Complete sequence of constant and 3′ noncoding regions of an immunoglobulin mRNA using the dideoxynucleotide method of RNA sequencing. Cell 15:1067–1075.

Hanken, J. 1983. Genetic variation in a dwarfed lineage, the Mexican salamander genus *Thorius* (Amphibia: Plethodontidae): Taxonomic, ecologic and evolutionary implications. Copeia 1983:1051–1073.

Harper, M. E. and Saunders, G. F. 1984. Localization of single-copy genes on human chromosomes by in situ hybridization of ^{3}H-probes and autoradiography, pp. 117–133. *In* R. S. Sparkes and F. F. de la Cruz (eds.), *Research Perspectives in Cytogenetics.* Univ. Park Press, Baltimore.

Harris, H. 1966. Enzyme polymorphism in man. Proc. Roy. Soc. London Ser. B 164:298–310.

Harris, H. and D. A. Hopkinson. 1976 et seq. *Handbook of Enzyme Electrophoresis in Human Genetics.* North-Holland, Amsterdam.

Harrison, R. G., D. M. Rand and W. C. Wheeler. 1987. Mitochondrial DNA variation in field crickets across a narrow hybrid zone. Mol. Biol. Evol. 4:144–158.

Harry, J. L. and D. A. Briscoe. 1988. Multiple paternity in the loggerhead turtle (*Caretta caretta*). J. Hered. 79:91–99.

Hartl, D. L. and A. G. Clark. 1989. *Principles of Population Genetics,* 2nd Ed. Sinauer, Sunderland, MA.

Hartman, B. K. and S. Udenfried. 1969. A method for immediate visualization of proteins in acrylamide gels and its use for preparation of antibodies to enzymes. Anal. Biochem. 30:391–394.

Hasegawa, M., Y. Iida, T. Yano, F. Takaiwa and M. Iwabuchi. 1985. Phylogenetic relationships among eukaryotic kingdoms inferred from ribosomal RNA sequences. J. Mol. Evol. 22:32–38.

Haslewood, G. A. D. 1967. *Bile Salts.* Methuen, London.

Hassouna, N., B. Michot and J.-P. Bachellerie. 1984. The complete nucleotide sequence of mouse 28S rRNA gene. Implications for the process of size increase of the large subunit rRNA in higher eukaryotes. Nucl. Acids Res. 12:3563–3583.

Haucke, H.-R. and G. Gellissen. 1988. Different mitochondrial gene orders among insects: Exchanged tRNA gene positions in the COII/COIII region between an orthopteran and a dipteran species. Curr. Genet. 14:471–476.

Haufler, C. H. 1987. Electrophoresis is modifying our concepts of evolution in homosporous pteridophytes. Am. J. Bot. 74:953–966.

Haufler, C. H. and J. S. Sweeney. 1989. Electrophoretic evidence that a reciprocal gene silencing mechanism can promote genetic diversity in polyploids. Am. J. Bot. 76(s):203.

Hauswirth, W. L. and P. J. Laipis. 1985. Transmission genetics of mammalian mitochondria: A molecular model and experimental evidence, pp. 49–59. *In* E. Quagliariello, E. C. Slater, F. Palmieri, C. Saccone and A. M. Kroon (eds.), *Achievements and Perspectives of Mitochondrial Research.* Elsevier, Amsterdam.

Hauswirth, W. W., L. O. Lim, B. Dujon and G. Turner. 1987. Methods for studying the genetics of mitochondria, pp. 171–282. *In* V. M. Darley-Usmar, D. Rickwood and M. T. Wilson (eds.), *Mitochondria: A Practical Approach.* IRL Press, Oxford.

Hay, R. J. 1979. Identification, separation and culture of mammalian tissue cells, pp. 143–318. *In* E. Reid (ed.), *Cell Populations, Methodology Surveys (B): Biochemistry,* Vol. 8. Wiley, New York.

Hay, R. J. and G. F. Gee. 1984. Procedures for collecting cell lines under field conditions, pp. 25–26. *In* H. C. Dessauer and M. S. Hafner (eds.), *Collections of Frozen Tissues: Value, Management, Field and Laboratory Procedures, and Directory of Exist-*

ing Collections. Association of Systematics Collections, Univ. Kansas Press, Lawrence.

Hayasaka, K., T. Gojobori and S. Horai. 1988. Molecular phylogeny and evolution of primate mitochondrial DNA. Mol. Biol. Evol. 5:626–644.

Healy, J. A. and M. F. Mulcahy. 1979. Polymorphic tetrameric superoxide dismutase in the pike *Esox lucius* L. (Pisces; Esocidae). Comp. Biochem. Physiol. 62B:563–565.

Hein, J. 1989a. A new method that simultaneously aligns and reconstructs ancestral sequences for any number of homologous sequences, when the phylogeny is given. Mol. Biol. Evol. 6:649–668.

Hein, J. 1989b. A tree reconstruction method that is economical in the number of pairwise comparisons used. Mol. Biol. Evol. 6:669–684.

Heinstra, P. W. H., W. J. M. Aben, W. Scharloo and G. E. W. Thorig. 1986. Alcohol dehydrogenase of *Drosophila melanogaster:* Metabolic differences mediated through cryptic allozymes. Heredity 57:23–29.

Helfman, D. M., J. C. Fiddes and D. Hanahan. 1987. Directional cDNA cloning in plasmid vectors by sequential addition of oligonucleotide linkers. Methods Enzymol. 152:349–359.

Henderson, A. S. 1982. Cytological hybridization to mammalian chromosomes. Int. Rev. Cytol. 76:1–46.

Henderson, N. S. 1965. Isozymes of isocitrate dehydrogenase: Subunit structure and intracellular location. J. Exp. Zool. 158:263–274.

Hendy, M. D. and D. Penny. 1982. Branch and bound algorithms to determine minimal evolutionary trees. Math. Biosci. 59:277–290.

Hennig, W. 1966. *Phylogenetic Systematics.* Univ. Illinois Press, Urbana.

Hereford, L. M. and R. Robash. 1977. Number and distribution of polyadenylated RNA sequences in yeast. Cell 10:453–462.

Herman, S. G. 1980. *The Naturalist's Field Journal.* Buteo Books, Vermillion, S.D.

Hernandez, J. L. and B. S. Weir. 1989. A disequilibrium coefficient approach to Hardy–Weinberg testing. Biometrics 45:53–70.

Hewitt, G. M. 1988. Hybrid zones—natural laboratories for evolutionary studies. Trends Ecol. Evol. 3:158–167.

Highton, R. 1979. A new cryptic species of salamander of the genus *Plethodon* from the southeastern United States (Amphibia: Plethodontidae). Brimleyana 1:31–36.

Highton, R., G. C. Maha and L. R. Maxson. 1989.

Biochemical evolution in the slimy salamanders of the *Plethodon glutinosus* complex in the eastern United States. Univ. Illinois Biol. Monogr.

Higuchi, R. G. and H. Ochman. 1989. Production of single-stranded DNA templates by exonuclease digestion following the polymerase chain reaction. Nucl. Acids Res. 17:5865.

Higuchi, R., B. Bowman, M. Freiberger, O. A. Ryder and A. C. Wilson. 1984. DNA sequences from the quagga, an extinct member of the horse family. Nature (London) 312:282–284.

Higuchi, R. G., L. A. Wrischnik, E. Oakes, M. George, B. Tong and A. C. Wilson. 1987. Mitochondrial DNA of the extinct quagga: Relatedness and postmortem change. J. Mol. Evol. 25:283–287.

Hilbish, T. J. and R. K. Koehn. 1985a. The physiological basis of natural selection at the LAP locus. Evolution 39:1302–1317.

Hilbish, T. J. and R. K. Koehn. 1985b. Dominance in physiological phenotypes and fitness at an enzyme locus. Science 229:52–54.

Hilbish, T. J., L. E. Deaton and R. K. Koehn. 1982. Effect of an allozyme polymorphism on regulation of cell volume. Nature (London) 298:688–689.

Hill, W. G. and B. S. Weir. 1988. Variances and covariances of squared linkage disequilibria. Theoret. Pop. Biol. 33:54–78.

Hillis, D. M. 1984. Misuse and modification of Nei's genetic distance. Syst. Zool. 33:238–240.

Hillis, D. M. 1985. Evolutionary genetics of the Andean lizard genus *Pholidobolus* (Sauria: Gymnophthalmidae): Phylogeny, biogeography, and a comparison of tree construction techniques. Syst. Zool. 34:109–126.

Hillis, D. M. 1987. Molecular versus morphological approaches to systematics. Annu. Rev. Ecol. Syst. 18:23–42.

Hillis, D. M. 1990. The phylogeny of amphibians: Current knowledge and the role of cytogenetics. *In* D. M. Green and S. K. Sessions (eds.), *Amphibian Cytogenetics and Evolution.* Academic Press, New York.

Hillis, D. M. and S. K. Davis. 1986. Evolution of ribosomal DNA: Fifty million years of recorded history in the frog genus *Rana.* Evolution 40:1275–1288.

Hillis, D. M. and S. K. Davis. 1987. Evolution of the 28S ribosomal RNA gene in anurans: Phylogenetic implications of length and restriction site variation. Mol. Biol. Evol. 4:117–125.

Hillis, D. M. and S. K. Davis. 1988. Ribosomal DNA:

Intraspecific polymorphism, concerted evolution, and phylogeny reconstruction. Syst. Zool. 32: 63–66.

Hillis, D. M. and M. T. Dixon. 1989. Vertebrate phylogeny: Evidence from 28S ribosomal DNA sequences, pp. 355–367. *In* B. Fernholm, K. Bremer, and H. Jörnvall (eds.), *The Hierarchy of Life. Proc. Nobel Symp. 70.* Elsevier, Amsterdam.

Hillis, D. M. and J. C. Patton. 1982. Morphological and electrophoretic evidence for two species of *Corbicula* (Bivalvia: Corbiculidae) in North America. Am. Midl. Natl. 108:74–80.

Hillis, D. M., M. T. Dixon and L. K. Ammerman. 1990. The relationships of coelacanths: Evidence from sequences of vertebrate ribosomal RNA genes. *In* J. A. Musiak and M. Bruton (eds.), *Biology and Evolution of Coelacanths.* Envir. Biol. Fishes (in press).

Hixson, J. E. and W. M. Brown. 1986. A comparison of the small ribosomal RNA genes from the mitochondrial DNA of the great apes and humans: Sequence, structure, evolution, and phylogenetic implications. Mol. Biol. Evol. 3:1–18.

Hoelzal, R. and G. A. Dover. 1987. Molecular techniques for examining genetic variation and stock identity in cetacean species. Rep. Int. Whale Comm.

Holmquist, R., M. M. Miyamoto and M. Goodman. 1988a. Analysis of higher-primate phylogeny from transversion differences in nuclear and mitochondrial DNA by Lake's methods of evolutionary parsimony and operator metrics. Mol. Biol. Evol. 5:217–236.

Holmquist, R., M. M. Miyamoto and M. Goodman. 1988b. Higher-primate phylogeny—Why can't we decide? Mol. Biol. Evol. 5:201–216.

Holsinger, K. E. and L. D. Gottlieb. 1988. Isozyme variability in the tetraploid *Clarkia gracilis* (Onagraceae) and its diploid relatives. Syst. Bot. 13: 1–6.

Honeycutt, R. L., S. W. Edwards, K. Nelson and E. Nevo. 1987. Mitochondrial DNA variation and the phylogeny of African mole rats (Rodentia: Bathyergidae). Syst. Zool. 36:280–293.

Hood, L. E., J. H. Wilson and W. B. Wood. 1974. *Molecular Biology of Eucaryotic Cells,* Vol. 1. Benjamin, Menlo Park, CA.

Hopkinson, D. A. 1975. The use of thiol reagents in the analysis of isozyme patterns, pp. 489–508. *In* C. L. Markert (ed.), *Isozymes.* Vol. 1. Academic Press, New York.

Houck, L. D., S. G. Tilley and S. J. Arnold. 1985.

Sperm competition in a plethodontid salamander: Preliminary results. J. Herpetol. 19:420–423.

Houde, P. and M. J. Braun. 1988. Museum collections as a source of DNA for studies of avian phylogeny. Auk 105:773–776.

Hsu, T. C. 1979. *Human and Mammalian Cytogenetics.* Springer-Verlag, Berlin.

Hsu, T. C. 1981. Polymorphism in human acrocentric chromosomes and the silver staining method for nucleolus organizer regions. Karyogram 7:45.

Hubby, J. L. and R. C. Lewontin. 1966. A molecular approach to the study of genic heterozygosity in natural populations. I. The number of alleles at different loci in *Drosophila pseudoobscura.* Genetics 54:577–594.

Hubby, J. L. and L. H. Throckmorton. 1965. Protein differences in *Drosophila.* II. Comparative species genetics and evolutionary problems. Genetics 52:203–215.

Hudspeth, M. E. S., D. S. Schumard, K. M. Tatti and L. I. Grossman. 1980. Rapid purification of yeast mitochondrial DNA in high yield. Biochim. Biophys. Acta. 610:221–228.

Hunt, J. A., T. J. Hall and R. J. Britten. 1981. Evolutionary distances in Hawaiian *Drosophila* measured by DNA reassociation. J. Mol. Evol. 17:361–367.

Hunt, W. G. and R. K. Selander. 1973. Biochemical genetics of hybridization in European house mice. Heredity 31:11–33.

Hunter, R. L. and C. L. Markert. 1957. Histochemical demonstration of enzymes separated by zone electrophoresis in starch gels. Science 125:1294–1295.

Hutchinson, M. N. and L. R. Maxson, 1987a. Biochemical studies on the relationships of the Gastric–brooding Frogs, genus *Rheobatrachus.* Amphibia/Reptilia 8:1–11.

Hutchinson, M. N. and L. R. Maxson. 1987b. Phylogenetic resolution among Australian tree frogs (Anura: Hylidae: Pelodryadinae): An immunological approach. Aust. J. Zool. 35:61–74.

Hutton, J. R. and J. G. Wetmur. 1973. Effect of chemical modification on the rate of renaturation of deoxyribonucleic acid: Deamination and glyoxalated deoxyribonucleic acid. Biochemistry 12:558–563.

Innis, M. A., D. H. Gelfand, J. J. Sninsky and T. J. White. 1990. *PCR Protocols: A Guide to Methods and Applications.* Academic Press, San Diego.

International Union of Biochemistry. Nomenclature Committee. 1984. *Enzyme Nomenclature, 1984.* Academic Press, Orlando, FL.

ISCN. 1981. An international system for human cytogenetic nomenclature—high resolution banding. Cytogenet. Cell Genet. 31:1–23.

Jackson, J. F. and J. A. Pounds. 1979. Comments on assessing the dedifferentiating effects of gene flow. Syst. Zool. 28:78–85.

Jacobs, H. T., J. W. Posakony, J. W. Grula, J. W. Roberts, J. H. Xin, R. J. Britten and E. H. Davidson. 1983. Mitochondrial DNA sequences in the nuclear genome of *Strongylocentrotus purpuratus*. J. Mol. Biol.165:609632.

Jansen, R. K. and J. D. Palmer. 1987a. Chloroplast DNA from lettuce and *Barnadesia* (Asteraceae): Structure, gene localization and characterization of a large inversion. Curr. Genet. 11:553–564.

Jansen, R. K. and J. D. Palmer. 1987b. A chloroplast DNA inversion marks an ancient evolutionary split in the sunflower family (Asteraceae). Proc. Natl. Acad. Sci. U.S.A. 84:5818–5822.

Jansen, R. K. and J. D. Palmer. 1988. Phylogenetic implications of chloroplast DNA restriction site variation in the Mutisieae (Asteraceae). Am. J. Bot. 75:751–764.

Jeanpierre, M. 1987. A rapid method for the purification of DNA from blood. Nucleic Acids Res. 15:9611.

Jeffreys, A. J. and D. B. Morton. 1987. DNA fingerprints of dogs and cats. Anim. Genet. 18:1–15.

Jeffreys, A. J., V. Wilson and S. L. Thein. 1985a. Hypervariable "minisatellite" regions in human DNA. Nature (London) 314:67–73.

Jeffreys, A. J., V. Wilson and S. L. Thein. 1985b. Individual-specific "fingerprints" of human DNA. Nature (London) 316:76–79.

Jeffreys, A. J., V. Wilson, R. Kelly, B. A. Taylor and G. Bulfield. 1987. Mouse DNA 'fingerprints': Analysis of chromosome localization and germ-line stability of hypervariable loci in recombinant inbred strains. Nucl. Acids Res. 15:2823–2836.

Jeffreys, A.J., N. J. Royle, V. Wilson and Z. Wong. 1988. Spontaneous mutation rates to new length alleles at tandem-repetitive hypervariable loci in human DNA. Nature (London) 332:278–281.

Jensen, U. and D. E. Fairbrothers (eds.) 1983. *Proteins and Nucleic Acids in Plant Systematics*. Springer-Verlag, New York.

Jiminez-Marin, D. and H. C. Dessauer. 1973. Protein phenotype variation in laboratory populations of*Rattus norvegicus*. Comp. Biochem. Physiol. 46B:487–492.

John, H., M. L. Birnsteil and K. W. Jones. 1969.

RNA-DNA hybrids at cytological levels. Nature (London) 223:582–587.

Johnson, A. G., F. M. Utter and H. O. Hodgins. 1970. Interspecific variation of tetrazolium oxidase in *Sebastodes* (rockfish). Comp. Biochem. Physiol. 37:281–285.

Johnson, G. B. 1976. Hidden alleles at the α-glycerophosphate locus in *Colias* butterflies. Genetics 83:149–167.

Johnson, G. B. 1977. Assessing electrophoretic similarity: The problem of hidden heterogeneity. Annu. Rev. Ecol. Syst. 8:309–328.

Johnson, G. B. 1979. Increasing the resolution of polyacrylamide gel electrophoresis by varying the degree of crosslinking. Biochem. Genet. 17:499–516.

Johnson, M. S. and R. Black. 1984. The Wahlund effect and the geographical scale of variation in the intertidal limpet *Siphonaria* sp. Marine Biol. 79:295–302.

Johnson, M. S. and R. F. Doolittle. 1986. A method for the simultaneous alignment of three or more amino acid sequences. J. Mol. Evol. 23:267–278.

Johnson, M. S., B. Clarke and J. Murray. 1977. Genetic variation and reproductive isolation in *Partula*. Evolution 31:116–126.

Johnson, N. K., R. M. Zink, G. F. Barrowclough and J. A. Marten. 1984. Suggested techniques for modern avian systematics. Wilson Bull. 96:543–560.

Johnson, N. K., R. M. Zink and J. A. Marten. 1988. Genetic evidence for relationships in the avian family Vireonidae. Condor 90:428–445.

Jones, C. S., H. Tegelstrom, D. S. Latchman and R. J. Berry. 1988. An improved rapid method for mitochondrial DNA isolation suitable for use in the study of closely related popualtions. Biochem. Genet. 26:83–88.

Jorgensen, R. A. and P. D. Cluster. 1988. Modes and tempos in the evolution of nuclear ribosomal DNA: New characters for evolutionary studies and new markers for genetic and population studies. Ann. Missouri Bot. Gard. 75:1238–1247.

Jukes, T. H. and C. R. Cantor. 1969. Evolution of protein molecules, pp. 21–132. *In* H. N. Munro (ed.), *Mammalian Protein Metabolism*. Academic Press, New York.

Kanehisa, M. 1984. Use of criteria for screening potential homologies in nucleic acid sequences. Nucl. Acids Res. 12:203–213.

Kaplan, J.-C. and E. Beutler. 1967. Electrophoresis of redcell NADH- and NADPH-diaphorases in normal subjects and patients with congenital methemo-globinemia. Biochem. Biophys. Res. Commun. 29:605–610.

Keilen, D. and Y. L. Wang. 1947. Stability of hemo-glo bin and certain non–erythrocytic enzymes *in vitro*. Biochem. J. 41:491–499.

Kempthorne, O. 1957. *An Introduction to Genetic Statistics*. Wiley, New York.

Kessler, L. G. and J. C. Avise. 1985a. Microgeograph-ic lineage analysis by mitochondrial genotype: Vari-ation in the cotton rat (*Sigmodon hispidis*). Evolu-tion 39:831–838.

Kessler, L. G. and J. C. Avise. 1985b. A comparative description of mitochondrial differentiation in se-lected avian and other vertebrate genera. Mol. Biol. Evol. 2:109–126.

Kettler, M. K. and G. S. Whitt. 1986. An apparent progressive and recurrent evolutionary restriction in tissue expression of a gene, the lactate dehydro-genase-C gene, within a family of bony fish (Sal-moniformes: Umbridae). J. Mol. Evol. 23:95–107.

Kettler, M. K., A. W. Ghent and G. S. Whitt. 1986. A comparison of phylogenies based on structural and tissue-expressional differences of enzymes in a family of teleost fishes (Salmoniformes: Umbridae). Mol. Biol. Evol. 3:485–498.

Kezer, J. and S. K. Sessions. 1979. Chromosome vari-ation in the plethodontid salamander, *Aneides fer-reus*. Chromosoma 71:65–80.

Kidd, K. K. and L. A. Sgaramella–Zonta. 1971. Phylo-genetic analysis: Concepts and methods. Am. J. Hum. Genet. 23:235–252.

Kilias, J. 1987. Protein characters as a taxonomic tool in lichen systematics. Bibl. Lichenol. 25:445–455.

Kimura, M. 1968. Evolutionary rate at the molecular level. Nature 217:624–626.

Kimura, M. 1980. A simple method for estimating evolutionary rate of base substitutions through comparative studies of nucleotide sequences. J. Mol. Evol. 16:111–120.

Kimura, M. 1981. Estimation of evolutionary dis-tances between homologous nucleotide sequences. Proc. Natl. Acad. Sci. U.S.A. 78:454–458.

Kimura, M. 1983a. The neutral theory of molecular evolution, pp. 208–233. *In* M. Nei and R. K. Koehn (eds.), *Evolution of Genes and Proteins*. Sin-auer, Sunderland, MA.

Kimura, M. 1983b. *The Neutral Theory of Molecular Evolution*. Cambridge Univ. Press, Cambridge.

Kimura, M. 1986. DNA and the neutral theory.

Phil. Trans. Roy. Soc. London Ser. B 312:343–354.

Kimura, M. and J. F. Crow. 1964. The number of alleles that can be maintained in a finite population. Genetics 49:725–738.

Kimura, M. and T. Ohta. 1972. On the stochastic modle for estimation of mutational distance be-tween homologous proteins. J. Mol. Evol. 2:87–90.

Kimura, M. and G. H. Weiss. 1964. The stepping stone model of population and the decrease of genetic correlation with distance. Genetics 49:561–576.

King, J. L. and T. H. Jukes. 1969. Non-Darwinian evolution. Science 164:788–798.

King, J. L. and T. Ohta. 1975. Polyallelic mutational equilibria. Genetics 79:681–691.

Kirsch, J. A. W., M. Springer, C. Krajewski, M. Archer, K. Aplin and A. W. Dickerman. 1990. DNA/DNA hybridization studies of the carnivorous marsupials. I. The intergeneric relationships of ban-dicoots (Marsupialia: Perameloidea). J. Mol. Evol. (in press).

Kishino, H. and M. Hasegawa. 1989. Evaluation of the maximum likelihood estimate of the evolution-ary tree topologies from DNA sequence data, and the branching order in Hominoidea. J. Mol. Evol. 29:170–179.

Kitto, G. B., P. M. Wasserman and N. O. Kaplan. 1966. Enzymatically active conformers of mito-chondrial malate dehydrogenase. Proc. Natl. Acad. Sci., U.S.A. 56:578–585.

Klebe, R. J. 1975. A simple method for the quantifica-tion of isozymes patterns. Biochem. Genet. 13:805–812.

Klein, J. 1982. *Immunology: The Science of Self-Nonself Discrimination*. Wiley, New York.

Kleppe, K., E. Ohtsuka, R. Kleppe, I. Molineux and H. G. Khorana. 1971. Studies on polynucleotides XCVI. Repair replication of short synthetic DNA's as catalyzed by DNA polymerases. J. Mol. Biol. 56:341–361.

Klotz, L. C. and R. L. Blanken. 1981. A practical method for calculating evolutionary trees from se-quence data. J. Theoret. Biol. 91:261–272.

Kluge, A. G. 1983. Cladistics and the classification of the great apes, pp. 151–177. *In* R. L. Ciochan and R. S. Corruccini (eds.), *New Interpretations of Ape and Human Ancestry*. Plenum, New York.

Kluge, A. G. 1984. The relevance of parsimony to phylogenetic inference, pp. 24–38. *In* T. Duncan and T. Stuessy (eds.), *Cladistics: Perspectives on the*

Reconstruction of Evolutionary History. Columbia Univ. Press, New York.

Kluge, A. G. 1988. Parsimony in vicariance biogeography: A quantitative method and a Greater Antillean example. Syst. Zool. 37:315–328.

Kluge, A. G. 1989. A concern for evidence and a phylogenetic hypothesis of relationships among *Epicrates* (Boidae, Serpentes). Syst. Zool. 38:7–25.

Kluge, A. G. and J. S. Farris. 1969. Quantitative phyletics and the evolution of anurans. Syst. Zool. 18:1–32.

Kluge, A. G. and R. E. Strauss. 1986. Ontogeny and systematics. Annu. Rev. Ecol. Syst. 16:247–268.

Knight, S. E. and D. M. Waller. 1987. Genetic consequences of outcrossing in the cleistogamous annual, *Impatiens capensis.* I. Population-genetic structure. Evolution 41:969–978.

Kobayashi, T., G. B. Milner, D. Teel and F. M. Utter. 1984. Genetic basis for electrophoretic variation of adenosine deaminase in chinook salmon. Trans. Am. Fish. Soc. 113:86–89.

Kocher, T. D. and R. D. Sage. 1986. Further genetic analyses of a hybrid zone between leopard frogs (*Rana pipiens* complex) in central Texas. Evolution 40:21–33.

Kocher, T. D. and T. J. White. 1989. Evolutionary analysis via PCR. *In* H. A. Erlich (ed.), *PCR Technology: Principles and Applications for DNA Amplification.* Stockton Press, New York.

Kocher, T. D., W. K. Thomas, A. Meyer, S. V. Edwards, S. Pääbo, F. X. Villablanca and A. C. Wilson. 1989. Dynamics of mitochondrial DNA evolution in animals: Amplification and sequencing with conserved primers. Proc. Natl. Acad. Sci. U.S.A., 86:6196–6200.

Koehn, R. K. 1978. Physiology and biochemistry of enzyme variation: The interface of ecology and population genetics, pp. 51–72. *In* P. Brussard (ed.), *Ecological Genetics: The Interface.* Springer, New York.

Koehn, R. K. and F. W. Immermann. 1981. Biochemical studies of aminopeptidase polymorphism in *Mytilus edulis.* I. Dependence of enzyme activity on season, tissue, and genotype. Biochem. Genet. 19:1115–1142.

Koehn, R. K. and J. F. Siebenaller. 1981. Biochemical studies of aminopeptidase polymorphism in *Mytilus edulis.* II. Dependence of reaction rate on physical factors and enzyme concentration. Biochem. Genet. 19:1143–1162.

Koehn, R. K., R. I. E. Newell and F. Immermann. 1980. Maintenance of an aminopeptidase allele fre-quency cline by natural selection. Proc. Natl. Acad. Sci. U.S.A. 77:5385–5389.

Koehn, R. K., W. J. Diehl and T. M. Scott. 1988. The differential contribution by individual enzymes of glycolysis and protein catabolism to the relationship between heterozygosity and growth rate in the coot clam, *Mulinia lateralis.* Genetics 118:121–130.

Kohne, D. E. 1970. Evolution of higher-organism DNA. Q. Rev. Biophys. 33:327–375.

Kohne, D. E. and R. J. Britten. 1971. Hydroxyapatite techniques for nucleic acid reassociation, pp. 500–512. *In* G. L. Cantoni and D. R. Davies (eds.), *Procedures in Nucleic Acid Research.* Harper and Row, New York.

Kohne, D. E., J. A. Chiscon and B. H. Hoyer. 1972. Evolution of primate DNA sequences. J. Hum. Evol. 1:627–644.

Kohne, D. E., S. A. Levison and M. J. Byers. 1977. Room temperature method for increasing the rate of DNA reassociation by many thousandfold: Thephenol emulsion reassociation technique. Biochemistry 16:5329–5341.

Kolodner, R. and K. K Temari. 1987. The molecular size and conformation of the chloroplast DNA from higher plants. Biochim. Biophys. Acta 402:372–390.

Koop, B., M. Goodman, P. Xu, K. Chan and J. L. Slightom. 1986. Primate η-globin DNA sequences and man's place among the great apes. Nature (London) 319:234–238.

Krajewski, C. 1989. Phylogenetic relationships among cranes (Gruiformes: Gruidae) based on DNA hybridization. Auk 106:603–618.

Kreitman, M. 1987. Molecular population genetics. Oxford Surv. Evol. Biol. 4:38–60.

Kreitman, M. and M. Aguade. 1986. Genetic uniformity in two populations of *Drosophila melanogaster* as revealed by filter hybridization of four-nucleotide-recognizing restriction enzyme digests. Proc. Natl. Acad. Sci. U.S.A. 83:3562–3566.

Küntzel, H. and H. G. Köchel. 1981. Evolution of rRNA and origin of mitochondria. Nature (London) 293:751–755.

Kuro-o, M., C. Ikebe and S. Kohno. 1986. Cytogenetic studies of Hynobiidae (Urodela) IV. DNA replication bands (R-banding) in the genus *Hynobius* and the banding karyotype of *Hynobius nigrescens* Stejneger. Cytogenet. Cell Genet. 43:14–18.

Kuro-o, M., C. Ikebe and S. Kohno. 1987. Cytogenetic studies of Hynobiidae (Urodela) VI. R-banding patterns in five pond-type *Hynobius* from Korea and Japan. Cytogenet. Cell Genet. 44:69–75.

Lacroix, J. C., R. Azzouz, D. Boucher, C. Abbadie, C. K. Pyne and J. Charlemagne. 1985. Monoclonal antibodies to lampbrush chromosome antigens of *Pleurodeles waltlii*. Chromosoma 92:69–80.

Laird, C., E. Jaffe, G. Karpen, M. Lamb and R. Nelson. 1987. Fragile sites in human chromosomes as regions of late-replicating DNA. Trends Genet. 3:274–281.

Laird, C. D., B. L. McConaughy and B. J. McCarthy. 1969. Rate of fixation of nucleotide substitutions in evolution. Nature (London) 224:149–154.

Lake, J. A. 1987. Rate-independent technique for analysis of nucleic acid sequences: Evolutionary parsimony. Mol. Biol. Evol. 4:167–191.

Lake, J. A. 1988. Origin of the eukaryotic nucleus determined by rate-invariant analysis of rRNA sequences. Nature (London) 331:184–186.

Lamb, T. and J. C. Avise. 1986. Directional introgression of mitochondrial DNA in a hybrid population of tree frogs: The influence of mating behavior. Proc. Natl. Acad. Sci. U.S.A. 83:2526–2530.

Lane, D. J., B. Pace, G. J. Olsen, D. A. Stahl, M. L. Sogin and N. R. Pace. 1985. Rapid determination of 16S ribosomal sequences for phylogenetic analyses. Proc. Natl. Acad. Sci. U.S.A. 82:6955–6959.

Langer, P. R., A. A. Waldrop and D. C. Ward. 1981. Enzymatic synthesis of biotin-labeled polynucleotides: Novel nucleic acid affinity probes. Proc. Natl. Acad. Sci. U.S.A. 78:6633–6637.

Langley, C. H., E. Montgomery and W, Quattlebaum. 1981. Restriction map variation in the ADH region of *Drosophila*. Proc. Natl. Acad. Sci. U.S.A. 79:5631–5635.

Lansman, R. A., R. O. Shade, J. F. Shapira and J. C. Avise. 1981. The use of restriction endonucleases to measure mitochondrial DNA sequence relatedness in natural populations. J. Mol. Evol. 17:214–226.

Lansman, R. A., J. C. Avise, C. F. Aquadro, J. F. Shapira and S. W. Daniel. 1983. Extensive genetic variation in mitochondrial DNAs among geographic populations of the deer mouse, *Peromyscus maniculatus*. Evolution 37:1–16.

Lanyon, S. 1985. Detecting internal inconsistencies in distance data. Syst. Zool. 34:397–403.

Larson, A. and R. Highton. 1978. Geographic protein variation and divergence in the salamanders of the *Plethodon welleri* group (Amphibia: Plethodontidae). Syst. Zool. 27:431–448.

Larson, A. and A. C. Wilson. 1989. Patterns of ribosomal RNA evolution in salamanders. Mol. Biol. Evol. 6:131–154.

Larson, A., D. B. Wake and K. P. Yanev. 1984. Measuring gene flow among populations having high levels of genetic fragmentation. Genetics 106:293–308.

Laurie-Ahlberg, C. C. and B. S. Weir. 1979. Allozymic variation and linkage disequilibrium in some laboratory populations of *Drosophila melanogaster*. Genetics 92:1295–1314.

Lavin, M., J. J. Doyle and J. D. Palmer. 1990. Evolutionary significance of the loss of the chloroplast DNA inverted repeat in the Leguminosae subfamily Papilionidae. Evolution 44 (in press).

Lawrence, J. G., D. E. Dykhuizen, R. F. DuBose and D. L. Hartl. 1989. Phylogenetic analysis using insertion sequence fingerprinting in *Escherichia coli*. Mol. Biol. Evol. 6:1–14.

Lawyer, F. C., S. Stoffel, R. K. Saiki, K. Myambo, R. Drummond and D. H. Gelfand. 1989. Isolation, characterization, and expression in *Escherichia coli* of the DNA polymerase gene from *Thermus aquaticus*. J. Biol. Chem. 264:6427–6437.

Learn, G. H., Jr. and B. A. Schaal. 1987. Population subdivision for ribosomal DNA repeat variants in *Clematis fremontii*. Evolution 41:433–438.

Leary, J. J., D. J. Brigati and D. C. Ward. 1983. Rapid and sensitive colorimetric method for visualizing biotin-labeled DNA probes hybridized to DNA or RNA immobilized on nitrocellulose: Bioblots. Proc. Natl. Acad. Sci. U.S.A. 80:4045–4049.

Leary, R. F., F. W. Allendorf and K. L. Knudsen. 1984. Major morphological effects of a regulatory gene: *Pgm1-t* in rainbow trout. Mol. Biol. Evol. 1:183–194.

Lebherz, H. G. 1983. On epigenetically generated isozymes ("pseudoisozymes") and their possible biological relevance, pp. 203–218. *In* M. C. Rattazzi, J. G. Scandalios and G. S. Whitt (eds.), *Isozymes: Current Topics in Biological and Medical Research*, Vol. 7. *Molecular Structure and Regulation*. Liss, New York.

Lee, M. R. and F. F. B. Elder. 1980. Yeast stimulation of bone marrow mitosis for cytogenetic investigations. Cytogenet. Cell Genet. 26:36–40.

Leone, C. A. 1964. *Taxonomic Biochemistry and Serology*. Ronald Press, New York.

Leone, C. A. 1968. The immunotaxonomic literature: The animal kingdom. Serol. Mus. Bull. 39:1–28.

LeQuesne, W. J. 1982. Compatibility analysis and its applications. Zool. J. Linn. Soc. 74:267–275.

Levan, A., D. Fredga and A. A. Sandberg. 1964. Nomenclature for centromeric position on chromosomes. Hereditas 52:201–220.

Levin, D. A. 1981. Dispersal versus gene flow in plants. Ann. Miss. Bot. Gard. 68:233–253.

Leviton, A. E., R. H. Gibbs, Jr., E. H. Heal and C. E. Dawson. 1985. Standards in herpetology and ichthyology: Part I. Standard symbolic codes for institutional resource collections in herpetology and ichthyology. Copeia 1985:802–832.

Lewin, B. M. 1987. *Genes,* 3rd Ed. Wiley, New York.

Lewin, R. 1988. Conflict over DNA clock results. Science 241:1598–1600.

Lewontin, R. C. 1974. *The Genetic Basis of Evolutionary Change.* Columbia Univ. Press, New York.

Lewontin, R. C. 1986. Population genetics. Annu. Rev. Genet. 19:81–102.

Lewontin, R. C. and C. C. Cockerham. 1959. The goodness–of–fit test for detecting natural selection in random mating populations. Evolution 13:561–564.

Lewontin, R. C. and J. Hubby. 1966. A molecular approach to the study of genic heterozygosity in natural populations. II. Amounts of variation and degree of heterozygosity in natural populations of *Drosophila pseudoobscura.* Genetics 54:595–609.

Li, C. C. 1988. Pseudo-random mating. In celebration of the 80th anniversary of the Hardy–Weinberg law. Genetics 119:731–737.

Li, W.-H. 1980. Rate of gene silencing at duplicate loci: A theoretical study and interpretation of data from tetraploid fishes. Genetics 95:237–258.

Li, W.-H. 1981. A simple method for constructing phylogenetic trees from distance matrices. Proc. Natl. Acad. Sci. U.S.A. 78: 1085–1089.

Li, W.-H. 1986. Evolutionary change of restriction cleavage sites and phylogenetic inference. Genetics 113:187–213.

Li, W.-H., C.-C. Luo and C.-I. Wu. 1985a. Evolution of DNA sequences, pp. 1–130. *In* R. MacIntyre (ed.), *Molecular Evolutionary Genetics.* Plenum, New York.

Li, W.-H., C.-I. Wu and C.-C. Luo. 1985b. A new method for estimating synonymous and nonsynonymous rates of nucleotide substitution considering the relative likelihood of nucleotide and codon changes. Mol. Biol. Evol. 2:150–174.

Li, W.-H., K. H. Wolfe, J. Sourdis and P. M. Sharp. 1987. Reconstruction of phylogenetic trees and estimation of divergence times under nonconstant rates of evolution. Cold Springs Harbor Symp. Quant. Biol. 52:847–856.

Libby, R. L. 1938. The photronreflectometer—an instrument for the measurement of turbid systems. J. Immunol. 34:71–73.

Lin, C. C., G. Shipmann, W. A. Kittrell and S. Ohno. 1969. The predominance of heterozygotes found in wild goldfish of Lake Erie at the gene locus for sorbitol dehydrogenase. Biochem. Genet. 3:603–607.

Linnaeus, C. 1758. *Systema Naturae,* 10th Ed. Stockholm.

Lint, D., J. Clayton, L. Postma and R. Lillie. 1988. Evolution of cetaceans: A serum albumin immunological and biochemical perspective. (Abstr. #33.21.36). XVI Intl. Cong. Genetics, Toronto.

Lipman, D. J. and W. R. Pearson. 1985. Rapid and sensitive protein similarity searches. Science 227:1435–1441.

Lipman, D. J., W. J. Wilbur, T. F. Smith and M. S. Waterman. 1984. On the statistical significance of nucleic acid similarities. Nucl. Acids Res. 12:215–226.

Loh, E. Y., J. F. Elliott, S. Cwirla, L. L. Lanier and M. M. Davis. 1989. Polymerase chain reaction with single-stranded specificity: Analysis of T cell receptor αchain. Science 243:217–220.

Long, E. H. and I. B. Dawid. 1980. Repeated genes in eukaryotes. Annu. Rev. Biochem. 49:727–764.

Loomis, W. F. 1988. *Four Billion Years: An Essay on the Evolution of Genes and Organisms.* Sinauer, Sunderland, MA.

Loveless, M. D. and J. L. Hamrick. 1984. Ecological determinants of genetic structure in plant populations. Annu. Rev. Ecol. Syst. 15:65–95.

Lowenstein, J. M. 1985a. Molecular approaches to the identification of species. Am. Sci. 73:541-547.

Lowenstein, J. M. 1985b. Radioimmune assay of mammoth tissue. Acta Zool. Fenn. 170:233–235.

Lowenstein, J. M. and O. A. Ryder. 1985. Immunological systematics of the extinct quagga (Equidae). Experientia 41:1192–1193.

Lowenstein, J. M., V. M. Sarich and B. J. Richardson. 1981. Albumin systematics of the extinct mammoth and Tasmanian wolf. Nature (London) 291:409–411.

Lumb, W. V. and E. W. Jones. 1984. *Anesthesia of Laboratory and Zoo Animals,* Chap. 18. *Veterinary Anesthesia.* 2nd Ed. Lea & Febriger, Philadelphia.

Lynch, M. 1988. Estimation of relatedness by DNA fingerprinting. Mol. Biol. Evol. 5:584–599.

Mabee, P. M. 1989. Assumptions underlying the use of ontogenetic sequences for determining character-state order. Trans. Am. Fish. Soc. 118:151–158.

MacArthur, R. H. and E. O. Wilson. 1963. An equilibrium theory of insular zoogeography. Evolution 17:373–387.

MacArthur, R. H. and E. O. Wilson. 1967. *The Theory of Island Biogeography.* Princeton Univ. Press, Princeton.

Macgregor, H. C. and S. K. Sessions. 1986. The biological significance of variation in satellite DNA and heterochromatin in newts of the genus *Triturus:* An evolutionary perspective. Phil. Trans. Roy. Soc. London Ser. B 312:243–259.

Macgregor, H. and S. Sherwood. 1979. The nucleolus organizers of *Plethodon* and *Aneides* located by in situ nucleic acid hybridization with *Xenopus* ^{3}H-ribosomal RNA. Chromosoma 72:271–280.

Macgregor H. and J. Varley. 1983. *Working with Animal Chromosomes.* Wiley, New York.

MacIntyre, R. J. 1976. Evolution and ecological value of duplicate genes. Annu. Rev. Ecol. Syst. 7:421–468.

MacIntyre, R. J. (ed.) 1985. *Molecular Evolutionary Genetics.* Plenum, New York.

MacIntyre, R. J., M. R. Dean and G. Batt. 1978. Evolution of acid phosphatase-1 in the genus *Drosophila.* Immunological studies. J. Molec. Evol. 12:121–142.

Maddison, W. P. 1989. Reconstructing character evolution on polytomous cladograms. Cladistics 5:365–377.

Maddison, W. P., M. J. Donoghue and D. R. Maddison. 1986. Outgroup analysis and parsimony. Syst. Zool. 33:83–103.

Malcolm, S., J. K. Cowell and B. D. Young. 1986. Specialist techniques in research and diagnostic clinical cytogenetics, pp. 197–226. *In* D. E. Rooney and B. H. Czepulkowski (eds.), *Human Cytogenetics.* IRL Press, Oxford.

Manchenko, G. P. 1988. Subunit structure of enzymes: Allozymic data. Isozyme Bull. 21:144–158.

Mancino, G., M. Ragghianti and S. Bucci-Innocenti. 1977. Cytotaxonomy and cytogenetics in European newt species, pp. 411–447. *In* D. H. Taylor and S. I. Guttman (eds.), *The Reproductive Biology of Amphibians.* Plenum, New York.

Maniatis, T., E. F. Fristch and J. Sambrook. 1982. *Molecular Cloning: A Laboratory Manual.* Cold Spring Harbor Publications, Cold Spring Harbor.

Manuelidis, L., P. R. Langer-Safer and D. C. Ward. 1982. High-resolution mapping of satellite DNA using biotin-labeled DNA probes. J. Cell Biol. 95:619–625.

Mao, S.-H. and B.-Y. Chen. 1982. Serological relationships of turtles and evolutionary implications. Comp. Biochem. Physiol. 71B:173–179.

Mao, S.-H., B.-Y. Chen, F.-Y. Yin and Y.-W. Guo. 1983. Immunotaxonomic relationships of sea snakes to terrestrial snakes. Comp. Biochem. Physiol. 74A:869–872.

Mao, S.-H., W. Frair, F.-Y. Yin and Y.-W. Guo. 1987. Relationships of some cryptodiran turtles as suggested by immunological cross-reactivity of serum albumins. Biochem. Syst. Ecol. 15:621–624.

Markert, C. L. 1983. Isozymes: Conceptual history and biological significance, pp. 1–17. *In* M. C. Rattazzi, J. G. Scandalios and G. S. Whitt (eds.), *Isozymes: Current Topics in Biological and Medical Research,* Vol. 7. *Molecular Structure and Regulation.* Liss, New York.

Markert, C. L. and F. Moller. 1959. Multiple forms of enzymes: Tissue, ontogenetic, and species-specific patterns. Proc. Natl. Acad. Sci. U.S.A. 45:753–763.

Markert, C. L., J. B. Shaklee and G. S. Whitt. 1975. Evolution of a gene. Science 189:102–114.

Marsden, J. E. and B. May. 1984. Feather pulp: A non-destructive sampling technique for electrophoretic studies of birds. Auk 101:173–175.

Martinson, H. G. 1973. The nucleic acid-hydroxyapatite interaction. II. Phase transitions in the deoxyribonucleic acid-hydroxyapatite system. Biochemistry 12:145–150.

Massaro, E. J. and C. L. Markert. 1968. Protein staining on starch gels. J. Histochem. Cytochem. 16:380–382.

Matson R. H. 1984. Applications of electrophoretic data in avian systematics. Auk 101:717–729.

Matson, R. H. 1989. Avian peptidase isozymes: Tissue distributions, substrate affinities, and assignment of homology. Biochem. Genet. 27:137–151.

Maure, R. R. 1978. Freezing mammalian embryos: A review of techniques. Theriogenology 9:45–68.

Maxam, A. M. and W. Gilbert. 1977. A new method for sequencing DNA. Proc. Natl. Acad. Sci. U.S.A. 74:560–564.

Maxam, A. M. and W. Gilbert. 1980. Sequencing end-labeled DNA with base-specific chemical cleavages. Meth. Enzymol. 65:499–559.

Maxson, L. R. 1981. Albumin evolution and its phylo

genetic implications in toads of the genus *Bufo*. II. Relationships among Eurasian *Bufo*. Copeia 1981:579–583.

Maxson, L. R. 1984. Molecular probes of phylogeny and biogeography in toads of the widespread genus *Bufo*. Molec. Biol. Evol. 1:345–356.

Maxson, L. R. and C. H. Daugherty. 1980. Evolutionary relationships of the monotypic toad family Rhinophrynidae: A biochemical perspective. Herpetologica 36:275–280.

Maxson, L. R. and J. D. Roberts. 1985. An immunological analysis of the phylogenetic relationships between two enigmatic frogs, *Myobatrachus* and *Arenophryne* J. Zool. (London) 207:289–300.

Maxson, L. R. and J. M. Szymura. 1984. Relationships among discoglossid frogs: An albumin perspective. Amphibia/Reptilia 5:245–252.

Maxson, L. R. and A. C. Wilson. 1975. Albumin evolution and organismal evolution in tree frogs (Hylidae). Syst. Zool. 24:1–15.

Maxson, L. R., R. Highton and D. B. Wake. 1979. Albumin evolution and its phylogenetic implications in the plethodontid salamander genera *Plethodon* and *Ensatina*. Copeia 1979:502–508.

Maxson, L. R., L. S. Ellis and A.-R. Song. 1981. Quantitative immunological studies of the albumins of North American squirrels, family Sciuridae. Comp. Biochem. Physiol. 68B:397–400.

Maxson, R. D. and L. R. Maxson 1986. Micro-complement fixation: A quantitative estimator of protein evolution. Mol. Biol. Evol. 3:375–88.

Mayr, E. 1969. *Principles of Systematic Zoology*. McGraw-Hill, New York.

Mayr, E. 1983. *The Growth of Biological Thought: Diversity, Evolution, and Inheritance*. Harvard Univ. Press, Cambridge, MA.

Mazur, P. 1970. Cryobiology: The freezing of biological systems. Science 168:939–949.

McBee, K., R. J. Baker and R. L. Honeycutt. 1987. Observations on rates of DNA degradation. Abstr. #87, Ann. Meet., Amer. Soc. Mammalogists, Albuquerque, NM.

McClenaghan, L. R., Jr., M. H. Smith and M. W. Smith. 1985. Biochemical genetics of mosquitofish. IV. Changes of allele frequencies through time and space. Evolution 39:451–460.

McCracken, G. F. and J. W. Bradbury. 1977. Paternity and genetic heterogeneity in the polygynous bat, *Phyllostomus hastatus*. Science 198:303–306.

McDonald, H. S. 1976. Methods for the physiological study of reptiles, pp. 19–126. *In* C. Gans and W. P. Dawson (eds.), *Biology of the Reptilia*, Vol. 5. Academic Press, New York.

McDonell, M. W., M. N. Simon and F. W. Studier. 1977. Analysis of restriction fragments of T7 DNA and determination of molecular weights by electrophoresis in neutral and alkaline gels. J. Mol. Biol. 110:119–146.

McGovern, M. and C. R. Tracy. 1981. Phenotypic variation in electromorphs previously considered to be genetic markers in *Microtus ochrogaster*. Oecologia 51:276–280.

McInnes, J. L., P. D. Vise, N. Habili and R. H. Symons. 1987. Chemical biotinylation of nucleic acids with photobiotin and their use as hybridization probes. Focus 9:1–4.

McKusick, V. A. 1988. *The Morbid Anatomy of the Human Genome*. Howard Hughes Medical Institute.

McLellan, T. 1984. Molecular charge and electrophoretic mobility in cetacean myoglobins of known sequence. Biochem. Genet. 22:181–200.

McLellan, T. and L. S. Inouye. 1986. The sensitivity of isoelectric focusing and electrophoresis in the detection of sequence differences in proteins. Biochem. Genet. 24:571–577.

McWright, C. G., J. J. Kearney and J. L. Mudd. 1975. Effect of environmental factors on starch gel electrophoretic patterns of human erythrocyte acid phosphatase, pp. 151–161. *In* G. Davis (ed.), *Forensic Science*. Am. Chem. Soc. Symp. Ser. 13, ACS, Washington, D. C.

Melchior, W. B. and P. H. Von Hippel. 1973. Alteration of the relative stability of dA-dT and dG-dC base pairs in DNA. Proc. Natl. Acad. Sci. U.S.A. 70:298–302.

Mellor, J. D. 1978. *Fundamentals of Freeze-Drying*. Academic Press, New York.

Menken, S. B. J. 1987. Is the extremely low heterozygosity level in *Yponomeuta rorellus* caused by bottlenecks? Evolution 41:630–637.

Merritt, R. B., J. F. Rogers and B. J. Kurz. 1978. Genetic variability in the longnose dace, *Rhinichthys cataractae*. Evolution 32:116–124.

Meyerowitz, E. M. and C. H. Martin. 1984. Adjacent chromosomal regions can evolve at very different rates: Evolution of the *Drosophila* 68C glue gene region. J. Mol. Evol. 20:251–264.

Micales, J. A., M. R. Bonde and G. L. Peterson. 1986. The use of isozyme analysis in fungal taxonomy and genetics. Mycotaxon 27:405–449.

Mickevich, M. F. and M. S. Johnson. 1976. Congruence between morphological and allozyme data in evolutionary inference and character evolution. Syst. Zool. 25:260–270.

Mihok, S., W. A. Fuller, R. P. Canham and E. C. McPhee. 1983. Genetic changes at the transferrin locus in the red-backed vole (*Clethrionomys gapperi*). Evolution 37:332–340.

Miller, H. 1987. Practical aspects of preparing phage and plasmid DNA: Growth, maintenance, and storage of bacteria and bacteriophage. Methods Enzymol. 152:145–170.

Miller, R. G. 1974. The jackknife—a review. Biometrika 61:1–15.

Minton, S. A. and S. K. Salanitro. 1972. Serological relationships among some colubrid snakes. Copeia 1972:246–252.

Mitchell, L. G. and C. R. Merril. 1989. Affinity generation of single-stranded DNA for dideoxy sequencing following the polymerase chain reaction. Analyt. Biochem. 178:239–242.

Miyamoto, M. M. 1981. Congruence among character sets in phylogenetic studies of the frog genus *Leptodactylus*. Syst. Zool. 30:281–290.

Miyamoto, M. M. 1983a. Biochemical variation in *Eleutherodactylus bransfordii*: Geographic patterns and cryptic species. Syst. Zool. 321:43–51.

Miyamoto, M. M. 1983b. Frogs of the *Eleutherodactylus rugulosus* group: A cladistic study of allozyme, morphological, and karyological data. Syst. Zool.32:109–124.

Miyamoto, M. M. 1985. Consensus cladograms and general classifications. Cladistics 1:186–189.

Miyamoto, M. M., J. L. Slightom and M. Goodman. 1987. Phylogenetic relationships of humans and African apes as ascertained from DNA sequences (7.1 kilobase pairs) of the ψη-globin region. Science 238:369–373.

Mizusawa, S., S. Nishimura and F. Seela. 1986. Improvement of the dideoxy chain termination method of DNA sequencing by use of deoxy-7-deaza-guanosine triphosphate in place of dGTP. Nucl. Acids Res. 14:1319–1324.

Moore, D. W. and T. L. Yates. 1983. Rate of protein inactivation in selected animals following death. J. Wildl. Manag. 47:1166–1169.

Moore, G. W. 1976. Proof for the maximum parsimony ("Red King") algorithm, pp. 117–137. *In* M. Goodman and R. E. Tashian (eds.), *Molecular Anthropology*, Plenum, New York.

Morden, C. W. and S. S. Golden. 1989. *psbA* genes indicate common ancestry of prochlorophytes and chloroplasts. Nature (London) 337:382–385.

Morgan, K. and C. Strobeck. 1979. Is intragenic re-

combination a factor in the maintenance of genetic variation in natural populations? Nature (London) 277:383–384.

Moritz, C. 1983. Parthenogenesis in the endemic Australian lizard *Heteronotia binoei* (Gekkonidae). Science 220:735737.

Moritz, C. 1987. Parthenogenesis in the tropical gekkonid lizard, *Nactus arnouxii* (Sauria: Gekkonidae). Evolution 41:1252–1266.

Moritz, C. and W. M. Brown. 1986. Tandem duplication of D-loop and ribosomal RNA sequences in lizard mitochondrial DNA. Science 233:1425–1427.

Moritz, C. and W. M. Brown. 1987. Tandem duplications in animal mitochondrial DNAs: Variation in incidence and gene content among lizards. Proc. Natl. Acad. Sci. U.S.A. 84:7183–7187.

Moritz, C., T. E. Dowling and W. M. Brown. 1987. Evolution of animal mitochondrial DNA: Relevance for population biology and systematics. Annu. Rev. Ecol. Syst. 18:269–292.

Moritz, C., M. Adams, S. Donnellan and P. Baverstock. 1989a. The origins and evolution of parthenogenesis in *Heteronotia binoei*: Genetic diversity among bisexual populations. Copeia 1989 (in press).

Moritz, C., S. Donnellan, M. Adams and P. R. Baverstock. 1989b. The origin and evolution of parthenogenesis in *Heteronotia binoei* (Gekkonidae): Extensive genotypic diversity among parthenogens. Evolution 43:994–1003.

Moritz, C., W. M. Brown, L. D. Densmore, J. W. Wright, D. Vyas, S. Donnellan, M. Adams and P. Baverstock. 1989c. Genetic diversity and the dynamics of hybrid parthenogenesis in *Cnemidophorus* (Teiidae) and *Heteronotia* (Gekkonidae), pp. 87–112. *In* R. M. Dawley and J. P. Bogart (eds.), *The Biology of Unisexual Vertebrates*. New York State Museum, Albany.

Morizot, D. C. and M. J. Siciliano. 1982. Linkage of two enzyme loci in fishes of the genus *Xiphophorus* (Poeciliidae). J. Hered. 73:163–167.

Morizot, D. C. and M. J. Siciliano. 1984. Gene mapping in fishes and other vertebrates, pp. 173–234. *In* B. J. Turner (ed.), *Evolutionary Genetics of Fishes*. Plenum, New York.

Morizot, D. C., J. A. Greenspan and M. J. Siciliano. 1983. Linkage group VI of fishes of the genus *Xiphophorus* (Poeciliidae): Assignment of genes coding for glutamine synthetase, uridine monophosphate kinase, and transferrin. Biochem. Genet. 21:1041–1049.

552   LITERATURE CITED

Moss, D. W. 1982. *Isoenzymes*. Chapman & Hall, New York.

Motro, U. and G. Thomson. 1982. On heterozygosity and the effective size of populations subject to size changes. Evolution 36:1059–1066.

Mueller, L. D. and F. J. Ayala. 1982. Estimation and interpretation of genetic distance in empirical studies. Genet. Res. 40:127–137.

Mulley, J. C. and B. D. H. Latter. 1980. Genetic variation and evolutionary relationships within a group of thirteen species of penaeid prawns. Evolution 34:904–916.

Mullis, K. B. and F. A. Faloona. 1987. Specific synthesis of DNA in vitro via a polymerase catalyzed chain reaction. Methods Enzymol. 155:335–350.

Muramatsu, T., S. Kan and M. Hiraishi. 1978. Isolation and characterization of lipoamide dehydrogenase from mackerel dark muscle. Comp. Biochem. Physiol. 61B:247–252.

Murphy, R. W. 1983a. Paleobiogeography and genetic differentiation of the Baja California herpetofauna. Occ. Pap. California Acad. Sci. 137:iv + 1–48.

Murphy, R. W. 1983b. The reptiles: Origin and evolution, pp. 130–158. *In* T. J. Case and M. L. Cody (eds.), *Island Biogeography in the Sea of Cortez.* Univ. of California Press, Berkeley.

Murphy, R. W. 1988. The problematic phylogenetic analysis of interlocus heteropolymer isozyme characters: A case study from sea snakes and cobras. Can. J. Zool. 66:2628–2633.

Murphy, R. W. and C. B. Crabtree. 1985a. Genetic relationships of the Santa Catalina Island rattleless rattlesnake, *Crotalus catalinensis* (Serpentes: Viperidae). Acta Zool. Mexicana (n.s.) 9:1–16.

Murphy, R. W. and C. B. Crabtree. 1985b. Evolutionary aspects of isozyme patterns, number of loci, and tissue-specific gene expression in the prairie rattlesnake, *Crotalus viridis viridis*. Herpetologica 41:451–470.

Murphy, R. W. and R. H. Matson. 1986. Gene expression in the tuatara, *Sphenodon punctatus*. N. Z. J. Zool. 13:573–581.

Murphy, R. W., W. E. Cooper, Jr. and W. S. Richardson. 1983. Phylogenetic relationships of the North American five-lined skinks, genus *Eumeces* (Sauria: Scincidae). Herpetologica 39:200–211.

Murphy, R. W., F. C. McCollum, G. C. Gorman and R. Thomas. 1984. Genetics of hybridizing populations of Puerto Rican *Sphaerodactylus*. J. Herpetol. 18:93–105.

Nadeau, J. H., J. Britton-Davidian, F. Bonhomme and L. Thaler. 1988. H-2 polymorphisms are more uniformly distributed than allozyme polymorphisms in natural populations of house mice. Genetics 118:131–140.

Nakamura, Y., M. Leppert, P. O'Connell, R. Wolff, T. Holm, M. Culver, C. Martin, E. Fujimoto, M. Hoff, E. Kumlin and R. White. 1987. Variable number of tandem repeat (VNTR) markers for human gene mapping. Science 235:1616–1622.

Nakanishi, M., A. C. Wilson, A. Nolan, G. C. Gorman and G. S. Bailey. 1969. Phenoxyethanol: Protein preservative for taxonomists. Science 163:681–683.

Nanney, D. L., R. M. Preparata, F. P. Preparata, E. B. Meyer and E. M. Simon. 1989. Shifting ditypic site analysis: Heuristics for expanding the phylogenetic range of nucleotide sequences in Sankoff analysis. J. Mol. Evol. 28:451–459.

Nardi, I., F. Andronico, S. De Lucchini and R. Batistoni. 1986. Cytogenetics of the European plethodontid salamanders of the genus *Hydromantes* (Amphibia, Urodela). Chromosoma 94:377–388.

Neale, D. B. and R. R. Sederoff. 1988. Inheritance and evolution of organelle genomes, pp. 251–264. *In* J. W. Hanover and D. E. Keathly (eds.), *Genetic Manipulation of Woody Plants*. Plenum, New York.

Neale, D. B., M. A. Saghai-Maroof, R. W. Allard, Q. Zhang and R. A. Jorgensen. 1988. Chloroplast DNA diversity in populations of wild and cultivated barley. Genetics 120:1105–1110.

Needleman, S. B. and C. D. Wunsch. 1970. A general method applicable to the search for similarities in the amino acid sequence of two proteins. J. Mol. Biol. 48:443–453.

Neff, N. A. 1986. A rational basis for a priori character weighting. Syst. Zool. 35:110–123.

Nei, M. 1972. Genetic distance between populations. Am. Natur. 106:283–292.

Nei, M. 1973. Analysis of gene diversity in subdivided populations. Proc. Natl. Acad. Sci. U.S.A. 70:3321–3323.

Nei, M. 1978. Estimation of average heterozygosity and genetic distance from a small number of individuals. Genetics 89:583–590.

Nei, M. 1987. *Molecular Evolutionary Genetics*. Columbia Univ. Press, New York.

Nei, M. and R. K. Chesser. 1983. Estimation of fixation indices and gene diversification. Ann. Human Genet. 47:253–259.

Nei, M. and T. Gojobori. 1986. Simple methods for estimating the numbers of synonymous and nonsynonymous nucleotide substitutions. Mol. Biol. Evol. 3:418–426.

Nei, M. and R. K. Koehn (eds.) 1983. *Evolution of Genes and Proteins.* Sinauer, Sunderland, MA.

Nei, M. and W.-H. Li. 1979. Mathematical model for studying genetic variation in terms of restriction endonucleases. Proc. Natl. Acad. Sci. U.S.A. 76: 5269–5273.

Nei, M. and F. Tajima. 1985. Evolutionary change of restriction cleavage sites and phylogenetic inference for man and apes. Mol. Biol. Evol. 2:189–205.

Nei, M., T. Maruyama and R. Chakraborty. 1975. The bottleneck effect and genetic variability in populations. Evolution 29:1–10.

Neigel, J. E. and J. C. Avise. 1986. Phylogenetic relationships of mitochondrial DNA under various demographic models of speciation, pp. 515–534. *In* E. Nevo and S. Karlin (eds.), *Evolutionary Processes and Theory.* Academic Press, New York.

Nelson, K., R. J. Baker and R. L. Honeycutt. 1987. Mitochondrial DNA and protein differentiation between hybridizing cytotypes of the white-footed mouse, *Peromyscus leucopus.* Evolution 41:864–872.

Nevo, E., A. Beiles and R. Ben-Shlomo. 1984. The evolutionary significance of genetic diversity: Ecological, demographic and life history correlates. Lec. Notes Biomath. 53:13–213.

Nichols, E., V. M. Chapman and F. H. Ruddle. 1973. Polymorphism and linkage for mannosephosphate isomerase in *Mus musculus.* Biochem. Genet. 8:47–53.

Nickrent, D. L. 1986. Genetic polymorphism in the morphologically reduced dwarf mistletoes (*Arceuthobium*, Viscaceae): An electrophoretic study. Am. J. Bot. 73:1492–1501.

Nickrent, D. L., S. I. Guttman and W. H. Eshbaugh. 1984. Biosystematic and evolutionary relationships among selected taxa of *Arceuthobium.* U. S. Dept. Agricult., Tech. Report RM-111.

Nuttall, G. H. F. 1904. *Blood Immunity and Blood Relationship.* Cambridge Univ. Press, Cambridge.

O'Brien, S. J., D. E. Wildt, D. Goldman, C. R. Merril and M. Bush. 1983. The cheetah is depauperate in genetic variation. Science 221:459–462.

O'Brien, S. J., W. G. Nash, D. E. Wildt, M. E. Bush and R. E. Benveniste. 1985a. A molecular solution to the riddle of the giant panda's phylogeny. Nature (London) 317:140–144.

O'Brien, S. J., M. E. Roelke, L. Marker, A. Newman, C. A. Winkler, D. Meltzer, L. Colly, J.F. Evermann, M. Bush and D. E. Wildt. 1985. Genetic basis for species vulnerability in the cheetah. Science 227:1428–1434.

O'Brien, S. J., D. E. Wildt, M. Bush, T. M. Caro, C. FitzGibbon, I. Aggundey and R.E. Leakey. 1987. East African cheetahs: Evidence for two population bottlenecks? Proc. Natl. Acad. Sci. U.S.A. 84:508–511.

Ochman, H., A. S. Gerber and D. L. Hartl. 1988. Genetic applications of an inverse polymerase chain reaction. Genetics 120:621–623.

O'Grady, R. T. and G. B. Deets. 1987. Coding multistate characters, with special reference to the use of parasites as characters of their hosts. Syst. Zool. 36:268–279.

Ohno, S. 1970. *Evolution by Gene Duplication.* Springer-Verlag, New York.

Ohno, S., C. Stenius, L. Christian and G. Schipmann. 1969. De novo mutation-like events observed at the 6PGD locus of the Japanese quail, and the principle of polymorphism breeding more polymorphism. Biochem. Genet. 3:417–428.

Ohta, T. 1977. Extension of neutral mutation drift hypothesis, pp. 148–167. *In* M. Kimura (ed.), *Molecular Evolution and Polymorphism.* National Institute of Genetics, Mishima.

Ohyama, K., H. Fukuzawa, T. Kohchi, T. Shirai, T. Sano, S. Sano, K. Umesono, Y. Shiki, M. Takeuchi, Z. Chang, S. Aota, H. Inokuchi and H. Ozeki. 1986. Complete nucleotide sequence of liverwort *Marchantia polymorpha* chloroplast DNA. Plant Mol. Biol. Rep. 4:148–175.

Olsen, G. J. 1987. Earliest phylogenetic branchings: Comparing rRNA-based evolutionary trees inferred with various techniques. Cold Spring Harbor Symp. Quant. Biol. 52:825–837.

Olsen, G. J. 1988. Phylogenetic analysis using ribosomal RNA. Methods Enzymol. 164:793–838.

Orosz, J. M. and J. G. Wetmur. 1974. In vitro iodination of DNA. Maximizing iodination while minimizing degradation: Use of buoyant density shifts for DNA-DNA hybrid isolation. Biochemistry 13:5467–5473.

Ouchterlony, O. 1958. Diffusion-in-gel methods for immunological analysis. Prog. Allergy 5:1.

Pääbo, S. 1985. Molecular cloning of ancient Egyptian mummy DNA. Nature (London) 314:644–645.

Pääbo, S. 1989. Ancient DNA: Extraction, character

ization, molecular cloning, and enzymatic amplification. Proc. Natl. Acad. Sci. U.S.A. 86:1939–1943.

Pääbo, S., J. A. Gifford and A. C. Wilson. 1988. Mitochondrial DNA sequences from a 7000-year-old brain. Nucl. Acid Res. 16:9775–9787.

Pääbo, S., R. Higuchi and A. C. Wilson. 1989. Ancient DNA and the polymerase chain reaction. J. Biol. Chem. 264: 9709–9712.

Pace, N. R., G. J. Olsen and C. R. Woese. 1986. Ribosomal RNA phylogeny and the primary lines of evolutionary descent. Cell 45:325–326.

Palmer, J. D. 1982. Physical and gene mapping of chloroplast DNA from *Atriplex triangularis* and *Cucumis sativa*. Nucl. Acids Res. 10:1593–1605.

Palmer, J. D. 1985a. Evolution of chloroplast and mitochondrial DNA in plants and algae, pp. 131–240. *In* R. J. MacIntyre (ed.), *Molecular Evolutionary Genetics*. Plenum, New York.

Palmer, J. D. 1985b. Comparative organization of chloroplast genomes. Annu. Rev. Genet. 19:325–354.

Palmer, J. D. 1986a. Isolation and structural analysis of chloroplast DNA. Methods Enzymol. 118:167–186.

Palmer, J. D. 1986b. Chloroplast DNA and phylogenetic relationships, pp. 63–80. *In* S. K. Dutta (ed.), *DNA Systematics*. CRC Press, Boca Raton, FL.

Palmer, J. D. 1987. Chloroplast DNA evolution and biosystematic uses of chloroplast DNA variation. Am. Natur. 130:S6–S29.

Palmer, J. D. and L. A. Herbon. 1988. Plant mitochondrial DNA evolves rapidly in structure, but slowly in sequence. J. Mol. Evol. 28:87–97.

Palmer, J. D., R. A. Jorgensen and W. F. Thompson. 1985. Chloroplast DNA variation and evolution in *Pisum*: Patterns of change and phylogenetic analysis. Genetics 109:195–213.

Palmer, J. D., B. Osorio, J. Aldrich and W. F. Thompson. 1987. Chloroplast DNA evolution among legumes: Loss of a large inverted repeat occurred prior to other sequence arrangements. Curr. Genet. 11:275–286.

Palmer, J. D., B. Osorio and W. F. Thompson. 1988a. Evolutionary significance of inversions in legume chloroplast DNA. Curr. Genet.14:75–89.

Palmer, J. D., R. K. Jansen, H. J. Michaels, M. W. Chase and J. R. Manhart. 1988b. Chloroplast DNA and plant phylogeny. Ann. Missouri Bot. Gard. 75:1180–1206.

Palva, T. K. and E. T. Palva. 1985. Rapid isolation of animal mitochondrial DNA by alkaline extraction. FEBS Lett. 192:267–270.

Pamilo, P. and M. Nei. 1988. Relationships between gene trees and species trees. Mol. Biol. Evol. 5:568–583.

Pardue, M. L. 1985. In situ hybridization, pp. 179–202. *In* B. D. Hames and S. J. Higgins (eds.), *Nucleic Acid Hybridisation: A Practical Approach*. IRL Press, Oxford.

Pardue, M. L. 1986. In situ hybridization to DNA of chromosomes and nuclei, pp. 111–137. *In* D. B. Roberts (ed.), *Drosophila: A Practical Approach*. IRL Press, Oxford.

Paris Conference. 1971. Standardization in human cytogenetics. Cytogenetics 11:313–362.

Parker, E. D., Jr. and R. K. Selander. 1976. The organization of genetic diversity in the parthenogenetic lizard *Cnemidophorus tesselatus*. Genetics 84:791–805.

Parkin, D. T. and S. R. Cole. 1985. Genetic differentiation and rates of evolution in some introduced populations of the House Sparrow, *Passer domesticus* in Australia and New Zealand. Heredity 54:15–23.

Pathak, S. and F. E. Arrighi. 1973. Loss of DNA following C-banding procedures. Cytogenet. Cell Genet. 12:414–422.

Patterson, C. (ed.) 1987. *Molecules and Morphology in Evolution: Conflict or Compromise?* Cambridge Univ. Press, Cambridge.

Patterson, C. 1988. Homology in classical and molecular biology. Mol. Biol. Evol. 5:603–625.

Patton, J. C. and J. C. Avise. 1983. An empirical evaluation of qualitative Hennigian analyses of protein electrophoretic data. J. Mol. Evol. 19:244–254.

Patton, J. L. and J. H. Feder. 1981. Microspatial genetic heterogeneity in pocket gophers: Non-random breeding and drift. Evolution 35:912–920.

Patton, J. L. and S. W. Sherwood. 1982. Genome evolution in pocket gophers (genus *Thomomys*) I. Heterochromatin variation and speciation potential. Chromosoma 85:149–162.

Patton, J. L. and S. W. Sherwood. 1983. Chromosome evolution and speciation in rodents. Annu. Rev. Ecol. Syst. 14:139–158.

Patton, J. L., M. F. Smith, R. D. Price and R. A. Hellenthal. 1984. Genetics of hybridization between the pocket gophers *Thomomys bottae* and *Thomomys townsendii* in northeastern California. Great Basin Natur. 44:431–440.

Penny, D. 1982. Towards a basis for classification:

The incompleteness of distance measures, incompatibility analysis and phenetic classification. J. Theoret. Biol. 96:129–142.

Penny, D. and M. D. Hendy. 1985. Testing methods of evolutionary tree construction. Cladistics 1:266–272.

Penny, D. and M. Hendy. 1986. Estimating the reliability of evolutionary trees. Mol. Biol. Evol. 3:403–417.

Pierson, E. D., V. M. Sarich, J. M. Lowenstein, M. J. Daniel and W. E. Rainey. 1986. A molecular link between the bats of New Zealand and South America. Nature (London) 323:60–63.

Pinkel, D., T. Straume and J. W. Gray. 1986. Cytogenetic analysis using quantitative, high-sensitivity, fluorescence hybridization. Proc. Natl. Acad. Sci. U.S.A. 83:2934–2938.

Pirrotta, V. 1986. Cloning Drosophila genes, pp. 83–110. In D. B. Roberts (ed.), Drosophila: A Practical Approach. IRL Press, Oxford.

Plante, Y., P. T. Boag and B. N. White. 1987. Nondestructive sampling of mitochondrial DNA from voles. Can. J. Zool. 65:175–180.

Ponath, P. D., D. M. Hillis and P. D. Gottlieb. 1989a. Structural and evolutionary comparisons of four alleles of the mouse immunoglobulin kappa chain gene, Igk-VSer. Immunogenetics 29:249–257.

Ponath, P. D., R. T. Boyd, D. M. Hillis and P. D. Gottlieb. 1989b. Structural and evolutionary comparisons of four alleles of the mouse Igk-J locus which encodes immunoglobulin kappa light chain joining (J_K) segments. Immunogenetics 29:389–396.

Powell, J. R. and A. Caccone. 1990. The TEACL method of DNA–DNA hybridization: Technical considerations. J. Mol. Evol.

Powell, J. R. and M. C. Zuniga. 1983. A simplified procedure for studying mtDNA polymorphisms. Biochem. Genet. 21:1051–1055.

Powell, J. R., A. Caccone, G. D. Amato and C. Yoon. 1986. Rates of nucleotide substitution in Drosophila mitochondrial DNA and nuclear DNA are similar. Proc. Natl. Acad. Sci. U.S.A. 83:9090–9093.

Powers, D. A., G. S. Greaney and A. R. Place. 1979. Physiological correlation between lactate dehydrogenase genotype and haemoglobin function in killifish. Nature (London) 277:240–241.

Prager, E. M. and A. C. Wilson. 1971a. The dependence of immunological cross-reactivity upon sequence resemblance among lysozymes. I. Micro

complement fixation studies. J. Biol. Chem. 246:5978–5989.

Prager, E. M. and A. C. Wilson. 1971b. The dependence of immunological cross-reactivity upon sequence resemblance among lysozymes. II. Comparison of precipitin and micro-complement fixation results. J. Biol. Chem. 246:7010–7017.

Prager, E. M. and A. C. Wilson. 1976. Congruency of phylogenies derived from different proteins. J. Mol. Evol. 9:45–57.

Prager, E. M., A. H. Brush, R. A. Nolan, M. Nakanishi and A. C. Wilson. 1974. Slow evolution of transferrin and albumin in birds according to microcomplement fixation analysis. J. Mol. Evol. 3:243–262.

Prager, E. M., A. C. Wilson, J. M. Lowenstein and V. M. Sarich. 1980. Mammoth albumin. Science 209:287–289.

Prensky, W. 1976. The radioiodination of RNA and DNA to high specific activities, pp. 121–152. In D. M. Prescott (ed.), Methods in Cell Biology. Academic Press, New York.

Prout, T. 1965. The estimation of fitness from genotypic frequencies. Evolution 19:546–551.

Qu, L. H., B. Michot and J.-P. Bachellerie. 1983. Improved methods for structure probing in large RNAs: A rapid heterologous sequencing approach is coupled to the direct mapping of nuclease accessible sites. Application to the 5' terminal domain of eukaryotic 28S rRNA. Nucl. Acids Res. 11:5903–5920.

Quellar, D. C., J. E. Strassmann and C. R. Hughes. 1988. Genetic relatedness in colonies of tropical wasps with multiple queens. Science 242:1155–1157.

Quinn, T. W. and B. N. White. 1987a. Analysis of DNA sequence variation, pp. 163–198. In F. Cooke and P. A. Buckley (eds.), Avian Genetics. Academic Press, London.

Quinn, T. W. and B. N. White. 1987b. Identification of restriction fragment length polymorphisms in genomic DNA of the lesser snow goose. Mol. Biol. Evol. 4:126–143.

Quinn, T. W., J. S. Quinn, F. Cooke and B. N. White. 1987. DNA marker analysis detects multiple maternity and paternity in single broods of the lesser snow goose (Anser caerulescens caerulescens). Nature 396:392–394.

Ragghianti, M., S. Bucci, G. Mancino, J. C. Lacroix, D. Boucher and J. Charlemagne. 1988. A novel approach to cytotaxonomic and cytogenetic studies in the genus *Triturus* using monoclonal antibodies to lampbrush chromosomes antigens. Chromosoma 97:134144.

Rainboth, W. J. and G. S. Whitt. 1974. Analysis of evolutionary relationships among shiners of the subgenus *Luxilus* (Teleostei, Cypriniformes, *Notropis*) with the lactate dehydrogenase and malate dehydrogenase isozyme systems. Comp. Biochem. Physiol. 49B:241–252.

Ramshaw, J. A. M., J. A. Coyne and R. C. Lewontin. 1979. The sensitivity of gel electrophoresis as a detector of genetic variation. Genetics 93:1019–1037.

Rand, D. M. and R. G. Harrison. 1986a. Ecological genetics of a mosaic hybrid zone: Mitochondrial, nuclear, and reproductive differentiation of crickets by soil type. Evolution 43:432–449.

Rand, D. M. and R. G. Harrison. 1986b. Mitochondrial DNA transmission genetics in crickets. Genetics 114:955–970.

Ranker, T. A. and A. F. Schnabel. 1986. Allozymic and morphological evidence for a progenitor-derivative species pair in *Camassia* (Liliaceae). Syst. Bot. 11:433–445.

Reeck, G. R., C. de Haen, D. C. Teller, R. F. Doolittle, W. M. Fitch, R. E. Dickerson, P. Chambon, A. D. McLachlan, E. Margoliash, T. H. Jukes and E. Zuckerkandl. 1987. "Homology" in proteins and nucleic acids: A terminology muddle and a way out of it. Cell 50:667.

Reed, K. C. and D. A. Mann. 1985. Rapid transfer of DNA from agarose gels to nylon membranes. Nucl. Acids Res. 13:7207–7221.

Remsen, J. V., Jr. 1977. On taking field notes. Am. Birds 31: 946–953.

Reynolds, J., B. S. Weir and C. C. Cockerham. 1983. Estimation of the coancestry coefficient: Basis for a short–term genetic distance. Genetics 105:767–779.

Richardson, B. J. 1981. The genetic structure of rabbit populations, pp. 37–52. *In* K. Myers and C.D. MacInnes (eds.), *Proceedings of the World Lagomorph Conference held in Guelph, Ontario, August, 1979*. Univ. of Guelph.

Richardson, B. J. 1983. Distribution of protein variation in skipjack tuna (*Katsumonus pelamis*) from the central and south–western Pacific. Aust. J. Mar. Freshw. Res. 34:231–251.

Richardson, B. J., P. R. Baverstock and M. Adams. 1986. *Allozyme Electrophoresis. A Handbook for Animal Systematics and Population Structure*. Academic Press, Sydney.

Rider, C. C. and C. B. Taylor. 1980. *Isoenzymes*. Chapman & Hall, London.

Ridgway, G. J., S. W. Sherburne and R. D. Lewis. 1970. Polymorphism in the esterases of Atlantic herring. Trans. Am. Fish. Soc. 99:147–151.

Rigby, P. W. J., M. Dieckmann, C. Rhodes and P. Berg. 1977. Labelling deoxyribonucleic acid to high specific activity in vitro by nicktranslation with DNA polymerase I. J. Mol. Biol. 113:237–251.

Riley, V. 1960. Adaptation of orbital bleeding technique to rapid serial blood studies. Proc. Soc. Exp. Biol. Med. 104:751–754.

Ritland, K. and M. T. Clegg. 1987. Evolutionary analysis of plant DNA sequences. Am. Natur. 130:s75–s100.

Roberts, J. W., S. A. Johnson, P. Kier, T. J. Hall, E. H. Davidson and R. J. Britten. 1985. Evolutionary conservation of DNA sequences expressed in sea urchin eggs and embryos. J. Mol. Evol. 22:99–107.

Roberts, L. 1989. Genome project under way, at last. Science 243:167–168.

Roberts, R. J. 1984. Restriction and modification enzymes and their recognition sequences. Nucl. Acids Res. 12:r167–r204.

Rogers, J. S. 1972. Measures of genetic similarity and genetic distance. Studies in Genet. VII. Univ. Texas Publ. 7213:145–153.

Rogers, J. S. 1984. Deriving phylogenetic trees from allele frequencies. Syst. Zool. 33:52–63.

Rogers, J. S. 1986. Deriving phylogenetic trees from allele frequencies: A comparison of nine genetic distances. Syst. Zool. 35:297–310.

Rogers, S. O. and A. J. Bendich. 1985. Extraction of DNA from milligram amounts of fresh, herbarium and mummified plant tissues. Plant Mol. Biol. 5:69–76.

Rogstad, S. H., J. C. Patton and B. A. Schaal. 1988. M13 repeat probe detects DNA minisatellite-like sequences in gymnosperm and angiosperm. Proc. Natl. Acad. Sci. U.S.A. 85:9176–9178.

Rollo, F. A., A. Amici, R. Salvi and A. Garbuglia. 1988. Short but faithful pieces of ancient DNA. Nature (London) 335:774.

Rooney, D. E. and B. H. Czepulkowski. 1986. *Human Cytogenetics*. IRL Press, Oxford.

Roose, M. L. and L. D. Gottlieb. 1976. Genetic and biochemical consequences of polyploidy in *Tragopogon*. Evolution 30:818–830.

Rosen, D. E. and D. G. Buth. 1980. Empirical evolutionary research versus neo-Darwinian speculation. Syst. Zool. 29:300–308.

Russell, F. E. 1980. *Snake Venom Poisoning.* Lippincott, Philadelphia.

Russell, F. E., J. A. Emery and T. E. Long. 1960. Some properties of rattlesnake venom following 26 years of storage. Proc. Soc. Exp. Biol. Med. 103:737–739.

Ryman, N. and F. Utter. (eds.) 1987. *Population Genetics and Fishery Management.* Univ. Washington Press, Seattle.

Ryman, N., F. W. Allendorf and G. Stahl. 1979. Reproductive isolation with little genetic divergence in sympatric populations of brown trout. Genetics 92:247–262.

Sackler, M. L. 1966. Xanthine oxidase from liver and duodenum of the rat: Histochemical localization and electrophoretic heterogeneity. J. Histochem. Cytochem. 14:326–333.

Sage, R. D. and R. K. Selander. 1979. Hybridization between species of the *Rana pipiens* complex in central Texas. Evolution 33:1069–1088.

Saghai-Moroof, M. A., K. M. Soliman, R. A. Jorgensen and R.W. Allard. 1984. Ribosomal DNA spacer-length polymorphisms in barley: Mendelian inheritance, chromosomal location, and population dynamics. Proc. Natl. Acad. Sci. U.S.A. 81:8014–8019.

Saiki, R. K., S. Scharf, F. Faloona, K. B. Mullis, G. T. Horu, H. A. Erlich and N. Arnheim. 1985. Enzymatic amplification of β-globin genomic sequences and restriction site analysis for diagnosis of sickle cell anemia. Science 230:1350–1354.

Saiki, R. K., D. H. Gelfand, S. Stoffel, S. J. Scharf, R. Higuchi, G. T. Horn, K. B. Mullis and H. A. Erlich. 1988. Primer-directed enzymatic amplification of DNA with a thermostable DNA polymerase. Science 239:487–491.

Saitou, N. and M. Nei. 1987. The neighbor-joining method: A new method for reconstructing phylogenetic trees. Mol. Biol. Evol. 4:406–425.

Salthe, S. N. and N. O. Kaplan. 1966. Immunology and rates of enzyme evolution in the amphibia in relation to the origins of certain taxa. Evolution 20:603–616.

Sambrook, J., E. F. Fritsch and T. Maniatis. 1989. *Molecular Cloning: A Laboratory Manual,* 2nd Ed. 3 Volumes. Cold Spring Harbor Laboratory, Cold Spring Harbor, NY.

Sanderson, M. J. 1989. Confidence limits on phylogenies: The bootstrap revisited. Cladistics 5:113–129.

Sanger, F., S. Nicklen and A. R. Coulson. 1977. DNA sequencing with chain-terminating inhibitors. Proc. Natl. Acad. Sci. U.S.A. 74:5463–5467.

Sankoff, D. 1975. Minimal mutation trees of sequences. SIAM J. Appl. Math. 28:35–42.

Sankoff, D. D. and R. J. Cedergren. 1983. Simultaneous comparison of three or more sequences related by a tree, pp. 253–263. *In* D. Sankoff and J. B. Kruskal (eds.), *Time Warps, String Edits, and Macromolecules: The Theory and Practice of Sequence Comparison.* Addison-Wesley, Reading, MA.

Sankoff, D., C. Morel and R. J. Cedergren. 1973. Evolution of 5S RNA and the non-randomness of base replacement. Nature (London) 245:232–234.

Sankoff, D., R. J. Cedergren and G. Lapalme. 1976. Frequency of insertion-deletion, transversion, and transition in the evolution of 5S ribosomal RNA. J. Mol. Evol. 7:133–149.

Sarich, V. M. 1977. Rates, sample sizes, and the neutrality hypothesis for electrophoresis in evolutionary studies. Nature (London) 265:24–28.

Sarich, V. M. 1985. Rodent macromolecular systematics, pp. 423–452, *In* W. P. Luckett and J.-L. Hartenberger (eds.), *Evolutionary Relationships among Rodents. A Multidisciplinary Analysis.* Plenum, New York.

Sarich, V. M. and J. E. Cronin. 1976. Molecular systematics of the primates, pp. 141–170. *In* M. Goodman and R. E. Tashian (eds.), *Molecular Anthropology.* Plenum, New York.

Sarich, V. M. and A. C. Wilson. 1966. Quantitative immunochemistry and the evolution of primate albumins: Micro-complement fixation. Science 154:1563–1566.

Sarich, V. M. and A. C. Wilson. 1967. Immunological time scale for hominid evolution. Science 158:1200–1203.

Sarich, V. M., C. W. Schmid and J. Marks. 1989. DNA hybridization as a guide to phylogenies: A critical analysis. Cladistics 5:3–32.

SAS Institute. 1985. *SAS User's Guide: Statistics,* Version 5. SAS Institute, Cary, North Carolina.

Sattath, S. and A. Tversky. 1977. Additive similarity trees. Psychometrika 42:319–345.

Scanlan, B. E., L. R. Maxson and W. E. Duellman. 1980. Albumin evolution in marsupial frogs (Hylidae: *Gastrotheca*). Evolution 34:222–229.

Schaal, B. A., W. J. Leverich and J. Nicto-Sotela. 1987. Ribosomal DNA variation in the native plant *Phlox divaricata*. Mol. Biol. Evol. 4:611–621.

Schaeffer, S. W. and C. F. Aquadro. 1987. Nucleotide

sequence of the alcohol dehydrogenase region of *Drosophila pseudoobscura:* Evolutionary change and evidence for an ancient duplication. Genetics 117:61–73.

Schafer, M. and W. Kunz. 1985. rDNA in *Locusta migratoria* is very variable: Two introns and extensive restriction site polymorphisms in the spacer. Nucl. Acids Res. 13:1251–1266.

Scherberg, N. H. and S. Refetoff. 1975. Radioiodine labeling of ribopolymers for special applications in biology, pp. 343– 359. *In* D. M. Prescott (ed.), *Methods in Cell Biology,* Vol. 10. Academic Press, New York.

Schleif, R. F. and P. C. Wensink. 1981. *Practical Methods in Molecular Biology.* Springer-Verlag, Berlin.

Schmid, M. and M. Guttenbach. 1988. Evolutionary diversity of reverse (R) fluorescent chromosome bands in vertebrates. Chromosoma 97:101–114.

Schmid M., J. Olert and C. Klett. 1979. Chromosome banding in Amphibia III. Sex chromosomes in *Triturus.* Chromosoma 71:29–55.

Schoen, D. J. 1982. Genetic variation and the breeding system of *Gilia achilleifolia.* Evolution 36:361–370.

Schwaner, T. D. and H. C. Dessauer. 1982. Comparative immunodiffusion survey of snake transferrins focused on the relationships of the Natricines. Copeia 1982: 541–549.

Schwaner, T. D., P. R. Baverstock, H. C. Dessauer and G. A. Mengden. 1985. Immunological evidence for the phylogenetic relationships of Australian elapid snakes, pp. 177–184. *In* G. Grigg, R. Shine and H. Ehmann (eds.), *Biology of Australasian Frogs and Reptiles.* Royal Zool. Soc., New South Wales.

Schwartz, M. K., J. S. Nisselbaum and O. Bodansky. 1963. Procedure for staining zones of activity of glutamic oxaloacetic transaminase following electrophoresis with starch gel. Am. J. Clin. Pathol. 40:103–106.

Schwartz, O. A. and K. B. Armitage. 1980. Genetic variation in social mammals: The marmot model. Science 207:665–667.

Schwartz, R. M. and M. O. Dayhoff. 1978. Origins of prokaryotes, eukaryotes, mitochondria, and chloroplasts: A perspective is derived from protein and nucleic acid sequence data. Science 199:395–403.

Schwert, G. W. 1957. Recovery of native bovine serum albumin after precipitation with trichloroacetic acid and solution in organic solvents. J. Am. Chem. Soc. 79:139–141.

Sears, B. B. 1980. The elimination of plastids during spermatogenesis and fertilization in the plant kingdom. Plasmid 4:233–255.

Seed, B., R. C. Parker and N. Davidson. 1982. Representation of DNA sequences in recombinant DNA libraries prepared by restriction enzyme partial digestion. Gene 19:201–209.

Selander, R. K., M. H. Smith, S. Y. Yang, W. E. Johnson and J. R. Gentry. 1971. Biochemical polymorphism and systematics in the genus *Peromyscus.* I. Variation in the old-field mouse (*Peromyscus polionotus*). Stud. Genet. VI. Univ. Texas Publ. 7103:49–90.

Selander, R. K., D. A. Caugant, H. Ochman, J. M. Musser, M. N. Gilmour and T. S. Whittam. 1986. Methods of multilocus enzyme electrophoresis for bacterial population genetics and systematics. Appl. Environ. Microbiol. 51:873–884.

Sellers, P. 1974. On the theory and computation of evolutionary distances. SIAM J. Appl. Math. 26:787–793.

Sensabaugh, G. F. 1982. Isozymes in forensic science, pp. 247–282. *In Isozymes: Current Topics in Biological and Medical Research,* Vol. 6. Liss, New York.

Sensabaugh, G. F., A. C. Wilson and P. L. Kirk. 1971a. Protein stability in preserved biological remains. I. Survival of biologically active proteins in an 8-year-old sample of dried blood. Int. J. Biochem. 2:545–557.

Sensabaugh, G. F., A. C. Wilson and P. L. Kirk. 1971b. Protein stability in preserved biological remains. II. Modification and aggregation of proteins in an 8-year-old sample of dried blood. Internat. J. Biochem. 2:558–568.

Separack, P., M. Slatkin and N. Arnheim. 1988. Linkage disequilibrium in human ribosomal genes: Implications for multigene family evolution. Genetics 119:943–949.

Sessions, S. K. 1982. Cytogenetics of diploid and triploid salamanders of the *Ambystoma jeffersonianum* complex. Chromosoma 84:599–621.

Sessions, S. K. and J. Kezer. 1987. Cytogenetic evolution in the plethodontid salamander genus *Aneides.* Chromosoma 95:17–30.

Sessions, S. K. and A. Larson. 1987. Developmental correlates of genome size in plethodontid salamanders and their implications for genome evolution. Evolution 41:1239–1251.

Shaklee, J. B. 1984. Genetic variation and population structure in the damselfish, *Stegastes fasciolatus,* throughout the Hawaiian Archipelago. Copeia 1984:629–640.

Shaklee, J. B. and C. P. Keenan. 1986. A practical laboratory guide to the techniques and methodology of electrophoresis and its application to fish fillet identification. CSIRO Marine Laboratories Publ. 177. Melbourne, Australia.

Shaklee, J. B. and C. S. Tamaru. 1981. Biochemical and morphological evolution of Hawaiian bonefishes (*Albula*). Syst. Zool. 30:125–146.

Shaklee, J. B. and G. S. Whitt. 1981. Lactate dehydrogenase isozymes of gadiform fishes: Divergent patterns of gene expression indicate a heterogeneous taxon. Copeia 1981:563–578.

Shaklee, J. B., K. L. Kepes and G. S. Whitt. 1973. Specialized lactate dehydrogenase isozymes: The molecular and genetic basis for the unique eye and liver LDHs of teleost fishes. J. Exp. Zool. 185:217–240.

Sharkey, M. J. 1989. A hypothesis-independent method of character weighting for cladistic analysis. Cladistics 5:63–86.

Shaw, C. R. 1965. Electrophoretic variation in enzymes. Science 149:936–943.

Shaw, C. R. and R. Prasad. 1970. Starch gel electrophoresis of enzymes—a compilation of recipes. Biochem. Genet. 4:297–330.

Shaw, D. D., A. D. Marchant, M. L. Arnold and N. Contreras. 1988. Chromosomal rearrangements, ribosomal genes and mitochondrial DNA: Contrasting patterns of introgression across a narrow hybrid zone, pp. 121–130. *In* P.E. Brandham and M.D. Bennett (eds.), *Kew Chromosome Conference*. III. Allen & Unwin, London.

Shaw, J., T. R. Meagher and P. Harley. 1987. Electrophoretic evidence of reproductive isolation between two varieties of the moss *Climacium americanum*. Heredity 59:337–343.

Sheldon, F. H. 1987. Rates of single-copy DNA evolution in herons. Mol. Biol. Evol. 4:56–69.

Sheldon, F. H. and A. H. Bledsoe. 1989. Indexes to the reassociation and stability of solution DNA hybrids. J. Mol. Evol. 29:328–343.

Shields, G. F. and A. C. Wilson. 1987. Calibration of mitochondrial DNA evolution in geese. J. Mol. Biol. 24:212–217.

Shinozaki, K., M. Ohme, M. Tanaka, T. Wakasugi, N. Hayashida, T. Matsubayashi, N. Zaita, J. Chunwongse, J. Obokata, K. Yamaguchi-Shinozaki, C. Ohto, K. Torazawa, B. Y. Meng, M. Sugita, H. Deno, T. Kamogashira, K. Yamada, J. Kusuda, F. Takaiwa, A. Kato, N. Tohdoh, H. Shimada and M.

Sugiura. 1986. The complete nucleotide sequence of tobacco chloroplast genome: Its gene organization and expression. EMBO J. 5:2043–2049.

Shochat, D. and H. C. Dessauer. 1981. Comparative immunological study of albumins of *Anolis* lizards of the Caribbean Islands. Comp. Biochem. Physiol. 68A:67–73.

Shoemaker, J. S. and W. M. Fitch. 1989. Evidence from nuclear sequences that invariable sites should be considered when sequence divergence is calculated. Mol. Biol. Evol. 6:270–289.

Shoshani, J., J. M. Lowenstein, D. A. Walz and M. Goodman. 1985. Proboscidean origins of mastodon and woolly mammoth demonstrated immunologically. Paleobiology 11:429–437.

Shows, T. B. and F. H. Ruddle. 1968. Function of the lactate dehydrogenase B gene in mouse erythrocytes: Evidence for control by a regulatory gene. Proc. Natl. Acad. Sci. U.S.A. 61:574.

Sibley, C. G. and J. E. Ahlquist. 1981a. The phylogeny and relationships of the ratite birds as indicated by DNA–DNA hybridization, pp. 301–335. *In* G. G. E. Scudder and J. L. Reveal (eds.), *Evolution Today*. Carnegie-Mellon Univ., Pittsburgh, PA.

Sibley, C. G. and J. E. Ahlquist. 1981b. Instructions for specimen preservation for DNA extraction: A valuable source of data for systematics. Assoc. Syst. Collections Newslett. 9:44–45.

Sibley, C. G. and J. E. Ahlquist. 1983. The phylogeny and classification of birds based on the data of DNA-DNA hybridization, pp. 245–292. *In* R. F. Johnston (ed.), *Current Ornithology*, Vol. 1. Plenum, New York.

Sibley, C. G. and J. Ahlquist. 1987a. Avian phylogeny reconstructed from comparisons of the genetic material, DNA, pp. 95–121. *In* C. Patterson (ed.), *Molecules and Morphology in Evolution: Conflict or Compromise?* Cambridge Univ. Press, Cambridge.

Sibley, C. G. and J. E. Ahlquist. 1987b. DNA hybridization evidence of hominoid phylogeny: Results from an expanded data set. J. Mol. Evol. 26:99–121.

Sibley, C. G., K. W. Corbin, J. E. Ahlquist and A. Ferguson. 1974. Birds, pp. 89–176. *In* C. A. Wright (ed.), *Biochemical and Immunological Taxonomy of Animals*. Academic Press, New York.

Sibley, C. G., J. E. Ahlquist and F. H. Sheldon. 1987. DNA hybridization and avian phylogenetics: Reply to Cracraft. Evol. Biol. 21:97–125.

Sibley, C. G., J. E. Ahlquist and B. L. Monroe Jr.

1988. A classification of living birds based on DNA–DNA hybridization studies. Auk 105:409–423.

Siciliano, M. J. and C. R. Shaw. 1976. Separation and visualization of enzymes on gels, pp. 185–209. *In* I. Smith (ed.), *Chromatographic and Electrophoretic Techniques,* Vol. 2. Wm. Heineman Medical Books, London.

Simmons, G. M., M. E. Kreitman, W. F. Quattlebaum and N. Miyashita. 1989. Molecular analysis of the alleles of alcohol dehydrogenase along a cline in *Drosophila melanogaster.* I. Maine, North Carolina, and Florida. Evolution 43:392–392.

Simon, C. M. 1979. Evolution of periodical cicadas: Phylogenetic inferences based upon allozyme data. Syst. Zool. 28:22–39.

Singh, G., N. Neckelmann and D. C. Wallace. 1987. Conformational mutations in human mitochondrial DNA. Nature 329:270–272.

Singh, R. S., R. C. Lewontin and A. A. Felton. 1976. Genetic heterogeneity within electrophoretic "alleles" of xanthine dehydrogenase in *Drosophila pseudoobscura.* Genetics 84:609–629.

Sites, J. W., Jr. and S. K. Davis. 1989. Phylogenetic relationships and molecular variability within and among six chromosome races of *Sceloporus grammicus* (Sauria, Iguanidae), based on nuclear and mitochondrial markers. Evolution 43:296–317.

Sites, J. W., Jr. and C. Moritz. 1987. Chromosome change and speciation revisited. Syst. Zool. 36:153–174.

Sites, J. W., Jr., J. W. Bickham, B. A. Pytel, I. F. Greenbaum and B. A. Bates. 1984. Biochemical characters and the reconstruction of turtle phylogenies: Relationships among batagurine genera. Syst. Zool. 33:137–158.

Sites, J. W., Jr., R. L. Bezy and P. Thompson. 1986. Nonrandom heteropolymer expression of lactate dehydrogenase isozymes in the lizard family Xantusiidae. Biochem. Syst. Ecol. 14:539–545.

Sites, J. W., Jr., D. M. Peccinini-Seale, C. Moritz, J. W. Wright and W. M. Brown. 1990. The evolutionary history of parthenogenetic *Cnemidophorus lemniscatus* (Sauria, Teiidae). I. Evidence for a hybrid origin. Evolution 44 (in press).

Slatkin, M. 1985. Gene flow in natural populations. Annu. Rev. Ecol. Syst. 16:393–430.

Slatkin, M. 1987. Gene flow and the geographic structure of natural populations. Science 236:787–792.

Slightom, J. L., T. W. Theisen, B. F. Koop and M. Goodman. 1987. Orangutan fetal globin genes.

Nucleotide sequences reveal multiple gene conversions during hominid phylogeny. J. Biol. Chem. 262:7472–7483.

Smith, C. A., J. M. Jordan and J. Vinograd. 1971. In vivo effects of intercalating drugs on the superhelix density of mitochondrial DNA isolated from human and mouse cells in culture. J. Mol. Biol. 59:255–272.

Smith, M. J., R. Nicholson, M. Stuerzl and A. Lui. 1982. Single copy DNA homology in sea stars. J. Mol. Evol. 18:92–101.

Smith, M. W., C. F. Aquadro, M. H. Smith, R. K. Chesser and W. J. Etges. 1982. *Bibliography of Electrophoretic Studies of Biochemical Variation in Natural Vertebrate Populations.* Texas Tech. Press, Lubbock.

Smith, T. F., M. S. Waterman and W. M. Fitch. 1981. Comparative biosequence metrics. J. Mol. Evol. 18:38–46.

Smith, T. F., M. S. Waterman and C. Burks. 1985. The statistical distribution of nucleic acid similarities. Nucl. Acids Res. 13:645–656.

Smithies, O. 1955. Zone electrophoresis in starch gels: Group variations in the serum proteins of normal individuals. Biochem. J. 61:629–641.

Sneath, P. H. A. and R. R. Sokal. 1973. *Numerical Taxonomy.* W. H. Freeman, San Francisco.

Sneath, P. H. A., M. J. Sackin and R. P. Amber. 1975. Detecting evolutionary incompatibilities from protein sequences. Syst. Zool. 24:311–332.

Snedecor, G. W. and W. G. Cochran. 1980. *Statistical Methods.* 7th ed. Iowa State Univ. Press, Ames.

Sober, E. 1983. Parsimony in systematics: Philosophical issues. Annu. Rev. Ecol. Syst. 14:335–357.

Sober, E. 1989. *Reconstructing the Past: Parsimony, Evolution, and Inference.* MIT Press, Cambridge, MA.

Solignac, M., J. Guermont, M. Monnerot and J.-C. Mounolou. 1984. Genetics of mitochondria in *Drosophila:* mtDNA inheritance in heteroplasmic strains of *D. mauritiana.* Mol. Gen. Genet. 197:183–88.

Soltis, D. E. and L. J. Rieseberg. 1986. Autopolyploidy in *Tolmiea menziesii* (Saxifragaceae): Genetic insights from enzyme electrophoresis. Am. J. Bot. 73:310–318.

Soltis, D. E., C. H. Haufler, D. C. Darrow and G. J. Gastony. 1983. Starch gel electrophoresis of ferns: A compilation of grinding buffers, gel and electrode buffers, and staining schedules. Am. Fern J. 73:9–27.

Soltis, D. E., P. S. Soltis and B. D. Ness. 1989a. Chloroplast DNA variation and multiple origins of autopolyploidy in *Heuchera micrantha* (Saxifragaceae). Evolution 43:650–656.

Soltis, D. E., T. A. Ranker and B. D. Ness. 1989b. Chloroplast DNA variation in a wild plant, *Tolmiea menziesii*. Genetics 121:819–826.

Southern, E. M. 1975. Detection of specific sequences among DNA fragments separated by gel electrophoresis. J. Mol. Biol. 98:503–517.

Spencer, D. F., M. N. Schnare and M. W. Gray. 1984. Pronounced structural similarities between the small ribosomal RNA genes of wheat mitochondria and *Escherichia coli*. Proc. Natl. Acad. Sci. U.S.A. 81:493–497.

Spencer, E. W., V. M. Ingram and C. Levinthal. 1966. Electrophoresis: An accident and some precautions. Science 152:1722–1723.

Spencer, N., D. A. Hopkinson and H. Harris. 1964. Phosphoglucomutase polymorphism in man. Nature (London) 204:742–745.

Spencer, N., D. A. Hopkinson and H. Harris. 1968. Adenosine deaminase polymorphism in man. Ann. Hum. Genet. London 32:9–14.

Spinella, D. G. and R. C. Vrijenhoek. 1982. Genetic dissection of clonally inherited genomes of *Poeciliopsis*. II. Investigation of a silent carboxylesterase allele. Genetics 100:279–286.

Springer, M. S. 1988. The Phylogeny of Diprotodontian Marsupials Based on Single-Copy DNA–DNA Hybridization and Craniodental Anatomy. Ph.D. Dissertation, Univ. of California, Riverside.

Springer, M. S. and J. A. W. Kirsch. 1989. Rates of single-copy DNA evolution in phalangeriform marsupials. Mol. Biol. Evol. 6:331–341.

Springer, M. S. and C. Krajewski. 1989. DNA hybridization in animal taxonomy: A critique from first principles. Q. Rev. Biol. 64:291–318.

Springer, M. S., J. A. W. Kirsch, K. Aplin and T. Flannery. 1990. DNA hybridization, cladistics, and the phylogeny of phalangerid marsupials. J. Mol. Evol. (in press).

St. Louis, V. L. and J. C. Barlow. 1988. Genetic differentiation among ancestral and introduced populations of the Eurasian tree sparrow (*Passer montanus*). Evolution 42:266–276.

Stahl, D. A., D. J. Lane, G. J. Olsen and N. R. Pace. 1984. Analysis of hydrothermal vent-associated symbionts by ribosomal RNA sequences. Science 224:409–411.

Stecher, P. G., M. Windholz, D. S. Leahy, D. M. Bolton and L. G. Eaton. 1968. *Merck Index,* 8th. Ed. Merck, Rahway, NJ.

Steffen, D. L., G. T. Cocks and A. C. Wilson. 1972. Micro-complement fixation in *Klebsiella* classification. J. Bacteriol. 110:803–808.

Steinemann, M., W. Pinsker and D. Sperlich. 1984. Chromosome homologies within the *Drosophila obscura* group probed by in situ hybridization. Chromosoma 91:46–53.

Steiner, W. W. M. and D. J. Joslyn. 1979. Electrophoretic techniques for the genetic study of mosquitoes. Mosquito News 39:35–54.

Stephen, W. P. 1974. Insects, pp. 303–349. *In* C. A. Wright (ed.), *Biochemical and Immunological Taxonomy of Animals*. Academic Press, New York.

Stoneking, M., B. May and J. Wright. 1981. Loss of duplicate gene expression in salmonids: Evidence for a null allele polymorphism at the duplicate aspartate aminotransferase loci in brook trout (*Salvelinus fontinalis*). Biochem. Genet. 19:1063–1077.

Stowell, R. E. (ed.), 1965. *Cryobiology*. Fed. Proc. 24:S1–S324.

Strobeck, C. and K. Morgan. 1978. The effect of intragenic recombination on the number of alleles in a finite population. Genetics 88:829–844.

Studier, J. A. and K. J. Keppler. 1988. A note on the neighbor-joining algorithm of Saitou and Nei. Mol. Biol. Evol. 5:729–731.

Sturtevant, A. H. and E. Novitski. 1941. The homologies of the chromosome elements in the genus *Drosophila*. Genetics 26:517–541.

Suzuki, H., K. Moriwaka and E. Nevo. 1987. Ribosomal DNA (rDNA) spacer polymorphism in mole rats. Mol. Biol. Evol. 4:602–610.

Swofford, D. L. 1981. On the utility of the distance Wagner procedure, pp. 25–43. *In* V. A. Funk and D. R. Brooks (eds.), *Advances in Cladistics. Proceedings of the First Meeting of the Willi Hennig Society*. New York Botanical Gardens, Bronx, NY.

Swofford, D. L. 1990. *PAUP: Phylogenetic Analysis Using Parsimony, Version 3.0*. Illinois Natl. Hist. Surv., Champaign, IL.

Swofford, D. L. and S. H. Berlocher. 1987. Inferring evolutionary trees from gene frequency data under the principle of maximum parsimony. Syst. Zool. 36:293–325.

Swofford, D. L. and W. P. Maddison. 1987. Reconstructing ancestral character states under Wagner parsimony. Math. Biosci. 87:199–229.

Swofford, D. L. and R. B. Selander. 1981. BIOSYS-1: A FORTRAN program for the comprehensive anal ysis of electrophoretic data in population genetics and systematics. J. Hered. 72:281–283.

Szymura, J. M. and N. H. Barton. 1986. Genetic analysis of a hybrid zone between the fire-bellied toads, *Bombina bombina* and *B. variegata,* near Cracow in southern Poland. Evolution 40:1141–1159.

Tabor, S. and C. C. Richardson. 1987. DNA sequence analysis with a modified bacteriophage T7 DNA polymerase. Proc. Natl. Acad. Sci. U.S.A. 84:4767–4771.

Tajima, F. and M. Nei. 1984. Estimation of evolutionary distance between nucleotide sequences. Mol. Biol. Evol. 1:269–285.

Takahata, N. and S. R. Palumbi. 1985. Extranuclear differentiation and gene flow in the finite island model. Genetics 109:441–457.

Tammar, A. R. 1974. Bile salts of Amphibia, pp. 67–76. *In* M. Florkin and B. T. Scheer (eds.), *Chemical Zoology.* Academic Press, New York.

Tateno, Y., M. Nei and F. Tajima. 1982. Accuracy of estimated phylogenetic trees from molecular data. I. Distantly related trees. J. Mol. Evol. 18:387–404.

Taylor, H. A., S. E. Riley, S. E. Parks and R. E. Stevenson. 1978. Longterm storage of tissue samples for cell culture. In Vitro 14:476–478.

Tegelstrom, H. 1986. Mitochondrial DNA in natural populations: An improved routine for the screening of genetic variation based on sensitive silverstaining. Electrophoresis 7:226–229.

Templeton, A. R. 1983a. Convergent evolution and non-parametric inferences from restriction fragment and DNA sequence data, pp. 151–179. *In* B. Weir (ed.), *Statistical Analysis of DNA Sequence Data.* Marcel Dekker, New York.

Templeton, A. R. 1983b. Phylogenetic inference from restriction endonuclease cleavage site maps with particular reference to the humans and apes. Evolution 37:221–244.

Templeton, A. R. 1987. Nonparametric inference from restriction cleavage sites. Mol. Biol. Evol. 4:315–319.

Templeton, A.R., K. Shaw, E. Routman and S. K. Davis. 1990. The genetic consequences of habitat fragmentation. Ann. Missouri Bot. Gard. (in press).

Tereba, A. and B. J. McCarthy. 1973. Hybridization of ^{125}I-labeled ribonucleic acid. Biochemistry 12:4675–4679.

Thorpe, J. P. 1982. The molecular clock hypothesis: Biochemical evaluation, genetic differentiation and systematics. Annu. Rev. Ecol. Syst. 13:139–168.

Tilley, S. G. 1981. A new species of *Desmognathus* (Amphibia: Caudata: Plethodontidae) from the southern Appalachian mountains. Occ. Pap. Mus. Zool. Univ. Michigan 695:1–23.

Tilley, S. G. and J. S. Hansman. 1976. Allozymic variation and occurrence of multiple inseminations in populations of the salamander *Desmognathus ochrophaeus.* Copeia 1976:734–41.

Tilley, S. G. and P. M. Schwerdtfeger. 1981. Electrophoretic variation in Appalachian populations of the *Desmognathus fuscus* complex (Amphibia: Plethodontidae). Copeia 1981:109–119.

Timmis, J. N. and N. S. Scott. 1984. Promiscuous DNA: Sequence homologies between DNA of separate organelles. Trends Biochem. Sci. 9:271–273.

Titus, T. A., D. M. Hillis and W. E. Duellman. 1989. Color polymorphism in neotropical treefrogs: An allozymic investigation of the taxonomic status of *Hyla favosa* Cope. Herpetologica 45:17–23.

Tjio, J. H. and A. Levan. 1956. The chromosome number of man. Hereditas 42:1–6.

Tobler, J. E. and E. H. Grell. 1978. Genetics and physiological expression of β-hydroxy acid dehydrogenase in *Drosophila.* Biochem. Genet. 16:333–342.

Turner, B. J. 1973. Genetic variation of mitochondrial aspartate aminotransferase in the teleost *Cyprinodon nevadensis.* Comp. Biochem. Physiol. 44B:89–92.

Turner, B. J. 1974. Genetic divergence of Death Valley pupfish species: Biochemical versus morphological evidence. Evolution 28:281–294.

Turner, B. J. 1980. A multiple slicer for starch gels. Isozyme Bull. 13:113.

Turner, B. J. 1984. Evolutionary genetics of artificial refugium populations of an endangered species, the desert pupfish. Copeia 1984:364–369.

Turner, B. J., R. R. Miller and E. M. Rasch. 1980. Significant differential gene duplication without ancestral tetraploidy in a genus of Mexican fish. Experientia 36:927–930.

Turner, B. J., J. S. Balsano, P. J. Monaco and E. M. Rasch. 1983. Clonal diversity and evolutionary dynamics in a diploid-triploid breeding complex of unisexual fishes (*Poecilia*). Evolution 37:798–809.

Turner, S., T. Burger-Wiersma, S. J. Giovannoni, L. R. Mur and N. R. Pace. 1989. The relationship of a prochlorophyte *Prochlorothrix hollandica* to green chloroplasts. Nature 337: 380–382.

Upholt, W. B. 1977. Estimation of DNA sequence divergence from comparison of restriction endonuclease digests. Nucl. Acids Res. 4:1257–65.

Utter, F., P. Aebersold and G. Winans. 1987. Interpreting genetic variation detected by electrophoresis, pp. 21–46. *In* N. Ryman and F. Utter (eds.), *Population Genetics and Fishery Management.* Univ. Washington Press, Seattle.

Uy, R. and F. Wold. 1977. Posttranslational covalent modification of proteins. Science 198:890–896.

VanlerBerghe, F., B. Dod, P. Boursot, M. Bellis and F. Bonhomme. 1986. Absence of Y-chromosome introgression across the hybrid zone between *Mus musculus* and *Mus domesticus.* Genet. Res. 48:191–197.

van Tets, P. and I. M. Cowan. 1966. Some sources of variation in the blood sera of deer (*Odocoileus*) as revealed by starch gel electrophoresis. Can. J. Zool. 44:631–647.

Varley, J. M., H. C. Macgregor, I. Nardi, C. Andrews and H. P. Erba. 1980. Cytological evidence of transcription of highly repeated DNA sequences during the lampbrush stage in *Triturus cristatus carnifex.* Chromosoma 80:289–307.

Vassart, G., M. Georges, R. Monsieur, H. Brocas, A. S. Lequarre and D. Christophe. 1987. A sequence in M13 phage detects hypervariable minisatellites in human and animal DNA. Science 235:683–684.

Vawter, L. and W. M. Brown. 1986. Nuclear and mitochondrial DNA comparisons reveal extreme rate variation in the molecular clock. Science 234:194–196.

Vrijenhoek, R. C. 1989. Genetic diversity and the ecology of asexual populations, pp. 175–197. *In* K. Wöhrmann and S. Jain (eds.), *Population Biology and Evolution.* Springer-Verlag, New York.

Vrijenhoek, R. C., M. E. Douglass and G. K. Meffe. 1985. Conservation genetics of endangered populations in Arizona. Science 229:400–402.

Wagner, D. B., G. R. Furnier, M. A. Saghai-Maroof, S. M. Williams, B. P. Dancik and R. W. Allard. 1987. Chloroplast DNA polymorphism in lodgepole and jack pines and their hybrids. Proc. Natl. Acad. Sci. U.S.A. 84:2097–2100.

Wagner, W. H. 1983. Reticulistics: The recognition of hybrids and their role in cladistics and classification, pp. 63–79. *In* N. I. Platnick and V. A. Funk (eds.), *Advances in Cladistics: Proceedings of the Second Meeting of the Willi Hennig Society.* Columbia Univ. Press, New York.

Wahlund, S. 1928. The combination of populations and the appearance of correlation examined from the standpoint of the study of heredity. Hereditas 11:65–106 (in German).

Wake, D. B. 1981. The application of allozyme evidence to problems in the evolution of morphology, pp. 257–270. *In* G. G. E. Scudder and J. L. Reveal (eds.), *Evolution Today.* Carnegie-Mellon Univ., Pittsburgh.

Wake, D. B. and A. Larson. 1987. Multidimensional analysis of an evolving lineage. Science 238:42–48.

Wake, D. B. and K. P. Yanev. 1986. Geographic variation in allozymes in a "ring species," the plethodontid salamander *Ensatina eschscholtzii* of western North America. Evolution 40:702–715.

Wake, D. B., G. Roth and M. H. Wake. 1983. On the problem of stasis in organismal evolution. J. Theoret. Biol. 101:211–224.

Wake, D. B., K. P. Yanev and C. W. Brown. 1986. Intraspecific sympatry in a "ring species," the plethodontid salamander *Ensatina eschscholtzii,* in southern California. Evolution 40:866–868.

Wallace, D. G. and D. Boulter. 1976. Immunological comparisons of higher plant plastocyanins. Phytochemistry 15:137–141.

Wallace, D. G., M.-C. King and A. C. Wilson. 1973. Albumin differences among ranid frogs: Taxonomic and phylogenetic implications. Syst. Zool. 22:1–13.

Walter, H., F. W. Selby and J. R. Fransisco. 1965. Altered electrophoretic mobilities of some erythrocytic enzymes as a function of their age. Nature (London) 208:76–77.

Ward, R. D., B. J. McAndrew and G. P. Wallis. 1979. Purine nucleoside phosphorylase variation in the brook lamprey, *Lampetra planeri* (Bloch) (Petromyzone, Agnatha): Evidence for a trimeric enzyme structure. Biochem. Genet. 17:251–256.

Ware, V. C., B. W. Tague, C. G. Clark, R. L. Gourse, R. C. Brand and S. A. Gerbi. 1983. Sequence analysis of 28S ribosomal DNA from the amphibian *Xenopus laevis.* Nucl. Acids Res. 11:7795–7817.

Waterman, M. S. 1984. General methods of sequence comparison. Bull. Math. Biol. 46:473–500.

Waterman, M. S., T. F. Smith and W. A. Beyer. 1976. Some biological sequence metrics. Adv. Math. 20:367–387.

Waterman, M. S., T. F. Smith, M. Singh and W. A. Beyer. 1977. Additive evolutionary trees. J. Theoret. Biol. 64:199–213.

Watson, P. F. (ed.) 1978. *Artificial Breeding of Non-Domestic Animals.* Symposium, Zool. Soc. London 43:1–376.

Watt, J. L. and G. S. Stephen. 1986. Lymphocyte culture for chromosome analysis, pp. 39–55. *In* D. E. Rooney and B. H. Czepulkowski (eds.), *Human Cytogenetics.* IRL Press, Oxford.

Watt, W. B. 1972. Intragenic recombination as a

source of population genetic variability. Am. Natur. 106:737–753.

Watt, W. B. 1977. Adaptation at specific loci. I. Natural selection on phosphoglucose isomerase of *Colias* butterflies: Biochemical and population aspects. Genetics 87:177–194.

Watt, W. B. 1983. Adaptation at specific loci. II. Demographic and biochemical elements in the maintenance of the *Colias* PGI polymorphism. Genetics 103:691–724.

Watt, W. B. 1985. Bioenergetics and evolutionary genetics: Opportunities for new synthesis. Am. Natur. 125:118–143.

Watt, W. B. 1986. Power and efficiency as indices of fitness in metabolic organization. Am. Natur. 127:629–653.

Watt, W. B., P. A. Carter and S. M. Blower. 1985. Adaptation at specific loci. IV. Differential mating success among glycolytic allozyme genotypes of *Colias* butterflies. Genetics 109:157–175.

Watt, W. B., P. A. Carter and K. Donohue. 1986. Females' choice of "good genotypes" as mates is promoted by an insect mating system. Science 233:1187–1190.

Weeden, N. F. 1983. Plastid isozymes, pp. 139–158. *In* S.D. Tanskey and T.J. Orton (eds.), *Isozymes in Plant Genetics and Breeding, Part A.* Elsevier, Amsterdam.

Weeks, D.P., N. Beerman and O. M. Griffith. 1986. A small scale five-hour procedure for isolating multiple samples of CsCl-purified DNA: Application to isolations from mammalian, insect, higher plant, algal, yeast, and bacterial sources. Anal. Biochem. 152:376–385.

Wegnez, M. 1987. Letter to the editor. Cell 51:516.

Weir, B. S. 1989. Sampling properties of gene diversity. pp. 23–43. *In* A. H. D. Brown, M. T. Clegg, A. L. Kahler and B. S. Weir (eds.), *Plant Population Genetics, Breeding, and Genetic Resources.* Sinauer, Sunderland, MA.

Weir, B. S. and C. C. Cockerham. 1984. Estimating F-statistics for the analysis of population structure. Evolution 38:1358–1370.

Weir, B. S. and C. C. Cockerham. 1989a. Complete characterization of disequilibrium at two loci, pp. 86–110. *In* M. W. Feldman (ed.), *Mathematical Evolutionary Theory.* Princeton Univ. Press, Princeton.

Weir, B. S. and C. C. Cockerham. 1989b. Analysis of disequilibrium coefficients, pp. 45–51. *In* W. G. Hill and T. F. C. Mackay (eds.), *Evolution and Animal Breeding: Reviews on Molecular and Quantitative Genetics Approaches in Honour of Alan Robertson.* Commonwealth Agricultural Bureaux, Slough, U. K.

Weisman, L. S., B. M. Krummel and A. C. Wilson. 1986. Evolutionary shift in the site of cleavage of prelysozyme. J. Biol. Chem. 261:2309–2313.

Werth, C. R. 1985. Implementing an isozyme laboratory at a field station. Virginia J. Sci. 36:53–76.

Werth, C. R. and M. D. Windham. 1987. A new model for speciation in polyploid pteridophytes resulting from reciprocal silencing of homoeologous genes. Am. J. Bot. 74:713–714.

Werth, C. R., S. I. Guttman and W. H. Eshbaugh. 1985a. Electrophoretic evidence of reticulate evolution in the Appalachian *Asplenium* complex. Syst. Bot. 10:184–192.

Werth, C. R., S. I. Guttman and W. H. Eshbaugh. 1985b. Recurring origins of allopolyploid species in *Asplenium.* Science 228:731–733.

Wetmur, J. G. and N. Davidson. 1968. Kinetics of renaturation of DNA. J. Mol. Biol. 31:349–370.

Wetton, J. H., R. E. Carter, D. T. Parkin and D. Walters. 1987. Demographic study of a wild house sparrow population by DNA fingerprinting. Nature (London) 327:147–149.

Wheeler, W. C. and R. L. Honeycutt. 1988. Paired sequence difference in ribosomal RNAs: Evolutionary and phylogenetic implication. Mol. Biol. Evol. 5:90–96.

White, M. J. D. 1973. *Animal Cytology and Evolution.* 3rd Ed. Cambridge Univ. Press, Cambridge.

White, T. J., N. Arnheim and H. A. Erlich. 1989. The polymerase chain reaction. Trends Genet. 5:185–189.

Whitt, G. S. 1970. Developmental genetics of the lactate dehydrogenase isozymes of fish. J. Exp. Zool. 175:1–36.

Whitt, G. S. 1981. Evolution of isozyme loci and their differential regulation, pp. 271–289. *In* G. G. E. Scudder and J. L. Reveal (eds.), *Evolution Today, Proceedings of the Second International Congress of Systematic and Evolutionary Biology.* Hunt Inst. Bot. Documentation, Carnegie-Mellon Univ., Pittsburgh, PA.

Whitt, G. S. 1983. Isozymes as probes and participants in developmental and evolutionary genetics, pp. 1–40. *In* M. C. Rattazzi, J. G. Scandalios and G. S. Whitt (eds.). *Isozymes: Current Topics in Biological and Medical Research, Vol. 10. Genetics and Evolution.* Liss, New York.

Whitt, G. S. 1987. Species differences in isozyme tissue patterns: Their utility for systematic and evolu-

tionary analyses, pp. 1–26. *In* M. C. Rattazzi, J. G. Scandalios and G. S. Whitt (eds.), *Isozymes: Current Topics in Biological and Medical Research, Vol. 15. Genetics, Development, and Evolution.* Liss, New York.

Whitt, G. S., J. B. Shaklee and C. L. Markert. 1975. Evolution of the lactate dehydrogenase isozymes of fishes, pp. 381–400. *In* C.L. Markert (ed.), *Isozymes IV: Genetics and Evolution.* Academic Press, New York.

Wiley, E. O. 1978. The evolutionary species concept reconsidered. Syst. Zool. 27:17–26.

Wiley, E. O. 1981. *Phylogenetics: The Theory and Practice of Phylogenetic Systematics.* Wiley Interscience, New York.

Wiley, E. O. 1988a. Vicariance biogeography. Annu. Rev. Ecol. Syst. 19:513–542.

Wiley, E. O. 1988b. Parsimony analysis and vicariance biogeography. Syst. Zool. 37:271–290.

Wilhelmi, R. W. 1942. The application of the precipitin technique to theories concerning the origin of vertebrates. Biol. Bull. 82:179–189.

Williams, P. L. and W. M. Fitch. 1989. Finding the minimal change in a given tree, pp. 453–470. *In* B. Fernholm, K. Bremer and H. Jörnvall (eds.), *The Hierarchy of Life.* Elsevier, Amsterdam.

Williams, S. M., R. DeSalle and C. Strobeck. 1985. Homogenization of geographical variants at the nontranscribed spacer of rDNA in *Drosophila mercatorum.* Mol. Biol. Evol. 2:338–346.

Williams, S. M., G. R. Furnier, E. Fuog and C. Strobeck. 1987. Evolution of the ribosomal DNA spacers of *Drosophila melanogaster:* Different patterns of variation on X and Y chromosomes. Genetics 116:225–232.

Williams, S. M., R. W. DeBry and J. L. Feder. 1988. A commentary on the use of ribosomal DNA in systematic studies. Syst. Zool. 37:60–63.

Wilson, A. C., V. M. Sarich and L. R. Maxson. 1974. The importance of gene rearrangement in evolution: Evidence from studies of rates of chromosomal, protein, and anatomical evolution. Proc. Natl. Acad. Sci. U.S.A. 71:3028–3030.

Wilson, A. C., G. L. Bush, S. M. Case and M.-C. King. 1975. Social structuring of mammalian populations and rate of chromosomal evolution. Proc. Natl. Acad. Sci. U.S.A. 72:5061–5065.

Wilson, A. C., S. S. Carlson and T. J. White. 1977. Biochemical evolution. Annu. Rev. Biochem. 46:473–639.

Wilson, A. C., R. L. Cann, S. M. Carr, M. George, Jr., U. B. Gyllensten, K. Helm-Bychowski, R. C. Hi-

guchi, S. R. Palumbi, E. M. Prager, R. D. Sage and M. Stoneking. 1985. Mitochondrial DNA and two perspectives on evolutionary genetics. Biol. J. Linn. Soc. 26:375–400.

Wilson, A. C., H. Ochman and E. M. Prager. 1987. Molecular time scale for evolution. Trends Genet. 3:241–247.

Wilson, E. O. 1985. Time to revive systematics. Science 230:1227.

Wilson, E. O. 1986. The value of systematics. Science 231:1057.

Wilson, F. R., G. S. Whitt and C. L. Prosser. 1973. Lactate dehydrogenase and malate dehydrogenase isozyme patterns in tissues of temperature acclimated goldfish (*Carassius auratus*). Comp. Biochem. Physiol. 46B:105–116.

Wilson, G. N., M. Knoller, L. L. Szyura and R. D. Schmickel. 1984. Individual and evolutionary variation of primate ribosomal DNA transcritpion initiation regions. Mol. Biol. Evol. 1:221–237.

Wolfe, K. H., W. H. Li and P. M. Sharp. 1987. Rates of nucleotide substitutions vary greatly among plant mitochondrial, chloroplast, and nuclear DNAs. Proc. Natl. Acad. Sci. U.S.A. 84:9054–9058.

Wolstenholme, D. R., D. O. Clary, J. L. MacFarlane, J. A. Wahleithner and L. Wilcox. 1985. Organization and evolution of invertebrate mitochondrial genomes, pp. 61–69. *In* E. Quagliariello, E. C. Slater, F. Palmieri, C. Saccone and A. M. Kroon (eds.), *Achievements and Perspectives of Mitochondrial Research.* Elsevier Press, Amsterdam.

Womack, J. E. 1983. Post-translational modification of enzymes: Processing genes, pp. 175–186. *In* M. C. Rattazzi, J. G. Scandalios and G. S. Whitt (eds.), *Isozymes: Current Topics in Biological and Medical Research, Vol. 7. Molecular Structure and Regulation.* Liss, New York.

Wong, C., C. E. Dowling, R. K. Saiki, R. G. Higuchi, H. A. Ehrlich and H. H. Kazazian, Jr. 1987. Characterization of β-thalassaemia mutations using direct genomic sequencing of amplified single copy DNA. Nature (London) 330:384–386.

Woodruff, D. S. 1989. Genetic anomalies associated with *Cerion* hybrid zones: The origin and maintenance of new electromorphic variants called hybrizymes. Biol. J. Linn. Soc. 36:281–294.

Woodruff, R. C. and J. N. Thompson. 1980. Hybrid release of mutator activity and the genetic structure of natural populations. Evol. Biol. 12:129–162.

Wright, C. A. 1974. *Biochemical and Immunological Taxonomy of Animals.* Academic Press, New York.

Wright, C. A. (ed.) 1978. *Biochemical and Immunological Taxonomy of Animals*. Academic Press, London.

Wright, D. A., C. M. Richards, J. S. Frost, A. M. Camozzi and B. J. Kunz. 1983. Genetic mapping in amphibians, pp. 287–311. *In* M.C. Rattazzi, J.G. Scandalios and G.S. Whitt (eds.), *Isozymes: Current Topics in Biological and Medical Research, Vol. 7. Molecular Structure and Regulation.* Liss, New York.

Wright, S. 1943. Isolation by distance. Genetics 28:114–138.

Wright, S. 1951. The genetical structure of populations. Ann. Eugen. 15:323–354.

Wright, S. 1978. *Evolution and the Genetics of Populations.* Vol. 4, *Variability in and among Natural Populations.* Univ. Chicago Press, Chicago.

Wrischnik, L. A., R. G. Higuchi, M. Stoneking, H. A. Erlich, N. Arnheim and A. C. Wilson. 1987. Length mutations in human mitochondrial DNA: Direct sequencing of enzymatically amplified DNA. Nucl. Acids Res. 15:529–542.

Wulf, J. H. and R. G. Cutler. 1975. Altered protein hypothesis of mammalian aging processes—I. Thermal stability of glucose-6-phosphate dehydrogenase in C57BL/6J mouse tissue. Exp. Gerontol. 10:101–117.

Yang, D., Y. Oyaizu, H. Oyaizu, G. J. Olsen and C. R. Woese. 1985. Mitochondrial origins. Proc. Natl. Acad. Sci. U.S.A. 82:4443–4447.

Yonenaga-Yassuda, Y., S. Kasahara, T. H. Chu and M. T. Rodrigues. 1988. High-resolution RBG-banding pattern in the genus *Tropidurus* (Sauria, Iguanidae). Cytogenet. Cell Genet. 48:68–71.

Youvan, D. C. and J. E. Hearst. 1979. Reverse transcriptase pauses at N^2-methylguanine during in vitro transcription of *Escherichia coli* 16S ribosomal RNA. Proc. Natl. Acad. Sci. U.S.A. 76:3571–3574.

Zhan, T. S., S. Pathak and J. C. Liang. 1984. Induction of G-bands in the chromosomes of *Melanoplus sanguinipes* (Orthoptera, Acrididae). Can. J. Genet. Cytol. 26:354–359.

Zimmer, E. A., S. L. Martin, S. M. Beverly, Y. W. Kan and A. C. Wilson. 1980. Rapid duplications and loss of genes coding for a chains of hemoglobin. Proc. Natl. Acad. Sci. U.S.A. 77:2158–2162.

Zimmer, E. A., C. J. Rivin and V. E. Walbot. 1981. A DNA isolation procedure suitable for most higher plant species. Plant Mol. Biol. Newslett. 2:93–96.

Zimmer E. A., R. K. Hamby, M. L. Arnold, D. A. Leblanc and E. C. Theriot. 1989. Ribosomal RNA phylogenies and flowering plant evolution, pp. 205–214. *In* B. Fernholm, K. Bremer and H. Jörnvall (eds.), *The Hierarchy of Life. Proc. Nobel Symp. 70.* Elsevier, Amsterdam.

Zuckerkandl, E. and L. Pauling. 1962. Molecular disease, evolution and genic heterogeneity, pp. 189–225. *In* M. Kasha and B. Pullman (eds.), *Horizons in Biochemistry.* Academic Press, New York.

Zurawski, G. and M. T. Clegg. 1987. Evolution of higher-plant chloroplast DNA-encoded genes: Implications for structure–function and phylogenetic studies. Annu. Rev. Plant Physiol. 38:391–418.

Index

Page numbers in **boldface** indicate instructions for preparing these stock solutions.

reassociation, 210–211
Stronglyocentrotus purpuratus, single-copy sequence polymorphism, 215
Subbands, isozyme electrophoresis, 64, 97, 98
"Subbed" slides, 181
Subbing solution, **202**
Subcloning, 322, 348–349
Subgenomic libraries, 321
Submarine horizontal gel rig, 290
Substitution(s)
  maximum likelihood phylogenies, 475
  patterns and effects on methods of phylogenetic tree inference, 472
  transition, 431, 474–475
  transversion, 431, 471, 472, 474–475
Substitution matrix, 477–478
Substoichiometric bands, 305–308
Subtree pruning and regrafting, 489
Succinate dehydrogenase (SUDH), 119
Sucrose step gradient, cpDNA isolation, 284–285
Sulfhydryl reagents, long-term storage, 37–38
Superimposed events, 429–432
Superoxide dismutase (SOD), 100, **119**
Sympatric species, 19
  boundaries, 20
Symplesiomorphic alleles, 58
Syringes, "bridge," 137
Systematics, 1, *see also* Molecular systematics

T_{25H}, 424
  derived from raw melting curve data, 242
T_{50H}, 208, 231, 424
  derived from raw melting curve data, 241
  estimation, 233, 234, 235
  genome size differences affecting, 247
  precision of measurements, 245–246
T_m, 206, 208, 231, 424
  analysis using S1 nuclease-TEACL assay, 229–231
  criterion of reassociation, 211
  derived from raw melting curve data, 241
  estimation, 233, 234
  precision of measurements, 245–246
  rapidly changing DNA effects, 239
  S1 nuclease, 212
  sequence divergence effects, 215
T_{mode}, 206, 231, 424
  derived from raw melting curve data, 239–240
  precision of measurements, 245
  rapidly changing DNA effects, 237, 239
TAE, *see* Tris-acetic acid-EDTA
Taq polymerase, 322–323, 329
Taq salts, 10×, **370**
TAT (tyrosine aminotransferase), **120**
Taxonomy, isozyme electrophoresis limits, 60–61
TBE, **317**
  10×, **370**
TBS, **202**
TCA, *see* Trichloroacetic acid
TCA/BSA (trichloroacetic acid/bovine serum albumin), **203**
Tcp-1 gene, 159
TE, *see* Tris-EDTA
TEACL, *see* Tetraethylammonium chloride

Temperature(s)
  incubation, 211
  melting, *see* Melting *entries*
Terminal length variation, 427
Terminal nodes, 414
Termination tubes, DNA sequencing reactions, 359–360
Tetraethylammonium chloride (TEACL), **249**
  DNA hybridization, 212–213
  melting curves combining advantages of hydroxy-apatite and, 213–214
  S1 nuclease combined with, 229–231
Tetraploidization, 59
Tetrazolium oxidase (TO), *see* Superoxide dismutase (SOD)
ThE, **317**
Thermal stability, reassociated hybrid DNA, *see* T_m
Thermus aquaticus, DNA polymerase, 322–323
Thiosulfate sulfurtransferase (TST), **120**
Thomomys bottae, paternity studies, 53
"3:1" fixative, 162
Three-level hierarchy, analysis of variance, 397
Three-point condition, 437–438
Thymine, 205
Time predictions, 508–514
Tissues, *see also* Specimens
  acquisition policies, 39
  biopsy, 34
  collection from animals, 30–32
  collection from plants, 32–33
  curatorial problems with collections, 39–41
  data base management of collections, 40
  deacquisition policies, 39–40
  differences in isozyme expression, 101
  disposition for long-term preservation, 39
  general collection procedures, 26–30
  guiding principles for collections, 40–41
  homogenization, 72, 74, 80
  packaging, 26–28
  preservation, 29–30
  storage, 35–36
  synoptic collections, 39–41
  thawing, 38
  transport from field to laboratory, 34–35
Titer, antiserum, 144–145
TM10, **203**
TO (tetrazolium oxidase), *see* Superoxide dismutase (SOD)
TPBS (Tris-phosphate-buffered saline), **203**
TPI (triose-phosphate isomerase), **120**
Tracer(s), 206, *see also* DNA hybridization
  estimation of fragment length, 224–225
  length of fragments, 219–220, 224–225
  paralogous sequences and, 247
  preparation by iodination, 225–227
  preparation with ^{32}P or ^3H, 220–221
  repeated elements in single-copy DNA used to make, 236
  self-reaction and repeat removal, 222–223
  short or degraded fragments, 236
  single-copy fractionation over hydroxyapatite, 223–224
Tracking dye, electrophoresis, 122